TODAY'S TECHNICIAN ™

CLASSROOM MANUAL

For Automotive Electricity and Electronics

SEVENTH EDITION

Barry Hollembeak

CENGAGE

Australia • Brazil • Mexico • Singapore • United Kingdom • United States

Today's Technician™: Classroom Manual for Automotive Electricity and Electronics, Seventh Edition

Barry Hollembeak

SVP, GM Skills & Global Product Management: Jonathan Lau

Product Director: Matthew Seeley

Senior Product Manager: Katie McGuire

Senior Director, Development: Marah Bellegarde

Senior Product Development Manager: Larry Main

Senior Content Developer: Mary Clyne

Product Assistant: Mara Ciacelli

Vice President, Marketing Services: Jennifer Ann Baker

Marketing Manager: Andrew Ouimet

Senior Production Director: Wendy Troeger

Production Director: Andrew Crouth

Senior Content Project Manager: Cheri Plasse

Senior Art Director: Jack Pendleton

Production Service/Composition: SPi Global

Cover image(s): Umberto Shtanzman/ Shutterstock.com

© 2019, 2015 Cengage Learning

ALL RIGHTS RESERVED. No part of this work covered by the copyright herein may be reproduced or distributed in any form or by any means, except as permitted by U.S. copyright law, without the prior written permission of the copyright owner.

For product information and technology assistance, contact us at **Cengage Customer & Sales Support, 1-800-354-9706**

For permission to use material from this text or product, submit all requests online at **www.cengage.com/permissions.** Further permissions questions can be e-mailed to **permissionrequest@cengage.com**

Library of Congress Control Number: 2017951704

Classroom Manual ISBN: 978-1-3376-1900-4
Package ISBN: 978-1-3376-1899-1

Cengage
20 Channel Center Street
Boston, MA 02210
USA

Cengage is a leading provider of customized learning solutions with employees residing in nearly 40 different countries and sales in more than 125 countries around the world. Find your local representative at **www.cengage.com.**

Cengage products are represented in Canada by Nelson Education, Ltd.

To learn more about Cengage Learning, visit **www.cengage.com**

Purchase any of our products at your local college store or at our preferred online store **www.cengagebrain.com**

Notice to the Reader

Publisher does not warrant or guarantee any of the products described herein or perform any independent analysis in connection with any of the product information contained herein. Publisher does not assume, and expressly disclaims, any obligation to obtain and include information other than that provided to it by the manufacturer. The reader is expressly warned to consider and adopt all safety precautions that might be indicated by the activities described herein and to avoid all potential hazards. By following the instructions contained herein, the reader willingly assumes all risks in connection with such instructions. The publisher makes no representations or warranties of any kind, including but not limited to, the warranties of fitness for particular purpose or merchantability, nor are any such representations implied with respect to the material set forth herein, and the publisher takes no responsibility with respect to such material. The publisher shall not be liable for any special, consequential, or exemplary damages resulting, in whole or part, from the readers' use of, or reliance upon, this material.

Printed in the United States of America
Print Number: 01   Print Year: 2017

CONTENTS

Thanks to the support the *Today's Technician*™ series has received from those who teach automotive technology, Cengage, the leader in automotive-related textbooks and learning solutions, is able to live up to its promise to provide new editions of the series every few years. We have listened and responded to our critics and fans and have presented this new updated and revised seventh edition. By revising this series on a regular basis, we can respond to changes in the industry, in technology, in the certification process, and the ever-changing needs of those who teach automotive technology.

We have also listened to instructors when they said that something was missing or incomplete in the previous edition. We have responded to those and have included the results in this seventh edition.

The *Today's Technician*™ series features textbooks and digital learning solutions that cover all mechanical and electrical systems of automobiles and light trucks. The individual titles correspond to the ASE (National Institute for Automotive Service Excellence) certification areas and are specifically correlated to the 2017 standards for Automotive Service Technicians, Master Service Technicians, as well as to the standards for Maintenance and Light Repair.

Additional titles include remedial skills and theories common to all of the certification areas and advanced or specific subject areas that reflect the latest technological trends. *Today's Technician: Automotive Electricity & Electronics, 7e* is designed to give students a chance to develop the same skills and gain the same knowledge that today's successful technician has. This edition also reflects the most recent changes in the guidelines established by the National Automotive Technicians Education Foundation (NATEF).

The purpose of NATEF is to evaluate technician training programs against standards developed by the automotive industry and recommend qualifying programs for certification (accreditation) by ASE. Programs can earn ASE certification upon the recommendation of NATEF. NATEF's national standards reflect the skills that students must master. ASE certification through NATEF evaluation ensures that certified training programs meet or exceed industry-recognized, uniform standards of excellence.

The technician of today and the future must know the underlying theory of all automotive systems, and be able to service and maintain those systems. Dividing the material into two volumes, a Classroom Manual and a Shop Manual, provides the reader with the information needed to begin a successful career as an automotive technician without interrupting the learning process by mixing cognitive and performance learning objectives into one volume.

The design of Cengage's *Today's Technician*™ series was based on features that are known to promote improved student learning. The design was further enhanced by a careful study of survey results, in which respondents were asked to value particular features. Some of these features can be found in other textbooks, while others are unique to this series.

Each Classroom Manual contains the principles of operation for each system and subsystem. The Classroom Manual also contains discussions on design variations of key components used by different vehicle manufacturers. It also looks into emerging technologies that will be standard or optional features in the near future. This volume is organized to build upon basic facts and theories. The primary objective of this volume is to allow the reader to gain an understanding of how each system and subsystem operates. This understanding is necessary to diagnose the complex automobiles of today and tomorrow.

Although the basics contained in the Classroom Manual provide the knowledge needed for diagnostics, diagnostic procedures appear only in the Shop Manual. An understanding of the underlying theories is also a requirement for competence in the skill areas covered in the Shop Manual.

A spiral-bound Shop Manual delivers hands-on learning experiences with step-by-step instructions for diagnostic and repair procedures. Photo Sequences are used to illustrate some of the common service procedures. Other common procedures are listed and are accompanied with fine line drawings and color photos that allow the reader to visualize and conceptualize the finest details of the procedure. This volume also contains the reasons for performing the procedures, as well as when that particular service is appropriate.

The two volumes are designed to be used together and are arranged in corresponding chapters. Not only are the chapters in the volumes linked together, but the contents of the chapters are also linked. The linked content is indicated by marginal callouts that refer the reader to the chapter and page where the same topic is addressed in the companion volume. This valuable feature saves users the time and trouble of searching the index or table of contents to locate supporting information in the other volume. Instructors will find this feature especially helpful when planning the presentation of material and when making reading assignments.

Both volumes contain clear and thoughtfully selected illustrations, many of which are original drawings or photos specially prepared for inclusion in this series. This means that art is a vital part of each textbook and not merely inserted to increase the number of illustrations.

—Jack Erjavec

HIGHLIGHTS OF THE NEW EDITION— CLASSROOM MANUAL

The text, photos, and illustrations in the seventh edition have been updated throughout to highlight the latest developments in automotive technology. In addition, some chapters have been combined. Although chapter 16 covers details associated with alternative powered vehicles, all pertinent information about hybrid vehicles is included in the main text that concerns relative topics. For example, the discussion of batteries in Chapter 5 includes coverage of HEV batteries and ultra-capacitors. Chapter 6 includes AC motor principles and the operation of the integrated starter/generator. Chapter 7 includes the HEV charging system, including regenerative braking and the DC/DC converter.

The flow of basic electrical to more complex electronic systems has been maintained. Chapters are arranged to enhance this flow and reduce redundancy.

Chapter 1 introduces the student to the automotive electrical and electronic systems with a general overview. This chapter emphasizes the interconnectivity of systems in today's vehicles, and describes the purpose and location of the subsystems, as well as the major components of the system and subsystems. The goal of this chapter is to establish a basic understanding for students to base their learning on. All systems and subsystems that are discussed in detail later in the text are introduced, and their primary purpose is described. Chapter 2 covers the underlying basic theories of electricity and includes discussion of Ohm's and Kirchhoff's laws. This is valuable to the student and the instructor because it covers the theories that other textbooks assume the reader knows. All related basic electrical theories are covered in this chapter.

Chapter 3 applies those theories to the operation of electrical and electronic components, and Chapter 4 covers wiring and the proper use of wiring diagrams. Emphasis is on using the diagrams to determine how the system works and how to use the diagram to isolate the problem.

The chapters that follow cover the major components of automotive electrical and electronic systems, such as batteries, starting systems and motor designs, charging systems, and basic lighting systems. This is followed by chapters that detail the functions of the body computer, input components, and vehicle communication networks. From here the student is guided into specific systems that utilize computer functions.

Current electrical and electronic systems are used as examples throughout the text. Most of these systems are discussed in detail. This includes computer-controlled interior and exterior lighting, night vision, adaptive lights, instrumentation, and electrical/electronic accessories. Coverage includes intelligent wiper, immobilizer, and adaptive cruise control systems. Chapter 15 details the passive restraint systems currently used.

HIGHLIGHTS OF THE NEW EDITION—SHOP MANUAL

Like the Classroom Manual, the Shop Manual is updated to match current trends. Service information related to the new topics covered in the Classroom Manual is also included in this manual. In addition, several new Photo Sequences are added. The purpose of these detailed photos is to show students what to expect when they perform the same procedure. They also help familiarize students with a system or type of equipment they may not encounter at school. Although the main purpose of the textbook is not to prepare someone to successfully pass an ASE exam, all the information required to do so is included in the textbook.

To stress the importance of safe work habits, Chapter 1 is dedicated to safety, and includes general HEV safety. As with the Classroom Manual, HEV system diagnosis is included within the main text. This provides the student with knowledge of safe system diagnosing procedures so they know what to expect as they further their training in this area. Included in this chapter are common shop hazards, safe shop practices, safety equipment, and the legislation concerning and the safe handling of hazardous materials and wastes.

Chapter 2 covers special tools and procedures. This chapter includes the use of isolation meters and expanded coverage of scan tools. In addition, a section on what it entails to be an electrical systems technician is included. This section covers relationships, completing the work order, and ASE certification. Another section emphasizes the importance of proper diagnostic procedures.

Chapter 3 leads the student through basic troubleshooting and service. This includes the use of various test equipment to locate circuit defects and how to test electrical and electronic components. Chapter 4 provides experience with wiring repairs along with extended coverage and exercises on using the wiring diagrams.

The remaining chapters have been thoroughly updated. The Shop Manual is cross-referenced to the Classroom Manual by using marginal notes. This provides students the benefit of being able to quickly reference the theory of the component or system that they are now working with.

Currently accepted service procedures are used as examples throughout the text. These procedures also serve as the basis for new job sheets that are included in the Shop Manual chapters.

CLASSROOM MANUAL

Features of this manual include the following:

Cognitive Objectives

These objectives outline the chapter's contents and identify what students should know and be able to do upon completion of the chapter. *Each topic is divided into small units to promote easier understanding and learning.*

Terms to Know List

A list of key terms appears immediately after the Objectives. Students will see these terms discussed in the chapter. Definitions can also be found in the Glossary at the end of the manual.

A Bit of History

This feature gives the student a sense of the evolution of the automobile. This feature not only contains nice-to-know information but also should spark some interest in the subject matter.

Margin Notes

The most important Terms to Know are highlighted and defined in the margins. Common trade jargon also appears in the margins and gives some of the common terms used for components. This helps students understand and speak the language of the trade, especially when conversing with an experienced technician.

Author's Notes

This feature includes simple explanations, stories, or examples of complex topics. These are included to help students understand difficult concepts.

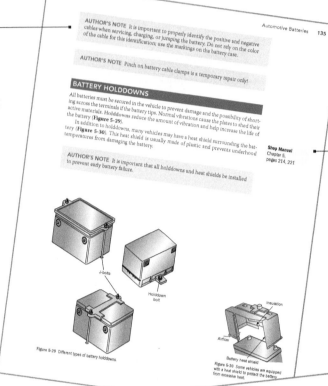

AUTHOR'S NOTE It is important to properly identify the positive and negative cables when servicing, charging, or jumping the battery. Do not rely on the color of the cable for this identification; use the markings on the battery case.

AUTHOR'S NOTE Pinch on battery cable clamps is a temporary repair only!

BATTERY HOLDDOWNS

All batteries must be secured in the vehicle to prevent damage and the possibility of short-ing across the terminals if the battery tips. Holddowns reduce the amount of vibration and help increase the life of the battery (**Figure 5-29**). Normal vibrations cause the plates to shed their active materials. Holddowns reduce the amount of vibration and help increase the life of the battery (**Figure 5-29**).
In addition to holddowns, many vehicles may have a heat shield surrounding the bat-tery (**Figure 5-30**). This heat shield is usually made of plastic and prevents underhood temperatures from damaging the battery.

AUTHOR'S NOTE It is important that all holddowns and heat shields be installed to prevent early battery failure.

Shop Manual
Chapter 5,
pages 214, 221

J-bolts

Holddown bolt

Figure 5-29 Different types of battery holddowns.

Insulation

Airflow

Battery heat shield
Figure 5-30 Some vehicles are equipped with a heat shield to protect the battery from excessive heat.

Cross-References to the Shop Manual

Reference to the appropriate page in the Shop Manual is given whenever necessary. Although the chapters of the two manuals are synchronized, material covered in other chapters of the Shop Manual may be fundamental to the topic discussed in the Classroom Manual.

Summary

Each chapter concludes with a summary of key points from the chapter. The key points are designed to help the reader review the chapter contents.

SUMMARY

- Through the use of gauges and indicator lights, the driver is capable of monitoring several engine and vehicle operating systems.
- The gauges include speedometer, odometer, tachometer, oil pressure, charging indicator, fuel level, and coolant temperature.
- The most common types of electromechanical gauges are the d'Arsonval, three coil, two coil, and air core.
- Computer-driven quartz swing needle displays are similar in design to the air-core electromagnetic gauges used in conventional analog instrument panels.
- All gauges require the use of a variable resistance sending unit. Styles of sending units include thermistors, piezoresistive sensors, and mechanical variable resistors.

- Digital instrument clusters use digital and linear displays to notify the driver of monitored system conditions.
- The most common types of displays used on electronic instrument panels are: light-emitting diodes (LEDs), liquid crystal displays (LCDs), vacuum fluorescent displays (VFDs), and a phos-phorescent screen that is the anode.
- A head-up display system displays visual images onto the inside of the windshield in the driver's field of vision.
- In the absence of gauges, important engine and vehicle functions are monitored by warning lamps. These circuits generally use an on/off switch–type sensor. The exception would be the use of voltage-controlled warning lights that use the principle of voltage drop.

Review Questions

Short-answer essays, fill in the blanks, and multiple-choice questions are found at the end of each chapter. These questions are designed to accurately assess the student's competence in the objectives stated at the beginning of the chapter.

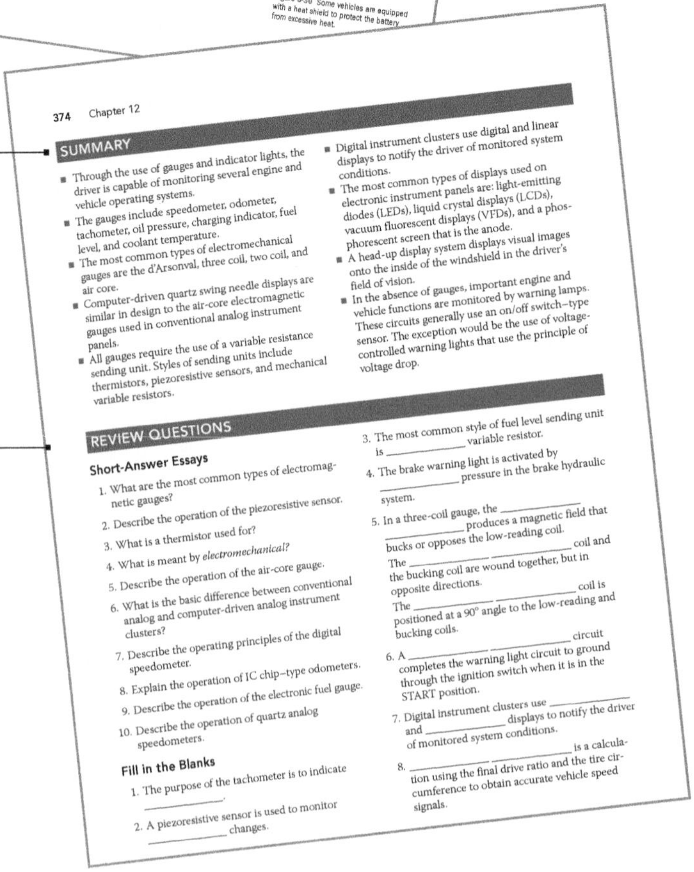

REVIEW QUESTIONS

Short-Answer Essays

1. What are the most common types of electromag-netic gauges?
2. Describe the operation of the piezoresistive sensor.
3. What is a thermistor used for?
4. What is meant by *electromechanical*?
5. Describe the operation of the air-core gauge.
6. What is the basic difference between conventional analog and computer-driven analog instrument clusters?
7. Describe the operating principles of the digital speedometer.
8. Explain the operation of IC chip–type odometers.
9. Describe the operation of the electronic fuel gauge.
10. Describe the operation of quartz analog speedometers.

Fill in the Blanks

1. The purpose of the tachometer is to indicate _____.
2. A piezoresistive sensor is used to monitor _____ changes.

3. The most common style of fuel level sending unit is _____ variable resistor.
4. The brake warning light is activated by _____ pressure in the brake hydraulic system.
5. In a three-coil gauge, the _____ produces a magnetic field that bucks or opposes the low-reading coil.
The _____ coil and the bucking coil are wound together, but in opposite directions.
The _____ coil is positioned at a 90° angle to the low-reading and bucking coils.
6. A _____ circuit completes the warning light circuit to ground through the ignition switch when it is in the START position.
7. Digital instrument clusters use _____ and _____ displays to notify the driver of monitored system conditions.
8. _____ is a calcula-tion using the final drive ratio and the tire cir-cumference to obtain accurate vehicle speed signals.

SHOP MANUAL

To stress the importance of safe work habits, the Shop Manual dedicates one full chapter to safety. Other important features of this manual include the following:

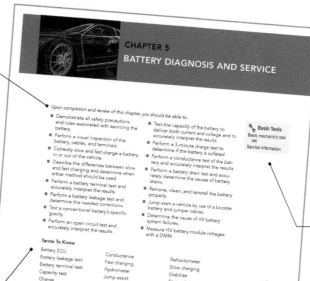

Performance-Based Objectives

These objectives define the contents of the chapter and define what the student should have learned upon completion of the chapter. These objectives also correspond with the list of required tasks for ASE certification. *Each ASE task is addressed.*

Although this textbook is not designed to simply prepare someone for the certification exams, it is organized around the ASE task list. These tasks are defined generically when the procedure is commonly followed and specifically when the procedure is unique for specific vehicle models. Imported- and domestic-model automobiles and light trucks are included in the procedures.

Terms to Know List

Terms in this list are also defined in the Glossary at the end of the manual.

Special Tools List

Whenever a special tool is required to complete a task, it is listed in the margin next to the procedure.

Cautions and Warnings

Throughout the text, warnings are given to alert the reader to potentially hazardous materials or unsafe conditions. Cautions are given to advise the student of things that can go wrong if instructions are not followed or if an unacceptable part or tool is used.

Basic Tools List

Each chapter begins with a list of the basic tools needed to perform the tasks included in the chapter.

Margin Notes

The most important Terms to Know are highlighted and defined in the margins. Common trade jargon also appears in the margins and gives some of the common terms used for components. This feature helps students understand and speak the language of the trade, especially when conversing with an experienced technician.

Photo Sequences

Many procedures are illustrated in detailed Photo Sequences. These detailed photographs show students what to expect when they perform particular procedures. They can also provide the student a familiarity with a system or type of equipment, which the school may not have.

Service Tips

Whenever a shortcut or special procedure is appropriate, it is described in the text. These tips generally describe common procedures used by experienced technicians.

Cross-References to the Classroom Manual

Reference to the appropriate page in the Classroom Manual is given whenever necessary. Although the chapters of the two manuals are synchronized, material covered in other chapters of the Classroom Manual may be fundamental to the topic discussed in the Shop Manual.

Customer Care

This feature highlights those little things a technician can do or say to enhance customer relations.

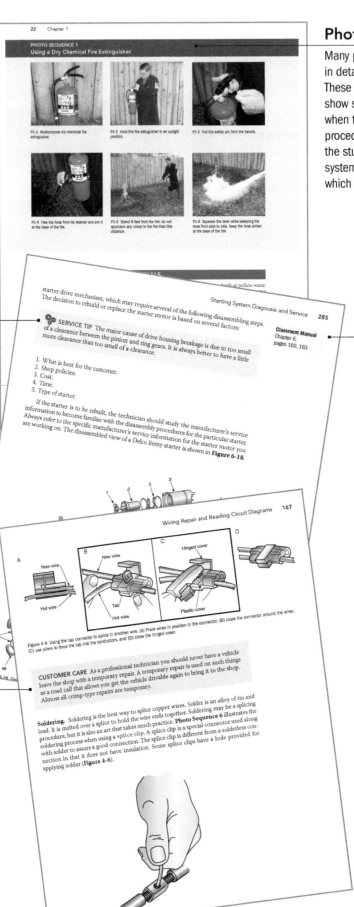

22   Chapter 1

PHOTO SEQUENCE 1
Using a Dry Chemical Fire Extinguisher

P1-1 Multipurpose dry chemical fire extinguisher.

P1-2 Hold the fire extinguisher in an upright position.

P1-3 Pull the safety pin from the handle.

P1-4 Free the hose from its retainer and aim it at the base of the fire.

P1-5 Stand 8 feet from the fire; do not approach any closer to the fire than this distance.

P1-6 Squeeze the lever while sweeping the hose from side to side. Keep the hose aimed at the base of the fire.

starter drive mechanism, which may require several of the following disassembling steps. The decision to rebuild or replace the starter motor is based on several factors:

Starting System Diagnosis and Service   285

SERVICE TIP The major cause of drive housing breakage is due to too small of a clearance between the pinion and ring gears. It is always better to have a little more clearance than too small of a clearance.

Classroom Manual
Chapter 6,
pages 160, 163

1. What is best for the customer.
2. Shop policies.
3. Cost.
4. Time.
5. Type of starter.

If the starter is to be rebuilt, the technician should study the manufacturer's service information to become familiar with the disassembly procedures for the particular starter. Always refer to the specific manufacturer's service information for the starter motor you are working on. The disassembled view of a Delco Remy starter is shown in **Figure 6-18**.

Wiring Repair and Reading Circuit Diagrams   167

Figure 4-5 Using the tap connector to splice in another wire. (A) Place wires in position in the connector, (B) close the connector around the wires, (C) use pliers to force the tab into the conductors, and (D) close the hinged cover.

CUSTOMER CARE As a professional technician you should never have a vehicle leave the shop with a temporary repair. A temporary repair is used on such things as a road call that allows you get the vehicle drivable again to bring it to the shop. Almost all crimp-type repairs are temporary.

Soldering. Soldering is the best way to splice copper wires. Solder is an alloy of tin and lead. It is melted over a splice to hold the wire ends together. Soldering may be a splicing procedure, but it is also an art that takes much practice. **Photo Sequence 6** illustrates the soldering process when using a splice clip. A splice clip is a special connector used along with solder to assure a good connection. The splice clip is different from a solderless connection in that it does not have insulation. Some splice clips have a hole provided for applying solder (**Figure 4-6**).

Case Studies

Beginning with Chapter 3, each chapter ends with a case study describing a particular vehicle problem and the logical steps a technician might use to solve the problem.

The vehicle owner complains that the brake lights do not light. He also says the dome light is not working. The technician verifies the problem and then checks the battery for good connections and tests the fusible links. All are in good condition.

A study of the wiring diagram indicates that the brake light and dome light circuits share the same fuse. It is also indicated that the ignition switch illumination light circuit is shared with these two circuits. A check of the ignition switch illumination light shows that it is not operating either. The technician checks the fuse that is identified in the wiring diagram. It is blown. When a replacement fuse is installed, the dome and brake lights work properly for three tests, and then the fuse blows again.

Upon further testing of the shared circuits, an intermittent short to ground is located in the steering column in the ignition switch illumination circuit. The technician solders in a repair wire to replace the damaged section. After all repairs are completed, a final test indicates proper operation of all circuits.

Ase-Style Review Questions

Each chapter contains ASE-style review questions that reflect the performance-based objectives listed at the beginning of the chapter. These questions can be used to review the chapter as well as to prepare for the ASE certification exam.

ASE-STYLE REVIEW QUESTIONS

1. Splicing copper wire is being discussed.
   Technician A says it is acceptable to use solderless connections.
   Technician B says acid core solder should not be used on copper wires.
   Who is correct?
   A. A only
   B. B only
   C. Both A and B
   D. Neither A nor B

2. Use of wiring diagrams is being discussed.
   Technician A says a wiring diagram is used to help determine related circuits.
   Technician B says the wiring
   exact locatio

4. Repairs to a twisted/shielded wire are being discussed.
   Technician A says a twisted/shielded wire carries high current.
   Technician B says because a twisted/shielded wire carries low current, any repairs to the wire must not increase the resistance of the circuit.
   Who is correct?
   A. A only
   B. B only
   C. B   d B
                nor B
              ssed.
              are
             g both

ASE Challenge Questions

Each technical chapter ends with five ASE challenge questions. These are not mere review questions; rather, they test students' ability to apply general knowledge to the contents of the chapter.

426   Chapter 9

ASE CHALLENGE QUESTIONS

1. *Technician A* says when diagnosing intermittent faults, it is good practice to substitute control modules to see if the problem goes away.
   Technician B says a circuit performance fault indicates that the continuity of the circuit is suspect.
   Who is correct?
   A. A only
   B. B only
   C. Both A and B
   D. Neither A nor B

2. The scan tool displays 5 volts for the ambient
   ature sensor. This indicates:
        or return circuit

4. *Technician A* says the MAP sensor reading with the key on, engine off should equal barometric pressure.
   Technician B says when the engine is started, the MAP sensor signal voltage should increase.
   Who is correct?
   A. A only
   B. B only
   C. Both A and B
   D. Neither A nor B

5. A vehicle with four-wheel ABS has a problem with the right rear wheel locking during heavy braking.
   Technician A says this could be caused by a bad speed sensor mounted at the wheel.
   Technician B says the speed sensor mounted at
        ould cause this problem.

Name _____

SHOP SAFETY SURVEY

As a professional technician, safety should be one of your first concerns. This job sheet will increase your awareness of shop safety items. As you survey your shop and answer the following questions, you will learn how to evaluate the safety of any workplace.

Date _____   Safety   41

Procedure

Your instructor will review your work at each Instructor Response point.

1. Before you begin to evaluate

JOB SHEET
1

Task Completed

Job Sheets

Located at the end of each chapter, job sheets provide a format for students to perform procedures covered in the chapter. A reference to the ASE task addressed by the procedure is included on the Job Sheet.

Basic Electrical Troubleshooting and Service   161

DIAGNOSTIC CHART 3-6
PROBLEM AREA:	Relays.
SYMPTOMS:	Component fails to turn on.
POSSIBLE CAUSES:	Faulty terminals to relay.
	Open in relay coil.
	Shorted relay coil.
	Burned relay high-current contacts.
	Internal open in relay high-current circuit.

DIAGNOSTIC CHART 3-7
PROBLEM AREA:	Relay.
SYMPTOMS:	Component fails to turn off.
POSSIBLE CAUSES:	Short across the relay terminals.
	Stuck relay high-current contacts.

DIAGNOSTIC CHART 3-8
PROBLEM AREA:	Stepped resistors.
SYMPTOMS:	Component fails to turn on.
POSSIBLE CAUSES:	Faulty terminal connections to the stepped resistor.
	Open in the input side of the resistor.
	Open in the output side of the resistor.

DIAGNOSTIC CHART 3-9
PROBLEM AREA:	Stepped resistors.
SYMPTOMS:	Component fails to operate at certain speeds or brightness.
POSSIBLE CAUSES:	Excessive resistance at the terminal connections to the stepped resistor.
	Open in one or more of the resistor's circuit.
	Short across one or more of the resistor's circuit.

DIAGNOSTIC CHART 3-10
PROBLEM AREA:	Variable resistors.
SYMPTOMS:	Component fails to turn on.
POSSIBLE CAUSES:	Faulty terminal connections to the variable resistor.
	Open in the input side of the resistor.
	Open in the output side of the resistor.
	Open in the return circuit of the resistor.
	Short to ground in the output circuit.
	Short to ground in the input circuit.

Diagnostic Charts

Some chapters include detailed diagnostic charts that list common problems and most probable causes. They also list a page reference in the Classroom Manual for better understanding of the system's operation and a page reference in the Shop Manual for details on the procedure necessary for correcting the problem.

SUPPLEMENTS

Instructor Resources

The *Today's Technician* series offers a robust set of instructor resources, available online at Cengage's Instructor Resource Center and on DVD. The following tools have been provided to meet any instructor's classroom preparation needs:

- An Instructor's Guide including lecture outlines, teaching tips, and complete answers to end-of-chapter questions.
- PowerPoint presentations with images and animations that coincide with each chapter's content coverage.
- Cengage Learning Testing Powered by Cognero® provides hundreds of test questions in a flexible, online system. You can choose to author, edit, and manage test bank content from multiple Cengage Learning solutions and deliver tests from your LMS, or you can simply download editable Word documents from the DVD or Instructor Resource Center.
- An Image Gallery includes photos and illustrations from the text.
- The Job Sheets from the Shop Manual are provided in Word format.
- End-of-Chapter Review Questions are provided in Word format, with a separate set of text rejoinders available for instructors' reference.
- To complete this powerful suite of planning tools, correlation guides are provided to the NATEF tasks and to the previous edition.

MINDTAP FOR AUTOMOTIVE ELECTRICITY & ELECTRONICS, 7E

NEW! The MindTap for *Automotive Electricity & Electronics, seventh edition,* features an integrated course offering a complete digital experience for the student and teacher. This MindTap is highly customizable and combines an enhanced ebook with videos, simulations, learning activities, quizzes, job sheets, and relevant DATO scenarios to help students analyze and apply what they are learning and help teachers measure skills and outcomes with ease.

- *A Guide:* Relevant activities combined with prescribed readings, featured multimedia, and quizzing to evaluate progress, will guide students from basic knowledge to analysis and application.

- *Personalized Teaching:* Teachers are able to control course content by removing, rearranging, or adding their own content to meet the needs of their specific program.

- *Promote Better Outcomes:* Through relevant and engaging content, assignments and activities, students are able to build the confidence they need to succeed. Likewise, teachers are able to view analytics and reports that provide a snapshot of class progress, time in course, engagement, and completion rates.

REVIEWERS

The author and publisher would like to extend special thanks to the following instructors for their contributions to this text:

Brett Baird
Salt Lake Community College
Riverton, UT

Timothy Belt
University of Northwestern Ohio
Lima, OH

Brian Brownfield
Kirkwood Community College
Cedar Rapids, IA

Michael Cleveland
Houston Community College
Houston, TX

Jason Daniels
University of Northwestern Ohio
Lima, OH

Kevin Fletcher
Asheville-Buncombe Technical
Community College
Asheville, NC

Jose Gonzalez
Automotive Training Center
Exton, PA

Jack D. Larmor
Baker College
Flint, MI

Gary Looft
Western Colorado Community College
Grand Junction, CO

Stanley D. Martineau
Utah State University
Price, UT

William McGrath
Moraine Valley College
Warrenville, IL

Tim Mulready
University of Northwestern Ohio
Lima, OH

Gary Norden
Elgin Community College
Elk Grove Village, IL

Michael Stiles
Indian River State College
Pierce, FL

David Tapley
University of Northwestern Ohio
Lima, OH

Omar Trinidad
S. Illinois University Carbondale
Carbondale, IL

Cardell Webster
Arapahoe Community College
Littleton, CO

David Young
University of Northwestern Ohio
Lima, OH

In addition, the author would like to extend a special thank-you and acknowledgment to Jerome "Doc" Viola and Arapahoe Community College in Littleton, Colorado, along with Phil Stiebler from Pro Chrysler Jeep Dodge Ram located in Denver, Colorado, for their valuable assistance with this textbook.

CHAPTER 1

INTRODUCTION TO AUTOMOTIVE ELECTRICAL AND ELECTRONIC SYSTEMS

Upon completion and review of this chapter, you should be able to understand and describe:

- The importance of learning automotive electrical systems.
- The role of electrical systems in today's vehicles.
- The interaction of the electrical systems.
- The purpose of the starting system.
- The purpose of the charging system.
- The role of the computer in today's vehicles.
- The purpose of vehicle communication networks.
- The purpose of various electronic accessory systems.
- The purpose of passive restraint systems.
- The purpose of alternate propulsion systems.

Terms To Know

Air bag systems	Electrical accessories	Neutral safety switch
Antitheft system	Fuel cell	Passive restraint systems
Automatic door locks (ADL)	Horn	Power door locks
Bus	Hybrid electric vehicle (HEV)	Power mirrors
Charging system	Ignition switch	Power windows
Computer	Keyless entry system	Starting system
Cruise control	Lighting system	Vehicle instrumentation systems
Easy exit	Memory seat	Voltage regulator
Electric defoggers	Multiplexing	Windshield wipers
Electric vehicle (EV)	Network	

INTRODUCTION

You are probably reading this book for one of two reasons. Either you are preparing yourself to enter the field of automotive service, or you are expanding your skills to include automotive electrical systems. In either case, congratulations on selecting one of the most fast-paced segments of the automotive industry. Working with electrical systems can be challenging, yet very rewarding; however, it can also be very frustrating at times.

For many people, learning electrical systems can be a struggle. It is my hope that I present the material to you in such a manner that you will not only understand electrical systems but also excel at it. There are many ways the theory of electricity can be explained, and many metaphors can be used. Some compare electricity to a water flow, while others explain it in a purely scientific fashion. Everyone learns differently. I am presenting electrical theory in a manner that I hope will be clear and concise. If you do not fully comprehend a concept, then it is important to discuss it with your instructor. Your instructor may be able to use a slightly different method of instruction to help you completely understand the concept. Electricity is somewhat abstract, so if you do have questions, be sure to ask your instructor.

WHY BECOME AN ELECTRICAL SYSTEM TECHNICIAN?

In the past, it was possible for technicians to work their entire careers and be able to almost completely avoid the vehicle's electrical systems. They would specialize in engines, steering/suspension, or brakes. Today there is not a system on the vehicle that is immune to the role of electrical circuits. Engine controls, electronic suspension systems, and antilock brakes are common on today's vehicles. Even electrical systems that were once thought of as being simple have evolved to computer controls. Headlights are now pulse-width modulated using high-side drivers and will automatically brighten and dim based on the light intensity of oncoming traffic. Today's vehicles are equipped with 20 or more computers, laser-guided cruise control, sonar park assist, infrared climate control, fiber optics, and radio frequency transponders and decoders. Simple systems have become more computer reliant. For example, the horn system on the 2010 Chrysler 300C uses three separate control modules and two bus networks to function. Even the tire and wheel assemblies have computers involved, with the addition of tire pressure monitoring systems!

Today's technician must possess a full and complete electrical background to succeed. The future will provide great opportunities for those technicians who have prepared themselves properly.

THE ROLE OF ELECTRICITY IN THE AUTOMOBILE

In the past, electrical systems were basically stand-alone. For example, the ignition system was responsible only for supplying the voltage needed to fire the spark plugs. Ignition timing was controlled by vacuum and mechanical advance systems. Today there are very few electrical systems that are still independent.

 A BIT OF HISTORY

Karl Benz of Mannheim, Germany, patented the world's first automobile on January 29, 1886. The vehicle was a three-wheeled automobile called the Benz Motorwagen. That same year Gottlieb Daimler built a four-wheeled vehicle. It was powered by a 1.5-horsepower engine that produced 50% more power than that of the Benz Motorwagen. The first automobile to be produced for sale in the United States was the 1896 Duryea.

Today, most manufacturers **network** their electrical systems together through computers. This means that information gathered by one system can be used by another. The result may be that a faulty component may cause several symptoms. Consider the following example. The wiper system can interact with the headlight system to turn on the headlights whenever the wipers are turned on. The wipers can interact with the vehicle speed sensor to provide for speed-sensitive wiper operation. The speed sensor may provide information to the antilock brake module. The antilock brake module can then share this information with the transmission control module, and the instrument cluster can receive vehicle speed information to operate the speedometer. If the vehicle speed sensor should fail, this could result in no antilock brake operation and a warning light turned on in the dash. But it could also result in the speedometer not functioning, the transmission not shifting, and the wipers not operating properly.

INTRODUCTION TO THE ELECTRICAL SYSTEMS

The purpose of this section is to acquaint you with the electrical systems that will be covered in this book. We will define the purpose of these systems.

> **AUTHOR'S NOTE** The discussion of the systems in this section of the chapter provides you with an understanding of their *main* purpose. Some systems have secondary functions. These will be discussed in detail in later chapters.

The Starting System

The **starting system** is a combination of mechanical and electrical parts that work together to start the engine. The starting system is designed to change the electrical energy, which is being supplied by the battery, into mechanical energy. For this conversion to be accomplished, a starter or cranking motor is used. The basic starting system includes the following components (**Figure 1-1**):

1. Battery.
2. Cable and wires.
3. Ignition switch.
4. Starter solenoid or relay.
5. Starter motor.
6. Starter drive and flywheel ring gear.
7. Starting safety switch.

The starter motor (**Figure 1-2**) requires large amounts of current (up to 400 amps [A]) to generate the torque needed to turn the engine. The conductors used to carry this amount of current (battery cables) must be large enough to handle the current. These cables may be close to 1/2 inch (12.7 mm) in diameter. It would be impractical to place a conductor of this size into the wiring harness to the ignition switch. To provide control

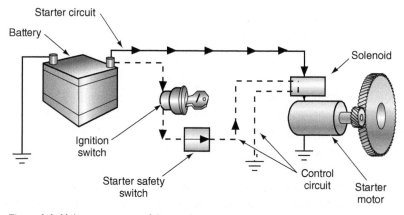

Figure 1-1 Major components of the starting system.

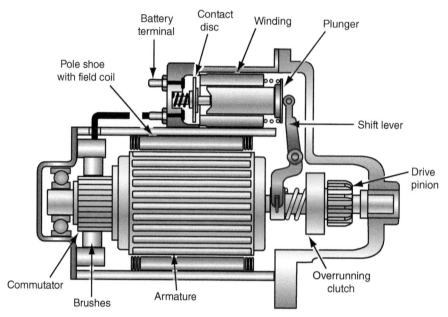

Figure 1-2 Starter motor.

of the high current, all starting systems contain some type of magnetic switch. There are two basic types of magnetic switches used: the solenoid and the relay.

The **ignition switch** is the power distribution point for most of the vehicle's primary electrical systems. The conventional ignition switch is spring loaded in the start position. This momentary contact automatically moves the contacts to the RUN position when the driver releases the key. All other ignition switch positions are detent positions.

The **neutral safety switch** is used on vehicles that are equipped with automatic transmissions. It opens the starter control circuit when the transmission shift selector is in any position except PARK or NEUTRAL. Vehicles that are equipped with automatic transmissions require a means of preventing the engine from starting while the transmission is in gear. Without this feature, the vehicle would lunge forward or backward once it is started, causing personal injury or property damage. The safety switch is connected into the starting system control circuit and is usually operated by the shift lever (**Figure 1-3**). When in the PARK or NEUTRAL position, the switch allows current to flow to the starter circuit. If the transmission is in a gear position, the switch prevents current flow to the starter circuit.

Figure 1-3 The neutral safety switch is usually attached to the transmission.

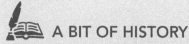

A BIT OF HISTORY

The Model T was called the first "people's car." Prior to its introduction by the Ford Motor Company in 1908, only the wealthy could afford to purchase an automobile. It was Henry Ford's desire to build a car for the masses. Although Henry Ford had no professional engineering education, he did possess a natural inclination toward mechanics. To keep production costs down, he used assembly-line production to manufacture the Model T. Henry Ford also introduced the moving conveyor belt into the assembly process, further accelerating production. The Model T was nicknamed Tin Lizzie because its body was made from lightweight sheet steel. The production of the Model T continued until 1927, with more than 16.5 million vehicles being produced. The electrical system was very simple and originally consisted of a flywheel magneto that produced low-voltage alternating current. This AC voltage was used to power a trembler coil that created a high-voltage current for use by the ignition system. The ignition pulse was passed to the timer (distributor) and directed to the proper cylinder. Ignition timing was adjusted manually via the spark advance lever that was mounted on the steering column. Moving the lever rotated the timer to advance or retard the ignition timing. Since the magneto may not produce sufficient current when starting the engine with the hand crank, a battery could be used to provide the required starting current. When electric headlights were introduced in 1915, the magneto was used to supply power for the lights and the horn.

Many vehicles that are equipped with manual transmissions use a similar type of safety switch. The start/clutch interlock switch is usually operated by movement of the clutch pedal (**Figure 1-4**).

The Charging System

The automotive storage battery is not capable of supplying the demands of the electrical systems for an extended period of time. Every vehicle must be equipped with a means of replacing the energy that is being drawn from the battery. A **charging system** is used to restore to the battery the electrical power that was used during engine starting. In addition, the charging system must be able to react quickly to high-load demands required of the electrical system. It is the vehicle's charging system that generates the current to operate all electrical accessories while the engine is running.

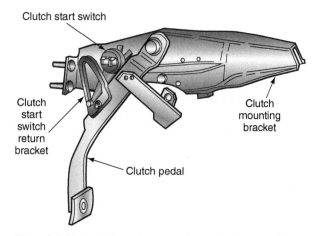

Figure 1-4 Most vehicles with a manual transmission use a clutch start switch.

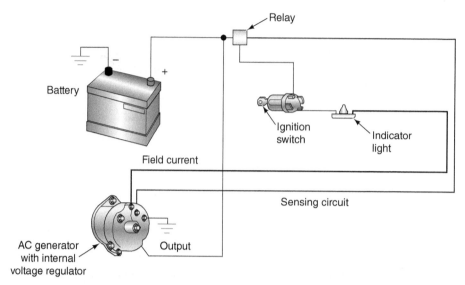

Figure 1-5 Components of the charging system.

The purpose of the charging system is to convert the mechanical energy of the engine into electrical energy to recharge the battery and run the electrical accessories. When the engine is first started, the battery supplies all the current required by the starting and ignition systems.

As illustrated in **Figure 1-5**, the entire charging system consists of the following components:

1. Battery.
2. Alternating current (AC) generator or direct current (DC) generator.
3. Drive belt.
4. Voltage regulator.
5. Charge indicator (lamp or gauge).
6. Ignition switch.
7. Cables and wiring harness.
8. Starter relay (some systems).
9. Fusible link (some systems).

A **voltage regulator** controls the output voltage of the AC generator, based on charging system demands, by controlling field current. The battery, and the rest of the electrical system, must be protected from excessive voltages. To prevent early battery and electrical system failure, regulation of the charging system is very important. Also, the charging system must supply enough current to run the vehicle's electrical accessories when the engine is running.

The Lighting System

The **lighting system** consists of all the lights used on the vehicle (**Figure 1-6**). This includes headlights, front and rear park lights, front and rear turn signals, side marker lights, daytime running lights, cornering lights, brake lights, back-up lights, instrument cluster backlighting, and interior lighting.

The lighting system of today's vehicles can consist of more than 50 light bulbs and hundreds of feet of wiring. Incorporated within these circuits are circuit protectors, relays, switches, lamps, and connectors. In addition, more sophisticated lighting systems use computers and sensors. Since the lighting circuits are largely regulated by federal laws, the systems are similar among the various manufacturers. However, there are variations that exist in these circuits.

Figure 1-6 Automotive lighting system.

With the addition of computer controls in the automobile, manufacturers have been able to incorporate several different lighting circuits or modify the existing ones. Some of the refinements that were made to the lighting system include automatic headlight washers, automatic headlight dimming, automatic on/off with timed-delay headlights, and illuminated entry systems. Some of these systems use sophisticated body computer–controlled circuitry and fiber optics.

Some manufacturers have included such basic circuits as turn signals into their body computer to provide for pulse-width dimming in place of a flasher unit. The body computer can also be used to control instrument panel lighting based on inputs that include if the side marker lights are on or off. By using the body computer to control many of the lighting circuits, the amount of wiring has been reduced. In addition, the use of computer control of these systems has provided a means of self-diagnosis in some applications.

Today, high-density discharge headlamps are becoming an increasingly popular option on many vehicles. These headlights provide improved lighting over conventional headlamps.

Vehicle Instrumentation Systems

Vehicle instrumentation systems (**Figure 1-7**) monitor the various vehicle operating systems and provide information to the driver about their correct operation. Warning devices also provide information to the driver; however, they are usually associated with an audible signal. Some vehicles use a voice module to alert the driver to certain conditions.

Electrical Accessories

Electrical accessories provide for additional safety and comfort. There are many electrical accessories that can be installed into today's vehicles. These include safety accessories such as the horn, windshield wipers, and windshield washers. Comfort accessories include the blower motor, electric defoggers, power mirrors, power windows, power seats, and power door locks.

Horns. A **horn** is a device that produces an audible warning signal (**Figure 1-8**). Automotive electrical horns operate by vibrating a diaphragm to produce a warning signal. This vibration of the diaphragm is repeated several times per second. As the diaphragm

Figure 1-7 The instrument panel displays various operating conditions.

Figure 1-8 Automotive horn.

vibrates, it causes a column of air that is in the horn to vibrate. The vibration of the column of air produces the sound.

Windshield Wipers. Windshield wipers are mechanical arms that sweep back and forth across the windshield to remove water, snow, or dirt (**Figure 1-9**). The operation of the wiper arms is through the use of a wiper motor. Most windshield wiper motors use permanent magnet fields, or electromagnetic field motors.

Electric Defoggers. Electric defoggers heat the rear window to remove ice and/or condensation. Some vehicles use the same circuit to heat the outside mirrors. When electrical current flows through a resistance, heat is generated. Rear window defoggers use this principle of controlled resistance to heat the glass. The resistance is through a grid that is baked on the inside of the glass (**Figure 1-10**). The system may incorporate a timer circuit that controls the relay.

Power Mirrors. Power mirrors are outside mirrors that are electrically positioned from the inside of the driver compartment. The electrically controlled mirror allows the driver to position the outside mirrors by use of a switch. The mirror assembly will use built-in motors.

Figure 1-9 Windshield wipers.

Figure 1-10 Rear window defogger grid.

Power Windows. Power windows are windows that are raised and lowered by use of electrical motors. The motor used in the power window system is capable of operating in forward and reverse rotations. The power window system usually consists of the following components:

1. Master control switch.
2. Individual control switches.
3. Individual window drive motors.

Power Door Locks. Electric power door locks use either a solenoid or a motor to lock and unlock the door. Typically, power door lock systems provide for the locking or unlocking of all doors with a single button push. This system may also incorporate automatic door locking, as discussed later.

Computers

A computer is an electronic device that stores and processes data and is capable of operating other devices (**Figure 1-11**). The use of computers on automobiles has expanded to include control and operation of several functions, including climate control, lighting

Figure 1-11 A control module is used to process data and operate different automotive systems.

circuits, cruise control, antilock braking, electronic suspension systems, and electronic shift transmissions. Some of these are functions of what is known as a body control module (BCM). Some body computer–controlled systems include direction lights, rear window defogger, illuminated entry, intermittent wipers, and other systems that were once thought of as basic.

A computer processes the physical conditions that represent information (data). The operation of the computer is divided into four basic functions:

1. Input.
2. Processing.
3. Storage.
4. Output.

Vehicle Communication Networks

All manufacturers now use a system of vehicle communications called **multiplexing (MUX)** to allow control modules to share information (**Figure 1-12**). Multiplexing provides the ability to use a single circuit to distribute and share data between several control modules throughout the vehicle. Because the data is transmitted through a single circuit, bulky wiring harnesses are eliminated.

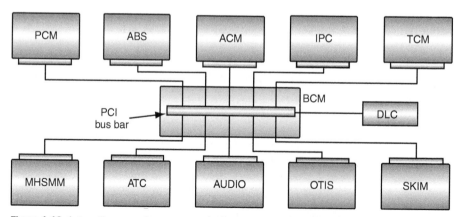

Figure 1-12 Automotive computers are networked together through multiplexing.

Vehicle manufacturers will use multiplexing systems to enable different control modules to share information. A MUX wiring system uses **bus** data links that connect each module. The term *bus* refers to the transporting of data from one module to another. Each module can transmit and receive digital codes over the bus data links. The signal sent from a sensor can go to any one of the modules and can be shared by the other modules.

Electronic Accessory Systems

With the growing use of computers, most systems can be controlled electronically. This provides for improved monitoring of the systems for proper operation and the ability to detect if a fault occurs. The systems that are covered in this book include the following:

Electronic Cruise Control Systems. **Cruise control** is a system that allows the vehicle to maintain a preset speed with the driver's foot off of the accelerator. Most cruise control systems are a combination of electrical and mechanical components.

> **Cruise control** systems are also referred to as *speed control*.

Memory Seats. The **memory seat** feature allows the driver to program different seat positions that can be recalled at the push of a button. The memory seat feature is an addition to the basic power seat system. Most memory seat systems share the same basic operating principles, the difference being in programming methods and the number of positions that can be programmed. Most systems provide for two seat positions to be stored in memory.

An **easy exit** feature may be an additional function of the memory seat that provides for easier entrance and exit of the vehicle by moving the seat all the way back and down. Some systems also move the steering wheel up and to full retract.

Electronic Sunroofs. Some manufacturers have introduced electronic control of their electric sunroofs. These systems incorporate a pair of relay circuits and a timer function into the control module. Motor rotation is controlled by relays that are activated based on signals received from the slide, tilt, and limit switches.

Antitheft Systems. The **antitheft system** is a deterrent system designed to scare off would-be thieves by sounding alarms and/or disabling the ignition system. **Figure 1-13** illustrates many of the common components that are used in an antitheft system. These components include the following:

1. An electronic control module.
2. Door switches at all doors.

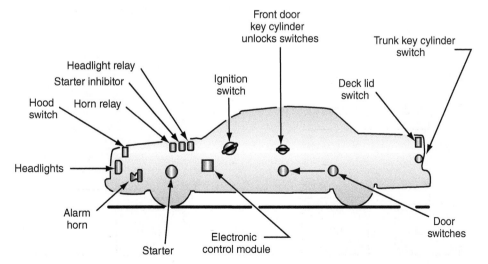

Figure 1-13 Typical components of an antitheft system.

3. Trunk key cylinder switch.
4. Hood switch.
5. Starter inhibitor relay.
6. Horn relay.
7. Alarm.

In addition, many systems incorporate the exterior lights into the system. The lights are flashed if the system is activated.

Some systems use ultrasonic sensors that will signal the control module if someone attempts to enter the vehicle through the door or window. The sensors can be placed to sense the parameter of the vehicle and sound the alarm if someone enters within the protected parameter distance.

Automatic Door Locks. **Automatic door locks (ADL)** is a passive system that locks all doors when the required conditions are met. Many automobile manufacturers are incorporating automatic door locks as an additional safety and convenience system. Most systems lock the doors when the gear selector is placed in DRIVE, the ignition switch in RUN, and all doors are shut. Some systems will lock the doors when the gear shift selector is passed through the REVERSE position, while others do not lock the doors unless the vehicle is moving 15 mph or faster.

In addition, the system may unlock one or all doors when the vehicle is stopped and the driver's door is opened. The sequence of which the doors are unlocked may be a programmable feature that vehicle owners can set to their preference.

The system may use the body computer or a separate controller to control the door lock relays. The controller (or body computer) takes the place of the door lock switches for automatic operation.

Keyless Entry. The **keyless entry system** allows the driver to unlock the doors or the deck lid (trunk) from outside of the vehicle without the use of a key. The main components of the keyless entry system are the control module, a coded-button keypad located on the driver's door (**Figure 1-14**), and the door lock motors.

Some keyless entry systems can be operated remotely. Pressing a button on a hand-held transmitter will allow operation of the system from distances of 25 to 50 feet (**Figure 1-15**).

Recently, most manufacturers have made available systems of remote engine starting and keyless start. These are usually designed into the function of the remote keyless entry system.

Figure 1-14 Keyless entry system keypad.

Figure 1-15 Remote keyless entry system transponder.

Passive Restraint Systems

Federal regulations have mandated the use of automatic **passive restraint systems** in all vehicles sold in the United States after 1990. Passive restraints are ones that operate automatically, with no action required on the part of the driver or occupant.

Air bag systems are on all of today's vehicles. The need to supplement the existing restraint system during frontal collisions has led to the development of the supplemental inflatable restraint (SIR) or air bag systems (**Figure 1-16**).

A typical air bag system consists of sensors, a diagnostic module, a clock spring, and an air bag module. **Figure 1-17** illustrates the typical location of the common components of the SIR system.

Alternate Propulsion Systems

Due to the increase in regulations concerning emissions and the public's desire to become less dependent on foreign oil, most major automotive manufacturers have developed alternative fuel or alternate power vehicles. Since the 1990s, most major automobile manufacturers have developed an **electric vehicle (EV)**. The primary advantage of an EV is a drastic reduction in noise and emission levels. General Motors introduced the EV1 electric car to the market in 1996. The original battery pack in this car contained twenty-six 12-volt (V) batteries that delivered electrical energy to a three-phase 102-kilowatt (kW) AC electric motor. The electric motor is used to drive the front wheels. The driving range is about 70 miles (113 km) of city driving or 90 miles (145 km) of highway driving.

With the improvements in battery technology and more user-friendly charging methods, the 100% electric vehicle is showing promise in some market segments. The Nissan Leaf (**Figure 1-18**) is an example of a 100% electric, zero emissions vehicle. This vehicle uses an 80 kW synchronous electric motor and a single speed constant ratio transmission. The lithium ion battery provides a driving range between 73 miles (117 km) and 109 miles (175 km) between charges.

EV battery limitation was a major stumbling block to most consumers. One method of improving the electric vehicle resulted in the addition of an on-board power generator that is assisted by an internal combustion engine, resulting in the **hybrid electric vehicle (HEV)**.

Basically, the HEV relies on power from the electric motor, the engine, or both (**Figure 1-19**). When the vehicle moves from a stop and has a light load, the electric

Stage 1

Stage 2

Stage 3

Figure 1-16 Air bag deployment sequence.

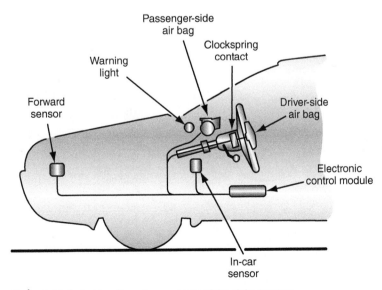

Figure 1-17 Typical location of components of the air bag system.

Figure 1-19 HEV power system.

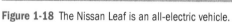

Figure 1-18 The Nissan Leaf is an all-electric vehicle.

© iStockphoto/Sjo

motor moves the vehicle. Power for the electric motor comes from stored energy in the battery pack. During normal driving conditions, the engine is the main power source. Engine power is also used to rotate a generator that recharges the storage batteries. The output from the generator may also be used to power the electric motor, which is run to provide additional power to the powertrain. A computer controls the operation of the electric motor, depending on the power needs of the vehicle. During full throttle or heavy load operation, additional electricity from the battery is sent to the motor to increase the output of the powertrain. **Fuel cell**–powered vehicles have a very good chance of becoming the drives of the future. They combine the reach of conventional internal combustion engines with high efficiency, low fuel consumption, and minimal or no pollutant emission. At the same time, they are extremely quiet. Because they work with regenerative fuel such as hydrogen, they reduce the dependence on crude oil and other fossil fuels.

A fuel cell–powered vehicle (**Figure 1-20**) is basically an electric vehicle. Like the electric vehicle, it uses an electric motor to supply torque to the drive wheels. The difference is that the fuel cell produces and supplies electric power to the electric motor instead of batteries. Most of the vehicle manufacturers and several independent laboratories are involved in fuel cell research and development programs. Many prototype fuel cell vehicles have been produced, with many being placed in fleets in North America and Europe.

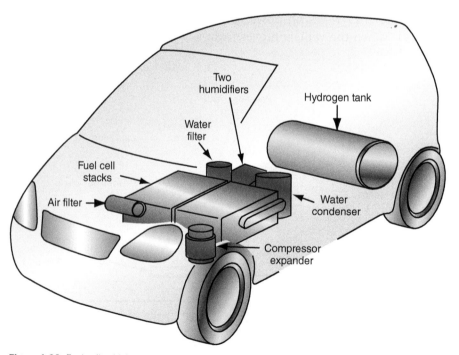

Figure 1-20 Fuel cell vehicle components.

SUMMARY

- The starting system is a combination of mechanical and electrical parts that work together to start the engine.
- The charging system replaces the electrical power used by the battery and provides current to operate the electrical accessories while the engine is running.
- The lighting system consists of all the lights used on the vehicle.
- Vehicle instrumentation systems monitor the various vehicle operating systems and provide information to the driver.
- Electrical accessories provide additional safety and comfort.
- Many of the basic electrical accessory systems have electronic controls added to them to provide additional features and enhancement.
- Computers are electronic devices that gather, store, and process data.
- Most vehicles use a multiplexing system to share information between computer systems.
- The memory seat feature allows the driver to program different seat positions that can be recalled at the push of a button.

- Some manufacturers have introduced electronic control of their electric sunroofs. These systems incorporate a pair of relay circuits and a timer function into the control module.
- Antitheft systems are deterrent systems designed to scare off would-be thieves by sounding alarms and/or disabling the ignition system.
- Automatic door locks are a passive system used to lock all doors when the required conditions are met. Many automobile manufacturers are incorporating the system as an additional safety and convenience feature.
- Passive restraints operate automatically, with no action required on the part of the driver or occupant.
- Electric vehicles powered by an electric motor run off a battery pack.
- The hybrid electric vehicle relies on power from the electric motor, the engine, or both.
- A fuel cell–powered vehicle is basically an electric vehicle, except that the fuel cell produces and supplies electric power to the electric motor instead of batteries.

REVIEW QUESTIONS

Short-Answer Essays

1. Describe your level of comfort concerning automotive electrical systems.

2. Explain why you feel it is important to understand the operation of the automotive electrical system.

3. Explain how the use of computers has changed the automotive electrical system.

4. Explain the difference between an electric vehicle and a fuel cell vehicle.

5. Explain the basics of HEV operation.

6. What is the purpose of the keyless entry system?

7. What safety benefits can be achieved from the automatic door lock system?

8. What is the purpose of the starting system?

9. What is the purpose of the charging system?

10. What is the function of the air bag system?

Fill in the Blanks

1. The ability of computers to share information with each other is _____.

2. Vehicle instrumentation systems _____ the various operating systems and provide information to the driver about their correct operation.

3. A _____ is an electronic device that stores and processes data.

4. The starting system is designed to change the _____ energy into mechanical energy.

5. The _____ _____ feature is an additional function of the memory seat that provides for easier entrance and exit of the vehicle.

6. The _____ _____ is the power distribution point for most of the vehicle's primary electrical systems.

7. The antitheft system is referred to as a _____ system.

8. The purpose of the charging system is to convert the _____ energy of the engine into _____ energy.

9. _____ restraints operate automatically, with no action required on the part of the driver.

10. The _____ _____ _____ uses an on-board power generator that is assisted by an internal combustion engine.

Multiple Choice

1. Electric vehicles power the motor by:
   A. A generator.   C. An engine.
   B. A battery pack.   D. None of the above.

2. The charging system:
   A. Provides electrical energy to operate the electrical system while the engine is running.
   B. Restores the energy to the battery after starting the engine.
   C. Uses the principle of magnetic induction to generate electrical power.
   D. All of the above.

3. The memory seat system:
   A. Operates separately of the power seat system.
   B. Requires the vehicle to be moving before the seat position can be recalled.
   C. Allows for the driver to program different seat positions that can be recalled at the push of a button.
   D. Can be equipped only on vehicles with manual position seats.

4. The following are true about the easy exit feature EXCEPT:
   A. It is an additional function of the memory seat.
   B. The driver's door is opened automatically.
   C. The seat is moved all the way back and down.
   D. The system may move the steering wheel up.

5. The following are components of the starting system EXCEPT:
   A. The flywheel ring gear.
   B. Neutral safety switch.
   C. Harmonic balancer.
   D. Battery.

6. Which of the following is the most correct statement?
   A. Automotive electrical systems are interlinked with each other.
   B. All automotive electrical systems function the same on every vehicle.
   C. Manufacturers are required by federal legislation to limit the number of computers used on today's vehicles.
   D. All of the above.

7. Automotive horns operate on the principle of:
   A. Induced voltage.
   B. Depletion zone bonding.
   C. Frequency modulation.
   D. Electromagnetism.

8. The purpose of multiplexing is to:
   A. Increase circuit loads to a sensor.
   B. Prevent electromagnetic interference.
   C. Allow computers to share information.
   D. Prevent multiple system failures from occurring.

9. The following are true about the air bag system EXCEPT:
   A. It is an active system.
   B. It is a supplemental system.
   C. It is mandated by the federal government.
   D. Deployment is automatic.

10. Alternate propulsion systems include:
    A. Electric vehicles.   C. Fuel cell vehicles.
    B. Hybrid vehicles.   D. All of the above.

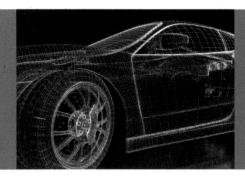

CHAPTER 2
BASIC THEORIES

Upon completion and review of this chapter, you should be able to:

- Explain the theories and laws of electricity.
- Describe the difference between insulators, conductors, and semiconductors.
- Define voltage, current, and resistance.
- Define and use Ohm's law correctly.
- Explain the difference between AC and DC currents.
- Define and illustrate series, parallel, and series-parallel circuits and the electrical laws that govern them.
- Explain the theory of electromagnetism.
- Explain the principles of induction.

Terms To Know

Alternating current (AC)	Electron theory	Photovoltaics (PV)
Ampere (A)	Equivalent series load	Power (P)
Atom	Ground	Protons
Balanced	Induction	Reluctance
Balanced atom	Insulator	Resistance
Circuit	Ion	Right-hand rule
Closed circuit	Kirchhoff's current law	Saturation
Conductor	Kirchhoff's voltage law	Self-induction
Continuity	Magnetic flux density	Semiconductors
Conventional theory	Mutual induction	Series circuit
Current	Neutrons	Series-parallel circuit
Cycle	Nucleus	Shell
Direct current (DC)	Ohms	Static electricity
Electromagnetic interference (EMI)	Ohm's law	Valence ring
	Open circuit	Voltage
Electromagnetism	Parallel circuit	Voltage drop
Electromotive force (EMF)	Permeability	Watt
Electron		

INTRODUCTION

The electrical systems used in today's vehicles can be very complicated (**Figure 2-1**). However, through an understanding of the principles and laws that govern electrical circuits, technicians can simplify their job of diagnosing electrical problems. In this chapter, you will learn the laws that dictate electrical behavior, how circuits operate, the difference

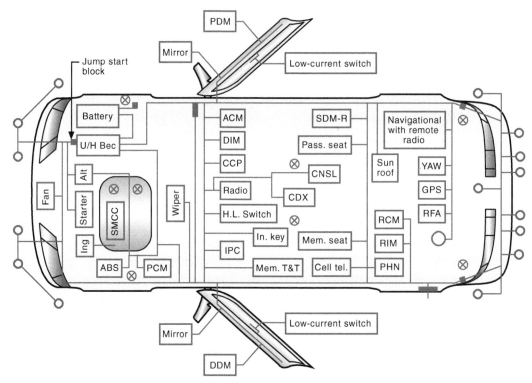

Figure 2-1 The electrical system of today's vehicle can be very complicated.

between types of circuits, and how to apply Ohm's law to each type of circuit. You will also learn the basic theories of semiconductor construction. Because magnetism and electricity are closely related, a study of electromagnetism and induction is included in this chapter.

BASICS OF ELECTRON FLOW

Because electricity is an energy form that cannot be seen, some technicians regard the vehicle's electrical system as being more complicated than it is. These technicians approach the vehicle's electrical system with some reluctance. It is important for today's technician to understand that electrical behavior is confined to definite laws that produce predictable results and effects. To facilitate the understanding of the laws of electricity, a short study of atoms is presented.

Atomic Structure

An **atom** is the smallest part of a chemical element that still has all the characteristics of that element. An atom is constructed of a fixed arrangement of **electrons** in orbit around a **nucleus**—much like planets orbiting the sun (**Figure 2-2**). Electrons are negatively

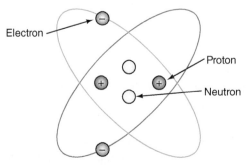

Figure 2-2 The composition of an atom.

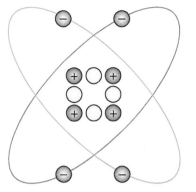 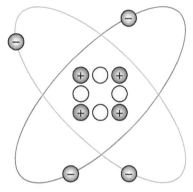 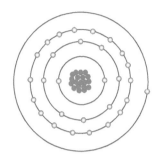

Figure 2-3 If the number of electrons and protons in an atom is the same, the atom is balanced.

Figure 2-4 Basic structure of a copper atom.

charged particles. The nucleus contains positively charged particles called **protons** and particles that have no charge, which are called **neutrons**. The protons and neutrons that make up the nucleus are tightly bound together. The electrons are free to move within their orbits at fixed distances around the nucleus. The attraction between the negative electrons and the positive protons causes the electrons to orbit the nucleus. All the electrons surrounding the nucleus are negatively charged, so they repel each other when they get too close. The electrons attempt to stay as far away from each other as possible without leaving their orbits.

Atoms attempt to have the same number of electrons as there are protons in the nucleus. This makes the atom **balanced** (**Figure 2-3**). To remain balanced, an atom will shed an electron or attract an electron from another atom. A specific number of electrons are in each of the electron orbit paths. The orbit closest to the nucleus has room for 2 electrons; the second orbit holds up to 8 electrons; the third holds up to 18; and the fourth and fifth hold up to 32 each. The number of orbits depends on the number of electrons the atom has. For example, a copper atom contains 29 electrons; 2 in the first orbit, 8 in the second, 18 in the third, and 1 in the fourth (**Figure 2-4**). The outer orbit, or **shell** as it is sometimes called, is referred to as the **valence ring**. In studying the laws of electricity, the only concern is with the electrons that are in the valence ring.

Since an atom seeks to be balanced, an atom that is missing electrons in its valence ring will attempt to gain other electrons from neighboring atoms. Also, if the atom has an excess number of electrons in its valence ring, it will try to pass them on to neighboring atoms.

Like charges repel each other; unlike charges attract each other.

Conductors and Insulators

To help explain why you need to know about these electrons and their orbits, let's continue to look at the atomic structure of copper. Copper is a metal and is the most commonly used **conductor** of electricity. A conductor is something that supports the flow of electricity through it. As stated earlier, the copper atom has 29 electrons and 29 protons, but there is only 1 electron in the valance ring. For the valence ring to be completely filled, it would require 32 electrons. Since there is only one electron, it is loosely tied to the atom and can be easily removed, making it a good conductor.

Copper, silver, gold, and other good conductors of electricity have only one or two electrons in their valence ring. These atoms can be made to give up the electrons in their valence ring with little effort.

Since electricity is the movement of electrons from one atom to another, atoms that have one to three electrons in their valence ring support electricity. They allow the electron to easily move from the valence ring of one atom to the valence ring of another atom. Therefore, if we have a wire made of millions of copper atoms, we have a good conductor of electricity. To have electricity, we simply need to add one electron to one of the copper atoms. That atom will shed the electron it had to another atom, which will shed its original electron to another, and so on. As the electrons move from atom to atom, a force is released. This force is what we use to light lamps, run motors, and so on. As long as we keep the electrons moving in the conductor, we have electricity.

Insulators are materials that don't allow electrons to flow through them easily. Insulators are atoms that have five to eight electrons in their valence ring. The electrons are held tightly around the atom's nucleus and they can't be moved easily. Insulators are used to prevent electron flow or to contain it within a conductor. Insulating material covers the outside of most conductors to keep the moving electrons within the conductor.

 A BIT OF HISTORY

Electricity was discovered by the Greeks over 2,500 years ago. They noticed that when amber was rubbed with other materials, it was charged with an unknown force that had the power to attract objects such as dried leaves and feathers. The Greeks called amber "elektron." The word *electric* is derived from this word and means "to be like amber."

In summary, the number of electrons in the valence ring determines whether an atom is a good conductor or an insulator. Some atoms are neither good insulators nor good conductors; these are called **semiconductors**. In short:

1. Three or fewer electrons—conductor.
2. Five or more electrons—insulator.
3. Four electrons—semiconductor.

ELECTRICITY DEFINED

Electricity is the movement of electrons from atom to atom through a conductor (**Figure 2-5**). Electrons are attracted to protons. Since we have excess electrons on one end of the conductor, we have many electrons being attracted to the protons. This attraction acts to push the electrons toward the protons. This push is normally called electrical pressure. The amount of electrical pressure is determined by the number of electrons that are attracted to protons. The electrical pressure or **electromotive force (EMF)** attempts to push an electron out of its orbit and toward the excess protons. If an electron is freed from its orbit, the atom acquires a positive charge because it now has one more proton than it has electrons. The unbalanced atom or **ion** attempts to return to its balanced state so it will attract electrons from the orbit of other **balanced atoms**. This starts a chain reaction as one atom captures an electron and another releases an electron. As this action continues to occur, electrons will flow through the conductor. A stream of free electrons forms and an electrical current is started. This does not mean a single electron travels the length of the insulator; it means the overall effect is electrons moving in one direction. All this happens at almost the speed of light. The strength of the electron flow is dependent on the potential difference or voltage.

The three elements of electricity are voltage, current, and resistance. How these three elements interrelate governs the behavior of electricity. Once the technician comprehends the laws that govern electricity, understanding the function and operation of the various automotive electrical systems is an easier task. This knowledge will assist the technician in diagnosis and repair of automotive electrical systems.

Voltage

Voltage can be defined as an electrical pressure (**Figure 2-6**) and is the electromotive force that causes the movement of the electrons in a conductor. In Figure 2-5, voltage is the force of attraction between the positive and the negative charges. An electrical

---

Random movement of electrons is not electric current; the electrons must move in the same direction.

It is often stated that the speed of light is 186,000 miles per second (299,000 kilometers per second). The actual speed of electricity depends on the composition of the wire and insulation and is slower than the speed of light due to the electrons bumping into each other and changing places.

An E can be used for the symbol to designate voltage (electromotive force). A V is also used as a symbol for voltage.

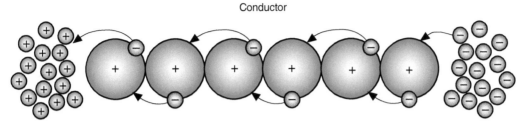

Figure 2-5 As electrons flow in one direction from one atom to another, an electrical current develops.

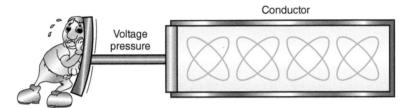

Figure 2-6 Voltage is the pressure that causes the electrons to move.

pressure difference is created when there is a mass of electrons at one point in the circuit, and a lack of electrons at another point in the circuit. In an automobile, the battery or generator is used to apply the electrical pressure.

The amount of pressure applied to a circuit is stated in the number of volts. If a voltmeter is connected across the terminals of an automobile battery, it may indicate 12.6 volts. This is actually indicating that there is a difference in potential of 12.6 volts. There is 12.6 volts of electrical pressure between the two battery terminals.

In a circuit that has current flowing, voltage will exist between any two points in that circuit (**Figure 2-7**). The only time voltage does not exist is when the potential drops to zero. In Figure 2-7 the voltage potential between points A and C and between points B and C is 12.6 volts. However, between points A and B the pressure difference is zero and the voltmeter will indicate 0 volts.

Shop Manual
Chapter 2,
pages 48, 54

One volt (V) is the amount of pressure required to move one ampere of current through one ohm of resistance.

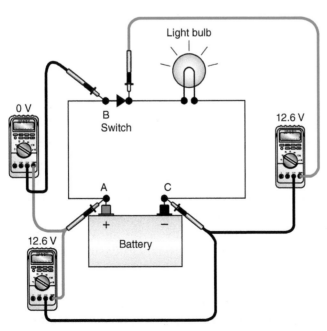

Figure 2-7 A simplified light circuit illustrating voltage potential.

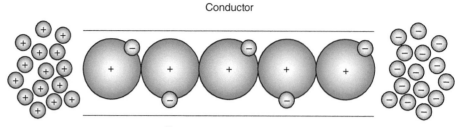

Conductor

6.25×10^{18} electrons per second = one ampere

Figure 2-8 The rate of electron flow is called current and is measured in amperes.

Current

Shop Manual
Chapter 2,
pages 48, 59

One **ampere (A)** represents the movement of 6.25×10^{18} electrons (or one coulomb) past one point in a conductor in one second.

The symbol for current is "I," which stands for intensity. Also "A" is used for amperage.

Current can be defined as the *rate* (intensity) of electron flow (**Figure 2-8**) and is measured in amperes. Current is a measurement of the electrons passing any given point in the circuit in 1 second. Because the flow of electrons is so fast, it would be impossible to physically see electron flow. However, the rate of electron flow can be measured. Current will increase as pressure (voltage) is increased, provided circuit resistance remains constant.

An electrical current will continue to flow through a conductor as long as the electromotive force is acting on the conductor's atoms and electrons. If a potential exists in the conductor, with a buildup of excess electrons at the end of the conductor farthest from the EMF and there is a lack of electrons at the EMF side, current will flow. The effect is called electron drift and accounts for the method in which electrons flow through a conductor.

An electrical current can be formed by the following forces: friction, chemical reaction, heat, pressure, magnetic induction, and photovoltaics (PV). Whenever electrons flow or drift in mass, an electrical current is formed. There are six laws that regulate this electrical behavior:

1. Like charges repel each other.
2. Unlike charges attract each other.
3. A voltage difference is created in the conductor when an EMF is acting on the conductor.
4. Electrons flow only when a voltage difference exists between the two points in a conductor.
5. Current tends to flow to ground in an electrical circuit as a return to source.
6. **Ground** is defined as the baseline when measuring electrical circuits and is the point of lowest voltage. Also, it is the return path to the source for an electrical circuit. The ground circuit used in most automotive systems is through the vehicle chassis and/or engine block. In addition, ground allows voltage spikes to be directed away from the circuit by absorbing them.

So far we have described current as the movement of electrons through a conductor. Electrons move because of a potential difference. This describes one of the common theories about current flow. The **electron theory** states that since electrons are negatively charged, current flows from the most negative to the most positive point within an electrical circuit. In other words, current flows from negative to positive. This theory is widely accepted by the electronic industry.

Another current flow theory is called the **conventional theory**. This theory describes current flow as being from positive to negative. The basic idea behind this theory is simply that although electrons move toward the protons, the energy or force that is released as the electrons move begins at the point where the first electron moved to the most positive charge. As electrons continue to move in one direction, the released energy moves in the opposite direction. This theory is the oldest theory and serves as the basis for most electrical diagrams.

Trying to make sense of it all may seem difficult. It is also difficult for scientists and engineers. In fact, another theory has been developed to explain the mysteries of current flow. This theory is called the hole-flow theory and is actually based on both electron theory and the conventional theory.

As a technician, you will find references to all of these theories. Fortunately, it really doesn't matter as long as you know what current flow is and what affects it. From this understanding, you will be able to figure out how the circuit basically works, how to test it, and how to repair it. In this text, we will present current flow as moving from positive to negative and electron flow as moving from negative to positive. Remember that current flow is the result of the movement of electrons, regardless of the theory.

A BIT OF HISTORY

The ampere is named after André Ampère, who in the late 1700s worked with magnetism and current flow to develop some foundations for understanding the behavior of electricity.

Resistance

The third component in electricity is **resistance**. Resistance is the opposition to current flow and is measured in **ohms** (Ω). In a circuit, resistance controls the amount of current. The size, type, length, and temperature of the material used as a conductor will determine its resistance. Devices that use electricity to operate (motors and lights) have a greater amount of resistance than the conductor.

Shop Manual
Chapter 2,
pages 48, 58

The **ohm** (Ω) is the unit of measurement for resistance of a conductor such that a constant current of one ampere in it produces a voltage of one volt between its ends.

The symbol for resistance is *R*.

A complete electrical **circuit** consists of the following: (1) a power source, (2) a load or resistance unit, and (3) conductors. Resistance (load) is required to change electrical energy to light, heat, or movement. There is resistance in any working device of a circuit, such as a lamp, motor, relay, coil, or other load component.

There are five basic characteristics that determine the amount of resistance in any part of a circuit:

1. The atomic structure of the material: The higher the number of electrons in the outer valence ring, the higher the resistance of the material.
2. The length of the conductor: The longer the conductor, the higher the resistance.
3. The diameter of the conductor: The smaller the cross-sectional area of the conductor, the higher the resistance.
4. Temperature: Normally an increase in temperature of the conductor causes an increase in the resistance.
5. Physical condition of the conductor: If the conductor is damaged by nicks or cuts, the resistance will increase because the conductor's diameter is decreased by these.

There may be unwanted resistance in a circuit. This could be in the form of a corroded connection or a broken conductor. In these instances, the resistance may cause the load component to operate at reduced efficiency or to not operate at all.

It does not matter if the resistance is from the load component or from unwanted resistance. There are certain principles that dictate its impact in the circuit:

1. Voltage always drops as current flows through the resistance.
2. An increase in resistance causes a decrease in current.
3. All resistances change the electrical energy into heat energy to some extent.

Shop Manual
Chapter 2, page 54

Voltage Drop Defined

Voltage drop occurs when current flows through a load component or resistance. Voltage drop is the amount of electrical energy that is converted as it pushes current flow through a resistance. Electricity is an energy. Energy cannot be created or destroyed but it can be changed from one form to another. As electrical energy flows through a resistance, it is converted to some other form of energy, usually heat energy. The amount of voltage drop over a resistance or load device is an indication of how much electrical energy is converted to another energy form. After a resistance, the voltage is lower than it was before the resistance.

Voltage drop can be measured by using a voltmeter (**Figure 2-9**). With current flowing through a circuit, the voltmeter may be connected in parallel over the resistor, wire, or component to measure voltage drop. The voltmeter indicates the amount of voltage potential between two points in the circuit. The voltmeter reading indicates the difference between the amount of voltage available to the resistor and the amount of voltage after the resistor.

There must be a voltage present for current to flow through a resistor. Kirchhoff's law basically states that the sum of the voltage drops in an electrical circuit will always equal source voltage. In other words, all of the source voltage is used by the circuit.

A BIT OF HISTORY

Gustav Kirchhoff was a German scientist who, in the 1800s, discovered two facts about the characteristics of electricity. One is called his voltage law, which states: "The sum of the voltage drops across all resistances in a circuit must equal the voltage of the source." His law on current states: "The sum of the currents flowing into any point in a circuit equals the sum of the currents flowing out of the same point." These laws describe what happens when electricity is applied to a load. Voltage drops while current remains constant; current does not drop.

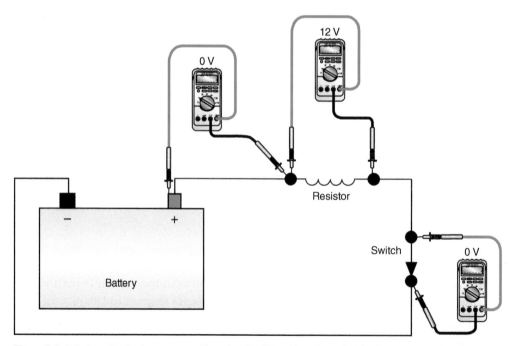

Figure 2-9 Using a voltmeter to measure voltage drop in different locations of a circuit.

ELECTRICAL LAWS

Electricity is governed by well-defined laws. The most fundamental of these are Ohm's law and Watt's law. Today's technician must understand these laws in order to completely grasp electrical theory.

Ohm's Law

Understanding **Ohm's law** is the key to understanding how electrical circuits work. Ohm's law defines the relationship between current, voltage, and resistance. The law states that it takes one volt of electrical pressure to push one ampere of electrical current through one ohm of electrical resistance. This law can be expressed mathematically as:

$$1 \text{ Volt} = 1 \text{ Ampere} \times 1 \text{ Ohm}$$

A BIT OF HISTORY

Georg S. Ohm was a German scientist in the 1800s who discovered that all electrical quantities are proportional to each other and therefore have a mathematical relationship.

This formula is most often expressed as follows: $E = I \times R$. E stands for electromotive force (electrical pressure or voltage), I stands for intensity (current or ampere), and R represents resistance. This formula is often used to find the amount of one electrical characteristic when the other two are known. As an example, if we have 2 amps of current and 6 ohms of resistance in a circuit, we must have 12 volts of electrical pressure.

$$E = 2 \text{ Amps} \times 6 \text{ Ohms} \qquad E = 2 \times 6 \qquad E = 12 \text{ Volts}$$

If we know the voltage and resistance but not the current of a circuit, we can quickly calculate it by using Ohm's law. Since $E = I \times R$, I would equal E divided by R. Let's supply some numbers to this. If we have a 12-volt circuit with 6 ohms of resistance, we can determine the amount of current in this way:

$$I = \frac{E}{R} \quad \text{or} \quad \frac{12 \text{ Volts}}{6 \text{ Ohms}} \quad \text{or} \quad I = 2 \text{ Amps}$$

The same logic is used to calculate resistance when voltage and current are known. $R = E/I$. One easy way to remember the formulas of Ohm's law is to draw a circle and divide it into three parts as shown in **Figure 2-10**. Simply cover the value you want to calculate. The formula you need to use is all that shows.

To show how easily this works, consider the 12-volt circuit in **Figure 2-11**. This circuit contains a 3-ohm light bulb. To determine the current in the circuit, cover the I in the circle to expose the formula $I = E/R$. Then plug in the numbers, $I = 12/3$. Therefore, the circuit current is 4 amperes.

To further explore how Ohm's law works, refer to **Figure 2-12**, which shows a simple circuit. If the battery voltage is 12 volts and the amperage is 24 amperes, the resistance of the lamp can be determined using $R = E/I$. In this instance, the resistance is 0.5 Ω. For another example, if the resistance of the bulb is 2 Ω and the amperage is 12, then voltage can be found using $E = I \times R$. In this instance, voltage would equal 24 volts. Now, if the circuit had 12 volts and the resistance of the bulb was 2 Ω, then amperage could be determined by $I = E/R$. In this case, amperage would be 6 amperes.

The resistance of a lamp used in an automotive application will change when current passes through it because its temperature changes.

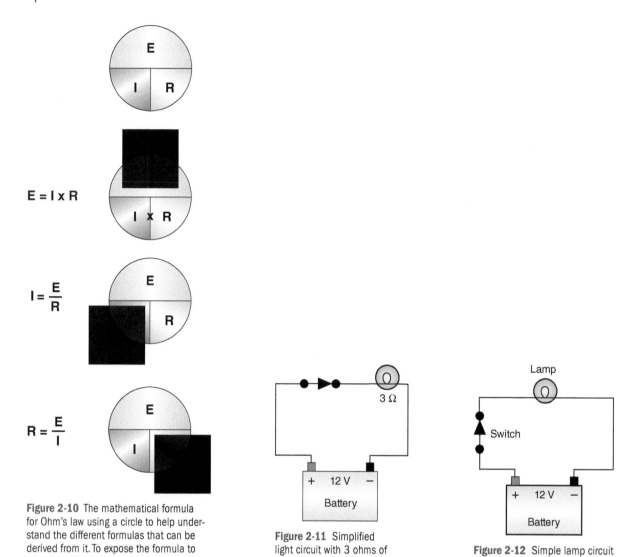

Figure 2-10 The mathematical formula for Ohm's law using a circle to help understand the different formulas that can be derived from it. To expose the formula to use, cover the unknown value.

Figure 2-11 Simplified light circuit with 3 ohms of resistance in the lamp.

Figure 2-12 Simple lamp circuit to help use Ohm's law.

Ohm's law is the basic law of electricity. It states that the amount of current in an electric circuit is inversely proportional to the resistance of the circuit and it is directly proportional to the voltage in the circuit. For example, if the resistance decreases and the voltage remains constant, the amperage will increase. If the resistance stays the same and the voltage increases, the amperage will also increase.

For example, refer to **Figure 2-13**; on the left side is a 12-volt circuit with a 3-ohm light bulb. This circuit will have 4 amps of current flowing through it. If a 1-ohm resistor is added to the same circuit (as shown to the right in Figure 2-13), total resistance is now 4 ohms. Because of the increased resistance, current dropped to 3 amps. The light bulb will be powered by less current and will be less bright than it was before the additional resistance was added.

Another point to consider is voltage drop. Before adding the 1-ohm resistor, the source voltage (12 volts) was dropped by the light bulb. With the additional resistance, the voltage drop of the light bulb decreased to 9 volts. The remaining 3 volts were dropped by the 1-ohm resistor. This can be proven by using Ohm's law. When the circuit current was 4 amps, the light bulb had 3 ohms of resistance. To find the voltage drop, we multiply the current by the resistance.

$$E = I \times R \quad \text{or} \quad E = 4 \times 3 \quad \text{or} \quad E = 12 \text{ Volts}$$

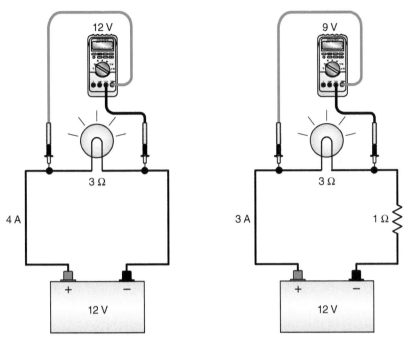

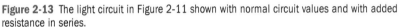

Figure 2-14 The mathematical formula for Watt's law.

Figure 2-13 The light circuit in Figure 2-11 shown with normal circuit values and with added resistance in series.

When the extra resistor was added to the circuit, the light bulb still had 3 ohms of resistance, but the current in the circuit decreased to 3 amps. Again, voltage drop can be determined by multiplying the current by the resistance.

$$E = I \times R \quad or \quad E = 3 \times 3 \quad or \quad E = 9\,Volts$$

The voltage drop of the additional resistor is calculated in the same way: $E = I \times R$ or $E = 3$ volts. The total voltage drop of the circuit is the same for both circuits. However, the voltage drop at the light bulb changed. This also would cause the light bulb to be dimmer. Ohm's law and its application will be discussed in greater detail later.

Horsepower ratings can be converted to electrical power rating using the conversion factor: 1 horsepower equals 746 watts.

Watt's Law

Power (P) is the rate of doing electrical work. Power is expressed in watts. A **watt** is equal to 1 volt multiplied by 1 ampere. There is another mathematical formula that expresses the relationship between voltage, current, and power. It is simply: $P = E \times I$ (**Figure 2-14**). Power measurements are measurements of the rate at which electricity is doing work.

A good example to demonstrate the concept of electrical power is light bulbs. Household light bulbs are sold by wattage. A 100-watt bulb is brighter and uses more electricity than a 60-watt bulb.

Referring to Figure 2-13, the light bulb in the circuit on the left has a 12-volt drop at 4 amps. We can calculate the power the bulb uses by multiplying the voltage and the current.

$$P = E \times I \quad or \quad P = 12 \times 4 \quad or \quad P = 48\,watts$$

The power output of the bulb is 48 watts. When the resistor was added to the circuit, the bulb dropped 9 volts at 3 amps. The power of the bulb is calculated in the same way as before.

$$P = E \times I \quad or \quad P = 9 \times 3 \quad or \quad P = 27\,watts$$

This bulb produced 27 watts of power, a little more than half of the original. It would be almost half as bright. The key to understanding what happened is to remember the light bulb didn't change; the circuit changed.

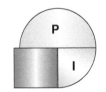

Another example of using Watt's law is to determine the amperage if an additional accessory is added to the vehicle's electrical system. If the accessory is rated at 75 watts, the amperage draw would be:

$$I = P/E = 75/12 = 6.25 \text{ amps}$$

This tells the technician that this circuit will probably require a 10-amp-rated fuse.

TYPES OF CURRENT

There are two classifications of electrical current flow: direct current (DC) and alternating current (AC). The type of current flow is determined by the direction it flows and by the type of voltage that drives it.

Direct Current

Direct current (DC) can be produced by a chemical reaction (such as in a battery) and has a current that is the same throughout the circuit and flows in the same direction (**Figure 2-15**). Voltage and current are constant if the switch is turned on or off. Most of the electrically controlled units in the automobile require direct current.

Alternating Current

The unrectified current produced within a AC generator is an example of alternating current found in the automobile. Circuits within the AC generator convert AC current to DC current.

Alternating current (AC) is produced anytime a conductor moves through a magnetic field. In an AC circuit, voltage and current do not remain constant. Alternating current changes directions from positive to negative. The voltage in an AC circuit starts at zero and rises to a positive value. Then it falls back to zero and goes to a negative value. Finally, it returns to zero (**Figure 2-16**). The AC voltage, shown in Figure 2-16, is called a sine wave. Figure 2-16 shows one **cycle**.

ELECTRICAL CIRCUITS

The portion of the circuit from the positive side of the source to the load component is called the insulated side or "hot" side of the circuit. The portion of the circuit that is from the load component to the negative side of the source is called the ground side of the circuit.

The electrical term **continuity** refers to the circuit being continuous. For current to flow, the electrons must have a continuous path from the source voltage to the load component and back to the source. A simple automotive circuit is made up of four parts:

1. Battery (power source).
2. Wires (conductors).
3. Load (light, motor, etc.).
4. Control device (switch).

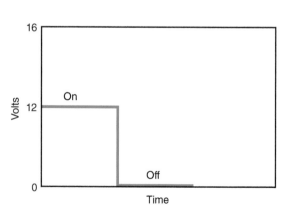

Figure 2-15 Direct current flow is in the same direction and remains constant on or off.

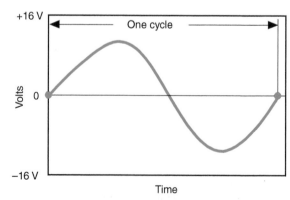

Figure 2-16 Alternating current does not remain constant.

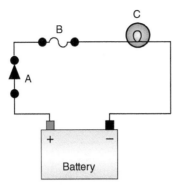

Figure 2-17 A basic electrical circuit including
(A) a switch, (B) a fuse, and (C) a lamp.

The basic circuit shown (**Figure 2-17**) includes a switch to turn the circuit on and off, a protection device (fuse), and a load. When the switch is turned to the ON position, the circuit is referred to as a **closed circuit**. When the switch is in the OFF position, the circuit is referred to as an **open circuit**. In this instance, with the switch closed, current flows from the positive terminal of the battery through the light and returns to the negative terminal of the battery. To have a complete circuit, the switch must be closed or turned on. The effect of opening and closing the switch to control electrical flow would be the same if the switch was installed on the ground side of the light.

There are three different types of electrical circuits: (1) the series circuit, (2) the parallel circuit, and (3) the series-parallel circuit.

> The electrical term **closed circuit** means that there are no breaks in the path and current will flow. **Open circuit** is used to mean that there will be no current flow since the path for electron flow is broken.

Series Circuit

A series **circuit** consists of one or more resistors (or loads) with only one path for current to flow. If any of the components in the circuit fails, the entire circuit will not function. All of the current that comes from the positive side of the battery must pass through each resistor and then back to the negative side of the battery.

The total resistance of a series circuit is calculated by simply adding the resistances together. As an example, refer to **Figure 2-18**. Here is a series circuit with three light bulbs; one bulb has 2 ohms of resistance and the other two have 1 ohm each. The total resistance of this circuit is 2 + 1 + 1 or 4 ohms.

The characteristics of a series circuit are as follows:

1. The total resistance is the sum of all resistances.
2. The current is the same at all points of the circuit (**Figure 2-19**).

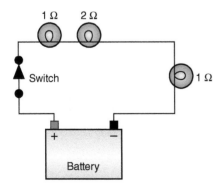

Figure 2-18 The total resistance in a series circuit is the sum of all resistances in the circuit.

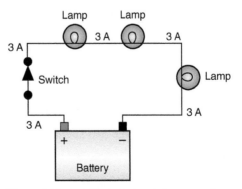

Figure 2-19 Regardless of where it is measured, amperage is the same at all points in a series circuit.

3. The voltage drop across each resistance will be equal if the resistance values are the same (**Figure 2-20**).
4. The voltage drop across each resistance will be different if the resistance values are different (**Figure 2-21**).
5. The sum of all voltage drops equals the source voltage.

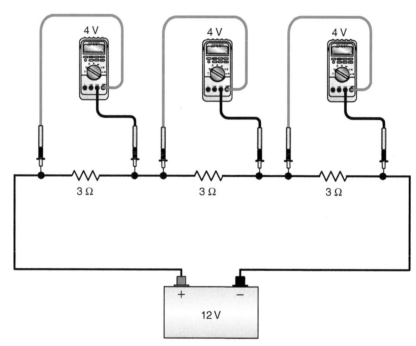

Figure 2-20 The voltage drop across each resistor in series will be the same if the resistance value of each is the same.

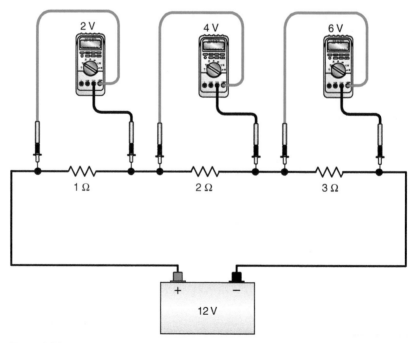

Figure 2-21 The voltage drop across each resistor in series will be different if the resistance value of each is different.

AUTHOR'S NOTE Over the years the symbols used for voltage, resistance, and current have changed. In the past the symbols were E for electromotive force (voltage), R for resistance, and I for intensity (current). More recently there has been a move to using the symbols V for voltage, R for resistance, and A for amps (current). As a result Ohm's law can be expressed as $E = I \times R$, or $V = A \times R$. Since illustrations used to illustrate Ohm's law use V and A, the formula will be expressed as $V = A \times R$.

To illustrate the laws of the series circuit, refer to **Figure 2-22**. The illustration labeled (A) is a simple 12-volt series circuit with a 2 Ω resistor. Using Ohm's law, it can be determined that the current is 6 amperes ($A = V/R$). Since the 2 Ω resistor is the only one in the circuit, all 12 volts are dropped across this resistor ($V = A \times R$).

In **Figure 2-22(B)**, an additional 4 Ω resistor is added in series to the existing 2 Ω resistor. Battery voltage is still 12 volts. Since this is a series circuit, total resistance is the sum of all of the resistances. In this case, total resistance is 6 Ω (4 Ω + 2 Ω). Using Ohm's law, total current through this circuit is 2 amperes ($A = V/R = 12/6 = 2$). In a series circuit, current is the same at all points of the circuit. No matter where current is measured in this example, the meter would read 2 amperes. This means that 2 amperes of current is flowing through each of the resistors. By comparing the amperage flow of the two circuits in Figure 2-22, you can see that circuit resistance controls (or determines) the amount of current flow. Understanding this concept is critical to performing diagnostics. Since this is a series circuit, adding resistance will decrease amperage draw.

Using Ohm's law, voltage drop over each resistor in the circuit can be determined. In this instance, V is the unknown value. Using $V = A \times R$ will determine the voltage drop over a resistance. Remember that amperage is the same throughout the circuit (2 amperes). The resistance value is the resistance of the resistor we are determining voltage drop for. For the 2 Ω resistor, the voltage drop would be $A \times R = 2 \times 2 = 4$ volts. Since the battery provides 12 volts and 4 volts are dropped over the 2 Ω resistor, 8 volts are left to be dropped by the 4 Ω resistor. To confirm this, $V = A \times R = 2 \times 4 = 8$. The sum of the voltage drops must equal the source voltage. Source voltage is 12 volts and the sum of the voltage drops is 4 volts + 8 volts = 12 volts.

These calculations work for all series circuits, regardless of the number of resistors in the circuit. Refer to **Figure 2-23** for an example of a series circuit with four

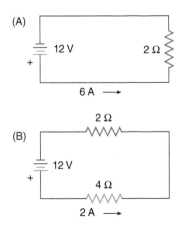

Figure 2-22 Circuit resistance controls, or determines, the amount of current flow.

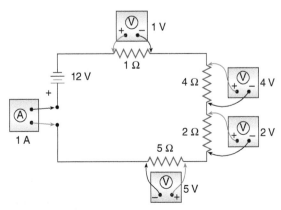

Figure 2-23 A series circuit used to demonstrate Ohm's law and voltage drop.

resistors. Total resistance is 12 Ω (1 Ω + 4 Ω + 2 Ω + 5 Ω). Total amperage is 1 amp (A = V/R = 12 volts/12 Ω = 1 amp). Voltage drop over each resistor would be calculated as follows:

1. Voltage drop over the 1 Ω resistor = A × R = 1 × 1 = 1 volt
2. Voltage drop over the 4 Ω resistor = A × R = 1 × 4 = 4 volts
3. Voltage drop over the 2 Ω resistor = A × R = 1 × 2 = 2 volts
4. Voltage drop over the 5 Ω resistor = A × R = 1 × 5 = 5 volts

$$\text{Total voltage drop} = 12 \text{ volts}$$

Parallel Circuit

The legs of a **parallel circuit** are also called parallel branches or shunt circuits.

In a **parallel circuit**, each path of current flow has separate resistances that operate either independently or in conjunction with each other (depending on circuit design). In a parallel circuit, current can flow through more than one parallel leg at a time (**Figure 2-24**). In this type of circuit, failure of a component in one parallel leg does not affect the components in other legs of the circuit.

The characteristics of a parallel circuit are as follows:

1. The voltage applied to each parallel leg is the same.
2. The voltage dropped across each parallel leg will be the same; however, if the leg contains more than one resistor, the voltage drop across each of them will depend on the resistance of each resistor in that leg.
3. The total resistance of a parallel circuit will always be less than the resistance of any of its legs.
4. The current flow through the legs will be different if the resistances are different.
5. The sum of the current in each leg equals the total current of the parallel circuit.

Figuring total resistance is a bit more complicated for a parallel circuit than a series circuit. Total resistance in a parallel circuit is always less than the lowest individual resistance because current has more than one path to flow. The method used to calculate total resistance depends on how many parallel branches are in the circuit, the resistance value of each branch, and personal preferences. Several methods of calculating total resistance are discussed. Choose the ones that work best for you.

If all resistances in the parallel circuit are equal, use the following formula to determine the total resistance:

$$R_T = \frac{\text{Value of one resistor}}{\text{Total number of branches}}$$

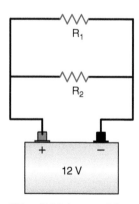

Figure 2-24 In a parallel circuit, there are multiple paths for current flow.

For example, if the parallel circuit, shown in Figure 2-24, had a 120 Ω resistor for R_1 and R_2, total circuit resistance would be

$$R_T = 120\,\Omega/2 \text{ branches} = 60\,\Omega$$

Note that total resistance is less than any of the resistances of the branches. If a third branch was added in parallel that also had 120 Ω resistance, total circuit resistance would be

$$R_T = 120\,\Omega/3 \text{ branches} = 40\,\Omega$$

Note what has happened when the third parallel branch was added. If more parallel resistors are added, more circuits are added, and the total resistance will decrease. With a decrease of total resistance, total amperage draw increases.

The total resistance of a parallel circuit with two legs or two paths for current flow can be calculated by using this formula:

$$R_T = \frac{R_1 \times R_2}{R_1 + R_2}$$

If the value of R_1 in Figure 2-24 was 3 ohms and R_2 had a value of 6 ohms, the total resistance can be found.

$$R_T = \frac{R_1 \times R_2}{R_1 + R_2} \quad \text{or} \quad R_T = \frac{3 \times 6}{3 + 6} \quad \text{or} \quad R_T = \frac{18}{9} \quad R_T = 2 \text{ ohms}$$

Based on this calculation, we can determine that the total circuit current is 6 amps (12 volts divided by 2 ohms). Using basic Ohm's law and a basic understanding of electricity, we can quickly determine other things about this circuit.

Each leg of the circuit has 12 volts applied to it; therefore, each leg must drop 12 volts. So the voltage drop across R_1 is 12 volts, and the voltage drop across R_2 is also 12 volts. Using the voltage drops, we can quickly find the current that flows through each leg. Since R_1 has 3 ohms and drops 12 volts, the current through it must be 4 amps. R_2 has 6 ohms and drops 12 volts and its current is 2 amps ($I = E/R$). The total current flow through the circuit is 4 + 2 or 6 amps. To calculate current in a parallel circuit, each shunt branch is treated as an individual circuit. To determine the branch current, simply divide the source voltage by the shunt branch resistance:

$$I = E/R$$

Referring to **Figure 2-25**, the total resistance of a circuit with more than two legs can be calculated with the following formula:

$$R_T = \frac{1}{\dfrac{1}{R_1} + \dfrac{1}{R_2} + \dfrac{1}{R_3} \cdots \dfrac{1}{R_n}}$$

Using Figure 2-25 and its resistance values, total resistance would be calculated by

$$R_T = \frac{1}{1/4 + 1/6 + 1/8}$$

This means total resistance is equal to the reciprocal of the sum of 1/4 + 1/6 + 1/8. The next step is to add 1/4 + 1/6 + 1/8. To do this, the least common denominator must be found. In this case, the least common denominator is 24, so the formula now looks like this:

$$\frac{1}{6/24 + 4/24 + 3/24}$$

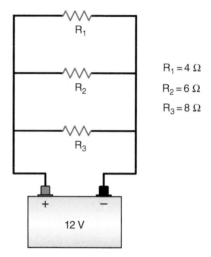

Figure 2-25 A parallel circuit with different resistances in each branch.

Now the fractions can be added together (remember to add only the numerator):

$$\frac{1}{13/24}$$

Since we are working with reciprocals, the formula now looks like this:

$$1 \times \frac{24}{13} = 1.85\,\Omega$$

Often it is much easier to calculate total resistance of a parallel circuit by using total current. Begin by determining the current through each leg of the parallel circuit; then add them together to find total current. Use basic Ohm's law to calculate the total resistance.

First, using the circuit illustrated in Figure 2-25, calculate the current through each branch:

1. Current through R_1 = E/R = $12/4$ = 3 amperes
2. Current through R_2 = E/R = $12/6$ = 2 amperes
3. Current through R_3 = E/R = $12/8$ = 1.5 amperes

Add all of the current flow through the branches together to get the total current flow:

Total amperage = $3 + 2 + 1.5 = 6.5$ amperes

Since this is a 12-volt system and total current is 6.5 amperes, total resistance is

$$R_t = 12\,\text{volts}/6.5\,\text{amps} = 1.85\,\Omega$$

This method can be mathematically expressed as follows:

$$R_T = V_T/A_T$$

Series-Parallel Circuits

The **series-parallel circuit** has some loads that are in series with each other and some that are in parallel (**Figure 2-26**). To calculate the total resistance in this type of circuit, calculate the **equivalent series loads** of the parallel branches first. Next, calculate the series resistance and add it to the equivalent series load. For example, if the parallel portion of the circuit has two branches with 4 Ω resistance each and the series portion has

The **equivalent series load**, or equivalent resistance, is the equivalent resistance of a parallel circuit plus the resistance in series and is equal to the equivalent resistance of a single load in series with the voltage source.

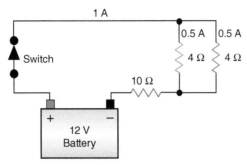

Figure 2-26 A series-parallel circuit with known resistance values.

a single load of 10 Ω, use the following method to calculate the equivalent resistance of the parallel circuit:

$$R_T = \frac{R_1 \times R_2}{R_1 + R_2} \quad \text{or} \quad \frac{4 \times 4}{4 + 4} \quad \text{or} \quad \frac{16}{8} \quad \text{or} \quad 2\,\text{ohms}$$

Then add this equivalent resistance to the actual series resistance to find the total resistance of the circuit.

$$2\,\text{ohms} + 10\,\text{ohms} = 12\,\text{ohms}$$

With the total resistance now known, total circuit current can be calculated. Because the source voltage is 12 volts, 12 is divided by 12 ohms.

$$I = E/R \quad \text{or} \quad I = 12/12 \quad \text{or} \quad I = 1\,\text{amp}$$

The current flow through each parallel leg is calculated by using the resistance of each leg and voltage drop across that leg. To do this, you must first find the voltage drops. Since all 12 volts are dropped by the circuit, we know that some are dropped by the parallel circuit and the rest by the resistor in series. We also know that the circuit current is 1 amp, the equivalent resistance value of the parallel circuit is 2 ohms, and the resistance of the series resistor is 10 ohms. Using Ohm's law we can calculate the voltage drop of the parallel circuit:

$$E = I \times R \quad \text{or} \quad E = 1 \times 2 \quad \text{or} \quad E = 2\,\text{volts}$$

Two volts are dropped by the parallel circuit. This means 2 volts are dropped by each of the 4-ohm resistors. Using our voltage drop, we can calculate our current flow through each parallel leg.

$$I = E/R \quad \text{or} \quad I = 2/4 \quad \text{or} \quad I = 0.5\,\text{amps}$$

Since the resistance on each leg is the same, each leg has 0.5 amps through it. If we did this right, the sum of the amperages will equal the total current of the circuit. It does: 0.5 + 0.5 = 1.

A slightly different series-parallel circuit is illustrated in **Figure 2-27**. In this circuit, a 2 Ω resistor is in series to a parallel circuit containing a 6 Ω and a 3 Ω resistor. To calculate total resistance, first find the resistance of the parallel portion of the circuit using $(6 \times 3)/(6 + 3) = 18/9 = 2$ ohms of resistance. Add this amount to the series resistance; $2 + 2 = 4$ ohms of total circuit resistance. Now total current can be calculated by $I = E/R = 12/4 = 3$ amperes. This means that 3 amperes is flowing through the series portion of the circuit. To figure how much amperage is in each of the parallel branches, the amount of applied voltage to each branch must be calculated. Since the series circuit has 3 amperes going through a 2 Ω resistor, the voltage drop over this resistor can be figured using $E = I \times R = 3 \times 2 = 6$ volts. Since the source voltage is 12 volts, this means that

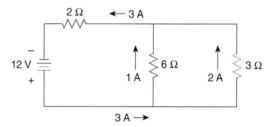

Figure 2-27 In a series-parallel circuit, the sum of the currents through the legs will equal the current through the series portion of the circuit.

6 volts are applied to each of the resistors in parallel (12 − 6 = 6 volts) and that 6 volts are dropped over each of these resistors. Current through each branch can now be calculated:

Current through the 6 Ω branch is I = E/R = 6/6 = 1 ampere

Current through the 3 Ω branch is I = E/R = 6/3 = 2 amperes

The sum of the current flow through the parallel branches should equal total current flow (1 + 2 = 3 amperes).

Based on what was just covered, the characteristics of a series-parallel circuit can be summarized as follows:

1. Total resistance is the sum of the resistance value of the parallel portion and the series resistance.
2. Voltage drop over the parallel branch resistance is determined by the resistance value of the series resistor.
3. Total amperage is the sum of the current flow through each parallel branch.
4. The amperage through each parallel branch is determined by the resistance in the branch.

It is important to realize that the actual or measured values of current, voltage, and resistance may be somewhat different than the calculated values. The change is caused by the effects of heat on the resistances. As the voltage pushes current through a resistor, the resistor heats up. The resistor changes the electrical energy into heat energy. This heat may cause the resistance to increase or decrease depending on the material it is made of. The best example of a resistance changing electrical energy into heat energy is a light bulb. A light bulb gives off light because the conductor inside the bulb heats up and glows when current flows through it.

Applying Ohm's Law

The primary importance of being able to apply Ohm's law is to predict what will happen if something else happens. Technicians use electrical meters to measure current, voltage, and resistance. When a measured value is not within specifications, you should be able to determine why. Ohm's law is used to do that.

Most automotive electrical systems are wired in parallel. Actually, the system is made up of a number of series circuits wired in parallel. This allows each electrical component to work independently of the others. When one component is turned on or off, the operation of the other components should not be affected.

Figure 2-28 illustrates a 12-volt circuit with one 3-ohm light bulb. The switch controls the operation of the light bulb. When the switch is closed, current flows and the bulb is lit. Four amps will flow through the circuit and the bulb.

I = E/R   or   I = 12/3   or   I = 4 amps

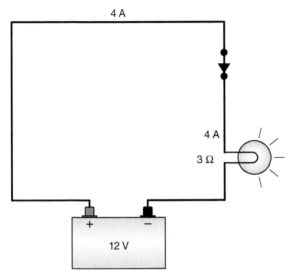

Figure 2-28 A simple light circuit.

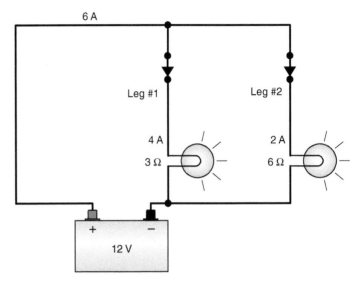

Figure 2-29 Two light bulbs wired in parallel.

Figure 2-29 illustrates the same circuit with a 6-ohm light bulb added in parallel to the 3-ohm light bulb. With the switch for the new bulb closed, 2 amps will flow through that bulb. The 3-ohm is still receiving 12 volts and has 4 amps flowing through it. It will operate in the same way and with the same brightness as it did before we added the 6-ohm light bulb. The only thing that changed was circuit current, which is now 6 (4 + 2) amps.

Leg #1 $I = E/R$ or $I = 12/3$ or $I = 4\,\text{amps}$

Leg #2 $I = E/R$ or $I = 12/6$ or $I = 2\,\text{amps}$

If the switch to the 3-ohm bulb is opened (**Figure 2-30**), the 6-ohm bulb works in the same way and with the same brightness as it did before we opened the switch. In this case, two things happened: the 3-ohm bulb no longer is lit and the total circuit current dropped to 2 amps.

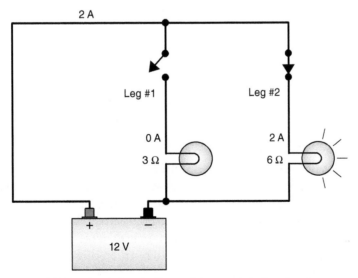

Figure 2-30 Two light bulbs wired in parallel; one switched on, the other switched off.

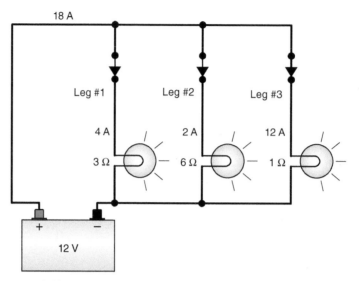

Figure 2-31 Three light bulbs wired in parallel.

Figure 2-31 is the same circuit as Figure 2-30 except a 1-ohm light bulb and switch was added in parallel to the circuit. With the switch for the new bulb closed, 12 amps will flow through that circuit. The other bulbs are working in the same way and with the same brightness as before. Again, total circuit resistance decreases so the total circuit current increases. Total current is now 18 amps:

Leg #1   I = E/R   or   I = 12/3   or   I = 4 amps

Leg #2   I = E/R   or   I = 12/6   or   I = 2 amps

Leg #3   I = E/R   or   I = 12/1   or   I = 12 amps

Total current = 4 + 2 + 12 or 18 amps

When the switch for any of these bulbs is opened or closed, the only things that happen are the bulbs turn either off or on and the total current through the circuit changes. Notice as we add more parallel legs, total circuit current goes up. There is a commonly used statement, "Current always takes the path of least resistance to ground." This statement is not totally correct. If this were a true statement, then parallel circuits would not work. However, as illustrated in the previous circuits, current flows to all of the bulbs regardless of the bulb's resistance. The resistances with lower values will draw higher currents, but all of the resistances will receive the current they allow. The statement is more accurate when expressed as, "Larger amounts of current will flow through lower resistances." Another way to state this would be, "The *majority* of the current flows the path of least resistance." This is very important to remember when diagnosing electrical problems.

From Ohm's law, we know that when resistance decreases, current increases. If we put a 0.6-ohm light bulb in place of the 3-ohm bulb (**Figure 2-32**), the other bulbs will work in the same way and with the same intensity as they did before. However, 20 amps of current will flow through the 0.6-ohm bulb. This will raise total circuit current to 34 amps. Lowering the resistance on the one leg of the parallel circuit greatly increases the current through the circuit. This high current may damage the circuit or components. It is possible that high current can cause wires to burn. In this case, the wires that would

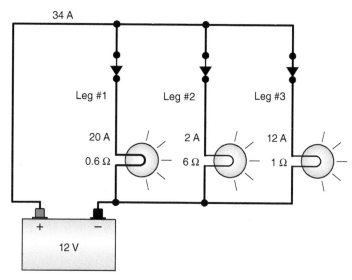

Figure 2-32 Parallel light circuit.

burn are the wires that would carry the 34 amps or the 20 amps to the bulb, not the wires to the other bulbs.

Leg #1   I = E/R   or   I = 12/0.6   or   I = 20 amps
Leg #2   I = E/R   or   I = 12/6   or   I = 2 amps
Leg #3   I = E/R   or   I = 12/1   or   I = 12 amps

Total current = 20 + 2 + 12 or 34 amps

Let's see what happens when we add resistance to one of the parallel legs. An increase in resistance should cause a decrease in current. In **Figure 2-33**, a 1-ohm resistor was added after the 1-ohm light bulb. This resistor is in series with the light bulb and the total resistance of that leg is now 2 ohms. The current through that leg is now 6 amps.

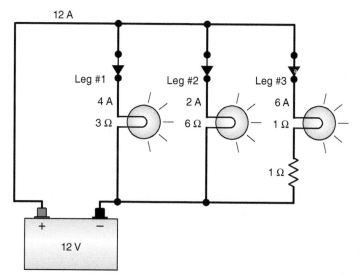

Figure 2-33 A parallel light circuit with one leg having a series resistance added.

Again, the other bulbs were not affected by the change. The only change to the whole circuit was in total circuit current, which now drops to 12 amps. The added resistance lowered total circuit current and changed the way the 1-ohm bulb works. This bulb will now drop only 6 volts. The remaining 6 volts will be dropped by the added resistor. The 1-ohm bulb will be much dimmer than before; its power rating dropped from 144 watts to 36 watts. Additional resistance causes the bulb to be dimmer. The bulb itself wasn't changed; only the resistance of that leg changed. The dimness is caused by the circuit, not the bulb.

Leg #1   I = E/R   or   I = 12/3   or   I = 4 amps

Leg #2   I = E/R   or   I = 12/6   or   I = 2 amps

Leg #3   I = E/R   or   I = 12/1+1   or   I = 12/2   or   I = 6 amps

Total current = 4 + 2 + 6 or 12 amps

Now let's see what happens when we add a resistance that is common to all of the parallel legs. In **Figure 2-34**, we added a 0.333-ohm resistor (0.333 was chosen to keep the math simple!) to the negative connection at the battery. This will cause the circuit's current to decrease; it will also change the operation of the bulbs in the circuit. The total resistance of the bulbs in parallel is 0.667 ohms.

$$R_T = \frac{1}{\frac{1}{3} + \frac{1}{6} + \frac{1}{1}} \quad \text{or} \quad R_T = \frac{1}{0.333 + 0.167 + 1} \quad \text{or} \quad R_T = \frac{1}{1.5} \quad \text{or} \quad R_T = 0.667 \text{ ohms}$$

The total resistance of the circuit is 1 ohm (0.667 + 0.333), which means the circuit current is now 12 amps. Because there will be a voltage drop across the 0.333-ohm resistor, each of the parallel legs will drop less than source voltage. To find the amount of voltage dropped by the parallel circuit, we multiply the amperage by the resistance. Twelve amps multiplied by 0.667 equals 8. So 8 volts will be dropped by the parallel circuit; the remaining 4 volts will be dropped by the 0.333 resistor. The amount of current

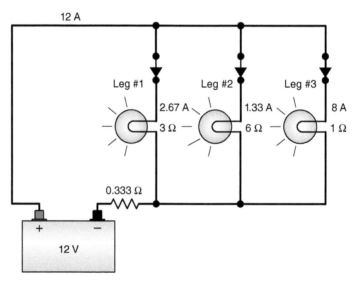

Figure 2-34 A parallel light circuit with a resistance in series to the whole circuit.

through each leg can be calculated by taking the voltage drop and dividing it by the resistance of the leg.

Leg #1 $I = E/R$ or $I = 8/3$ or $I = 2.667$ amps

Leg #2 $I = E/R$ or $I = 8/6$ or $I = 1.333$ amps

Leg #3 $I = E/R$ or $I = 8/1$ or $I = 8$ amps

Total circuit current $= 2.667 + 1.333 + 8$ or 12 amps

The added resistance affected the operation of all the bulbs, because it was added to a point that was common to all of the bulbs. All of the bulbs would be dimmer and circuit current would be lower.

KIRCHHOFF'S LAWS

Ohm's law is stated as it takes one volt to push one ampere through one ohm of resistance. Ohm's law is the principal law of electricity and is used to determine electrical values. However, there are times when Ohm's law would be difficult to use in determining electrical values. This is true in circuits that do not have clearly defined series or parallel connections. Also, Ohm's law can be difficult to use if the circuit has more than one power source. In these instances, Kirchhoff's laws are used.

Kirchhoff stated two laws that described voltage and current relationships in an electric circuit. The first law, known as **Kirchhoff's voltage law**, states that the algebraic sum of the voltage sources and voltage drops in a closed circuit must equal zero. This law is actually the rule that states that the sum of the voltage drops in a series circuit must equal the source voltage.

The second law, known as **Kirchhoff's current law**, states that the algebraic sum of the currents entering and leaving a point must equal zero. This is the rule that states that total current flow in a parallel circuit will be the sum of the currents through all the circuit branches.

Kirchhoff's Current Law

The concept of Kirchhoff's current law is the fact that if more current entered a particular point than left that point, a charge would have to develop at that point. In the parallel circuit illustrated (**Figure 2-35**), if 4 amperes of current flow through R_1 to the junction at point A, and 6 amperes of current flow through R_2 to point A, then the sum of the two

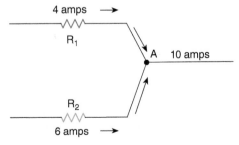

Figure 2-35 The sum of the currents entering and leaving a point in the circuit must equal zero.

currents is 10 amperes leaving point A. Since Kirchhoff's current law states that the algebraic sum of the currents must equal zero, to use this law the current entering a point is considered to be positive and the current leaving a point is considered to be negative. Since the currents flowing through R_1 and R_2 are entering point A, they are considered positive. The current leaving point A is considered negative. In this example then:

$$+ 4\,A + 6\,A - 10\,A = 0\,A$$

> **AUTHOR'S NOTE** The figures used in the following explanation will have a source voltage of 120 volts. This amount is used so the example will be easier to understand and to keep the values high enough to prevent having to use fractions.

Figure 2-36 illustrates a more complex series-parallel circuit. Ohm's law tells us that this circuit will have a total of 2 amperes flowing through it since there is a total of 60 ohms of resistance. Since 2 amperes is flowing through R_1, 32 volts will be dropped and the applied voltage at point B will be 88 volts. With 2 amperes flowing through R_3, the voltage drop over this resistor will be 40 volts. With 88 volts applied to point B and R_3 dropping 40 volts, that leaves 48 volts to be dropped over R_2 and the parallel circuit of R_4 through R_6. This means that points B to E will have a current of 0.8 amperes and the parallel branch will have 1.2 amperes.

Notice what occurs to the current at point B. Two amperes is flowing into point B from R_1. At this point, the current splits, with part flowing to R_2 and part to the circuit containing resistors R_4 through R_6. The current entering point B is considered positive, and the two currents leaving point B are considered negative. The equation would look like this:

$$+ 2\,A - 0.8\,A - 1.2\,A = 0\,A$$

At point E, there is 0.8 ampere of current entering from the circuit with resistor R_2 and 1.2 amperes of current entering from the circuit with resistors R_4 through R_6. Two amperes of current leaves point E and flows through R_3.

$$+ 0.8\,A + 1.2\,A - 2\,A = 0\,A$$

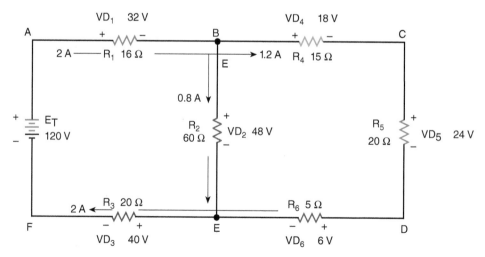

Figure 2-36 Illustration of use of Kirchhoff's laws.

Kirchhoff's Voltage Law

Kirchhoff's voltage law states that the algebraic sum of the voltages around any closed loop must equal zero. To be able to calculate the sum, first it must be established which end of the resistive element is positive and which is negative. Referring back to **Figure 2-36**, we will use the conventional current flow of positive to negative. Therefore, the point at which current enters a resistor is marked positive, and the point where current leaves the resistor is marked negative.

AUTHOR'S NOTE It does not matter if you use the conventional theory of current flow or the electron theory; you only need to be consistent. The sum of voltage drops will be the same regardless of which current flow assumption you use.

The circuit illustrated in **Figure 2-36** has three separate closed loops. Closed loop ACDF contains the voltage drops VD_1, VD_4, VD_5, VD_6, VD_3, and E_T. E_T is the source and must be included in the equation. The voltage drops for this loop are as follows:

$$+ VD_1 + VD_4 + VD_5 + VD_6 + VD_3 - E_T = 0$$
$$+ 32\,V + 18\,V + 24\,V + 6\,V + 40\,V - 120\,V = 0\,V$$

The positive or negative sign for each number is determined by the assumed direction of current flow. In this example, it is assumed that current leaves point A and returns to point A. Current leaving point A enters R_1 at the positive side. Therefore, the voltage is considered to be positive (+32 V). The same is true for R_4, R_5, R_6, and R_3. However, the current enters the voltage source at the negative side so ET is assumed to be negative.

Closed loop ABEF contains voltage drops VD_1, VD_2, VD_3, and ET. Current will leave point A and return to point A through R_1, R_2, R_3, and the voltage source. The voltage drops are as follows:

$$+ VD_1 + VD_2 + VD_3 - VD_T = 0$$
$$+ 32\,V + 48\,V + 40\,V - 120\,V = 0\,V$$

Closed loop BCDE contains voltage drops VD_4, VD_5, VD_6, and VD_2. Current leaves point B and returns to point B, flowing through R_4, R_5, R_6, and R_2.

$$+ VD_4 + VD_5 + VD_{6S} - VD_{26} = 0$$
$$+ 18\,V + 24\,V + 6\,V - 48\,V = 0\,V$$

MAGNETISM PRINCIPLES

Magnetism is a force that is used to produce most of the electrical power in the world. It is also the force used to create the electricity to recharge a vehicle's battery, make a starter work, and produce signals for various operating systems. A magnet is a material that attracts iron, steel, and a few other materials. Because magnetism is closely related to electricity, many of the laws that govern electricity also govern magnetism.

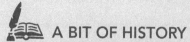

A BIT OF HISTORY

The force of a magnet was first discovered over 2,000 years ago by the Greeks. They noticed that a type of stone, now called magnetite, was attracted to iron. During the Dark Ages, people believed that evil spirits caused the strange powers of magnetite.

There are two types of magnets used on automobiles: permanent magnets and electromagnets. Permanent magnets are magnets that do not require any force or power to keep their magnetic field. Electromagnets depend on electrical current flow to produce and, in most cases, keep their magnetic field.

Magnets

All magnets have polarity. A magnet that is allowed to hang free will align itself north and south. The end facing north is called the north-seeking pole and the end facing south is called the south-seeking pole. Like poles will repel each other and unlike poles will attract each other. These principles are shown in **Figure 2-37**. The magnetic attraction is the strongest at the poles.

Magnetic flux density is a concentration of the lines of force (**Figure 2-38**). A strong magnet produces many lines of force and a weak magnet produces fewer lines of force. Invisible lines of force leave the magnet at the north pole and enter again at the south pole. While inside the magnet, the lines of force travel from the south pole to the north pole (**Figure 2-39**).

The field of force (or magnetic field) is all the space, outside the magnet, that contains lines of magnetic force. Magnetic lines of force penetrate all substances; there is no known insulation against magnetic lines of force. The lines of force may be deflected only by other magnetic materials or by another magnetic field.

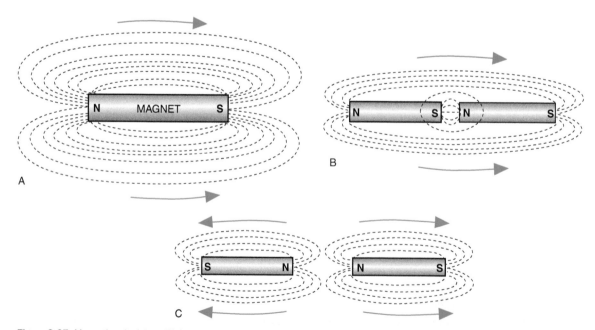

Figure 2-37 Magnetic principles: (A) All magnets have poles, (B) unlike poles attract each other, and (C) like poles repel.

Figure 2-38 Iron filings indicate the lines of magnetic flux.

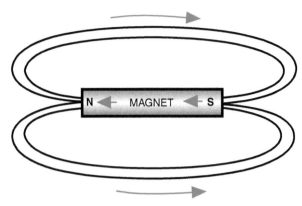

Figure 2-39 Lines of force through the magnet.

Electromagnetism

Electromagnetism uses the theory that whenever an electrical current flows through a conductor, a magnetic field is formed around the conductor (**Figure 2-40**). The number of lines of force and the strength of the magnetic field produced will be in direct proportion to the amount of current flow.

The direction of the lines of force is determined by the **right-hand rule**. Using the conventional theory of current flow being from positive to negative, the right hand is used to grasp the wire, with the thumb pointing in the direction of current flow. The fingers will point in the direction of the magnetic lines of force (**Figure 2-41**).

André Marie Ampère noted that current flowing in the same direction through two nearby wires will cause the wires to attract each other. Also, he observed that if current flow in one of the wires is reversed, the wires will repel each other. In addition, he found that if a wire is coiled with current flowing through the wire, the same magnetic field that surrounds a straight wire combines to form one larger magnetic field. This magnetic field has true north and south poles (**Figure 2-42**). Looping the wire doubles the flux density where the wire is running parallel to itself. **Figure 2-43** shows how these lines of force will join and add to each other.

The north pole can be determined in the coil by use of the right-hand rule. Grasp the coil with the fingers pointing in the direction of current flow (+ to −) and the thumb will point toward the north pole (**Figure 2-44**).

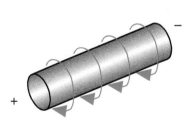

Figure 2-40 A magnetic field surrounds a conductor that has current flowing through it.

Figure 2-41 Right-hand rule to determine direction of magnetic lines of force.

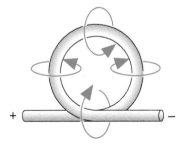

Figure 2-42 Looping the conductor increases the magnetic field.

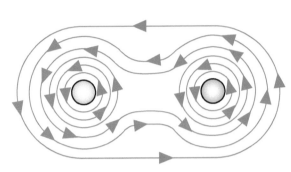

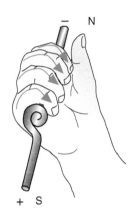

Figure 2-43 Lines of force join together and attract each other.

Figure 2-44 Right-hand rule to determine magnetic poles.

Permeability is the term used to indicate the magnetic conductivity of a substance compared with the conductivity of air. The greater the permeability, the greater the magnetic conductivity and the easier a substance can be magnetized or the more attracted it is to a magnet.

As more loops are added, the fields from each loop will join and increase the flux density (**Figure 2-45**). To make the magnetic field even stronger, an iron core can be placed in the center of the coil (**Figure 2-46**). The soft iron core has high **permeability** and low **reluctance**, which provides an excellent conductor for the magnetic field to travel through the center of the wire coil.

The strength of an electromagnetic coil is affected by the following factors:

1. The amount of current flowing through the wire.
2. The number of windings or turns.
3. The size, length, and type of core material.
4. The direction and angle at which the lines of force are cut.

The strength of the magnetic field is measured in ampere-turns:

Ampere-turns = amperes × number of turns

Reluctance is the term used to indicate a material's resistance to the passage of flux lines. Highly reluctant materials are not attracted to magnets.

The magnetic field strength is measured by multiplying the current flow in amperes through a coil by the number of complete turns of wire in the coil. For example, in **Figure 2-47**, a 1,000-turn coil with 1 ampere of current would have a field strength of 1,000 ampere-turns. This coil would have the same field strength as a coil with 100 turns and 10 amperes of current.

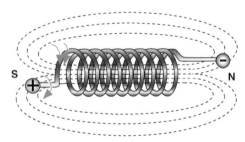

Figure 2-45 Adding more loops of wire increases the magnetic flux density.

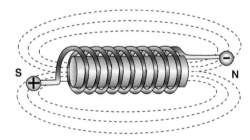

Figure 2-46 The addition of an iron core concentrates the flux density.

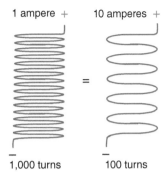

Figure 2-47 Magnetic field strength is determined by the amount of amperage and the number of coils.

THEORY OF INDUCTION

Electricity can be produced by magnetic **induction**. Magnetic induction occurs when a conductor is moved through the magnetic lines of force (**Figure 2-48**) or when a magnetic field is moved across a conductor. A difference of potential is set up between the ends of the conductor and a voltage is induced. This voltage exists only when the magnetic field or the conductor is in motion.

The induced voltage can be increased by either increasing the speed in which the magnetic lines of force cut the conductor or increasing the number of conductors that are cut. It is this principle that is behind the operation of all ignition systems, starter motors, and charging systems.

A common induction device is the ignition coil. As the current increases, the coil will reach a point of **saturation**. This is the point at which the magnetic strength eventually levels off and where current will no longer increase as it passes through the coil. The magnetic lines of force, which represent stored energy, will collapse when the applied voltage is removed. When the lines of force collapse, the magnetic energy is returned to the wire as electrical energy.

Mutual induction is used in ignition coils where a rapidly changing magnetic field in the primary windings creates a voltage in the secondary winding (**Figure 2-49**).

A desirable induction is called a **mutual induction**.

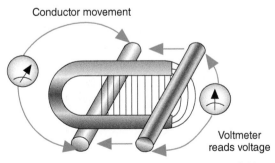

Figure 2-48 Moving a conductor through a magnetic field induces an electrical potential difference.

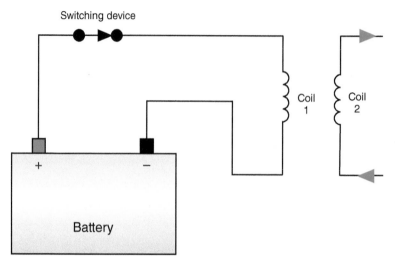

Figure 2-49 A mutual induction is used to create an electrical current in coil 2 if the current flow in coil 1 is turned off.

Self-induction is also referred to as counter EMF (CEMF) or as a voltage spike.

If voltage is induced in the wires of a coil when current is first connected or disconnected, it is called **self-induction**. The resulting current is in the opposite direction of the applied current and tends to reduce the magnetic force. Self-induction is governed by Lenz's law, which states that an induced current flows in a direction opposite the magnetic field that produces it.

Self-induction is generally not wanted in automotive circuits. For example, when a switch is opened, self-induction tends to continue to supply current in the same direction as the original current because as the magnetic field collapses, it induces voltage in the wire. According to Lenz's law, voltage induced in a conductor tends to oppose a change in current flow. Self-induction can cause an electrical arc to occur across an opened switch. The arcing may momentarily bypass the switch and allow the circuit that was turned off to operate for a short period of time. The arcing will also burn the contacts of the switch.

Self-induction is commonly found in electrical components that contain a coil or an electric motor. To help reduce the arc across contacts, a capacitor or clamping diode may be connected to the circuit. The capacitor will absorb the high-voltage arcs and prevent arcing across the contacts. Diodes are semiconductors that allow current flow in only one direction. A clamping diode can be connected in parallel to the coil and will prevent current flow from the self-induction coil to the switch. The capacitor and clamping diode will be discussed in Chapter 3.

Magnetic induction is also the basis for a generator and many of the sensors on today's vehicles. In a generator, a magnetic field rotates inside a set of conductors. As the magnetic field crosses the wires, a voltage is induced. The amount of voltage induced by this action depends on the speed of the rotating field, the strength of the field, and the number of conductors the field cuts through. This principle will be discussed in greater detail in Chapter 7.

Magnetic sensors are used to measure speeds, such as engine, vehicle, and shaft speeds. These sensors typically use a permanent magnet. Rotational speed is determined by the passing of blades or teeth in and out of the magnetic field. As a tooth moves in and out of the magnetic field, the strength of the magnetic field is changed and a voltage signal is induced. This signal is sent to a control device, where it is interpreted. This principle is discussed in greater detail in Chapter 10.

PHOTOVOLTAICS

Photovoltaics (PV) is one of the forces that can be used to generate electrical current by converting solar radiation through the use of solar cells. Solar cells are used to make up solar panels that convert energy from the sun into a direct current electricity. Some automotive manufacturers are looking to this energy source as a means to power some accessories and to recharge batteries.

EMI SUPPRESSION

Electromagnetic interference (EMI) is an undesirable creation of electromagnetism whenever current is switched on and off. As manufacturers began to increase the number of electronic components and systems in their vehicles, the problem of EMI had to be controlled. The low-power integrated circuits used on modern vehicles are sensitive to the signals produced as a result of EMI. EMI is produced as current in a conductor is turned on and off. EMI is also caused by **static electricity** that is created by friction. The friction is a result of tires contacting the road, or of fan belts contacting the pulleys.

Static electricity is electricity that is not in motion.

EMI can disrupt the vehicle's computer systems by inducing false messages to the computer. The computer requires messages to be sent over circuits in order to communicate with other computers, sensors, and actuators. If any of these signals are disrupted, the engine and/or accessories may turn off.

EMI can be suppressed by any one of the following methods:

1. Adding a resistance to the conductors. This is usually done to high-voltage systems, such as the secondary circuit of the ignition system.
2. Connecting a capacitor in parallel and a choke coil in series with the circuit.
3. Shielding the conductor or load components with a metal or metal-impregnated plastic.
4. Increasing the number of paths to ground by using designated ground circuits. This provides a clear path to ground that is very low in resistance.
5. Adding a clamping diode in parallel to the component.
6. Adding an isolation diode in series to the component.

SUMMARY

- An atom is constructed of a complex arrangement of electrons in orbit around a nucleus. If the number of electrons and protons is equal, the atom is balanced or neutral.
- A conductor allows electricity to easily flow through it.
- An insulator does not allow electricity to easily flow through it.
- Electricity is the movement of electrons from atom to atom. In order for the electrons to move in the same direction, an electromotive force (EMF) must be applied to the circuit.

- The electron theory defines electron flow as motion from negative to positive.
- The conventional theory of current flow states that current flows from a positive point to a less positive point.
- Voltage is defined as an electrical pressure and is the difference between the positive and negative charges.
- Current is defined as the rate of electron flow and is measured in amperes. Amperage is the amount of electrons passing any given point in the circuit in 1 second.

- Resistance is defined as opposition to current flow and is measured in ohms (Ω).
- Ohm's law defines the relationship between current, voltage, and resistance. It is the basic law of electricity and states that the amount of current in an electric circuit is inversely proportional to the resistance of the circuit and is directly proportional to the voltage in the circuit.
- Wattage represents the measure of power (P) used in a circuit. Wattage is measured by using Watt's law formula, which defines the relationship between amperage, voltage, and wattage.
- Direct current results from a constant voltage and a current that flows in one direction.
- In an alternating current circuit, voltage and current do not remain constant. AC current changes direction from positive to negative and negative to positive.
- For current to flow, the electrons must have a complete path from the source voltage to the load component and back to the source.
- The series circuit provides a single path for current flow from the electrical source through all the circuit's components and back to the source.

- A parallel circuit provides two or more paths for current to flow.
- A series-parallel circuit is a combination of the series and parallel circuits.
- The equivalent series load is the total resistance of a parallel circuit plus the resistance of the load in series with the voltage source.
- Voltage drop is caused by a resistance in the circuit that reduces the electrical pressure available after the resistance.
- Kirchhoff's voltage law states that the algebraic sum of the voltage sources and voltage drops in a closed circuit must equal zero. This law is actually the rule that states that the sum of the voltage drops in a series circuit must equal the source voltage.
- Kirchhoff's current law states that the algebraic sum of the currents entering and leaving a point must equal zero. This is the rule that states that total flow of current in a parallel circuit will be the sum of the currents through all the circuit branches.

REVIEW QUESTIONS

Short-Answer Essays

1. List and define the three elements of electricity.
2. Explain the basic principles of Ohm's law.
3. List and describe the three types of circuits.
4. Explain the principle of electromagnetism.
5. Describe the principle of induction.
6. Describe the basics of electron flow.
7. Define the two types of electrical current.
8. Describe the difference between insulators, conductors, and semiconductors.
9. Define "voltage drop."
10. What does the measurement of "watt" represent?

Fill in the Blanks

1. _____ are negatively charged particles. The nucleus contains positively charged particles called _____ and particles that have no charge called _____.

2. A _____ allows electricity to easily flow through it. An _____ does not allow electricity to easily flow through it.

3. For the electrons to move in the same direction, there must be an _____ applied.

4. The _____ _____ of current flow states that current flows from a positive point to a less positive point.

5. Resistance is defined as _____ to current flow and is measured in _____.

6. In a series circuit, the voltage drop across a resistance is determined by the _____ value.

7. Kirchhoff's voltage law states that the _____ _____ _____ in an electrical circuit will always _____ available voltage at the source.

8. The _____ of all the resistors in series is the total resistance of that series circuit.

9. _____ is defined as an electrical pressure.

10. _____ is defined as the rate of electron flow.

Multiple Choice

1. Which of the following methods can be used to form an electrical current?

   A. Magnetic induction.

   B. Chemical reaction.

   C. Heat.

   D. All of the above.

   E. None of the above.

2. In a parallel circuit:

   A. Total resistance is the sum of all of the resistances in the circuit.

   B. Total resistance is less than the lowest resistor.

   C. Amperage will decrease as more branches are added.

   D. All of the above.

3. All of the following concerning voltage drop are true EXCEPT:

   A. All of the voltage from the source must be dropped before it returns to the source.

   B. Corrosion is not a contributor to voltage drop.

   C. Voltage drop is the conversion of electrical energy into another energy form.

   D. Voltage drop can be measured with a voltmeter.

4. All of the following concerning voltage are true EXCEPT:

   A. Voltage is the electrical pressure that causes electrons to move.

   B. Voltage will exist between any two points in a circuit unless the potential drops to zero.

   C. Voltage is $A \times R$.

   D. In a series circuit, voltage is the same at all points in the circuit.

5. Wattage is:

   A. A measure of the total electrical work being performed per unit of time.

   B. Expressed as $P = R \times A$.

   C. Both A and B.

   D. Neither A nor B.

6. Which answer is incorrect concerning a series circuit?

   A. Total resistance is the sum of all of the resistances in the circuit.

   B. Total resistance is less than the lowest resistor.

   C. Amperage will decrease as more resistance is added.

   D. All of the above.

7. Which statement about electrical currents is correct?

   A. Alternating current can be stored in a battery.

   B. Alternating current is produced from a voltage and current that remain constant and flow in the same direction.

   C. Direct current is used for most electrical systems on the automobile.

   D. Direct current changes directions from positive to negative.

8. Induction:

   A. Is the magnetic process of producing a current flow in a wire without any actual contact to the wire.

   B. Exists when the magnetic field or the conductor is in motion.

   C. All of the above.

   D. None of the above.

9. All of the following statements are true EXCEPT:

   A. If the resistance increases and the voltage remains constant, the amperage will increase.

   B. Ohm's law can be stated as $A = V/R$.

   C. If voltage is increased, amperage will increase.

   D. An open circuit does not allow current flow.

10. Which of the following statements is correct?

   A. An insulator is capable of supporting the flow of electricity through it.

   B. A conductor is not capable of supporting the flow of electricity.

   C. All of the above.

   D. None of the above.

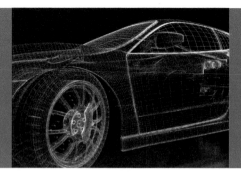

CHAPTER 3

ELECTRICAL AND ELECTRONIC COMPONENTS

Upon completion and review of this chapter, you should be able to:

- Describe the common types of electrical system components used and how they affect the electrical system.
- Explain the operation of the electrical controls, including switches, relays, and variable resistors.
- Explain the basic function of capacitors.
- Describe the basic operating principles of electronic components.
- Explain the use of electronic components in the circuit.

- Explain the purpose of circuit protection devices. Describe the most common types in use.
- Define circuit defects, including opens, shorts, grounds, and excessive resistance.
- Explain the effects that each type of circuit defect has on the operation of the electrical system.

Terms To Know

Anode	Fixed resistors	Pole
Avalanche diodes	Forward-biased	Positive plate
Base	Fuse	Positive temperature coefficient (PTC) thermistors
Bimetallic strip	Fuse block	
Bipolar	Fusible link	Potentiometer
Buzzer	Ganged switch	Protection device
Capacitance	Hole	Relay
Capacitor	Integrated circuit (IC)	Reverse-biased
Cathode	ISO relays	Rheostat
Circuit breaker	Light-emitting diode (LED)	Short
Clamping diode	Load device	Shorted circuit
Collector	Maxi-fuse	Solenoid
Covalent bonding	Normally closed (NC) switch	Sound generator
Crystal	Normally open (NO) switch	Stepped resistor
Darlington pair		Throw
Depletion-type FET	N-type material	Thyristor
Dielectric	Open	Transistor
Diode	Overload	Turn-on voltage
Electrostatic field	Peak reverse voltage (PRV)	Variable resistor
Emitter	Photodiode	Wiper
Enhancement-type FET	Phototransistor	Zener diode
Field-effect transistor (FET)		Zener voltage

INTRODUCTION

In this chapter you will be introduced to electrical and electronic components. These components include circuit protection devices, switches, relays, variable resistors, diodes, and different forms of transistors. Today's technician must comprehend the operation of these components and the ways they affect electrical system operation. With this knowledge, the technician will be able to accurately and quickly diagnose many electrical failures.

To be able to properly diagnose the components and circuits, the technician must be able to use the test equipment that is designed for electrical system diagnosis. This chapter discusses the various types of test equipment used for diagnosing electrical systems. You will learn the appropriate equipment to use to locate the fault based on the symptoms. In addition, the various types of defects that cause the system to operate improperly are discussed.

ELECTRICAL COMPONENTS

Electrical circuits require different components depending on the type of work they do and how they are to perform it. A light may be wired directly to the battery, but it will remain on until the battery drains. A switch will provide for control of the light circuit. However, if variable dimming of the light is required, a rheostat is also needed.

There are several electrical components that may be incorporated into a circuit to achieve the desired results from the system. These components include switches, relays, solenoids, buzzers, various types of resistors, and capacitors.

Switches

A switch is the most common means of providing control of electrical current flow to an accessory (**Figure 3-1**). A switch can control the on/off operation of a circuit or direct the flow of current through various circuits. The contacts inside the switch assembly carry the current when they are closed. When they are open, current flow is stopped.

Shop Manual
Chapter 3, page 121

A **normally open (NO) switch** will not allow current flow when it is in its rest position. The contacts are open until they are acted on by an outside force that closes them to complete the circuit. A **normally closed (NC) switch** will allow current flow when it is in its rest position. The contacts are closed until they are acted on by an outside force that opens them to stop current flow.

The simplest type of switch is the single-**pole**, single-**throw** (SPST) switch (**Figure 3-2**). This switch controls the on/off operation of a single circuit. The most common type of SPST switch design is the hinged pawl. The pawl acts as the contact and changes position as directed to open or close the circuit.

The term **pole** refers to the number of input circuits.

Some SPST switches are designed to be a momentary contact switch. This switch usually has a spring that holds the contacts open until an outside force is applied and closes them. The horn switch on most vehicles is of this design.

The term **throw** refers to the number of output circuits.

Some electrical systems may require the use of a single-pole, double-throw switch (SPDT). The dimmer switch used in the headlight system is usually an SPDT switch. This switch has one input circuit with two output circuits. Depending on the position of the contacts, voltage is applied to the high-beam circuit or to the low-beam circuit (**Figure 3-3**).

Figure 3-1 Common types of switches used in the automotive electrical system.

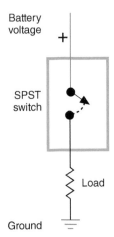

Figure 3-2 A simplified illustration of an SPST switch.

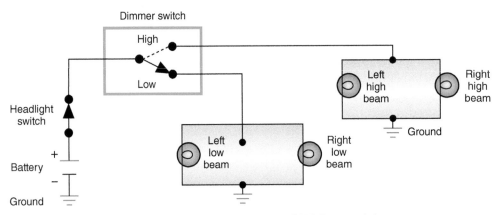

Figure 3-3 A simplified schematic of a headlight system using an SPDT dimmer switch.

One of the most complex switches is the **ganged switch**. This type of switch is commonly used as an ignition switch. In **Figure 3-4**, the five wipers are all ganged together and will move together. Battery voltage is applied to the switch from the starter relay terminal. When the ignition key is turned to the START position, all wipers move to the "S" position. Wipers D and E will complete the circuit to ground to test the instrument panel warning lamps. Wiper B provides battery voltage to the ignition coil. Wiper C supplies battery voltage to the starter relay and the ignition module. Wiper A has no output.

> **AUTHOR'S NOTE** The dashed lines used in the switch symbol indicate that the wipers of the switch move together.

Once the engine starts, the wipers are moved to the RUN position. Wipers D and E are moved out of contact with any output terminals. Wiper A supplies battery voltage to the comfort controls and turn signals, wiper B supplies battery voltage to the ignition coil and other accessories, and wiper C supplies battery voltage to other accessories. The

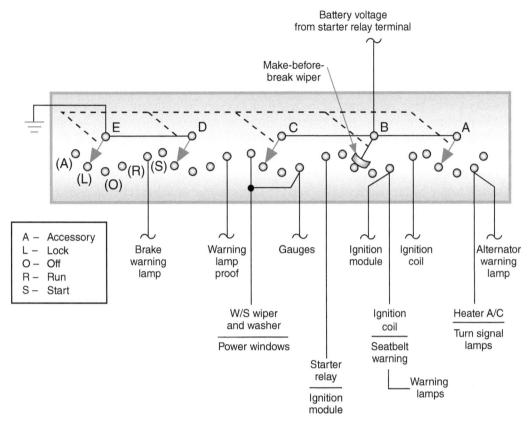

Figure 3-4 Illustration of an ignition switch.

jumper wire between terminals A and R of wiper C indicates that those accessories listed can be operated with the ignition switch in the RUN or ACC position.

Mercury switches are used by many vehicle manufacturers to detect motion. This switch uses a capsule that is partially filled with mercury and has two electrical contacts located at one end. If the switch is constructed as a normally open switch, the contacts are located above the mercury level (**Figure 3-5**). Mercury is an excellent conductor of electricity. If the capsule is moved so the mercury touches both of the electrical contacts, the circuit is completed (**Figure 3-6**). This type of switch is used to illuminate the engine compartment when the hood is opened. While the hood is shut, the capsule is tilted in a

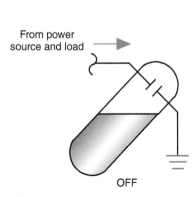

Figure 3-5 A mercury switch in the open position. The mercury is not covering the points.

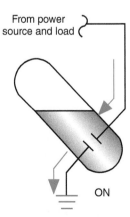

Figure 3-6 When the mercury switch is tilted, the mercury covers the points and closes the circuit.

position such that the mercury is not able to complete the circuit. Once the hood opened, the capsule tilts with the hood, the mercury completes the circuit, and the light turns on.

Relays

Shop Manual
Chapter 3, page 123

The collapsing of the magnetic field around the coil produces voltage spikes. To prevent the voltage spike from damaging sensitive electronic components, relays may be equipped with a clamping diode or a resistor installed in parallel to the coil. See page 69.

Some circuits utilize electromagnetic switches called **relays** (**Figure 3-7**). The coil in the relay has a very high resistance; thus, it will draw very low current. This low current is used to produce a magnetic field that will close the contacts. Normally open relays have their points closed by the electromagnetic field, and normally closed relays have their points opened by the magnetic field. The contacts are designed to carry the high current required to operate the load component. When current is applied to the coil, the contacts close and heavy battery current flows to the load component that is being controlled.

The illustration (**Figure 3-8**) shows a relay application in a horn circuit. Battery voltage is applied to the coil. Because the horn switch is a normally open–type switch, the current flow to ground is open. Pushing the horn switch will complete the circuit, allowing current flow through the coil. The coil develops a magnetic field, which closes the contacts. With the contacts closed, battery voltage is applied to the horn (which is grounded). Used in this manner, the horn relay becomes a control of the high current necessary to blow the horn. The control circuit may be wired with very small diameter wire because it will have low current flowing through it. The control unit may have only 0.25 ampere flowing through it, and the horn may require 24 or more amperes.

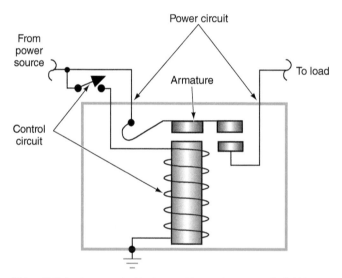

Figure 3-7 A relay uses electrical current to create a magnetic field to draw the contact point closed.

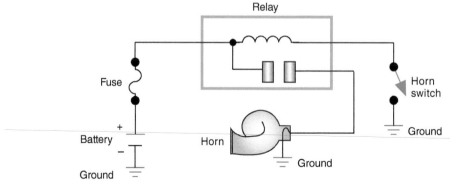

Figure 3-8 A relay can be used in the horn circuit to reduce the required size of the conductors installed in the steering column.

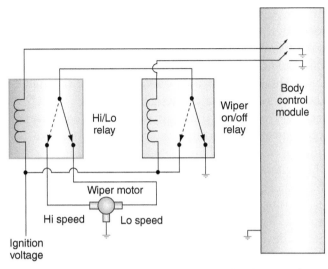

Figure 3-9 Using a relay as a diverter to control Hi/Lo wiper operation. The Hi/Lo relay diverts current to the different brushes of the wiper motor.

Relays can also be used as a circuit diverter (**Figure 3-9**). In this example, the Hi/Lo wiper relay will direct current flow to either the high-speed brush or the low-speed brush of the wiper motor to control wiper speeds.

The coil side of the relay is rated based on its voltage. In the automotive industry this means they are typically rated for around 12 volts. The contact side of the relay is rated based on its current-carrying capabilities. Typically, automotive relays are 20, 30, 35, or 40 amp rated. It is important to use the correct relay if you need to replace it. The relays can also be NO type or NC type. They may have different amperage requirements based on whether they are NC or NO relays.

ISO Relays. **ISO relays** conform to the specifications of the International Standards Organization (ISO) for common size and terminal patterns (**Figure 3-10**). The terminals are identified as 30, 87a, 87, 86, and 85. Terminal 30 is usually connected to battery

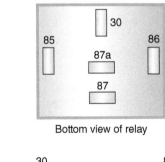

Bottom view of relay

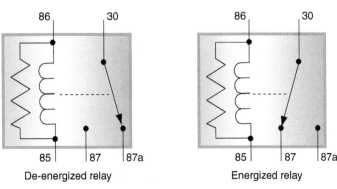

De-energized relay   Energized relay

Figure 3-10 ISO relay terminal identification.

voltage. This source voltage can be either switched (on or off by some type of switch) or connected directly to the battery. Terminal 87a is connected to terminal 30 when the relay is de-energized. Terminal 87 is connected to terminal 30 when the relay is energized. Terminal 86 is connected to battery voltage (switched or unswitched) to supply current to the electromagnet. Finally, terminal 85 provides ground for the electromagnet. Once again, the ground can be switched or unswitched.

Some manufacturers may name the relay terminals based on their function. Terminal 86 is named "Trigger" since it is the voltage supply terminal that will activate the coil. Terminal 85 is identified as "Ground." This terminal may be connected directly to a ground location, or may be switched. If the relay does not have a clamping diode, terminals 86 and 85 can be interchanged and the terminal functions are reversed.

Terminal 30 is identified as "Input" since it is connected to the supply voltage for the switch side of the relay. Terminals 87 and 87a are both named "Output" since either one can direct current to a device depending on the activation state of the relay.

Solenoids

A **solenoid** is an electromagnetic device that operates in the same way as a relay; however, a solenoid uses a movable iron core. Solenoids can do mechanical work, such as switching electrical, vacuum, and liquid circuits. The iron core inside the coil of the solenoid is spring loaded (**Figure 3-11**). When current flows through the coil, the magnetic field created around the coil attracts the core and moves it into the coil. This example shows a solenoid that works as a high-amperage relay in that it uses low current to control high current. This would be a common type of starter motor solenoid.

A typical example of a solenoid designed to perform a mechanical function is an electric door lock or trunk release system that mechanically moves the locking mechanism. To do work, the core is attached to a mechanical linkage, which causes something to move. When current flows through the coil, it moves the core, and ultimately moves the linkage. When current flow through the coil stops, the spring pushes the core back to its original position.

Solenoids may also switch a circuit on or off, in addition to causing a mechanical action. Such is the case with some starter motor–mounted solenoids. These devices move the starter gear in and out of mesh with the flywheel. At the same time, they complete the circuit from the battery to the ignition circuit. Both of these actions are necessary to start an engine.

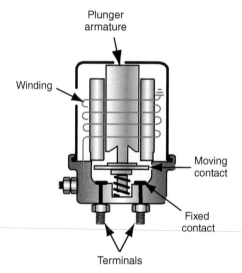

Figure 3-11 Typical electromechanical solenoid.

AUTHOR'S NOTE There are probably more solenoids on the vehicle than you may realize. For example, each fuel injector is really nothing more than a solenoid. As the magnetic field builds, a pintle is lifted off its seat and fuel is allowed to flow past the pintle, through the orifices and into the intake manifold.

Buzzers

A **buzzer**, or **sound generator**, is sometimes used to warn the driver of possible safety hazards by emitting an audio signal (such as when the seat belt is not buckled). A buzzer is similar in construction to a relay except for the internal wiring (**Figure 3-12**). The coil is supplied current through the normally closed contact points. When voltage is applied to the buzzer, current flows through the contact points to the coil. When the coil is energized, the contact arm is attracted to the magnetic field. As soon as the contact arm is pulled down, the current flow to the coil is opened, and the magnetic field is dissipated. The contact arm then closes again, and the circuit to the coil is closed. This opening and closing action occurs very rapidly. It is this movement that generates the vibrating signal and the buzzing sound.

Resistors

All circuits require resistance in order to operate. If the resistance performs a useful function, it is referred to as the **load device**. However, resistance can also be used to control current flow and as sensing devices for computer systems. There are several types of resistors that may be used within a circuit. These include fixed resistors, stepped resistors, and variable resistors.

Fixed Resistors. **Fixed resistors** are usually made of carbon or oxidized metal (**Figure 3-13**). These resistors have a set resistance value and are used to limit the amount of current flow in a circuit. Like any other electrical component, fixed resistors must be matched to the circuit. Resistors not only have different resistance values but are also rated based on their wattage. The resistance value can be determined by the color bands on the

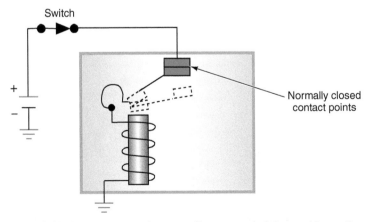

Figure 3-12 A buzzer reacts to the current flow to open and close rapidly, creating a noise.

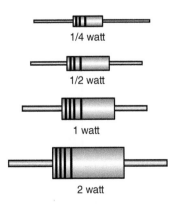

Figure 3-13 Fixed resistors.

protective shell (**Figure 3-14**). Usually there are four or five color bands. When there are four bands, the first two are the digit bands, the third is the multiplier, and the fourth is the tolerance. On a resistor with five bands, the first three are digit bands.

For example, if the resistor has four color bands of yellow, black, brown, and gold, the resistance value is determined as follows:

The first color band (yellow) gives the first-digit value of 4.
The second color band (black) gives the second-digit value of 0.
The digit value is now 40. Multiply this by the value of the third band. In this case, brown has a value of 10, so the resistor should have 400 ohms of resistance (40 × 10 = 400).
The last band gives the tolerance. Gold equals a tolerance range of ±5%.

Shop Manual
Chapter 3, page 126

Stepped Resistors. A **stepped resistor** has two or more fixed resistor values. The stepped resistor can have an integral switch or a switch wired in series. A stepped resistor is commonly used to control electrical motor speeds (**Figure 3-15**). By changing the position of the switch, resistance increases or decreases within the circuit. If the current flows through a low resistance, a higher current flows to the motor and its speed increases. If the switch is placed in the low-speed position, additional resistance is added to the circuit. Less current flows to the motor, which causes the motor to operate at a reduced speed.

A stepped resistor is also used to convert digital to analog signals in a computer circuit. This is accomplished by converting the on/off digital signals into a continuously variable analog signal.

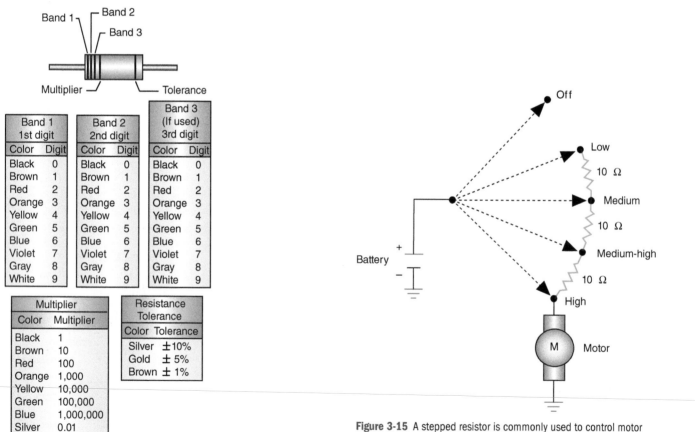

Band 1		Band 2		Band 3 (If used)	
1st digit		2nd digit		3rd digit	
Color	Digit	Color	Digit	Color	Digit
Black	0	Black	0	Black	0
Brown	1	Brown	1	Brown	1
Red	2	Red	2	Red	2
Orange	3	Orange	3	Orange	3
Yellow	4	Yellow	4	Yellow	4
Green	5	Green	5	Green	5
Blue	6	Blue	6	Blue	6
Violet	7	Violet	7	Violet	7
Gray	8	Gray	8	Gray	8
White	9	White	9	White	9

Multiplier	
Color	Multiplier
Black	1
Brown	10
Red	100
Orange	1,000
Yellow	10,000
Green	100,000
Blue	1,000,000
Silver	0.01
Gold	0.1

Resistance Tolerance	
Color	Tolerance
Silver	±10%
Gold	± 5%
Brown	± 1%

Figure 3-14 Resistor color code chart.

Figure 3-15 A stepped resistor is commonly used to control motor speeds. The total resistance of the switch is 30 Ω in the low position, 20 Ω in the medium position, 10 Ω in the medium-high position, and 0 Ω in the high position.

Variable Resistors.   **Variable resistors** provide for an infinite number of resistance values within a range. The most common types of variable resistors are rheostats and potentiometers. A **rheostat** is a two-terminal variable resistor used to regulate the strength of an electrical current. A rheostat has one terminal connected to the fixed end of a resistor and a second terminal connected to a movable contact called a **wiper** (**Figure 3-16**). By changing the position of the wiper on the resistor, the amount of resistance can be increased or decreased. The most common use of the rheostat is in the instrument panel lighting switch. As the switch knob is turned, the instrument lights dim or brighten depending on the resistance value.

A **potentiometer** is a three-wire variable resistor that acts as a voltage divider to produce a continuously variable output signal proportional to a mechanical position. When a potentiometer is installed into a circuit, one terminal is connected to a power source at one end of the wound resistor. The second wire is connected to the opposite end of the wound resistor and is the ground return path. The third wire is connected to the wiper contact (**Figure 3-17**). The wiper senses a variable voltage drop as it is moved over the resistor. Because the current always flows through the same amount of resistance, the total voltage drop measured by the potentiometer is very stable. For this reason, the potentiometer is a common type of input sensor for the vehicle's on-board computers.

Shop Manual
Chapter 3, page 126

Capacitors

Some automotive electrical systems use a capacitor or condenser to store electrical charges (**Figure 3-18**). A capacitor uses the theory of **capacitance** to temporarily store electrical energy. Capacitance (C) is the ability of two conducting surfaces to store voltage. The two surfaces must be separated by an insulator.

A **capacitor** does not consume any power; however, it will store and release electrical energy. All of the voltage stored in the capacitor is returned to the circuit when the capacitor discharges. Because the capacitor stores voltage, it will also absorb voltage changes in the circuit. By providing for this storage of voltage, damaging voltage spikes can be controlled. Capacitors are also used to reduce radio noise.

A capacitor is made by wrapping two conductor strips around an insulating strip. The insulating strip, or **dielectric**, prevents the plates from coming in contact while keeping them very close to each other. The dielectric can be made of insulator material such as ceramic, glass, paper, plastic, or even the air between the two plates. A capacitor blocks direct current. A small amount of current enters the capacitor and charges it.

Shop Manual
Chapter 3, page 128

The insulator in a capacitor is called a **dielectric**. The dielectric can be made of some insulator material such as ceramic, glass, paper, plastic, or even the air between the two plates.

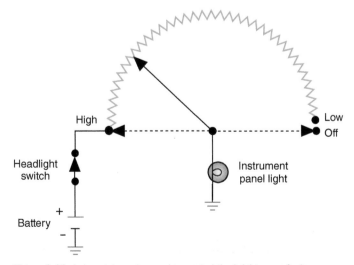

Figure 3-16 A rheostat can be used to control the brightness of a lamp.

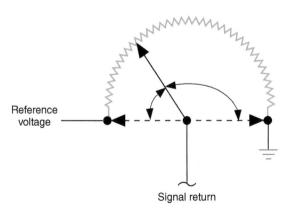

Figure 3-17 A potentiometer is used to send a signal voltage from the wiper.

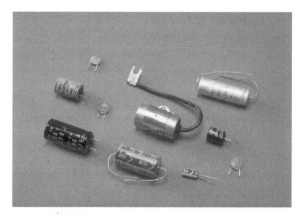

Figure 3-18 Capacitors that can be used in automotive electrical circuits.

Most capacitors are connected in parallel across the circuit (**Figure 3-19**). Capacitors operate on the principle that opposite charges attract each other and that there is a potential voltage between any two oppositely charged points. When the switch is closed, the protons at the positive battery terminal will attract some of the electrons on one plate of the capacitor away from the area near the dielectric material. As a result, the atoms of the **positive plate** are unbalanced because there are more protons than electrons in the atom. This plate now has a positive charge because of the shortage of electrons (**Figure 3-20**). The positive charge of this plate will attract electrons on the other plate. The dielectric keeps the electrons on the negative plate from crossing over to the positive plate, resulting in a storage of electrons on the negative plate (**Figure 3-21**). The movement of electrons to the negative plate and away from the positive plate is an electrical current.

Current will flow through the capacitor until the voltage charges across the capacitor and across the battery are equalized. Current flow through a capacitor is only the effect of the electron movement onto the negative plate and away from the positive plate. Electrons do not actually pass through the capacitor from one plate to another. The charges on the plates do not move through the **electrostatic field**. They are stored on the plates as static electricity.

When the charges across the capacitor and battery are equalized, there is no potential difference and no more current will flow through the capacitor (**Figure 3-22**). Current will now flow through the load components in the circuit (**Figure 3-23**).

When the switch is opened, current flow from the battery through the resistor is stopped. However, the capacitor has a storage of electrons on its negative plate. Because

The plate connected to the positive battery terminal is the **positive plate**.

An atom that has fewer electrons than protons is said to be a positive ion.

The field that is between the two oppositely charged plates is called the **electrostatic field**.

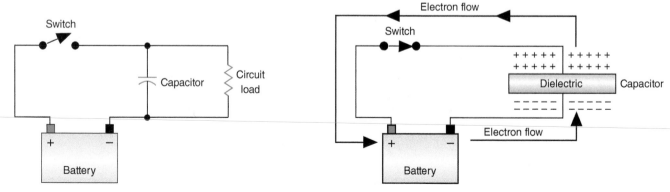

Figure 3-19 A capacitor connected to a circuit.

Figure 3-20 The positive plate sheds its electrons.

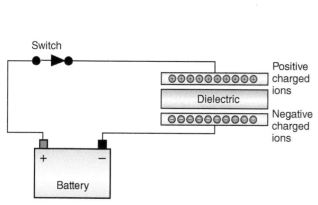

Figure 3-21 The electrons will be stored on the negative plate.

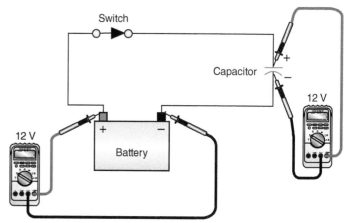

Figure 3-22 A capacitor when it is fully charged.

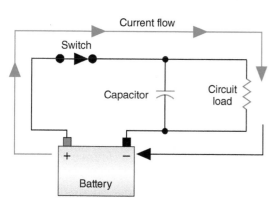

Figure 3-23 Current flow with a fully charged capacitor.

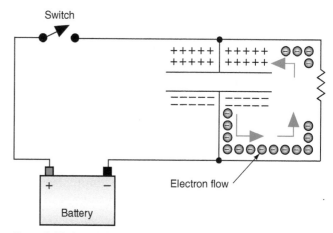

Figure 3-24 Current flow with the switch open and the capacitor discharging.

the negative plate of the capacitor is connected to the positive plate through the resistor, the capacitor acts as the source. The capacitor will discharge the electrons through the resistor until the atoms of the positive plate and negative plate return to a balanced state (**Figure 3-24**).

In the event that a high-voltage spike occurs in the circuit, the capacitor will absorb the additional voltage before it is able to damage the circuit components. A capacitor can also be used to stop current flow quickly when a circuit is opened (such as in the ignition system). It can also store a high-voltage charge and then discharge it when a circuit needs the voltage (such as in some air bag systems).

Capacitors are rated in units called farads. A one-farad capacitor connected to a one-volt source will store 6.28×10^{18} electrons. A farad is a large unit and most commonly used capacitors are rated in picofarad (a trillionth of a farad) or microfarads (a millionth of a farad). In addition, the capacitor has a voltage rating that is determined by how much voltage can be applied to it without the dielectric breaking down. The maximum voltage rating and capacitance determine the amount of energy a capacitor holds. The voltage rating is related to the strength and thickness of the dielectric. The voltage rating increases with increasing dielectric strength and the thickness of the dielectric. The capacitance increases with the area of the plates and decreases with the thickness of the dielectric.

ELECTRONIC COMPONENTS

Because a semiconductor material can operate as both a conductor and an insulator, it is very useful as a switching device. How a semiconductor material works depends on the way current flows, or tries to flow, through it.

As discussed in Chapter 2, electrical materials are classified as conductors, insulators, or semiconductors. Semiconductors include diodes, transistors, and silicon-controlled rectifiers. These semiconductors are often called solid-state devices because they are constructed of a solid material. The most common materials used in the construction of semiconductors are silicon or germanium. Both of these materials are classified as a **crystal**, since they have a definite atomic structure.

Silicon and germanium have four electrons in their outer orbits. Because of their crystal-type structure, each atom shares an electron with four other atoms (**Figure 3-25**). As a result of this **covalent bonding**, each atom will have eight electrons in its outer orbit. All the orbits are filled and there are no free electrons; thus, the material (as a category of matter) falls somewhere between conductor and insulator.

Perfect crystals are not used for manufacturing semiconductors. They are doped with impurity atoms. This doping adds a small percentage of another element to the crystal. The doping element can be arsenic, antimony, phosphorus, boron, aluminum, or gallium.

If the crystal is doped by using arsenic, antimony, or phosphorus, the result is a material with free electrons (**Figure 3-26**). Materials such as arsenic have five electrons, which leaves one electron left over. This doped material becomes negatively charged and is referred to as an **N-type material**. Under the influence of an EMF, it will support current flow.

If boron, aluminum, or gallium is added to the crystal, a P-type material is produced. Materials like boron have three electrons in their outermost orbit. Because there is one fewer electron, there is an absence of an electron that produces a **hole** (**Figure 3-27**) and becomes positively charged.

By putting N-type and P-type materials together in a certain order, solid-state components are built that can be used for switching devices, voltage regulators, electrical control, and so on.

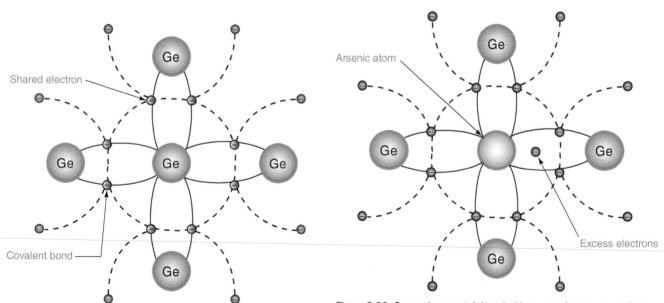

Figure 3-25 Crystal structure of germanium.

Figure 3-26 Germanium crystal doped with an arsenic atom to produce an N-type material.

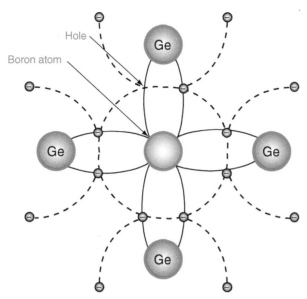

Figure 3-27 Germanium crystal doped with a boron atom to produce a P-type material.

Diodes

A **diode** is an electrical one-way check valve that will allow current to flow in one direction only. A diode is the simplest semiconductor device. It is formed by joining P-type semiconductor material with N-type material. The N (negative) side of a diode is called the **cathode** and the P (positive) side the **anode** (**Figure 3-28**). The point where the cathode and anode join together is called the PN junction. The outer shell of the diode will have a stripe painted around it. This stripe designates which end of the diode is the cathode.

When a diode is made, the positive holes from the P region and the negative charges from the N region are drawn toward the junction. This is because of the attraction of unlike charges to each other. Electrons from the N material will cross over and fill holes from the P material along the PN junction. This forms a depletion zone. When the charges cross over, the two halves are no longer balanced and the diode builds up a network of internal charges. In the depletion zone, the material is an insulator since all of the holes are filled and there are no free electrons. When the diode is incorporated within a circuit and a voltage is applied, the internal characteristics change. If the diode is **forward-biased**, there will be current flow (**Figure 3-29**). In this state, the depletion zone will become smaller as the electrons of the N region and the holes of the P region move across the junction. The negative region pushes electrons across the junction as the free electrons in the N region are repelled by the negative charge from the battery and are attracted to the P region of the diode. The repelling of like charges of the positive side of the battery and the P region of the diode results in the holes in the P region moving toward the N region. Since the depletion zone is reduced, when the diode is forward-biased, the diode acts as a conductor.

If the diode is **reverse-biased**, there will be no current flow (**Figure 3-30**). The negative region will attract the positive holes away from the junction and the positive region will attract electrons away. Since the electrons and holes are moving in the wrong direction, the depletion zone is enlarged and makes the diode act as an insulator.

When the diode is forward-biased, it will have a small voltage drop across it. On standard silicon diodes, this voltage is usually about 0.6 volt. This is referred to as the **turn-on voltage**.

Shop Manual
Chapter 3, page 128

Forward-biased means that a positive voltage is applied to the P-type material and a negative voltage to the N-type material.

Reverse-biased means that a positive voltage is applied to the N-type material and a negative voltage to the P-type material.

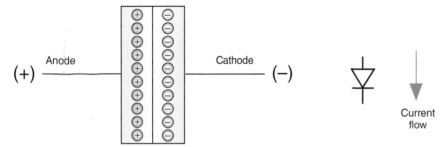

Figure 3-28 A diode and its symbol.

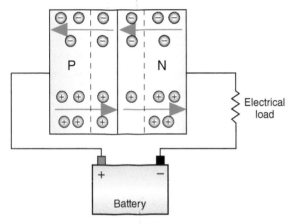

Figure 3-29 Forward-biased voltage causes current flow.

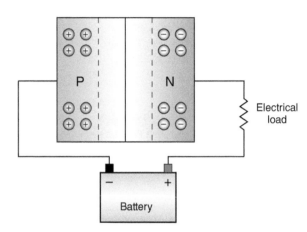

Figure 3-30 Reverse-biased voltage prevents current flow.

Diode Identification and Ratings. Since there are many types of diodes that can vary in size from very small to those capable of withstanding 250 amperes, a semiconductor identification system was developed to identify and rate diodes (**Figure 3-31**). This system is also used to identify transistors and many other special semiconductor devices. The identification system uses a combination of numbers and letters to identify different types of semiconductor devices.

The first two characters of the system identify the component. The first number indicates the number of junctions in the semiconductor device. Since this number is one less than the number of active elements, a 1 designates a diode, a 2 designates a transistor,

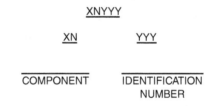

XNYYY

XN   YYY

COMPONENT   IDENTIFICATION
                      NUMBER

X - NUMBER OF SEMICONDUCTOR JUNCTIONS
N - A SEMICONDUCTOR
YYY - IDENTIFICATION NUMBER (ORDER OR REGISTRATION NUMBER)
      ALSO INCLUDES SUFFIX LETTER (IF APPLICABLE) TO INDICATE

1. MATCHING DEVICES
2. REVERSE POLARITY
3. MODIFICATION

EXAMPLE - 1N345A (AN IMPROVED VERSION OF THE
            SEMICONDUCTOR DIODE TYPE 345)

Figure 3-31 Standard semiconductor identification markings.

and a 3 designates a tetrode (a four-element transistor). The letter "N" that follows the first number indicates that the device is a semiconductor.

The last series of characters that follow the "N" is a serialized identification number. This number may also contain a suffix letter after the third digit. Common suffix letters that are used include "M" to describe matching pairs of separate semiconductor devices and "R" to indicate reverse polarity. In addition, suffix letters are used to indicate modified versions of the device. For example, a semiconductor diode designated as type 1N345A signifies a two-element diode (1) of semiconductor material (N) that is an improved version (A) of type 345.

To distinguish the anode from the cathode side of the diode, manufacturers generally code the cathode end of the diode. The identification can be a colored band or a dot, a "k," "+," "cath," or an unusual shape such as raised edge or taper. Some manufacturers use standard color code bands on the cathode side to not only identify the cathode end of the diode but identify the diode by number.

Zener Diodes

As stated, if a diode is reverse-biased it will not conduct current. However, if the reverse voltage is increased, a voltage level will be reached at which the diode will conduct in the reverse direction. This voltage level is referred to as **zener voltage**. At this point the diode is able to conduct, but will try to limit the voltage dropped across it to this zener voltage. Reverse current can destroy a simple PN-type diode due to the buildup of heat. The point when the PN junction of a diode begins to breakdown is called the **peak reverse voltage (PRV)**. However, the diode can be doped with materials that will withstand reverse current.

Shop Manual
Chapter 3, page 129

 A BIT OF HISTORY

The first diodes were vacuum tube devices (also known as thermionic valves). Arrangements of electrodes were surrounded by a vacuum within a glass envelope that appeared similar to light bulbs. This arrangement of a filament and plate to create a diode was invented by John Ambrose Fleming in 1904. Current flow through the filament results in generation of heat. When heated, electrons are emitted into the vacuum. These electrons are electrostatically drawn to a positively charged outer metal plate (anode). Since the plate is not heated, the electrons will not return to the filament, even if the charge on the plate is made negative.

A **zener diode** is designed to operate in reverse bias at the PRV region. At the point that PRV is reached, a large current flows in reverse bias. Since the voltage drop across the zener is limited to the voltage level, the voltage output is prevented from climbing any higher. This makes the zener diode an excellent component for regulating voltage. If the zener diode is rated at 15 volts, it will not conduct in reverse bias when the voltage is below 15 volts. At 15 volts it will conduct. The function of the zener diode is to limit the voltage across the load and the voltage will not increase over 15 volts.

Figure 3-32 shows a simplified circuit that has a zener diode in it to provide a constant voltage level to the instrument gauge. In this example, the zener diode is connected in series with the resistor and in parallel to the gauge. If the voltage to the gauge must be limited to 7 volts, the zener diode used would be rated at 7 volts. The zener diode maintains a constant voltage drop, and the total voltage drop in a series circuit must equal the amount of source voltage; thus, voltage that is greater than the zener voltage must be dropped over the resistor. Even though source voltage may vary (as a normal result of the charging system), causing different currents to flow through the resistor and zener diode, the voltage that the zener diode drops remains the same.

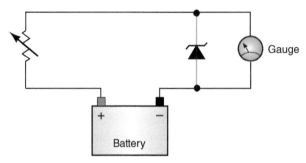

Figure 3-32 Simplified instrument gauge circuit that uses a zener diode to maintain a constant voltage to the gauge. Note the symbol used for a zener diode.

The zener breaks down when system voltage reaches 7 volts. At this point, the zener diode conducts reverse current, causing an additional voltage drop across the resistor. The amount of voltage to the instrument gauge will remain at 7 volts because the zener diode makes the resistor drop the additional voltage to maintain this limit.

Here we see the difference between the standard diode and the zener diode. When the zener diode is reverse-biased, the zener holds the available voltage to a specific value.

Avalanche Diodes

Shop Manual
Chapter 3, page 130

Avalanche diodes are specialized diodes that are used as "relief valves" to protect electrical systems from excess voltages. Similar to zener diodes in operation, avalanche diodes conduct in the reverse direction when the reverse-bias voltage exceeds the breakdown voltage. However, breakdown is caused by the avalanche effect. This occurs when the reverse electric field moves across the PN junction and causes a wave of ionization (like an avalanche), leading to a large current. Avalanche diodes are designed to break down at a well-defined reverse voltage without being destroyed.

Avalanche diodes are commonly used in automobile AC generators (alternators) to protect against voltage surges that can damage computer systems. Typically, these will go into avalanche at about 21 to 29 volts and will conduct the current to ground instead of into the vehicle's electrical system. Voltages higher than the zener voltage will result in the diode conducting constantly in the reverse-bias direction. This prevents the output voltage of the AC generator from going above the rating of the avalanche diode.

Light-Emitting Diodes

Shop Manual
Chapter 3, page 130

In the discussion on diodes it was mentioned that when the diode is forward-biased there is an interaction of electrons and holes that occurs. A side effect of this interaction is the generation of light. A **light-emitting diode (LED)** has a small lens built into it so this light can be seen when current flows through it (**Figure 3-33**). When the LED is forward-biased, the holes and electrons combine and current is allowed to flow through it. The energy generated is released in the form of light. The light from an LED is not heat energy

Figure 3-33 (A) A light-emitting diode uses a lens to emit the generated light. (B) Symbol for LED.

as is the case with other lights. It is electrical energy. Because of this, LEDs last longer than light bulbs. It is the material used to make the LED that will determine the color of the light emitted, and the turn-on voltage.

Similar to standard silicon diodes, the LED has a constant turn-on voltage. However, this turn-on voltage is usually higher than that of standard diodes. The turn-on voltage defines the color of the light; 1.2 volts correspond to red, 2.4 volts to yellow.

Photodiodes

A **photodiode** also allows current to flow in one direction only. However, the direction of current flow is opposite to that in a standard diode. Reverse current flow occurs only when the diode receives a specific amount of light. These types of diodes can be used in automatic headlight systems.

Clamping Diodes

Whenever the current flow through a coil (such as used in a relay or solenoid) is discontinued, a voltage surge or spike is produced. This surge results from the collapsing of the magnetic field around the coil. The movement of the field across the windings induces a very high-voltage spike, which can damage electronic components as it flows through the system. In some circuits, a capacitor can be used as a shock absorber to prevent component damage from this surge. In today's complex electronic systems, a **clamping diode** is commonly used to prevent the voltage spike. By installing a clamping diode in parallel with the coil, a bypass is provided for the electrons during the time that the circuit is open (**Figure 3-34**).

> A **clamping diode** is nothing more than a standard diode; the term *clamping* refers to its function.

An example of the use of clamping diodes is on some air-conditioning compressor clutches. Because the clutch operates by electromagnetism, opening the clutch coil circuit produces a voltage spike. If this voltage spike was left unchecked, it could damage the vehicle's on-board computers. The installation of the clamping diode prevents the voltage spike from reaching the computers. The clamping diode must be connected to the circuit in reverse bias.

Relays may also be equipped with a clamping diode. However, some use a resistor to dissipate the voltage spike. The two types of relays are not interchangeable.

Transistors

A **transistor** is a three-layer semiconductor. It is typically used as a very fast switching device. The word *transistor* is a combination of two words, *transfer* and *resist*. The transistor is used to control current flow in the circuit (**Figure 3-35**). It can be used to allow a predetermined amount of current flow or to resist this flow.

> **Shop Manual**
> Chapter 3, page 130

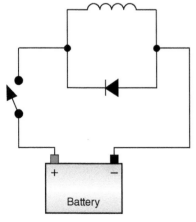

Figure 3-34 A clamping diode in parallel to a coil prevents voltage spikes when the switch is opened.

Figure 3-35 Transistors that are used in automotive applications.

Transistors are made by combining P-type and N-type materials in groups of three. The two possible combinations are NPN (**Figure 3-36**) and PNP (**Figure 3-37**).

The three layers of the transistor are designated as **emitter**, **collector**, and **base**. The emitter is the outside layer of the forward-biased diode that has the same polarity as the circuit side to which it is applied. The arrow on the transistor symbol refers to the emitter lead and points in the direction of positive current flow and to the N material. The collector is the outside layer of the reverse-biased diode. The base is the shared middle layer. Each of these different layers has its own lead for connecting to different parts of the circuit. In effect, a transistor is two diodes that share a common center layer. When a transistor is connected to the circuit, the emitter-base junction will be forward-biased and the collector-base junction will be reverse-biased.

In the NPN transistor, the emitter conducts current flow to the collector when the base is forward-biased. The transistor cannot conduct unless the voltage applied to the base leg exceeds the emitter voltage by approximately 0.7 volt. This means both the base and collector must be positive with respect to the emitter. With less than 0.7 volt applied to the base leg (compared to the voltage at the emitter), the transistor acts as an opened switch. When the voltage difference is greater than 0.7 volt at the base, compared to the emitter voltage, the transistor acts as a closed switch (**Figure 3-38**).

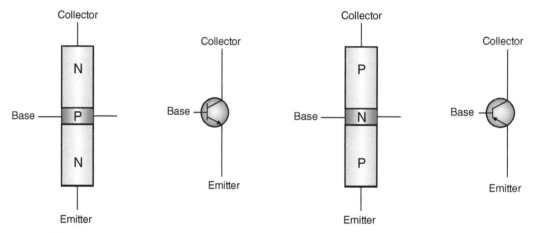

Figure 3-36 An NPN transistor and its symbol.   **Figure 3-37** A PNP transistor and its symbol.

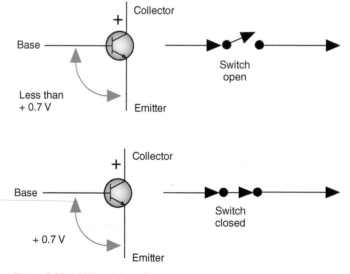

Figure 3-38 NPN transistor action.

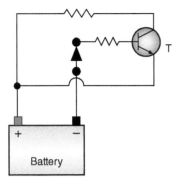

Figure 3-39 NPN transistor with reverse-biased voltage applied to the base. No current flow.

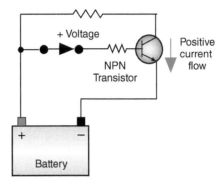

Figure 3-40 NPN transistor with forward-biased voltage applied to the base. Current flows.

When an NPN transistor is used in a circuit, it normally has a reverse bias applied to the base-collector junction. If the emitter-base junction is also reverse-biased, no current will flow through the transistor (**Figure 3-39**). If the emitter-base junction is forward-biased (**Figure 3-40**), current flows from the emitter to the base. Because the base is a thin layer and a positive voltage is applied to the collector, electrons flow from the emitter to the collector.

In the PNP transistor, current will flow from the emitter to the collector when the base leg is forward-biased with a voltage that is more negative than that at the emitter (**Figure 3-41**). For current to flow through the emitter to the collector, both the base and the collector must be negative with respect to the emitter.

> **AUTHOR'S NOTE** Current flow through transistors is always based on hole and/ or electron flow.

Current can be controlled through a transistor. Thus, transistors can be used as a very fast electrical switch. It is also possible to control the amount of current flow through the collector. This is because the output current is proportional to the amount of current through the base leg.

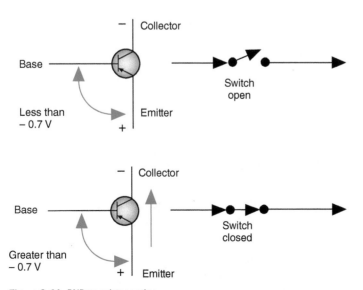

Figure 3-41 PNP transistor action.

A transistor has three operating conditions:

1. **Cutoff:** When reverse-biased voltage is applied to the base leg of the transistor. In this condition the transistor is not conducting and no current will flow.
2. **Conduction:** Bias voltage difference between the base and the emitter has increased to the point that the transistor is switched on. In this condition the transistor is conducting. Output current is proportional to that of the current through the base.
3. **Saturation:** This occurs when the collector to emitter voltage is reduced to near zero by a voltage drop across the collector's resistor.

These types of transistors are called **bipolar** because they have three layers of silicon; two of these layers are the same. Another type of transistor is the **field-effect transistor (FET)**. The FET's leads are listed as source, drain, and gate. The source supplies the electrons and is similar to the emitter in the bipolar transistor. The drain collects the current and is similar to the collector. The gate creates the electrostatic field that allows electron flow from the source to the drain. It is similar to the base.

While electrons are flowing from the source to the drain (electron theory), positive charges are flowing from the drain to the source (conventional theory).

The FET transistor does not require a constant bias voltage. A voltage needs to be applied to the gate terminal to get electron flow from the source to the drain. The source and drain are constructed of the same type of doped material. They can be either N-type or P-type materials. The source and drain are separated by a thin layer of either N-type or P-type material opposite the gate and drain.

Using **Figure 3-42**, if the source voltage is held at 0 volts and 6 volts are applied to the drain, no current will flow between the two. However, if a lower positive voltage is applied to the gate, the gate forms a capacitive field between it and the channel. The voltage of the capacitive field attracts electrons from the source, and current will flow through the channel to the higher positive voltage of the drain.

A BIT OF HISTORY

The transistor was developed by a team of three American physicists: Walter Houser Brattain, John Bardeen, and William Bradford Shockley. They announced their achievement in 1948. These physicists won the Nobel Prize in physics for this development in 1956.

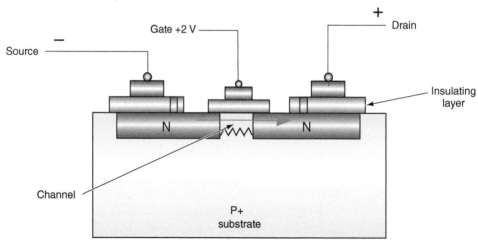

Figure 3-42 An FET uses a positive voltage to the gate terminal to create a capacitive field to allow electron flow.

This type of FET is called an **enhancement-type FET** because the field effect improves current flow from the source to the drain. This operation is similar to that of a NO switch. A **depletion-type FET** is like a NC switch, whereas the field effect cuts off current flow from the source to the drain.

Transistor Amplifiers

A transistor can be used in an amplifier circuit to amplify the voltage. This is useful when using a very small voltage for sensing computer inputs but needing to boost that voltage to operate an accessory (**Figure 3-43**). The waveform showing the small signal voltage that is applied to the base leg of a transistor may look like that shown in **Figure 3-44A**. The waveform showing the corresponding signal through the collector will be inverted (**Figure 3-44B**). Three things happen in an amplified circuit:

1. The amplified voltage at the collector is greater than that at the base.
2. The input current increases.
3. The pattern has been inverted.

Some amplifier circuits use a **Darlington pair**, which is two transistors that are connected together. The first transistor in a Darlington pair is used as a preamplifier to produce a large current to operate the second transistor (**Figure 3-45**). The second

Shop Manual
Chapter 3, page 131

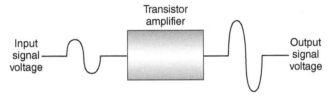

Figure 3-43 A simplified amplifier circuit.

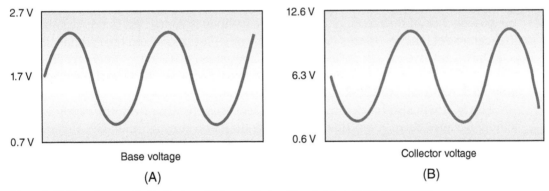

Figure 3-44 The voltage applied to the base (A) is amplified and inverted through the collector (B).

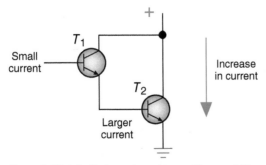

Figure 3-45 A Darlington pair used to amplify current. T1 acts as a preamplifier that creates a larger base current for T2, which is the final amplifier that creates a larger current.

transistor is isolated from the control circuit and is the final amplifier. The second transistor boosts the current to the amount required to operate the load component. The Darlington pair is utilized by most control modules used in electronic ignition systems.

Phototransistors

A **phototransistor** is a transistor that is sensitive to light. In a phototransistor, a small lens is used to focus incoming light onto the sensitive portion of the transistor (**Figure 3-46**). When light strikes the transistor, holes and free electrons are formed. These increase current flow through the transistor according to the amount of light. The stronger the light intensity, the more current that will flow. This type of phototransistor is often used in automatic headlight dimming circuits.

Thyristors

A **thyristor** is a semiconductor switching device composed of alternating N and P layers. It can be used to rectify current from AC to DC, and to control power to light dimmers, motor speed controls, solid-state relays, and other applications where power control is needed.

The most common type of thyristor used in automotive applications is the silicon-controlled rectifier (SCR). Like the transistor, the SCR has three legs. However, it consists of four regions arranged PNPN (**Figure 3-47**). The three legs of the SCR are called the anode (or P-terminal), the cathode (or N-terminal), and the gate (one of the center regions).

The SCR requires only a trigger pulse (not a continuous current) applied to the gate to become conductive. Current will continue to flow through the anode and cathode as long as the voltage remains high enough, or until gate voltage is reversed.

The SCR can be connected to a circuit in either the forward or the reverse direction. Using Figure 3-47 of a forward-direction connection, the P-type anode is connected to the positive side of the circuit and the N-type cathode is connected to the negative side. The center PN junction blocks current flow through the anode and the cathode.

Once a positive voltage pulse is applied to the gate, the SCR turns on. Even if the positive voltage pulse is removed, the SCR will continue to conduct. If a negative voltage pulse is applied to the gate, the SCR will no longer conduct.

The SCR will also block any reverse current from flowing from the cathode to the anode. Because current can flow in only one direction through the SCR, the SCR can rectify AC current to DC current.

Figure 3-46 Phototransistor.

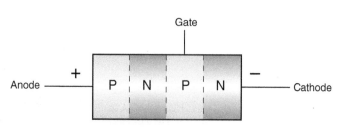

Figure 3-47 A forward-direction SCR.

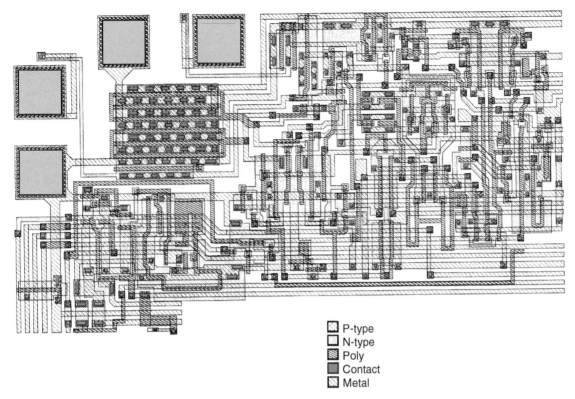

P-type
N-type
Poly
Contact
Metal

Figure 3-48 An enlarged illustration of an integrated circuit with thousands of transistors, diodes, resistors, and capacitors. Actual size can be less than 1/4 inch (6.35 mm) square.

Integrated Circuits

An **integrated circuit (IC)** is a complex circuit of thousands of transistors, diodes, resistors, capacitors, and other electronic devices that are formed onto a tiny silicon chip (**Figure 3-48**). As many as 30,000 transistors can be placed on a chip that is 1/4 inch (6.35 mm) square.

Integrated circuits are constructed by photographically reproducing circuit patterns onto a silicon wafer. The process begins with a large-scale drawing of the circuit. This drawing can be room size. Photographs of the circuit drawing are reduced until they are the actual size of the circuit. The reduced photographs are used as a mask. Conductive P-type and N-type materials, along with insulating materials, are deposited onto the silicon wafer. The mask is placed over the wafer and selectively exposes the portion of material to be etched away or the portions requiring selective deposition. The entire process of creating an IC chip takes over 100 separate steps. Out of a single wafer 4 inches (101.6 mm) in diameter, thousands of ICs can be produced.

The small size of the integrated chip has made it possible for vehicle manufacturers to add several computer-controlled systems to the vehicle without taking up much space. Also, a single computer is capable of performing several functions.

CIRCUIT PROTECTION DEVICES

Most automotive electrical circuits are protected from high current flow that would exceed the capacity of the circuit's conductors and/or loads. Excessive current results from a decrease in the circuit's resistance. Circuit resistance will decrease when too many components are connected in parallel or when a component or wire becomes shorted. A short is an undesirable, low-resistance path for current flow. When the circuit's current reaches a predetermined level, most circuit **protection devices** open and stop current flow in the circuit. This action prevents damage to the wires and the circuit's components.

Shop Manual
Chapter 3, page 116

Fuses

The most commonly used circuit protection device is the **fuse** (**Figure 3-49**). A fuse is a replaceable element that contains a metal strip that will melt when the current flowing through it exceeds its rating. The thickness of the metal strip determines the rating of the fuse. When the metal strip melts, excessive current is indicated. The cause of the **overload** must be found and repaired; then a new fuse of the same rating should be installed. The most commonly used automotive fuses are rated from 3 to 30 amps.

There are three basic types of fuses: glass or ceramic fuses, blade-type fuses, and bullet or cartridge fuses. Glass and ceramic fuses are found mostly on older vehicles. Sometimes, however, you can find them in a special holder connected in series with a circuit. Glass fuses are small glass cylinders with metal caps. The metal strip connects the two caps. The rating of the fuse is normally marked on one of the caps.

Blade-type fuses are flat plastic units and are available in six different physical sizes (**Figure 3-50**). The plastic housing is formed around two male blade-type connectors. The metal strip connects these connectors inside the plastic housing. The rating of these fuses is on top of the plastic housing and the plastic is color coded (**Figure 3-51**).

The low-profile mini, MICRO 2, and MICRO 3 fuses are the latest fuse types used in the automotive industry. All three of these fuse designs offer space savings. The low-profile mini fuse is a shorter version of the mini fuse and uses a different terminal design. The MICRO 3 fuse has three terminals and two fuse elements (**Figure 3-52**). It is essentially two fuses in one housing. The center terminal is common to both outputs. The MICRO 3 fuse offers additional circuit protection in less space.

Cartridge-type fuses are made of plastic or ceramic material. They have pointed ends and the metal strip rounds from end to end. This type of fuse is much like a glass fuse except the metal strip is not enclosed.

> Excess current flow in a circuit is called an **overload**.

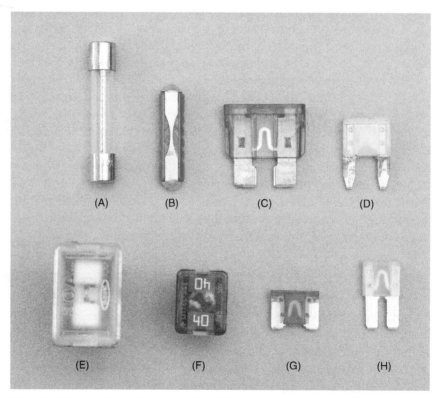

Figure 3-49 Common fuses (A) glass cartridge, (B) ceramic, (C) blade (auto-fuse), (D) mini, (E) maxi, (F) "F" type, (G) low-profile mini, and (H) MICRO.

Auto-fuse

Current Rating	Color
3	Violet
5	Tan
7.5	Brown
10	Red
15	Blue
20	Yellow
25	Natural
30	Green

Blade Size	Blade Group	Dimensions L × W × H	Common Ratings (Maximum Amperage)
Micro 2	APT, ATR	9.1 × 3.8 × 15.3 mm	5, 7.5, 10, 15, 20, 25, 30
Micro 3	ATL	14.4 × 4.2 × 18.1 mm	5, 7.5, 10, 15
Low-Profile Mini	APS, ATT	10.9 × 3.81 × 8.73 mm	2, 3, 4, 5, 7.5, 10, 15, 20, 25, 30
Mini	APM, ATM	10.9 × 3.6 × 16.3 mm	2, 3, 4, 5, 7.5, 10, 15, 20, 25, 30
Regular	APR, ATC (Closed), ATO (open), ATS[1]	19.1 × 5.1 × 18.5 mm	0.5, 1, 2, 3, 4, 5, 7.5, 10, 15, 20, 25, 30, 35, 40
Maxi	APX	29.2 × 8.5 × 34.3 mm	20, 25, 30, 35, 40, 50, 60, 70, 80, 100, 120

Figure 3-50 Fuse size and group classifications.

Maxi-fuse

Current Rating	Color
20	Yellow
30	Green
40	Amber
50	Red
60	Blue
70	Brown
80	Natural

Mini-fuse

Current Rating	Color
5	Tan
7.5	Brown
10	Red
15	Blue
20	Yellow
25	White
30	Green

Figure 3-51 Color coding for blade-type fuses. An auto-fuse is a standard blade-type fuse.

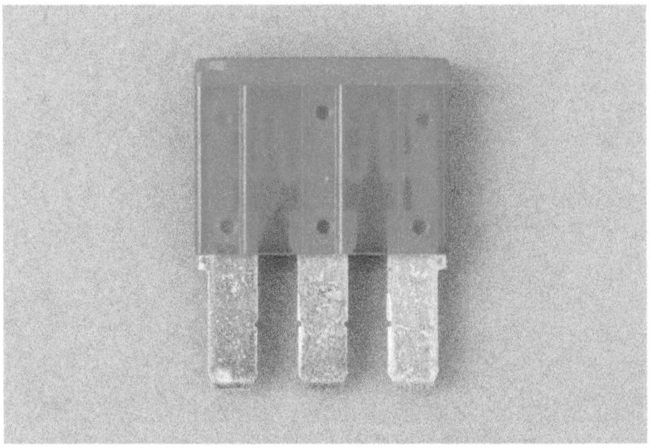

Figure 3-52 The MICRO 3 fuse protects two circuits.

Figure 3-53 Fuse boxes are normally located under the dash or in the engine compartment.

Fuses are typically located in a central **fuse block** or power distribution box. However, fuses may also be found in relay boxes and electrical junction boxes. Power distribution boxes are normally located in the engine compartment and house fuses and relays. A common location for a fuse box is under the instrumental panel (**Figure 3-53**). The fuse box may also be located behind kick panels, in the glove box, in the engine compartment, or in a variety of other places on the vehicle. Fuse ratings and the circuits they protect are normally marked on the cover of the fuse or power distribution box. Of course, this information can also be found in the vehicle owner's manual and the service information.

A "blown" fuse is identified by a burned-through metal wire in the capsule (**Figure 3-54**).

A fuse is connected in series with the circuit. Normally the fuse is located before all of the loads of the circuit (**Figure 3-55**). However, it may be placed before an individual load (**Figure 3-56**).

When adding accessories to the vehicle, the correct fuse rating must be selected. Use the power formula to determine the correct fuse rating (watts / volts = amperes). The fuse selected should be rated slightly higher than the actual current draw to allow for current surges (5% to 10%).

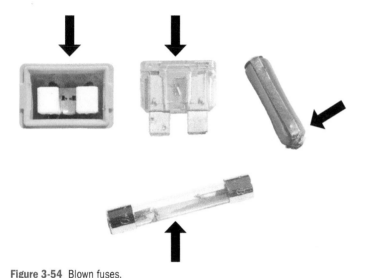

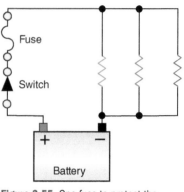

Figure 3-55 One fuse to protect the entire parallel circuit.

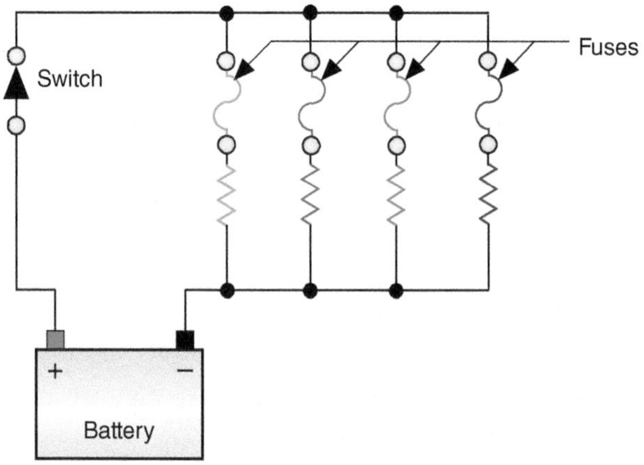

Figure 3-54 Blown fuses.

Figure 3-56 Fuses used to protect each branch of a parallel circuit.

Figure 3-57 Fusible links located near the battery.

Fusible Links

Shop Manual
Chapter 4, page 176

Fusible links are made of meltable conductor material with a special heat-resistant insulation (**Figure 3-57**). When there is an overload in the circuit, the conductor link melts and opens the circuit. To properly test a fusible link, use an ohmmeter or continuity tester. A vehicle may have one or several fusible links to provide protection for the main power wires before they are divided into smaller circuits at the fuse box. The fusible links are usually located at a main connection near the battery or starter solenoid. The current capacity of a fusible link is determined by its size. A fusible link is usually four wire sizes smaller (four numbers larger) than the circuit it protects. The smaller the wire, the larger its number. A circuit that uses 14-gauge wire would require an 18-gauge fusible link for protection.

AUTHOR'S NOTE Some GM vehicles have the fusible link located at the main connection near the starter motor.

AUTHOR'S NOTE A "blown" fusible link is usually identified by bubbling of the insulator material around the link.

Maxi-Fuses

In place of fusible links, many manufacturers use a **maxi-fuse**. A maxi-fuse looks similar to a blade-type fuse except it is larger and has a higher current capacity. It is also referred to as a cartridge fuse. By using maxi-fuses, manufacturers are able to break down the electrical system into smaller circuits. If a fusible link burns out, many of the vehicle's electrical systems may be affected. By breaking down the electrical system into smaller circuits and installing maxi-fuses, the consequence of a circuit defect will not be as severe as it would have been with a fusible link. In place of a single fusible link, there may be many maxi-fuses, depending on how the circuits are divided. This makes the technician's job of diagnosing a faulty circuit much easier.

Maxi-fuses are used because they are less likely to cause an underhood fire when there is an overload in the circuit. If the fusible link is burned in two, it is possible that the "hot" side of the link can come into contact with the vehicle frame and the wire can catch on fire.

Today many manufacturers are replacing maxi-fuses with "F"-type fuses. These are smaller versions of the maxi-fuses.

Circuit Breakers

A circuit that is susceptible to an overload on a routine basis is usually protected by a **circuit breaker**. A circuit breaker uses a **bimetallic strip** that reacts to excessive current (**Figure 3-58**). When an overload or circuit defect occurs that causes an excessive amount of current draw, the current flowing through the bimetallic strip causes it to heat. As the strip heats, it bends and opens the contacts. Once the contacts are opened, current can no longer flow. With no current flowing, the strip cools and closes again. If the excessive current cause is still in the circuit, the breaker will open again. The circuit breaker will continue to open and close as long as the overload is in the circuit. This type of circuit breaker is self-resetting or "cycled." Some circuit breakers require manual resetting by pressing a button, while others must be removed from the power to reset (**Figure 3-59**).

An example of the use of a circuit breaker is in the power window circuit. Because the window is susceptible to jams due to ice buildup on the window, a current overload is possible. If this should occur, the circuit breaker will heat up and open the circuit before the window motor is damaged. If the operator continues to attempt to operate the power window, the circuit breaker will open and close until the cause of the jam is removed.

A **bimetallic strip** consists of two different types of metals. One strip will react more quickly to heat than the other, which causes the strip to flex in proportion to the amount of current flow.

Shop Manual
Chapter 3, page 117

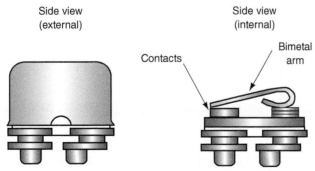

Figure 3-58 The circuit breaker uses a bimetallic strip that opens if current draw is excessive.

A

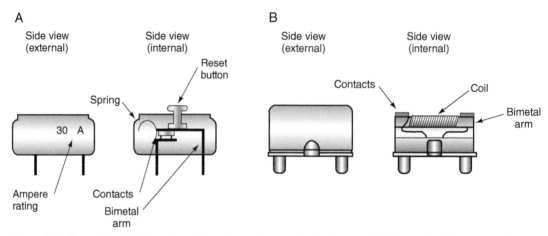

B

Figure 3-59 Non-cycling circuit breakers. (A) can be reset by pressing the button, while (B) requires being removed from the power to reset.

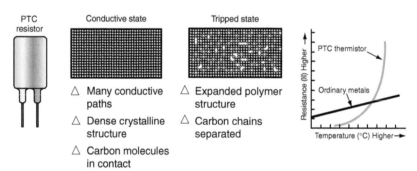

Figure 3-60 PTC operation.

Shop Manual
Chapter 3, page 119

PTCs as Circuit Protection Devices

Automotive manufacturers must utilize means of providing reliable circuit protection, yet at the same time reduce vehicle weight and cost. When fuses are used to protect multiple circuits, this can result in large, heavy, and complex wiring assemblies. The use of polymer, **positive temperature coefficient (PTC) thermistors** provides a lightweight alternative to protecting the circuit against shorts or overload conditions. A PTC thermistor increases in resistance as temperature increases. Because of its design, a PTC thermistor has the ability to trip (increase resistance to the point it becomes the load device in the circuit) during an over-current condition and reset after the fault is no longer present (**Figure 3-60**).

Conductive polymers consist of specially formulated plastics and various conductive materials. At normal temperatures, the plastic materials form a crystalline structure. The structure provides a low-resistance conductive chain. The resistance is so low that it does not affect the operation of the circuit. However, if the current flow increases above the trip threshold, the additional heat causes the crystalline structure to change to an amorphous state. In this condition, the conductive paths separate, causing a rapid increase in the resistance of the PTC. The increased resistance reduces the current flow to a safe level.

Shop Manual
Chapter 3, page 106

CIRCUIT DEFECTS

All electrical problems can be classified as being one of three types: an open, short, or high resistance. Each one of these will cause a component to operate incorrectly or not at all. Understanding what each of these problems will do to a circuit is the key to proper diagnosis of any electrical problem.

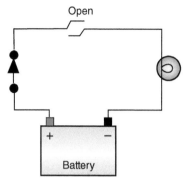

Figure 3-61 An open circuit stops all current flow.

Open

An **open** is simply a break in the circuit (**Figure 3-61**). An open is caused by turning a switch off, a break in a wire, a burned-out light bulb, a disconnected wire or connector, or anything that opens the circuit. When a circuit is open, current does not flow and the component doesn't work. Because there is no current flow, there are no voltage drops in the circuit. Source voltage is available everywhere in the circuit up to the point at which it is open. Source voltage is even available after a load, if the open is after that point.

Opens caused by a blown fuse will still cause the circuit not to operate, but the cause of the problem is the excessive current that blew the fuse. Nearly all other opens are caused by a break in the continuity of the circuit. These breaks can occur anywhere in the circuit.

Shorts

A **short** results from an unwanted path for current. **Shorted circuits** cause an increase in current flow by bypassing part of the normal circuit path. This increased current flow can burn wires or components.

An example of a shorted circuit could be found in a faulty coil. The windings within a coil are insulated from each other; however, if this insulation breaks down, a copper-to-copper contact is made between the turns. Since part of the windings will be bypassed, this reduces the number of windings in the coil through which current will flow. This results in the effectiveness of the coil being reduced. Also, since the current bypasses a portion of the normal circuit resistance, current flow is increased and excess heat can be generated.

Another example of a shorted circuit is if the insulation of two adjacent wires breaks down and allows a copper-to-copper contact (**Figure 3-62**). If the short is between points

Shop Manual
Chapter 3, page 106

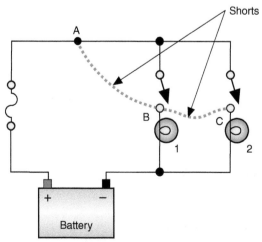

Figure 3-62 A short circuit can be a copper-to-copper contact between two adjacent wires.

A and B, light 1 would be on all the time. If the short is between points B and C, both lights would illuminate when either switch is closed.

Another example is shown in **Figure 3-63**. With the two wires shorted together, the horn will sound every time the brake pedal is depressed. Also, if the horn switch is pressed, the brake lights will come on.

Shop Manual
Chapter 3, page 110

Another type of electrical defect is a short to ground. A short to ground allows current to flow an unintentional path to ground (**Figure 3-64**). To see what happens in a circuit

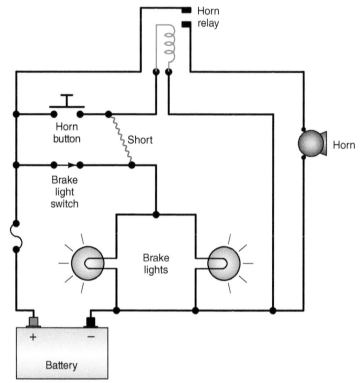

Figure 3-63 A wire-to-wire short.

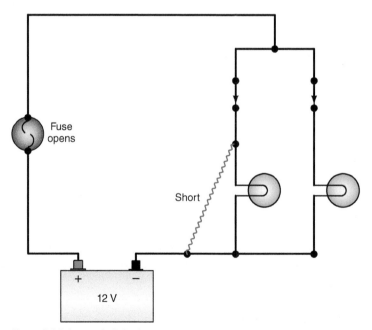

Figure 3-64 A grounded circuit.

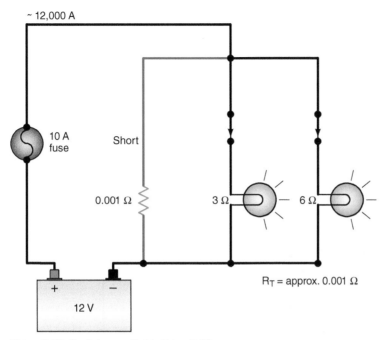

Figure 3-65 Ohm's law applied to Figure 3-63.

that has a short to ground, refer to **Figure 3-65**. If normal resistance of the two bulbs is 3 ohms and 6 ohms, since they are in parallel, the total circuit resistance is 2 ohms. The short makes a path from the power side of one bulb to the return path, and to the battery. The short creates a low-resistance path. If the low-resistance path has a resistance value of 0.001 ohms, it is possible to calculate what would happen to the current in this circuit.

The short becomes another leg in the parallel circuit. Since the total resistance of a parallel circuit is always lower than the lowest resistance, we know the total resistance of the circuit is now less than 0.001 ohms. Using Ohm's law, we can calculate the current flowing through the circuit.

$$A = V/R \quad \text{or} \quad A = 12/0.001 \quad \text{or} \quad A = 12{,}000 \text{ Amps}$$

Needless to say, it would take a large wire to carry that amount of amperage. Our 10-amp fuse would melt quickly when the short occurred. This would protect the wires and light bulbs.

High Resistance

High-resistance problems occur when there is unwanted resistance in the circuit. The high resistance can come from a loose connection, corroded connection, corrosion in the wire, wrong size wire, and so on. Since the resistance becomes an additional load in the circuit, the effect is that the load component, with reduced voltage and current applied, operates with reduced efficiency. An example would be a taillight circuit with a load component (light bulb) that is rated at 50 watts. To be fully effective, this bulb must draw 4.2 amperes at 12 volts (A = P/V). This means a full 12 volts should be applied to the bulb. If resistance is present at other points in the circuit, some of the 12 volts will be dropped. With less voltage (and current) being available to the light bulb, the bulb will illuminate with less intensity.

Shop Manual
Chapter 3, page 112

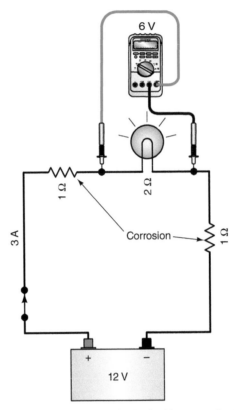

Figure 3-66 A simple light circuit with unwanted resistance.

Figure 3-66 illustrates a light circuit with unwanted resistance at the power feed for the bulb and at the negative battery terminal. When the circuit is operating properly, the 2-ohm light bulb will have 6 amps of current flowing through it and drop 12 volts. With the added resistance, the current is reduced to 3 amps and the bulb drops only 6 volts. As a result, the bulb's illumination is very dim.

SUMMARY

- A switch can control the on/off operation of a circuit or direct the flow of current through various circuits.
- A normally open switch will not allow current flow when it is in its rest position. A normally closed switch will allow current flow when it is in its rest position.
- A relay is a device that uses low current to control a high-current circuit.
- A buzzer is sometimes used to warn the driver of possible safety hazards by emitting an audio signal (such as when the seat belt is not buckled).
- A stepped resistor has two or more fixed resistor values. It is commonly used to control electrical motor speeds.
- A variable resistor provides for an infinite number of resistance values within a range. A rheostat is a

two-terminal variable resistor used to regulate the strength of an electrical current. A potentiometer is a three-wire variable resistor that acts as a voltage divider to produce a continuously variable output signal proportional to a mechanical position.
- Capacitance is the ability of two conducting surfaces to store voltage.
- A diode is an electrical one-way check valve that will allow current to flow in one direction only.
- Forward bias means that a positive voltage is applied to the P-type material and negative voltage to the N-type material. Reverse bias means that positive voltage is applied to the N-type material and negative voltage is applied to the P-type material.
- A transistor is a three-layer semiconductor that is commonly used as a very fast switching device.

- An integrated circuit is a complex circuit of thousands of transistors, diodes, resistors, capacitors, and other electronic devices that are formed onto a tiny silicon chip.
- The protection device is designed to turn off the system it protects. This is done by creating an open (like turning off a switch) to prevent a complete circuit.
- Fuses are rated by amperage. Never install a larger rated fuse into a circuit than the one that was

designed by the manufacturer. Doing so may damage or destroy the circuit.
- An open circuit is a circuit in which there is a break in continuity.
- A shorted circuit is a circuit that allows current to bypass part of the normal path.
- A short to ground is a condition that allows current to return to ground before it has reached the intended load component.

REVIEW QUESTIONS

Short-Answer Essays

1. Describe the use of three types of semiconductors.

2. What types of mechanical variable resistors are used on automobiles?

3. Define what is meant by opens, shorts, grounds, and excessive resistance.

4. Explain the effects that each type of circuit defect will have on the operation of the electrical system.

5. Explain the purpose of a circuit protection device.

6. Describe the most common types of circuit protection devices.

7. Describe the common types of electrical system (nonelectronic) components used and how they affect the electrical system.

8. Explain the basic concepts of capacitance.

9. Explain the difference between normally open (NO) and normally closed (NC) switches.

10. Explain the differences between forward-biasing and reverse-biasing a diode.

Fill in the Blanks

1. Never install a larger rated _____ into a circuit than the one that was designed by the manufacturer.

2. A _____ can control the on/off operation of a circuit or direct the flow of current through various circuits.

3. A normally _____ switch will not allow current flow when it is in its rest position. A normally _____ switch will allow current flow when it is in its rest position.

4. An _____ is a complex circuit of many transistors, diodes, resistors, capacitors, and other electronic devices that are formed onto a tiny silicon chip.

5. When a _____ voltage is applied to the P-type material of a diode and _____ voltage is applied to the P-type material, the diode is reverse-biased. When a _____ voltage is applied to the N-type material of a diode and _____ voltage is applied to the P-type material, the diode is forward-biased.

6. A _____ is used in electronic circuits as a very fast switching device.

7. A _____ is an electrical one-way check valve that will allow current to flow in one direction only.

8. A _____ is an electromechanical device that uses low current to control a high-current circuit.

9. A _____ is a three-wire variable resistor that acts as a voltage divider. A_____ is a two-terminal variable resistor used to regulate the strength of an electrical current.

10. The _____ requires only a trigger pulse applied to the gate to become conductive.

Multiple Choice

1. All of the following are true concerning electrical shorts, **EXCEPT**:
   A. A short can add a parallel leg to the circuit, which lowers the entire circuit's resistance.
   B. A short can result in a blown fuse.
   C. A short decreases amperage in the circuit.
   D. A short bypasses the circuit's intended path.

2. A "blown" fusible link is identified by:
   A. A burned-through metal wire in the capsule.
   B. A bubbling of the insulator material around the link.
   C. All of the above.
   D. None of the above.

3. All of the statements concerning circuit components are true, **EXCEPT**:
   A. A switch can control the on/off operation of a circuit.
   B. A switch can direct the flow of current through various circuits.
   C. A relay can be an SPDT-type switch.
   D. A potentiometer changes voltage drop due to the function of temperature.

4. Which of the following statements is/are correct?
   A. A zener diode is an excellent component for regulating voltage.
   B. A reverse-biased diode lasts longer than a forward-biased diode.
   C. The switches of a transistor last longer than those of a relay.
   D. Both A and C.

5. The light-emitting diode (LED):
   A. Emits light when it is reverse-biased.
   B. Has a variable turn-on voltage.
   C. Has a light color that is defined by the materials used to construct the diode.
   D. Has a turn-on voltage that is usually less than standard diodes.

6. Which of the following is the correct statement?
   A. An open means there is continuity in the circuit.
   B. A short bypasses a portion of the circuit.
   C. High-amperage draw indicates an open circuit.
   D. High resistance in a circuit increases current flow.

7. Transistors:
   A. Can be used to control the switching on/off of a circuit.
   B. Can be used to amplify voltage.
   C. Control high current with low current.
   D. All of the above.

8. All of the following are true, **EXCEPT**:
   A. Voltage drop can cause a lamp in a parallel circuit to burn brighter than normal.
   B. Excessive voltage drop may appear on either the insulated or the grounded return side of a circuit.
   C. Increased resistance in a circuit decreases current.
   D. A diode is used as an electrical one-way check valve.

9. Which statement is correct concerning diodes?
   A. Diodes are aligned to allow current flow in one direction only.
   B. Diodes can be used to rectify DC voltages into AC voltages.
   C. The stripe is on the anode side of the diode.
   D. Normal turn-on voltage of a standard diode is 1.5 volts.

10. A capacitor:
    A. Consumes electrical power.
    B. Induces voltage.
    C. Both A and B.
    D. Neither A nor B.

CHAPTER 4

WIRING AND CIRCUIT DIAGRAMS

Upon completion and review of this chapter, you should be able to understand and describe:

- When single-stranded or multistranded wire should be used.
- The use of resistive wires in a circuit.
- The construction of spark plug wires.
- How wire size is determined by the American Wire Gauge (AWG) and metric methods.
- How to determine the correct wire gauge to be used in a circuit.

- How temperature affects resistance and wire size selection.
- The purpose and use of printed circuits.
- Why wiring harnesses are used and how they are constructed.
- The purpose of wiring diagrams.
- The common wiring diagram electrical symbols that are used.
- The purpose of the component locator.

Terms To Know

Block diagrams	Ground straps	Splice
Common connection	Primary wiring	Stranded wire
Component locator	Printed circuit boards	Tracer
Electrical symbols	Schematic	Wiring diagram
Gauge	Secondary wiring	Wiring harness

INTRODUCTION

Today's vehicles have a vast amount of electrical wiring that, if laid end to end, could stretch for half a mile or more. Today's technician must be proficient at reading wiring diagrams in order to sort through this great maze of wires. Trying to locate the cause of an electrical problem can be quite difficult if you do not have a good understanding of wiring systems and diagrams.

In this chapter, you will learn how wiring harnesses are made (**Figure 4-1**), how to read the wiring diagram, how to interpret the symbols used, and how terminals are used. This will reduce the amount of confusion you may experience when repairing an electrical circuit. It is also important to understand how to determine the correct type and size of wire to carry the anticipated amount of current. It is possible to cause an electrical problem by simply using the wrong gauge size of wire. A technician must understand the three factors that cause resistance in a wire—length, diameter, and temperature—to perform repairs correctly.

AUTOMOTIVE WIRING

Primary wiring is the term used for conductors that carry low voltage. The insulation of primary wires is usually thin. **Secondary wiring** refers to wires used to carry high voltage, such as ignition spark plug wires. Secondary wires have much thicker insulation than primary wires.

Shop Manual
Chapter 4, page 164

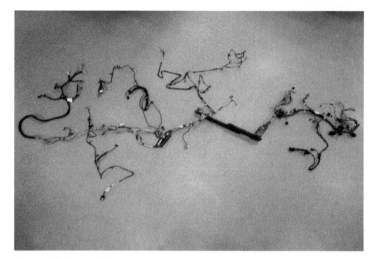

Figure 4-1 Portion of the vehicle's wiring harness.

Most of the primary wiring conductors used in the automobile are made of several strands of copper wire wound together and covered with a polyvinyl chloride (PVC) insulation (**Figure 4-2**). Copper has low resistance and can be connected to easily by using crimping connectors or soldered connections. Other types of conductor materials used in automobiles include silver, gold, aluminum, and tin-plated brass.

AUTHOR'S NOTE Copper is used mainly because of its low cost and availability.

Stranded wire means the conductor is made of several individual wires that are wrapped together. Stranded wire is used because it is very flexible and has less resistance than solid wire of the same size. This is because electrons tend to flow on the outside surface of conductors. Since there is more surface area exposed in a stranded wire (each strand has its own surface), there is less resistance in the stranded wire than in the solid wire (**Figure 4-3**). The PVC insulation is used because it can withstand temperature extremes and corrosion. PVC insulation is also capable of withstanding battery acid, antifreeze, and gasoline. The insulation protects the wire from electrical shorts, corrosion, and other environmental factors.

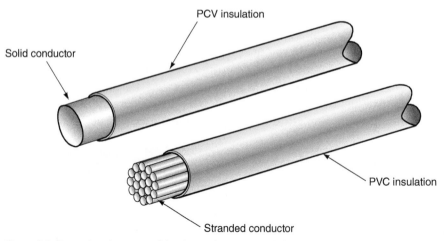

Figure 4-2 Comparison between solid and stranded primary wire.

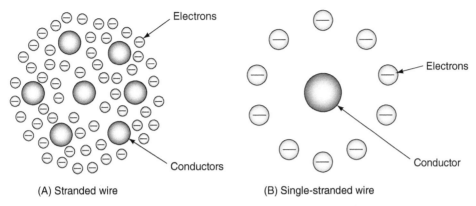

(A) Stranded wire   (B) Single-stranded wire

Figure 4-3 Stranded wire provides flexibility and more surface area for electron flow than a single-stranded solid wire.

AUTHOR'S NOTE General Motors has used single-stranded aluminum wire in limited applications where no flexing of the wire is expected. For example, it was used in some taillight circuits.

Wire Sizes

Consideration must be given for some margin of safety when selecting wire size. There are three major factors that determine the proper size of wire to be used:

1. The wire must have a large enough diameter, for the length required, to carry the necessary current for the load components in the circuit to operate properly.
2. The wire must be able to withstand the anticipated vibration.
3. The wire must be able to withstand the anticipated amount of heat exposure.

Wire size is based on the diameter of the conductor. The larger the diameter, the less the resistance. There are two common size standards used to designate wire size: American Wire Gauge (AWG) and metric.

The AWG standard assigns a **gauge** number to the wire based on the diameter of the conductor not including the insulation. The higher the number, the smaller the wire diameter. For example, 20-gauge wire is smaller in diameter than 10-gauge wire (**Table 4-1**). Most electrical systems in the automobile use 14-, 16-, or 18-gauge wire. Some high-current circuits will also use 10- or 12-gauge wire. Most battery cables are 2-, 4-, or 6-gauge cable.

Both wire diameter and wire length affect resistance. Sixteen-gauge wire is capable of conducting 20 amps (A) for 10 feet with minimal voltage drop. However, if the current is to be carried for 15 feet, 14-gauge wire would be required. If 20 amps were required to be carried for 20 feet, then 12-gauge wire would be required. The additional wire size is needed to prevent voltage drops in the wire. **Table 4-2** lists the wire size required to carry a given amount of current for different lengths.

Another factor that affects wire resistance is temperature. An increase in temperature creates a similar increase in resistance. A wire may have a known resistance of 0.03 ohm (Ω) per 10 feet at 70°F. When exposed to

Table 4-1 Gauge and wire size chart.

American Wire Gauge Sizes	
Gauge Size	**Conductor Diameter (inch)**
20	0.032
18	0.040
16	0.051
14	0.064
12	0.081
10	0.102
8	0.128
6	0.162
4	0.204
2	0.258
1	0.289
0	0.325
2/0	0.365
4/0	0.460

Table 4-2 The distance the current must be carried is a factor in determining the correct wire gauge to use.

Total Approximate Circuit Amperes	Wire Gauge (for Length in Feet)								
12 V	3	5	7	10	15	20	25	30	40
1.0	18	18	18	18	18	18	18	18	18
1.5	18	18	18	18	18	18	18	18	18
2	18	18	18	18	18	18	18	18	18
3	18	18	18	18	18	18	18	18	18
4	18	18	18	18	18	18	18	16	16
5	18	18	18	18	18	18	18	16	16
6	18	18	18	18	18	18	16	16	16
7	18	18	18	18	18	18	16	16	14
8	18	18	18	18	18	16	16	16	14
10	18	18	18	18	16	16	16	14	12
11	18	18	18	18	16	16	14	14	12
12	18	18	18	18	16	16	14	14	12
15	18	18	18	18	14	14	12	12	12
18	18	18	16	16	14	14	12	12	10
20	18	18	16	16	14	12	10	10	10
22	18	18	16	16	12	12	10	10	10
24	18	18	16	16	12	12	10	10	10
30	18	16	16	14	10	10	10	10	10
40	18	16	14	12	10	10	8	8	6
50	16	14	12	12	10	10	8	8	6
100	12	12	10	10	6	6	4	4	4
150	10	10	8	8	4	4	2	2	2
200	10	8	8	6	4	4	2	2	1

temperatures of 170°F, the resistance may increase to 0.04 ohm per 10 feet. Wires that are to be installed in areas that experience high temperatures, as in the engine compartment, must be of a size such that the increased resistance will not affect the operation of the load component. Also, the insulation of the wire must be capable of withstanding the high temperatures.

In the metric system, wire size is determined by the cross-sectional area of the wire. Metric wire size is expressed in square millimeters (mm²). In this system, the smaller the number, the smaller the wire conductor. The approximate equivalent wire size of metric to AWG is shown in **Table 4-3**.

Ground Straps

Shop Manual
Chapter 4, page 177

Usually there is not a direct metal to metal connection between the powertrain components and the vehicle chassis. The engine, transmission, and axle assemblies are supported by rubber mounts or bushings. The rubber acts as an insulator so any electrical components such as actuators or sensors that are mounted to the powertrain components will not have a completed circuit back to the vehicle's battery. This is especially true if the negative battery cable is attached to the vehicle's chassis instead of the engine block. **Ground straps** between

Table 4-3 Approximate AWG to metric equivalents.

Metric Size (mm²)	AWG (Gauge) Size	Ampere Capacity
0.5	20	4
0.8	18	6
1.0	16	8
2.0	14	15
3.0	12	20
5.0	10	30
8.0	8	40
13.0	6	50
19.0	4	60

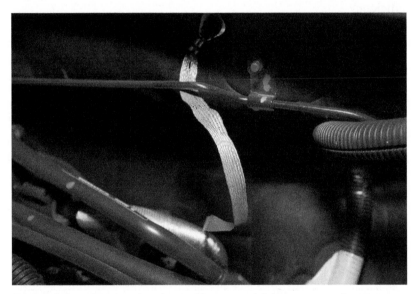

Figure 4-4 Ground straps are used to provide a return path for components that are insulated from the chassis.

the powertrain components and the vehicle's chassis are used to complete the return path to the battery (**Figure 4-4**). In addition, ground straps suppress electromagnetic induction (EMI) and radiation by providing a low-resistance circuit ground path.

AUTHOR'S NOTE Ground straps are also referred to as bonding straps. Ground straps can be installed in various locations. Some of the most common locations are:

- Engine to bulkhead or fender.
- Across the engine mounts.
- Radio chassis to instrument panel frame.
- Air-conditioning evaporator valve to the bulkhead.

The ground strap can be a large gauge insulated-type cable or a braided strap. Even on vehicles with the battery negative cable attached to the engine block, ground straps are used to connect between the engine block and the vehicle chassis. The additional ground cable ensures a good, low-resistance ground path between the engine and the chassis.

Ground straps are also used to connect sheet metal parts such as the hood, fender panels, and the exhaust system even though there is no electrical circuit involved. In these cases, the strap is used to suppress EMI since the sheet metal could behave as a large capacitor. The air space between the sheet metal forms an electrostatic field and can interfere with any computer-controlled circuits that are routed near the sheet metal.

Terminals and Connectors

Shop Manual
Chapter 4,
pages 165, 174

To perform the function of connecting the wires from the voltage source to the load component reliably, terminal connections are used. Today's vehicles can have as many as 500 separate circuit connections. The terminals used to make these connections must be able to perform with very low voltage drop. Terminals are constructed of either brass or steel. Steel terminals usually have a tin or lead coating. Some steel terminals have a gold plating to prevent corrosion. A loose or corroded connection can cause an unwanted voltage drop resulting in poor operation of the load component. For example, a connector used in a light circuit that has as little as 10% voltage drop (1.2 volts [V]) may result in a 30% loss of lighting efficiency.

Terminals can be either crimped or soldered to the conductor. The terminal makes the electrical connection, and it must be capable of withstanding the stress of normal vibration. **Figure 4-5** shows several different types of terminals used in the automotive electrical system. In addition, the following connectors are used on the automobile:

1. **Molded connector:** These connectors usually have one to four wires that are molded into a one-piece component (**Figure 4-6**). Although the male and female connector halves separate, the connector itself cannot be taken apart.

Shop Manual
Chapter 4, page 175

2. **Multiple-wire, hard-shell connector:** These connectors usually have a hard, plastic shell that holds the connecting terminals of separate wires (**Figure 4-7**). The wire terminals can be removed from the shell to be repaired.

Shop Manual
Chapter 4, page 175

3. **Bulkhead connectors:** These connectors are used when several wires must pass through the bulkhead (**Figure 4-8**).

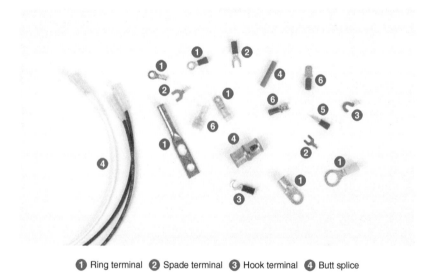

① Ring terminal ② Spade terminal ③ Hook terminal ④ Butt splice
⑤ Snap plug terminal ⑥ Quick disconnect terminal

Figure 4-5 Examples of primary wire terminals and connectors used in automotive applications.

Figure 4-6 Molded connectors cannot be disassembled to replace damaged terminals or to test.

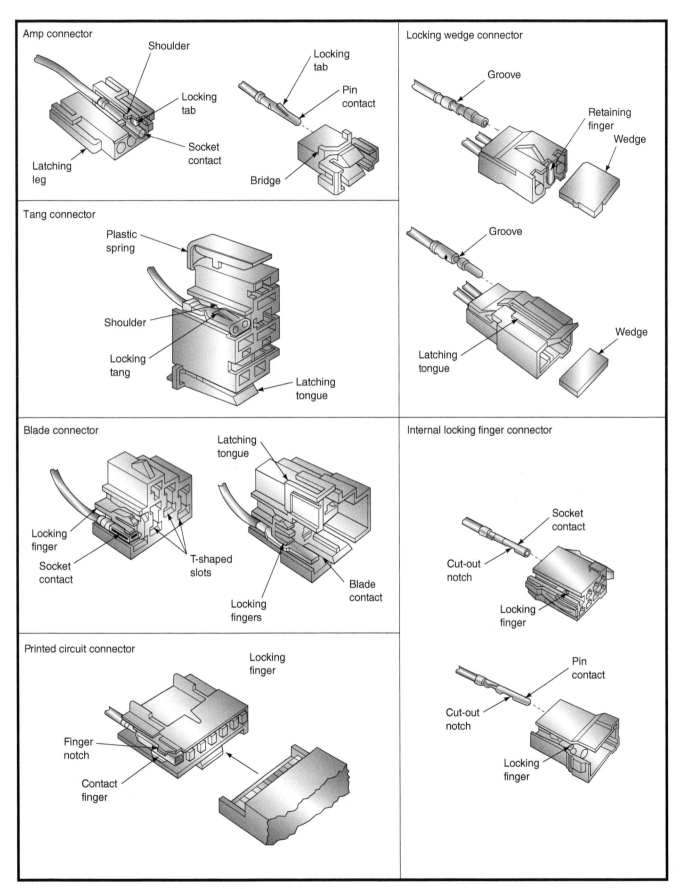

Figure 4-7 Multiple-wire, hard-shell connectors.

Figure 4-8 Bulkhead connector.

Figure 4-9 Weather-pack connector is used to prevent connector corrosion.

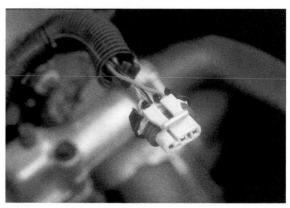

Figure 4-10 Metri-pack connector.

Shop Manual
Chapter 4, page 175

Shop Manual
Chapter 4, page 177

4. **Weather-pack connectors:** These connectors have rubber seals on the terminal ends and on the covers of the connector half (**Figure 4-9**). They are used on computer circuits to protect the circuit from corrosion, which may result in a voltage drop.
5. **Metri-pack connectors:** These are like the weather-pack connectors but do not have the seal on the cover half (**Figure 4-10**).
6. **Heat shrink–covered butt connectors:** Recommended for air bag applications by some manufacturers. Other manufacturers allow NO repairs to the circuitry, while still others require silver-soldered connections.

To reduce the number of connectors in the electrical system, a **common connection** can be used (**Figure 4-11**). Common connections are used to share a source of power or a common ground and are often called a **splice**. If there are several electrical components that are physically close to each other, a single common connection (splice) eliminates using a separate connector for each wire.

Printed Circuits

Printed circuit boards are used to simplify the wiring of the circuits they operate. Other uses of printed circuit boards include the inside of radios, computers, and some voltage regulators. Most instrument panels use printed circuit boards as circuit conductors.

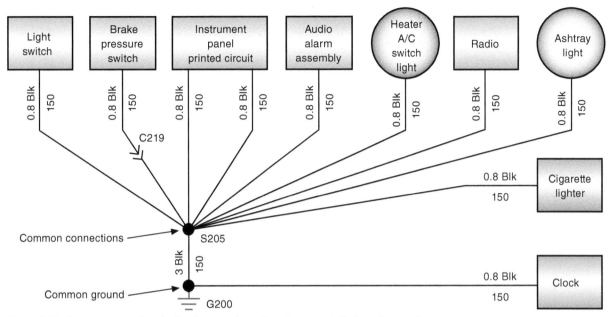

Figure 4-11 Common connections (splices) are used to reduce the amount of wire and connectors.

A printed circuit is made of a thin phenolic or fiberglass board that has copper (or some other conductive material) deposited on it. Portions of the conductive metal are then etched or eaten away by acid. The remaining strips of conductors provide the circuit path for the instrument panel illumination lights, warning lights, indicator lights, and gauges of the instrument panel (**Figure 4-12**). The printed circuit board is attached to the back of the instrument panel housing. An edge connector joins the printed circuit board to the vehicle wiring harness.

Whenever it is necessary to perform repairs on or around the printed circuit board, it is important to follow these precautions:

1. When replacing light bulbs, be careful not to cut or tear the surface of the printed circuit board.
2. Do not touch the surface of the printed circuit with your fingers. The acid present in normal body oils can damage the surface.

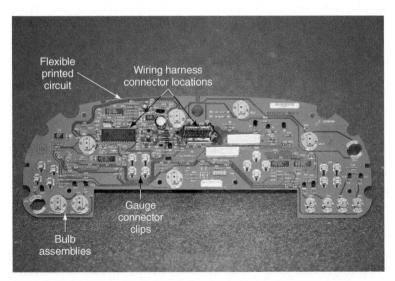

Figure 4-12 Printed circuits eliminate bulky wires behind the instrument panel.

3. If the printed circuit board needs to be cleaned, use a commercial cleaning solution designed for electrical use. If this solution is not available, it is possible to clean the board by *lightly* rubbing the surface with an eraser.

 A BIT OF HISTORY

The printed circuit board was developed in 1947 by the British scientist J. A. Sargrove to simplify the production of radios.

Wiring Harness

Most manufacturers use **wiring harnesses** to reduce the number of loose wires hanging under the hood or dash of an automobile. The wiring harness provides for a safe path for the wires of the vehicle's lighting, engine, and accessory components. The wiring harness is made by grouping insulated wires and wrapping them together. The wires are bundled into separate harness assemblies that are joined together by connector plugs. The multiple-pin connector plug may have more than 60 individual wire terminals.

There are several complex wiring harnesses in a vehicle, in addition to the simple harnesses. The engine compartment harness and the under-dash harness are examples of complex harnesses (**Figure 4-13**). Lighting circuits usually use a simpler harness (**Figure 4-14**). A complex harness serves many circuits, while a simple harness services only a few circuits. Some individual circuit wires may branch out of a complex harness to other areas of the vehicle.

Most wiring harnesses use a flexible plastic or mesh nylon conduit to provide for quick wire installation (**Figure 4-15**). The conduit has a seam that can be opened to accommodate the installation or removal of wires from the harness. The seam will close once the wires are installed, and will remain closed even if the conduit is bent.

In areas around the exhaust manifolds, the wires may be exposed to excessive heat. To protect the wires of the harness, a special heat reflective conduit is typically used (**Figure 4-16**).

The conduit is commonly referred to as the *wire loom* or *corrugated loom*.

Wiring Protective Devices

Often overlooked, but very important to the electrical system, are proper wire protection devices (**Figure 4-17**). These devices prevent damage to the wiring by maintaining proper wire routing and retention. Special clips, retainers, straps, and supplementary insulators provide additional protection to the conductor over what the insulation itself is capable of providing. Whenever the technician must remove one of these devices to perform a repair, it is important that the device be reinstalled to prevent additional electrical problems.

Whenever it is necessary to install additional electrical accessories, try to support the primary wire in at least 1-foot intervals. If the wire must be routed through the frame or body, use rubber grommets to protect the wire.

Wiring Diagrams

Shop Manual
Chapter 4, page 179

One of the most important tools for diagnosing and repairing electrical problems is a **wiring diagram**. A wiring diagram is an electrical schematic that shows a representation of actual electrical or electronic components (by use of symbols) and the wiring of the vehicle's electrical systems. These diagrams identify the wires and connectors from each

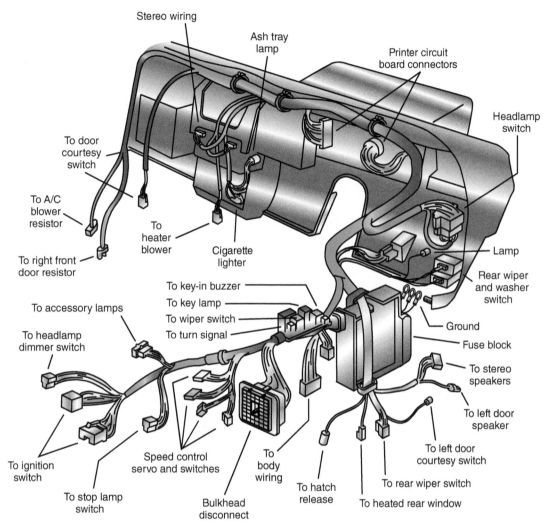

Stereo wiring

Ash tray
lamp

Printer circuit
board connectors

Headlamp
switch

To door
courtesy
switch

To A/C
blower
resistor

To
heater
blower

Cigarette
lighter

To right front
door resistor

Lamp

Rear wiper
and washer
switch

To key-in buzzer

To accessory lamps

To key lamp

To wiper switch

To turn signal

Ground

Fuse block

To headlamp
dimmer switch

To stereo
speakers

To left door
speaker

To ignition
switch

Speed control
servo and switches

To
body
wiring

To left door
courtesy switch

To stop lamp
switch

Bulkhead
disconnect

To hatch
release

To rear wiper switch

To heated rear window

Figure 4-13 Complex wiring harness.

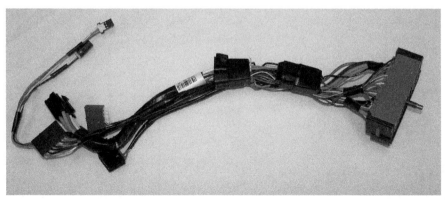

Figure 4-14 Simple wiring harness.

circuit on a vehicle. They also show where different circuits are interconnected, where they receive their power, where the ground is located, and the colors of the different wires. All of this information is critical to proper diagnosis of electrical problems. Some wiring diagrams also give additional information that helps you understand how a circuit

Figure 4-15 Flexible conduit used to make wiring harnesses.

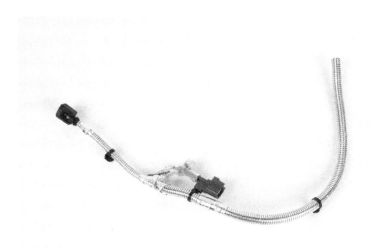

Figure 4-16 Special heat reflective conduit.

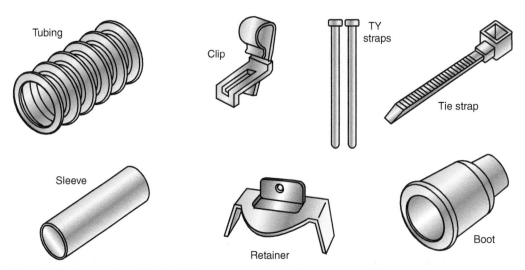

Tubing

Clip

TY straps

Tie strap

Sleeve

Retainer

Boot

Figure 4-17 Common wire protection devices.

operates and how to identify certain components (**Figure 4-18**). Wiring diagrams do not explain how the circuit works; this is where your knowledge of electricity is required. By using the wiring diagrams, service information, and your electrical knowledge, you should be able to ascertain the proper operating design of the system. Also, you need to determine how electrical systems interact with each other.

A wiring diagram can show the wiring of the entire vehicle or a single circuit (**Figure 4-19**). These single-circuit diagrams are also called **block diagrams**. Wiring diagrams of the entire vehicle tend to look more complex and threatening than block diagrams. However, once you simplify the diagram to only those wires, connectors, and components that belong to an individual circuit, they become less complex and more valuable.

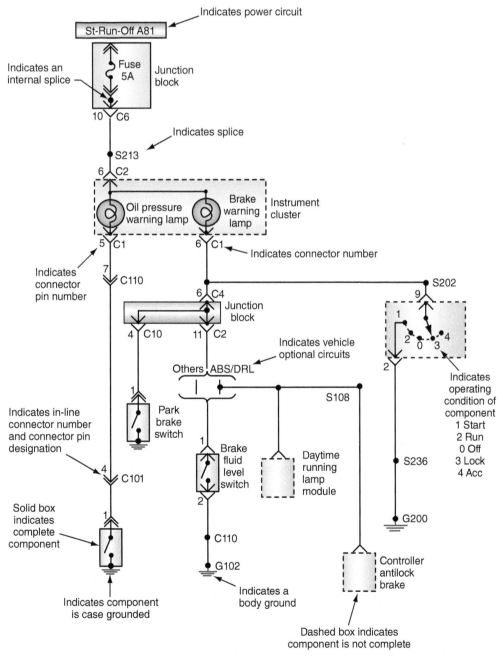

Figure 4-18 Wiring diagrams provide the technician with necessary information to accurately diagnose the electrical systems.

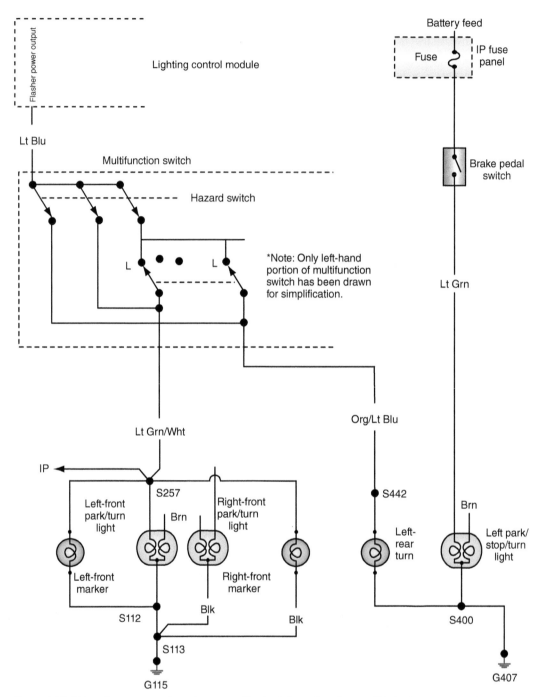

Figure 4-19 Wiring diagram illustrating only one specific circuit for easier reference. This is also known as a block diagram.

Wiring diagrams show the wires, connections to switches and other components, and the type of connector used throughout the circuit. Total vehicle wiring diagrams are normally spread out over many pages of service information. Some are displayed on a single large sheet of paper that folds out of the manual. A system wiring diagram is actually a portion of the total vehicle diagram. The system and all related circuitry are shown on a single page. System diagrams are often easier to use than vehicle diagrams simply because there is less information to sort through.

Remember that electrical circuits need a complete path in order to work. A wiring diagram shows the insulated side of the circuit and the point of ground. Also, when lines (or wires) cross on a wiring diagram, this does not mean they connect. If wires are connected, there will be a connector or a dot at the point where they cross. Most wiring diagrams do not show the location of the wires, connectors, or components in the vehicle. Some have location reference numbers displayed by the wires. After studying the wiring diagram, you will know what you are looking for. Then you move to the car to find it.

In addition to entire vehicle and system-specific wiring diagrams, there are other diagrams that may be used to diagnose electricity problems. An electrical **schematic** shows how the circuit is connected. It does not show the colors of the wires or their routing. Schematics are what have been used so far in this book. They display a working model of the circuit. These are especially handy when trying to understand how a circuit works. Schematics are typically used to show the internal circuitry of a component or to simplify a wiring diagram. One of the troubleshooting techniques used by good electrical technicians is to simplify a wiring diagram into a schematic.

Electrical Symbols

Most wiring diagrams do not show an actual drawing of the components. Rather, they use **electrical symbols** to represent the components. Often the symbol illustrates the basic operation of the component. Many different symbols have been used in wiring diagrams through the years. **Table 4-4** shows some of the commonly used symbols. You need to be familiar with all of the symbols; however, you don't need to memorize all of the variations. Wiring diagram manuals include a legend that helps you interpret the symbols.

 A BIT OF HISTORY

Early service manuals had hand drawn and labeled wiring diagrams. They also had drawings of the actual components. As more and more electrical components were added to cars, this became impractical. Soon schematic symbols replaced the component drawings.

Color Codes and Circuit Numbering

Nearly all of the wires in an automobile are covered with colored insulation. These colors are used to identify wires and electrical circuits. The color of the wires is indicated on the wiring diagram. Some wiring diagrams also include circuit numbers. These numbers, or letters and numbers, help identify a specific circuit. Both types of coding make it easier to diagnose electrical problems. Unfortunately, not all manufacturers use the same method of wire identification. **Table 4-5** shows common color codes and their abbreviations. Most wiring diagrams list the appropriate color coding used by the manufacturer. Make sure you understand what color the code is referring to before looking for a wire.

AUTHOR'S NOTE Some manufacturers use "L" to denote a blue wire. They do this to prevent confusion when using "B" which can mean either black or blue. The "L" is used regardless of the shade or darkness of the blue wire.

In most color codes, the first group of letters designates the base color of the insulation. If a second group of letters is used, it indicates the color of the **tracer**. For example, a wire designated as WH/BLK would have a white base color with a black tracer. A tracer is a thin or dashed line of a different color than the base color of the insulation.

Table 4-4 Common electrical and electronic symbols used in wiring diagrams.

COMPONENT	SYMBOL	COMPONENT	SYMBOL
Ammeter		Jack, Coaxial	
AND Gate		Jack, Phono	
Antenna		Lamp, Neon	
Battery		Male Contact	
Capacitor		Microphone	
Capacitor, Variable		Motor, One speed	
Cell		Motor, Reversible	
Circuit Breaker/PTC device		Motor, two Speed	
Clockspring		Multiple connectors	
Coaxial Cable		Nand Gate	
Coil		Negative Voltage Connection	
Crystal, Piezoelectric		NOR Gate	
Diode		Operational Amplifier	
Diode, Gunn		OR Gate	
Diode, Light-Emitting		Outlet, Utility, 117-V	
Diode, Photosensitive		Oxygen Sensor	
Diode, Photovoltaic		Page Reference	(BW-30-10)
Diode, Zener		Piezoelectric Cell	
Dual Filament Lamp		Photocell, Tube	
Female Contact		Plug, Utility, 117-V	
Fuse		Positive Voltage Connection	
Fusible link		Potentiometer	
Gauge		Probe, Radio-Frequency	
Ground, Chassis		Rectifier, Semiconductor	
Ground, Earth		Relay, DPDT	
Heating element		Relay, DPST	
Hot Bar	BATT AO	Relay, SPDT	
Inductor, Air-Core		Relay, SPST	
Inductor, Iron-Core		Resistor	
In-Line Connectors		Resonator	
Integrated Circuit		Rheostat, Variable Resistor, Thermistor	
Inverter		Shielding	-----

Table 4-4   *(continued)*

COMPONENT	SYMBOL
Signal Generator	
Single Filament Lamp	
Sliding Door Contact	
Solenoid	
Solenoid Valve	
Speaker	
Splice, External	
Splice, Internal	
Splice, Internal (Incompleted)	
Switch, Closed	
Switch, DPDT	
Switch, DPST	
Switch, Ganged	
Switch, Momentary-Contact	
Switch, Open	
Switch, Resistive Multiplex	
Switch, Rotary	
Switch, SPDT	
Switch, SPST	
Terminals	
Test Point	
Thermocouple	
Thyristor	
Tone Generator	
Transformer, Air-Core	

COMPONENT	SYMBOL
Transformer, Iron-Core	
Transformer, Tapped Primary	
Transformer, Tapped Secondary	
Transistor, Bipolar, npn	
Transistor, Bipolar, pnp	
Transistor, Field-Effect, N-Channel	
Transistor, Field-Effect, P-Channel	
Transistor, Metal-Oxide, Dual-Gate	
Transistor, Metal-Oxide, Single-Gate	
Transistor, Photosensitive	
Transistor, Unijunction	
Tube, Diode	
Tube, Pentode	
Tube, Photomultiplier	
Tube, Tetrode	
Tube, Triode	
Unspecified Component	
Voltmeter	
Wattmeter	
Wire Destination in Another Cell	
Wire Origin & Destination within Cell	
Wires	
Wires, Connected, Crossing	
Wires, Not Connected, Crossing	

Table 4-5 Common color codes used in automotive applications.

Color		Abbreviations	
Black	BLK	BK	B
Blue (Dark)	BLU DK	DB	DK BLU
Blue	BLU	B	L
Blue (Light)	BLU LT	LB	LT BLU
Brown	BRN	BR	BN
Glazed	GLZ	GL	
Gray	GR A	GR	G
Green (Dark)	GRN DK	DG	DK GRN
Green (Light)	GRN LT	LG	LT GRN
Maroon	MAR	M	
Natural	NAT	N	
Orange	ORN	O	ORG
Pink	PNK	PK	P
Purple	PPL	PR	
Red	RED	R	RD
Tan	TAN	T	TN
Violet	VLT	V	
White	WHT	W	WH
Yellow	YEL	Y	YL

Ford uses four methods of color coding its wires (**Figure 4-20**):

1. Solid color.
2. Base color with a stripe (tracer).
3. Base color with hash marks.
4. Base color with dots.

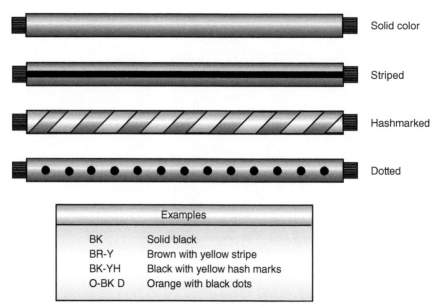

Solid color

Striped

Hashmarked

Dotted

Examples	
BK	Solid black
BR-Y	Brown with yellow stripe
BK-YH	Black with yellow hash marks
O-BK D	Orange with black dots

Figure 4-20 Four methods that Ford uses to color code circuit wires.

Chrysler uses a numbering method to designate the circuits on the wiring diagram (**Figure 4-21**). The circuit identification, wire gauge, and color of the wire are included in the wire number. Chrysler identifies the main circuits by using a main circuit identification code that corresponds to the first letter in the wire number (**Table 4-6**).

General Motors uses numbers that include the wire gauge in metric millimeters, the wire color, circuit number, splice number, and ground identification (**Figure 4-22**). In this

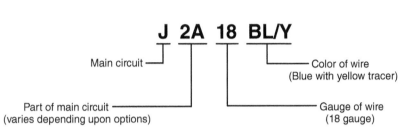

Figure 4-21 Chrysler's wiring code identification.

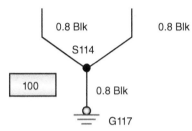

Figure 4-22 GM's method of circuit and wire identification.

Table 4-6 Chrysler's circuit identification codes.

Circuit Identification Code Chart	
Circuit	**Function**
A	Battery feed
B	Brake controls
C	Climate controls
D	Diagnostic circuits
E	Dimming illumination circuits
F	Fused circuits
G	Monitoring circuits (gauges)
H	Multiple
I	Not used
J	Open
K	Powertrain control module
L	Exterior lighting
M	Interior lighting
N	Multiple
O	Not used
P	Power option (battery feed)
Q	Power options (ignition feed)
R	Passive restraint
S	Suspension/steering
T	Transmission/transaxle/transfer case
U	Open
V	Speed control, wiper/washer
W	Wipers
X	Audio systems
Y	Temporary
Z	Grounds

example, the circuit is designated as 100, the wire size is 0.8 mm², the insulation color is black, the splice is numbered S114, and the ground is designated as G117.

Most manufacturers also number connectors, terminals, splices, and grounds for identification. The numbers correspond to their general location within the vehicle. The following is typical identification numbers:

100–199	Engine compartment forward of the dash panel
200–299	Instrument panel area
300–399	Passenger compartment
400–499	Deck area
500–599	Left front door
600–699	Right front door
700–799	Left rear door
800–899	Right rear door
900–999	Deck lid or hatch

As mentioned, manufacturers use differing ways to illustrate wiring diagrams. As an example of reading import wiring diagrams, consider the system that has been used by Toyota. One way Toyota differs from methods we have discussed is that it relies heavily on ID numbers. When looking at a relay or junction block component, the ID number will be located in an oval next to the terminal identifiers (**Figure 4-23**). These ID numbers indicate which relay or junction block the component is located in.

Components and parts not part of the junction or relay block also use ID numbers made up of letters and numbers (**Figure 4-24**). Usually the letter matches the first letter of the component's name. In this example the component ID number is "I1." The "I" is the first letter of idle air control valve. The ID number crosses over to the connector diagram and to the parts location table.

The ID numbers used for in-line connectors is different than that used for components. The connector ID number is located within the illustration of the connector (**Figure 4-25**). Also, notice the splice point identifier. The letter used for in-line connectors and splices is the same. The ID numbers will begin with "E" for Engine, "I" for Instrument cluster, and "B" for Body. These ID numbers correspond to how you use the location table.

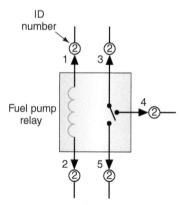

Figure 4-23 The ID number used with junction and relay block components identifies which block it is located in.

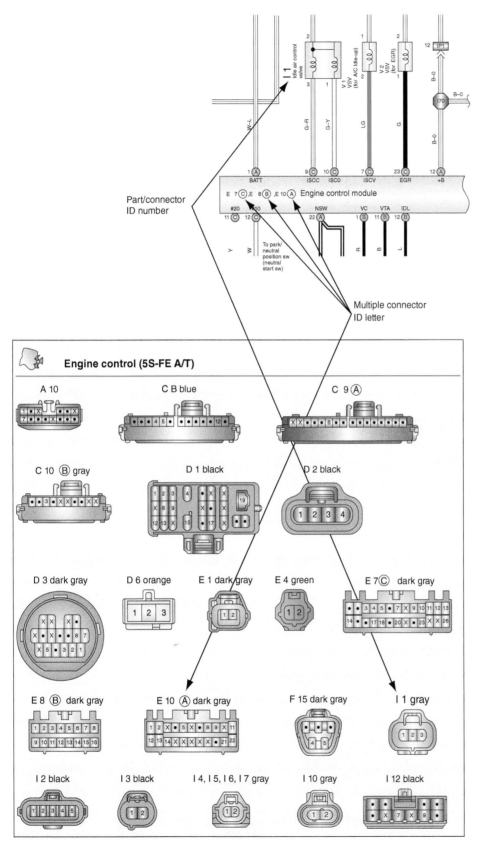

Part/connector
ID number

Multiple connector
ID letter

Figure 4-24 The component ID number is used to cross-reference to additional information.

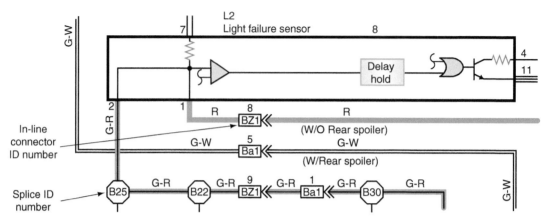

Figure 4-25 In-line connector and splice ID numbers.

Grounds are identified by an inverted triangle with a two-letter identifier (**Figure 4-26**). Like the in-line connector and splice identifiers, the first letter identifies which harness the ground is located in. Again, the ground ID is used to refer to the location chart to determine where it is located on the vehicle.

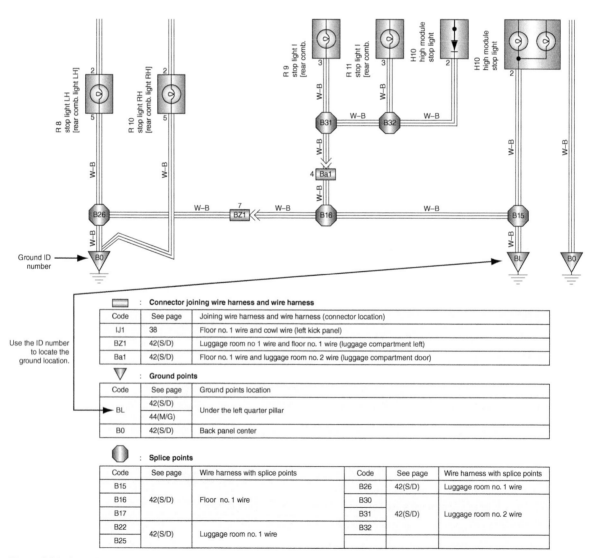

Figure 4-26 Ground point ID.

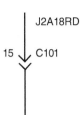

Figure 4-27
The wiring diagram shows the red wire going through cavity 15 of the C101 connector. Now you need to find this cavity in the actual connector.

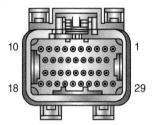

Figure 4-28 The service information will provide cavity numbering orientation. Pay close attention to which way the cavities are counted.

Today most manufacturers have converted to some form of electronic or on-line formats for wiring diagrams. However, they usually use the same basic format as they did for their paper manuals, the difference being the use of hyperlinks or other methods to navigate throughout the circuit layout.

When the wires pass through a bulk connector, you must properly identify the correct wire for the circuit you are diagnosing. Sometimes on larger bulk connectors, there may be several cavities that are filled with the same color of wires; thus just using the color codes may not be sufficient. For example, the wiring diagram indicates the red wire goes through cavity 15 of the C101 connector (**Figure 4-27**). Once this connector is located, use the service manual information to determine the orientation of the connector (**Figure 4-28**). Orientation may be different between the male and female sides of the connector. Since the view is from the terminal side of the connector as opposed to the harness side, the numbers are counted from the right to the left. Some connectors may have molded numbers at the end of each row or in the corners of the connector to help orient the connector. Once you know which way to count, simply start at a numbered connector and count until you reach the desired cavity.

Component Locators

The wiring diagrams in most service information may not indicate the exact physical location of the components of the circuit. In another section of the service information, or in a separate manual, a **component locator** is provided to help find where a component is installed in the vehicle. The component locator may use both drawings and text to lead the technician to the desired component (**Figure 4-29**).

Many electrical components may be hidden behind kick panels, dashboards, fender wells, and under seats. The use of a component locator will save the technician time in finding the suspected defective unit.

Component locators are also called installation diagrams.

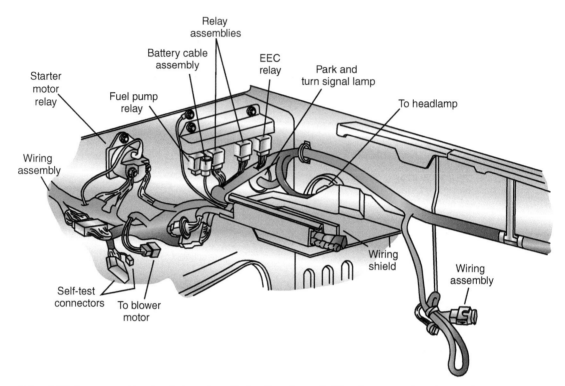

Figure 4-29 A component locator diagram shows the location of several different components and connectors.

SUMMARY

- Most of the primary wiring conductors used in the automobile are made of several strands of copper wire wound together and covered with a polyvinyl chloride (PVC) insulation.
- Stranded wire is used because of its flexibility and current flows on the surface of the conductors. Because there is more surface area exposed in a stranded wire, there is less resistance in the stranded wire than in the solid wire.
- There are three major factors that determine the proper size of wire to be used: (1) the wire must have a large-enough diameter—for the length required—to carry the necessary current for the load components in the circuit to operate properly; (2) the wire must be able to withstand the anticipated vibration; and (3) the wire must be able to withstand the anticipated amount of heat exposure.
- Wire size is based on the diameter of the conductor.
- Factors that affect the resistance of the wire include the conductor material, wire diameter, wire length, and temperature.

- Ground straps are used to complete the return path to the battery between components that are insulated. They are also used to suppress electromagnetic induction (EMI) and radiation.
- Terminals can be either crimped or soldered to the conductor. The terminal makes the electrical connection, and it must be capable of withstanding the stress of normal vibration.
- Printed circuit boards are used to simplify the wiring of the circuits they operate. A printed circuit is made of a thin phenolic or fiberglass board that has copper (or some other conductive material) deposited on it.
- A wire harness is an assembled group of wires that branch out to the various electrical components. It is used to reduce the number of loose wires hanging under the hood or dash. It provides for a safe path for the wires of the vehicle's lighting, engine, and accessory components.
- The wiring harness is made by grouping insulated wires and wrapping them together. The wires are bundled into separate harness assemblies that are joined together by connector plugs.

- A wiring diagram shows a representation of actual electrical or electronic components and the wiring of the vehicle's electrical systems.
- The technician's greatest helpmate in locating electrical problems is the wiring diagram. Correct use of the wiring diagram will reduce the amount of time a technician needs to spend tracing the wires in the vehicle.
- In place of actual pictures, a variety of electrical symbols are used to represent the components in the wiring diagram.
- Color codes and circuit numbers are used to make tracing wires easier.
- In most color codes, the first group of letters designates the base color of the insulation. If a second group of letters is used, it indicates the color of the tracer.
- A component locator is used to determine the exact location of several of the electrical components.

REVIEW QUESTIONS

Short-Answer Essays

1. Explain the purpose of wiring diagrams.

2. Explain how wire size is determined by the American Wire Gauge (AWG) and metric methods.

3. Explain the purpose and use of printed circuits.

4. Explain the purpose of the component locator.

5. Explain when single-stranded or multistranded wire should be used.

6. Explain how temperature affects resistance and wire size selection.

7. List the three major factors that determine the proper size of wire to be used.

8. List and describe the different types of terminal connectors used in the automotive electrical system.

9. What is the difference between a complex and a simple wiring harness?

10. Describe the methods the three domestic automobile manufacturers use for wiring code identification.

Fill in the Blanks

1. There is _____ resistance in the stranded wire than in the solid wire.

2. _____ complete the return path to the battery between components that are insulated.

3. Wire size is based on the _____ of the conductor.

4. In the AWG standard, the _____ the number, the smaller the wire diameter.

5. An increase in conductor temperature creates an _____ in resistance.

6. _____ connectors are used on computer circuits to protect the circuit from corrosion.

7. _____ are used to prevent damage to the wiring by maintaining proper wire routing and retention.

8. A wiring diagram is an electrical schematic that shows a _____ of actual electrical or electronic components (by use of symbols) and the _____ of the vehicle's electrical systems.

9. In most color codes, the first group of letters designates the _____ of the insulation. The second group of letters indicates the color of the _____.

10. A _____ is used to determine the exact location of several of the electrical components.

Multiple Choice

1. Automotive wiring is being discussed.

   Technician A says most primary wiring is made of several strands of copper wire wound together and covered with insulation.

   Technician B says the types of conductor materials used in automobiles include copper, silver, gold, aluminum, brass, and tin-plated steel.

   Who is correct?

   A. A only

   B. B only

   C. Both A and B

   D. Neither A nor B

2. Stranded wire use is being discussed.

   Technician A says there is less exposed surface area for electron flow in a stranded wire.

   Technician B says there is more resistance in the stranded wire than in the same gauge solid wire.

   Who is correct?

   A. A only
   B. B only
   C. Both A and B
   D. Neither A nor B

3. A Chrysler wiring diagram designation of M2 14 BK/YL identifies the main circuit as being:

   A. Climate control.
   B. Interior lighting.
   C. Wipers.
   D. None of the above.

4. Ground straps are used to:

   A. Provide a return path to the battery between insulated components.
   B. Suppress electromagnetic induction.
   C. Both A and B.
   D. Neither A nor B.

5. The selection of the proper size of wire to be used is being discussed.

   Technician A says the wire must be large enough, for the length required, to carry the amount of current necessary for the load components in the circuit to operate properly.

   Technician B says temperature has little effect on resistance and it is not a factor in wire size selection.

   Who is correct?

   A. A only
   B. B only
   C. Both A and B
   D. Neither A nor B

6. Terminal connectors are being discussed.

   Technician A says good terminal connections will resist corrosion.

   Technician B says the terminals can be either crimped or soldered to the conductor.

   Who is correct?

   A. A only
   B. B only
   C. Both A and B
   D. Neither A nor B

7. Wire routing is being discussed.

   Technician A says to install additional electrical accessories that are necessary to support the primary wire in at least 10-foot intervals.

   Technician B says if the wire must be routed through the frame or body, use metal clips to protect the wire.

   Who is correct?

   A. A only
   B. B only
   C. Both A and B
   D. Neither A nor B

8. Printed circuit boards are being discussed.

   Technician A says printed circuit boards are used to simplify the wiring of the circuits they operate.

   Technician B says care must be taken not to touch the board with bare hands.

   Who is correct?

   A. A only
   B. B only
   C. Both A and B
   D. Neither A nor B

9. Wiring harnesses are being discussed.

   Technician A says a wire harness is an assembled group of wires that branches out to the various electrical components.

   Technician B says most under-hood harnesses are simple harnesses.

   Who is correct?

   A. A only
   B. B only
   C. Both A and B
   D. Neither A nor B

10. Wiring diagrams are being discussed.

    Technician A says wiring diagrams give the exact location of the electrical components.

    Technician B says a wiring diagram will indicate what circuits are interconnected, where circuits receive their voltage source, and what color of wires are used in the circuit.

    Who is correct?

    A. A only
    B. B only
    C. Both A and B
    D. Neither A nor B

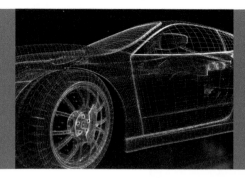

CHAPTER 5
AUTOMOTIVE BATTERIES

Upon completion and review of this chapter, you should be able to understand and:

- Describe the purposes of a battery.
- Describe the construction of conventional, maintenance-free, hybrid, and recombination batteries.
- Describe the main elements of a battery.
- Describe the chemical action that occurs to produce current in a battery.
- Describe the chemical reaction that occurs in a battery during cycling.
- Describe the differences, advantages, and disadvantages between battery types.
- Describe the function of HEV batteries.
- Describe the operation and purpose of ultra-capacitors.

- Describe the different types of battery terminals used.
- Describe the methods used to rate batteries.
- Determine the correct battery to be installed into a vehicle.
- Explain the effects of temperature on battery performance.
- Describe the different loads or demands placed upon a battery during different operating conditions.
- Explain the major reasons for battery failure.
- Define battery-related terms such as *deep cycle*, *electrolyte solution*, and *gassing*.

Terms To Know

Absorbed glass mat (AGM) battery	Gassing	Radial grid
Ampere-hour rating	Grid	Recombination battery
Battery cables	Grid growth	Regenerative braking
Cell element	Holddowns	Reserve capacity
Cold cranking rating	Hybrid battery	Reserve-capacity rating
Contactors	Hydrometer	Specific gravity
Cranking amps (CA)	Integrated starter generator (ISG)	Terminals
Deep cycling	Maintenance-free battery	Ultra-capacitors
Electrochemical	Material expanders	Valve-regulated lead-acid (VRLA) battery
Electrolyte	Memory effect	
Energy density	Radial	

INTRODUCTION

An automotive battery (**Figure 5-1**) is an **electrochemical** device capable of storing chemical energy that can be converted to electrical energy. *Electrochemical* refers to the chemical reaction of two dissimilar materials in a chemical solution that results in electrical current.

Shop Manual
Chapter 5, page 211

Figure 5-1 Typical automotive 12-volt battery.

When the battery is connected to an external load, such as a starter motor, an energy conversion occurs that results in an electrical current flowing through the circuit. Electrical energy is produced in the battery by the chemical reaction that occurs between two dissimilar plates that are immersed in an electrolyte solution. The automotive battery produces direct current (DC) electricity that flows in only one direction.

When discharging the battery (current flowing from the battery), the battery changes chemical energy into electrical energy. It is through this change that the battery releases stored energy. During charging (current flowing through the battery from the charging system), electrical energy is converted into chemical energy. As a result, the battery can store energy until it is needed.

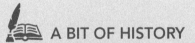 **A BIT OF HISTORY**

Buick first introduced the storage battery as standard equipment in 1906.

The automotive battery has several important functions, including the following:

1. It operates the starting motor, ignition system, electronic fuel injection, and other electrical devices for the engine during cranking and starting.
2. It supplies all the electrical power for the vehicle accessories whenever the engine is not running or when the vehicle's charging system is not working.
3. It furnishes current for a limited time whenever electrical demands exceed charging system output.
4. It acts as a stabilizer of voltage for the entire automotive electrical system.
5. It stores energy for extended periods of time.

AUTHOR'S NOTE The battery does not store energy in electrical form. The battery stores energy in chemical form.

The largest demand placed on the battery occurs when it must supply current to operate the starter motor. The amperage requirements of a starter motor may be over several hundred amperes. This requirement is also affected by temperatures, engine size, and engine condition.

After the engine starts, the vehicle's charging system works to recharge the battery and to provide the current to run the electrical systems. Alternating current (AC) generators can have a maximum output of 60 to 250 amps (A). This is usually enough to operate all of the vehicle's electrical systems and meet the demands of these systems. However, under some conditions (such as the engine idling), generator output is below its maximum rating. If there are enough electrical accessories turned on during this time (heater, wipers, headlights radio, etc.), the demand may exceed the AC generator output. During this time, the battery must supply the additional current. Electrical loads that are still present when the ignition switch is in the OFF position are called key-off or parasitic loads.

Even with the ignition switch turned off, there are electrical demands placed on the battery. Clocks, memory seats, engine computer memory, body computer memory, and electronic sound system memory are all examples of key-off loads. The total current draw of key-off loads is usually less than 50 mA.

In the event that the vehicle's charging system fails, the battery must supply all of the current necessary to run the vehicle. The amount of time a battery can be discharged at a certain current rate until the voltage drops below a specified value is referred to as **reserve capacity**. Most batteries will supply a reserve capacity of 25 amps for approximately 120 minutes before discharging too low to keep the engine running.

The amount of electrical energy that a battery is capable of producing depends on the type, size, weight, active area of the plates, and the amount of sulfuric acid in the electrolyte solution.

The chemical molecular formula for sulfuric acid is H_2SO_4.

In this chapter, you will study the design and operation of different types of batteries currently used in automobiles. These include conventional batteries, maintenance-free batteries, hybrid batteries, recombination batteries, absorbed glass mat batteries, and valve-regulated batteries. In addition, high-voltage batteries used in electric and electric hybrid vehicles (HEVs) are discussed.

CONVENTIONAL BATTERIES

A conventional battery is constructed of seven basic components:

1. Positive plates.
2. Negative plates.
3. Separators.
4. Case.
5. Plate straps.
6. Electrolyte.
7. Terminals.

The condition of the battery should be the first test performed on a vehicle with an electrical problem. Without proper battery performance, the entire electrical system is affected.

The difference between three-year and five-year batteries is the quantity of **material expanders** used in the construction of the plates and the number of plates used to build a cell. Material expanders are fillers that can be used in place of the active materials. They are used to keep the manufacturing costs low.

A plate, either positive or negative, starts with a **grid**. Grids are generally made of lead alloys, usually antimony. About 5% to 6% antimony is added to increase the strength of the grid. The grid is the frame structure with connector tabs at the top. The grid has

Lead dioxide is composed of small grains of particles. This gives the plate a high degree of porosity, allowing the electrolyte to penetrate the plate.

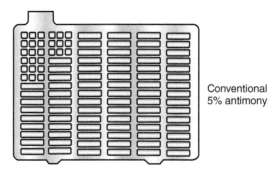

Figure 5-2 Conventional battery grid.

horizontal and vertical grid bars that intersect at right angles (**Figure 5-2**). An active material made from ground lead oxide, acid, and material expanders is pressed into the grid in paste form. The positive plate is given a "forming charge" that converts the lead oxide paste into lead dioxide. The negative plate is given a "forming charge" that converts the paste into sponge lead.

The negative and positive plates are arranged alternately in each **cell element** (**Figure 5-3**). Each cell element can consist of 9 to 13 plates. The positive and negative plates are insulated from each other by separators made of microporous materials. The construction of the element is completed when all of the positive plates are connected to each other and all of the negative plates are connected to each other. The connection of the plates is by plate straps (**Figure 5-4**).

A typical 12-volt (V) automotive battery is made up of six cells connected in series (**Figure 5-5**). This means the positive side of a cell element is connected to the negative side of the next cell element. This is repeated throughout all six cells. By connecting the cells in series, the current capacity of the cell and cell voltage remain the same. The six cells produce 2.1 volts each. Wiring the cells in series produces the 12.6 volts required by the automotive electrical system. The plate straps provide a positive cell connection and a negative cell connection. The cell connection may be one of three

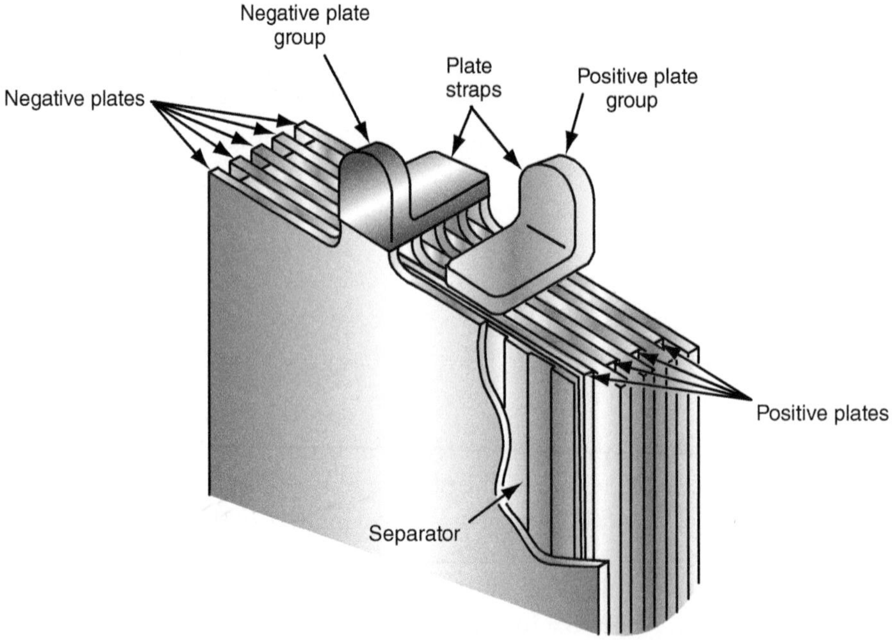

> Usually, negative plate groups contain one more plate than positive plate groups to help equalize the chemical activity.

Figure 5-3 A battery cell consists of alternate positive and negative plates.

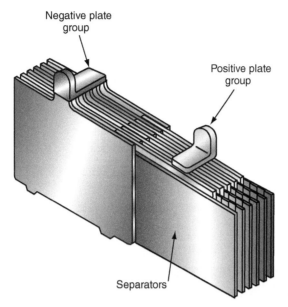

Figure 5-4 Construction of a battery element.

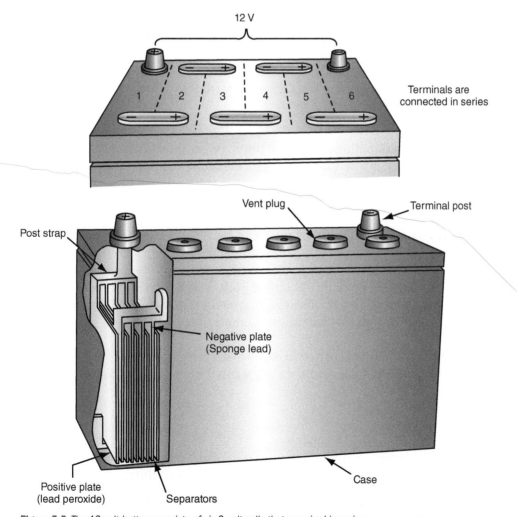

Figure 5-5 The 12-volt battery consists of six 2-volt cells that are wired in series.

Intercell connections

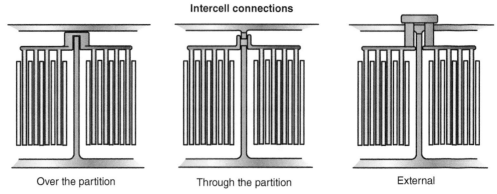

Over the partition   Through the partition   External

Figure 5-6 The cell elements can be connected using one of three intercell connection methods.

Figure 5-7 The vents of a conventional battery allow the release of gases.

Many batteries have envelope-type separators that retain active materials near the plates.

The most common connection used to connect cell elements is through the partition. It provides the shortest path and the least resistance.

Lead acid batteries are also called flooded batteries.

Shop Manual
Chapter 5, page 219

types: over the partition, through the partition, or external (**Figure 5-6**). The cell elements are submerged in a cell case filled with **electrolyte** solution. Electrolyte consists of sulfuric acid diluted with water. The electrolyte solution used in automotive batteries consists of 64% water and 36% sulfuric acid, by weight. Electrolyte is both conductive and reactive.

The battery case is made of polypropylene, hard rubber, and plastic base materials. The battery case must be capable of withstanding temperature extremes, vibration, and acid absorption. The cell elements sit on raised supports in the bottom of the case. By raising the cells, chambers are formed at the bottom of the case that traps the sediment that flakes off the plates. If the sediment is not contained in these chambers, it could cause a conductive connection across the plates and short the cell. The case is fitted with a one-piece cover.

Because the conventional battery releases hydrogen gas when it is being charged, the case cover will have vents. The vents are located in the cell caps of a conventional battery (**Figure 5-7**).

Chemical Action

Activation of the battery is through the addition of electrolyte. This solution causes the chemical actions to take place between the lead dioxide of the positive plates and the sponge lead of the negative plates. The electrolyte is also the carrier that moves electric current between the positive and negative plates through the separators.

The automotive battery has a fully charged **specific gravity** of 1.265 corrected to 80°F(27°C). Therefore, a specific gravity of 1.265 for electrolyte means it is 1.265 times heavier than an equal volume of water. As the battery discharges, the specific gravity of the electrolyte decreases because the electrolyte becomes more like water. The specific gravity of a battery can give you an indication of how charged a battery is.

Shop Manual
Chapter 5,
pages 216, 220, 226

Specific gravity is the weight of a given volume of a liquid divided by the weight of an equal volume of water. Water has a specific gravity of 1.000.

Fully charged:	1.265 specific gravity
75% charged:	1.225 specific gravity
50% charged:	1.190 specific gravity
25% charged:	1.155 specific gravity
Discharged:	1.120 or lower specific gravity

These specific gravity values may vary slightly according to the design of the battery. However, regardless of the design, the specific gravity of the electrolyte in all batteries will decrease as the battery discharges. Temperature of the electrolyte will also affect its specific gravity. All specific gravity specifications are based on a standard temperature of 80°F(27°C)). When the temperature is above that standard, the specific gravity is lower. When the temperature is below that standard, the specific gravity increases. Therefore, all specific gravity measurements must be corrected for temperature. A general rule to follow is to add 0.004 for every 10°F(5.5°C) above 80°F(27°C) and subtract 0.004 for every 10°F(5.5°C) below 80°F(27°C).

In operation, the battery is being partially discharged and then recharged. This represents an actual reversing of the chemical action that takes place within the battery. The constant cycling of the charge and discharge modes slowly wears away the active materials on the cell plates. This action eventually causes the battery plates to sulfate. The battery must be replaced once the sulfation of the plates has reached the point of insufficient active plate area.

In the charged state, the positive plate material is essentially pure lead dioxide, PbO_2. The active material of the negative plates is spongy lead, Pb. The electrolyte is a solution of sulfuric acid, H_2SO_4, and water. The voltage of the cell depends on the chemical difference between the active materials.

Figure 5-8 shows what happens to the plates and electrolyte during discharge. The electrolyte contains negatively charged sulfate ions (SO_4^{2-}) and positively charged

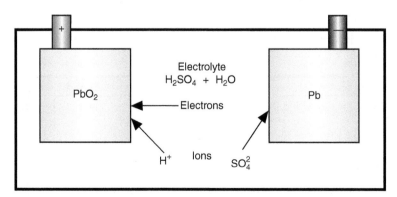

Figure 5-8 Chemical action that occurs inside of the battery during the discharge cycle.

hydrogen ions (H^+). When a load is applied to the battery, the sulfate ions move to the negative plates and give up their negative charge. Any remaining sulfate combines with the sulfuric acid on the plates to form lead sulfate $(PbSO_4)$. The lead sulfate becomes an electrical insulator. As the electrons accumulate an electric field is created that attracts hydrogen ions and repels sulfate ions, leading to a double-layer near the surface. The hydrogen ions screen the charged electrode from the solution, which limits further reactions unless charge is allowed to flow out of electrode

As the electrons are attracted to the positive plates, the oxygen (O_2) joins with the hydrogen ions from the electrolyte to form water (H_2O). The water dilutes the strength of the electrolyte and decreases the number of ions. Also, the lead sulfate formation on the positive and negative plates reduces the area of active material available to the ions.

The result of discharging is changing the PbO_2 of the positive plates and the Pb of the negative plates into lead sulfate $(PbSO_4)$. Because the plates are the same, the battery does not have an electrical potential.

📖 A BIT OF HISTORY

Lead acid batteries date back to 1859. Alexander Graham Bell used a primitive battery to make his first local call in 1876. Once it was learned how to recharge the lead acid batteries, they were installed into the automobile. These old-style batteries could not hold a charge very well and it was believed that placing the battery on a concrete floor made them discharge faster. Although this fable has no truth to it, the idea has hung on for years.

The charge cycle is exactly the opposite (**Figure 5-9**). The generator creates an excess of electrons at the negative plates, attracting the positive hydrogen ions to them. The hydrogen combines with the sulfate to revert it back to sulfuric acid (H_2SO_4) and lead. When most of the sulfate have been converted, the hydrogen rises from the negative plates. The water in the electrolyte splits into hydrogen and oxygen. The oxygen reacts with the lead sulfate on the positive plates to revert it back into lead dioxide. When the reaction is about complete, the oxygen bubbles rise from the positive plates. This puts the plates and the electrolyte back in their original form and the cell is charged.

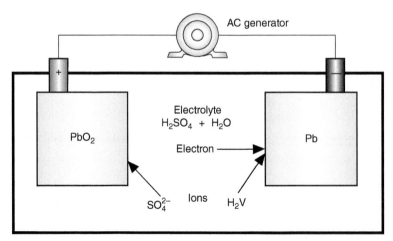

Figure 5-9 Chemical action inside of the battery during the charge cycle.

MAINTENANCE-FREE BATTERIES

In a **maintenance-free battery** there is no provision for the addition of water to the cells since the battery is sealed (**Figure 5-10**). This battery type contains cell plates made of a slightly different compound than what is in a conventional battery. The plate grids contain calcium, cadmium, or strontium to reduce **gassing** and self-discharge. Gassing is the conversion of the battery water into hydrogen and oxygen gas. This process is also called electrolysis. The antimony used in conventional batteries is not used in maintenance-free batteries because it increases the breakdown of water into hydrogen and oxygen and because of its low resistance to overcharging. The use of calcium, cadmium, or strontium reduces the amount of vaporization that takes place during normal operation. The grid may be constructed with additional supports to increase its strength and to provide a shorter path, with less resistance, for the current to flow to the top tab (**Figure 5-11**).

Each plate is wrapped and sealed on three sides by an envelope design separator. The envelope is made from microporous plastic. By enclosing the plate in an envelope, the plate is insulated and reduces the shedding of the active material from the plate.

The battery is sealed except for a small vent so the electrolyte and vapors cannot escape (**Figure 5-12**). An expansion or condensation chamber allows the water to condense and drain back into the cells. Because the water cannot escape from the battery, it is not

> Maintenance-free batteries are also called lead calcium batteries.

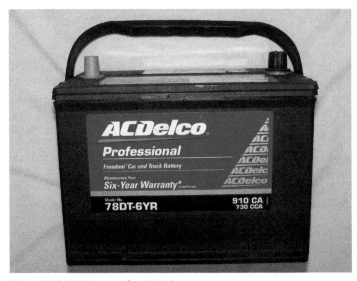

Figure 5-10 Maintenance-free batteries.

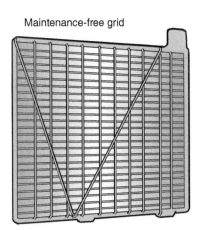

Maintenance-free grid

Calcium or strontium alloy:
• Adds strength.
• Cuts gassing up to 97%.
• Resists overcharge.

Figure 5-11 Maintenance-free battery grids with support bars give increased strength and faster electrical delivery.

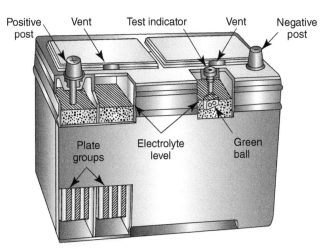

Figure 5-12 Construction of a maintenance-free battery.

necessary to add water to the battery on a periodic basis. Containing the vapors also reduces the possibility of corrosion and discharge through the surface because of electrolyte on the surface of the battery. Vapors leave the case only when the pressure inside the battery is greater than atmospheric pressure.

> **AUTHOR'S NOTE** If electrolyte and dirt are allowed to accumulate on the top of the battery case, it may create a conductive connection between the positive and negative terminals, resulting in a constant discharge on the battery.

Shop Manual
Chapter 5,
pages 214, 220

Some maintenance-free batteries have a built-in **hydrometer** to indicate the state of charge (**Figure 5-13**). A hydrometer is a test instrument that is used to check the specific gravity of the electrolyte to determine the battery's state of charge. If the dot that is at the bottom of the hydrometer is green, then the battery is fully charged (more than 65% charged). If the dot is black, the battery state of charge is low. If the battery does not have a built-in hydrometer, it cannot be tested with a hydrometer because the battery is sealed.

> **AUTHOR'S NOTE** It is important to remember that the built-in hydrometer is only an indication of the state of charge for one of the six cells of the battery and should not be used for testing purposes.

Many manufacturers have revised the maintenance-free battery to a low-maintenance battery, in that the caps are removable for testing and electrolyte level checks. Also, the grid construction contains about 3.4% antimony. To decrease the distance and resistance of the path that current flows in the grid, and to increase its strength, the horizontal and vertical grid bars do not intersect at right angles (**Figure 5-14**).

The advantages of maintenance-free batteries over conventional batteries include:

1. A larger reserve of electrolyte above the plates.
2. Increased resistance to overcharging.
3. Longer shelf life (approximately 18 months).
4. Ability to be shipped with electrolyte installed, reducing the possibility of accidents and injury to the technician.
5. Higher cold cranking amps rating.

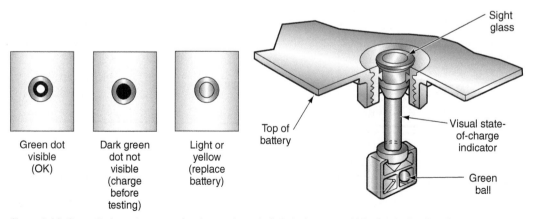

Green dot visible (OK)

Dark green dot not visible (charge before testing)

Light or yellow (replace battery)

Top of battery

Sight glass

Visual state-of-charge indicator

Green ball

Figure 5-13 One cell of a maintenance-free battery has a built-in hydrometer, which gives indication of overall battery condition.

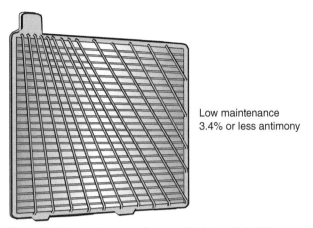

Figure 5-14 Low-maintenance battery grid with vertical grid bars intersecting at an angle.

The major disadvantages of the maintenance-free battery include:

1. **Grid growth** when the battery is exposed to high temperatures.
2. Inability to withstand **deep cycling**.
3. Low reserve capacity.
4. Faster discharge by parasitic loads.
5. Shorter life expectancy.

Grid growth is a condition where the grid grows little metallic fingers that extend through the separators and short out the plates.

HYBRID BATTERIES

AUTHOR'S NOTE The following discussion on hybrid batteries refers to a battery type and not to the batteries that are used in hybrid electric vehicles (HEVs).

Deep cycling is to discharge the battery to a very low state of charge before recharging it.

The **hybrid battery** combines the advantages of the low-maintenance and maintenance-free batteries. The hybrid battery can withstand at least six deep cycles and still retain 100% of its original reserve capacity. The grid construction of the hybrid battery consists of approximately 2.75% antimony alloy on the positive plates and a calcium alloy on the negative plates. This allows the battery to withstand deep cycling while retaining reserve capacity for improved cranking performance. Also, the use of antimony alloys reduces grid growth and corrosion. The lead calcium has less gassing than conventional batteries.

Grid construction differs from other batteries in that the plates have a lug located near the center of the grid. In addition, the vertical grid bars are arranged in a **radial** pattern (**Figure 5-15**). By locating the lug near the center of the grid and using the **radial grid** design, the current has less resistance and a shorter path to follow to the lug (**Figure 5-16**). This means the battery is capable of providing more current at a faster rate.

Radial means branching out from a common center.

The separators used are constructed of glass with a resin coating. The glass separators offer low electrical resistance with high resistance to chemical contamination. This type of construction provides for increased cranking performance and battery life. Some manufacturers are using fiberglass separators.

RECOMBINATION BATTERIES

A recent variation of the automobile battery is the **recombination battery** (**Figure 5-17**). The recombination battery is sometimes called a gel-cell battery. It does not use a liquid electrolyte. Instead, it uses separators that hold a gel-type material. The separators are

Shop Manual
Chapter 5, page 218

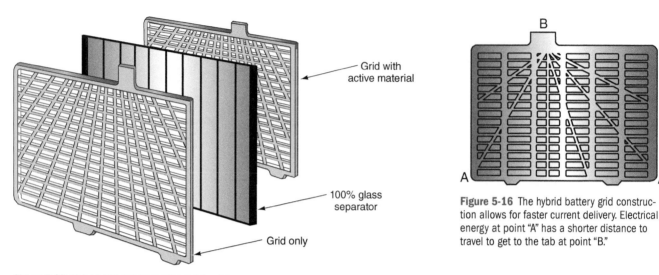

Figure 5-15 Hybrid grid and separator construction.

Figure 5-16 The hybrid battery grid construction allows for faster current delivery. Electrical energy at point "A" has a shorter distance to travel to get to the tab at point "B."

Figure 5-17 The recombination battery is one of the most recent advances in the automotive battery.

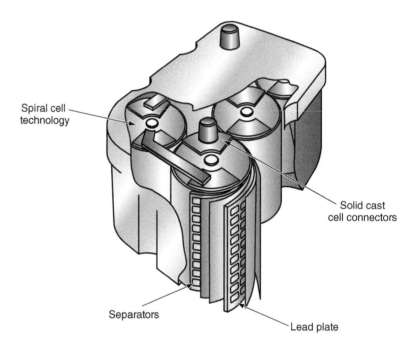

Figure 5-18 Construction of the recombination battery cells.

placed between the grids and have very low electrical resistance. The spiral design provides a larger plate surface area than that in conventional batteries (**Figure 5-18**). In addition, the close plate spacing results in decreased resistance. Because of this design, output voltage and current in recombination batteries are higher than that in conventional batteries. The extra amount of available voltage (approximately 0.6 volt) assists in cold-weather starting. Also, gassing is virtually eliminated and the battery can recharge faster.

The following are some other safety features and advantages of the recombination battery:

1. Contains no liquid electrolyte. If the case is cracked, no electrolyte will spill.
2. Can be installed in any position, including upside down.
3. Is corrosion free.
4. Has very low maintenance because there is no electrolyte loss.
5. Can last as much as four times longer than conventional batteries.
6. Can withstand deep cycling without damage.
7. Can be rated over 800 cold cranking amperes.

Recombination batteries recombine the oxygen gas that is normally produced on the positive plates with the hydrogen given off by the negative plates. This recombination of oxygen and hydrogen produces water (H_2O) and replaces the moisture in the battery. The electrolyte solution of the recombination battery is absorbed into the separators.

The oxygen produced by the positive plates is trapped in the cell by special pressurized sealing vents. The oxygen gases then travel to the negative plates through small fissures in the gelled electrolyte. There are between one and six one-way safety valves in the top of the battery. The safety valves are necessary for maintaining a positive pressure inside of the battery case. This positive pressure prevents oxygen from the atmosphere from entering the battery and causing corrosion. Also, the safety valves must release excessive pressure that may be produced if the battery overcharges.

Absorbed Glass Mat Batteries

A variation of the recombination battery is the **absorbed glass mat (AGM) battery**. Instead of using a gel, AGM batteries hold their electrolyte in a moistened fiberglass matting. The matting is sandwiched between the battery's lead plates. The plates are made of high-purity lead and are tightly compressed into six cells. Separation of the plates is done by acid-permeated vitreous separators that act as sponges to absorb acid. Each cell is enclosed in its own cylinder within the battery case. This results in a sealed battery.

During normal discharging and charging of the battery, the hydrogen and oxygen sealed within the battery are captured and recombined to form water within the electrolyte. This process of recombining hydrogen and oxygen eliminates the need to add water to the battery.

Typical of recombination batteries, AGM batteries are not easily damaged due to vibrations or impact. AGM batteries also have short recharging times and low internal resistance, which increases output.

Hybrid vehicles also use 12-volt AGM batteries for operation of the vehicle accessories. Because the 12-volt battery is not used to crank the engine, it typically has very low energy capacity.

Valve-Regulated Batteries

All recombination batteries are classified as valve-regulated batteries since they have one-way safety valves that control the internal pressure of the battery case. The valve will open to relieve any excessive pressure within the battery, but at all other times the valve closes and seals the battery. A **valve-regulated lead-acid (VRLA) battery** is another variation of the recombination battery. Each cell contains a one-way check valve in the vent. In addition, the VRLA battery immobilizes the electrolyte. Because the electrolyte is immobilized, the oxygen produced at the positive plates is defused. Within the VRLA battery, the oxygen produced on the positive plate is absorbed by the negative plate. This causes a decrease in the amount of hydrogen produced at the negative plate. The small amount of hydrogen that is produced is combined with the oxygen to produce water that is returned to the electrolyte.

The VRLA uses a plate construction with a base of lead-tin-calcium alloy. The active material of one of the plates is porous lead dioxide, while the active material of the other plate is spongy lead. Sulfuric acid is used as the electrolyte. The sulfuric acid is absorbed into plate separators made of a glass-fiber fabric.

HIGH-VOLTAGE BATTERIES

Automotive manufacturers have been investigating the use of high-voltage batteries for over 20 years. The first step was the use of 42-volt systems. Today, high-voltage batteries are required by electric-drive vehicles. This section discusses high-voltage battery systems and their progression.

42-Volt Systems

The first production vehicle that used the 42-volt system was the 2002 Toyota Crown Sedan, which was sold only in the Japanese market. Although the use of 42-volt systems is very limited now, the technology learned has been applied to the electric hybrid systems.

The system is called 12 volt/14 volt because the battery is 12 volts but the charging system delivers 14 volts. The 42-volt system is actually a 36-volt/42-volt system.

The increased number of electrical systems finding their way onto today's vehicles is taxing the capabilities of the 12-volt/14-volt electrical system. Electronic content in vehicles has been rising at a rate of about 6% per year. It has been estimated that by the end of the decade the electronic content will be about 40% of the total cost of a high-line vehicle. The electrical demands on the vehicle will have risen from about 500 watts in 1970 to about 10,000 watts by 2020.

AUTHOR'S NOTE The reason for using 42 volts is based on the current 12-volt system. Today's batteries are rated at 12 volts but actually store about 14 volts. In addition, the charging system produces about 14 to 15 volts when the engine is running. With the engine running, the primary source of electrical power is the charging system. This means that the automobile's electrical system is actually a 14-volt system. Forty-two volts represent three 12-volt batteries, but since they actually hold a 14-volt charge, the system is considered 42 volts (3 times 14 volts equals 42 volts).

An additional benefit that may be derived from the use of a 42-volt system is that it allows manufacturers to electrify most of the inefficient mechanical and hydraulic systems that are currently used. The new technology will allow electromechanical intake and exhaust valve control, active suspension, electrical heating of the catalytic converters, electrically operated coolant and oil pumps, electric air-conditioning compressor, brake-by-wire, steer-by-wire, and so on to be utilized. Studies have indicated that as these mechanical systems are replaced, fuel economy will increase by about 10% and emissions will decrease.

A BIT OF HISTORY

In 1955, automobile manufacturers started to move from 6-volt electrical systems to the present 12-volt system. The change was due to the demand for increased power to accommodate a greater number of electrical accessories. In 1955, the typical car wiring harness weighed 8 to 10 pounds and required approximately 250 to 300 watts. In 1990, the typical car wiring harness weighed 15 to 20 pounds and required over 1,000 watts. In 2000, the typical car wiring increased to weigh between 22 and 28 pounds and required over 1,800 watts. The conventional 14-volt generator is capable of producing a maximum output of 2,000 watts.

Additional fuel savings can also be realized due to the more efficient charging system used for the 42-volt system. Current 14-volt generators have an average efficiency across the engine speed range band of less than 60%. This translates to about 0.5 gallon (1.9 liters) of fuel for 65 miles (104.6 km) of driving to provide a continuous electrical load of 1,000 watts. With a 42-volt generator, the fuel consumption can be reduced to an equivalent of up to 15% in fuel savings.

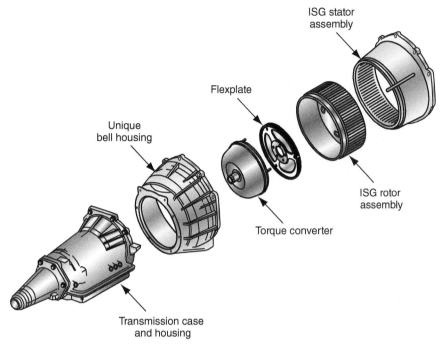

Figure 5-19 The ISG is usually located between the engine and the transmission in the bell housing.

As simple as it may seem to convert from a 12-volt/14-volt system to a 36-volt/42-volt system, many challenges need to be overcome. It is not as simple as adding a higher-voltage output generator and expecting the existing electrical components to work. One of the biggest hurdles is the light bulb. Current 12-volt lighting filaments can't handle 42 volts. The dual-voltage systems can be used as a step toward full 42-volt implantation. However, dual-voltage systems are expensive to design. Another aspect of the 42-volt system is service technician training to address aspects of arcing, safety, and dual-voltage diagnostics.

Arcing is perhaps the greatest challenge facing the design and use of the 42-volt system. In fact, some manufacturers have abandoned further research and development of the system because of the problem with arcing. In the conventional 14-volt system, the power level is low enough that it is almost impossible to sustain an arc. Since there isn't enough electrical energy involved, the arcs collapse quickly and there is less heat buildup. Electrical energy in an arc at 42 volts is significantly greater and is sufficient to maintain a steady arc. The arc from a 42-volt system can reach a temperature of 6000°F (3,316°C).

One of the newest technologies to emerge from the research and development of the 42-volt system is the **integrated starter generator (ISG)**. Although this technology was not actually used on a 42-volt system, development was a result of this system. The ISG is one of the key contributors to the hybrid's fuel efficiency due to its ability to automatically stop and restart the engine under different operating conditions. A typical hybrid vehicle uses an electric induction motor or ISG between the engine and the transmission (**Figure 5-19**). The ISG performs many functions such as fast, quiet starting, automatic engine stops/starts to conserve fuel, recharging the vehicle batteries, smoothening driveline surges, and providing **regenerative braking**. These features will be discussed in greater detail in later chapters.

Regenerative braking is a method of capturing the vehicle's kinetic energy while it slows down. The energy is used to recharge the batteries or a capacitor.

HEV Batteries

As discussed earlier, lead-acid batteries are the most commonly used batteries in the automotive industry. By connecting the batteries in series to each other, they can provide

Figure 5-20 The Nissan Leaf is an example of the growing EV market.

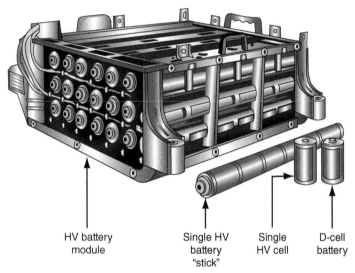

HV battery module   Single HV battery "stick"   Single HV cell   D-cell battery

Figure 5-21 HV battery constructed of cylindrical cells.

Figure 5-22 Battery pack made of several prismatic cells.

Shop Manual
Chapter 5, page 237

NiCad batteries are also referred to as NiCd.

Energy density refers to the amount of energy that is available for a given amount of space.

high-enough voltages to power some electric vehicles (EVs). For example, the first-generation General Motors' EV used twenty-six 12-volt lead-acid batteries connected in series to provide 312 volts. The downside of this arrangement is that the battery pack weighed 1,310 pounds (595 kg). In addition, distance that could be traveled between battery recharges was 55 to 95 miles (88 to 153 km). The next-generation EV used nickel-metal hydride (NiMH) batteries. These provided for a slightly longer traveling range between recharges. Hybrid technology is accelerating battery technology to the point that battery-powered EVs are becoming more popular (**Figure 5-20**).

The battery pack in a hybrid vehicle is typically made up of several cylindrical cells (**Figure 5-21**) or prismatic cells (**Figure 5-22**). These battery packs are often called high-voltage (HV) batteries. There are several types of HV batteries being used or in development.

Nickel–Cadmium (NiCad) Batteries. NiCad cells may have a future role in hybrid and electric vehicles because of several advantages that it has. These include being able to withstand many deep cycles, low cost of production, and long service life. NiCad batteries perform very well when high-energy boosts are required.

The negatives associated with the use of NiCad batteries include the following points: they use toxic metals, have low **energy density**, need to be recharged if they have not been used for a while, and suffer from the **memory effect**. The memory effect refers to the battery not being able to be fully recharged because it "remembers" its previous charge level. This results in a low battery charge due to a battery that is not completely discharged before it is recharged. For example, if the battery is consistently being recharged after it is only discharged 50%, the battery will eventually accept and hold only a 50% charge and not accept any higher charge.

The cathode (positive) electrode in a NiCad cell is made of fiber mesh covered with nickel hydroxide. The anode (negative) electrode is a fiber mesh that is covered with cadmium. The electrolyte is aqueous potassium hydroxide (KOH). The KOH is a conductor of ions and has little involvement in the chemical reaction process. During discharge, ions travel from the anode, through the KOH, and on to the cathode. During charging, the opposite occurs. Each cell produces 1.2 volts.

Nickel-Metal Hydride (NiMH) Batteries.

NiMH batteries are very quickly replacing nickel–cadmium batteries since they are more environment friendly. They also have more capacity than the NiCad battery since they have a higher energy density. However, they have a lower current capacity when placed under a heavy load. Currently, the NiMH is the most common HV battery used in the hybrid vehicle.

The issue facing HEV manufacturers is that the NiMH battery has a relatively short service life. Service life suffers as a result of the battery being subjected to several deep cycles of charging and discharging over its lifetime. In addition, NiMH cells generate heat while being charged and they require long charge times. Because of the service life issue, most batteries used in HEVs have an eight-year warranty.

The cathode electrode of the NiMH battery is a fiber mesh that contains nickel hydroxide. The anode electrode is made of hydrogen-absorbing metal alloys. The most commonly used alloys are compounds containing two to three of the following metals: titanium, vanadium, zirconium, nickel, cobalt, manganese, and aluminum. The amount of hydrogen that can be accumulated and stored by the alloy is far greater than the actual volume of the alloy.

The cathode and anode electrodes are separated by a sheet of fine fibers saturated with an aqueous and alkaline electrolyte, KOH. The components of the cell are typically placed in a metal housing and then the unit is sealed. There is a safety vent that allows high pressures to escape, if needed.

Under load the cell discharges and the hydrogen moves from the anode to the cathode electrode. Since the electrolyte supports only the ion movement from one electrode to the other, it has no active role in the chemical reaction. This means that the electrolyte level does not change because of the chemical reaction. When the cell is recharged, hydrogen moves from the cathode to the anode electrode.

The cells can be constructed either cylindrical or prismatic. Both designs are currently being used in today's hybrid vehicles. The prismatic design requires less storage space but has less energy density than the cylindrical design.

A 300-volt battery is constructed of 240 cells that produce 1.2 volts each. The cells are made into a module, with each module having six cells. Each module is actually a self-contained 7.2-volt battery. The modules are connected in series to create the total voltage (**Figure 5-23**).

A service disconnect is used to disable the HV system if repairs or service to any part of the system is required (**Figure 5-24**). This service connector provides two functions that are used to separate the HV battery pack into two separate batteries, with approximately 150 volts each. First, when the service disconnect is lifted up, it opens a high-voltage interlock loop (HVIL); then when the service disconnect is fully removed, it opens the high-voltage connector. When the HVIL is open, **contactors** should open. Contactors

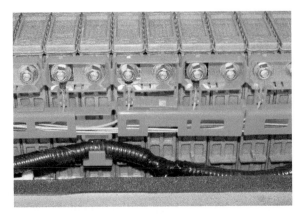

Figure 5-23 Cell module connections.

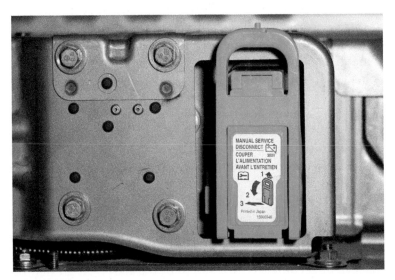

Figure 5-24 Service disconnect plug.

are heavy-duty relays that are connected to the positive and negative sides of the HV battery. The contactors are normally open and require a 12-volt supply to keep them closed. When the service disconnect is lifted and the HVIL is opened, the voltage supply to the contactors is interrupted and the contactors should open. However, if arcing has occurred that may have welded the contacts of the contactors together, the circuit will not be opened. This will result in a diagnostic trouble code being set.

Lithium-Ion (Li-Ion) Batteries.

Rechargeable lithium-based batteries are very similar in construction to the nickel-based batteries just discussed. Positives associated with the use of lithium batteries include high energy density, limited memory effect, and their environment-friendly property. The negatives are that lithium is considered an alkali metal and oxidizes very rapidly in air and water, which makes lithium highly flammable and slightly explosive when exposed to air and water. Lithium metal is also corrosive.

AUTHOR'S NOTE Lithium is the lightest metal and provides the highest energy density of all known metals.

The anode electrode of the Li-ion battery is made of graphite (a form of carbon). The cathode mostly comprises graphite and a lithium alloy oxide. Due to the safety issues associated with lithium metal, the Li-ion battery uses a variety of lithium compounds. A manganese Li-ion battery has been developed for use in hybrid vehicles that has the potential of lasting twice as long as the NiMH battery.

The electrolyte is a lithium salt mixed in a liquid. Polyethylene membranes are used to separate the plates inside the cells and, in effect, separate the ions from the electrons. The membranes have extremely small pores that allow the ions to move within the cell.

As with most other rechargeable cells, ions move from the anode to the cathode when the cell is providing electrical energy, and during recharging, the ions move back from the cathode to the anode.

Lithium-Polymer (Li-Poly) Batteries.   The lithium-polymer battery is nearly identical to the Li-ion battery and shares the same electrode construction. The difference is in the lithium salt electrolyte. The Li-Poly cell holds the electrolyte in a thin solid, polymer composite (polyacrylonitrile) instead as a liquid. The solid polymer electrolyte is not flammable.

The dry polymer electrolyte does not conduct electricity. Instead, it allows ions to move between the anode and cathode. The polymer electrolyte also serves as the separator between the plates. Since the dry electrode has very high resistance, it is unable to provide bursts of current for heavy loads. The efficiency can be increased by increasing the cell temperature above 140°F (60°C). The voltage of the Li-Poly cell is about 4.23 volts when fully charged.

 A BIT OF HISTORY

The Toyota Prius Hybrid was the first automobile to use a bank of ultra-capacitors. The ultra-capacitors store energy captured during deceleration and braking and release that energy to assist the engine during acceleration. The energy in the capacitors is also used to restart the engine during the stop/start sequence.

ULTRA-CAPACITORS

Ultra-capacitors are capacitors constructed to have a large electrode surface area and a very small distance between the electrodes. Unlike conventional capacitors that use a dielectric, the ultra-capacitors use an electrolyte (**Figure 5-25**). It also stores electrical energy at the boundary between the electrodes and the electrolyte. Although an ultra-capacitor is an electrochemical device, no chemical reactions are involved in the storing of electrical energy. This means that the ultra-capacitor remains an electrostatic device. The design of the ultra-capacitor increases its capacitance capabilities to as much as 5,000 farads.

Ultra-capacitors are used in many present-day hybrid vehicles and in some experimental fuel cell EVs because of their ability to quickly discharge high voltages and then be quickly recharged. This makes them ideal for increasing boost to electrical motors during times of acceleration or heavy loads. Ultra-capacitors are also very good at absorbing the energy from regenerative braking.

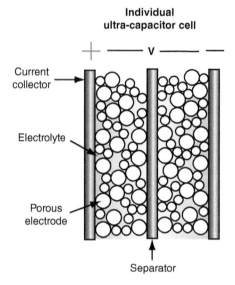

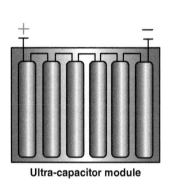

Figure 5-25 Ultra-capacitor cell construction.

BATTERY TERMINALS

Shop Manual
Chapter 5, page 219

Terminals provide a means of connecting the battery plates to the vehicle's electrical system. All automotive batteries have two terminals. One terminal is a positive connection; the other is a negative connection. The battery terminals extend through the cover or the side of the battery case. The following are the most common types of battery terminals (**Figure 5-26**):

1. **Post or top terminals:** Used on most automotive batteries. The positive post is typically larger than the negative post to prevent connecting the battery in reverse polarity. In addition, the terminals are identified using positive and negative symbols.
2. **Side terminals:** Positioned in the side of the container near the top. These terminals are threaded and require a special bolt to connect the cables. Polarity identification is by positive and negative symbols.
3. **L terminals:** Used on specialty batteries and some imports. Polarity identification is by positive and negative symbols.

BATTERY RATINGS

Shop Manual
Chapter 5, page 224

Battery capacity ratings are established by the Battery Council International (BCI) in conjunction with the Society of Automotive Engineers. Battery cell voltage depends on the types of materials used in constructing the battery. Current capacity depends on several factors:

1. The size of the cell plates. The larger the surface area of the plates, the more chemical action that can occur. This means a greater current is produced.

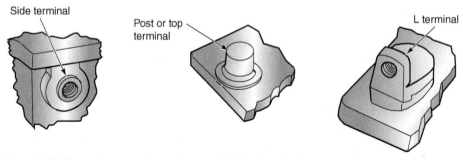

Figure 5-26 The most common types of automotive battery terminals.

2. The weight of the positive and negative plate active materials.
3. The weight of the sulfuric acid in the electrolyte solution.

The battery's current capacity rating is an indication of its ability to deliver cranking power to the starter motor and of its ability to provide reserve power to the electrical system. The commonly used current capacity ratings are explained in the following sections.

Ampere-Hour Rating

The **ampere-hour rating** is the amount of steady current that a fully charged battery can supply for 20 hours at 80°F (26.7°C) without the terminal voltage falling below 10.5 volts. For example, if a battery can be discharged for 20 hours at a rate of 4.0 amps before its terminal voltage reads 10.5 volts, it would be rated at 80 ampere-hours. This method of battery rating is not widely used.

Cold Cranking Rating

Cold cranking rating is the most common method of rating automotive batteries. It is determined by the load, in amperes, that a battery is able to deliver for 30 seconds at 0°F (−17.7°C) without terminal voltage falling below 7.2 volts (1.2 volts per cell) for a 12-volt battery. The cold cranking rating is given in total amperage and is identified as 300 CCA, 400 CCA, 500 CCA, and so on. Some batteries are rated as high as 1,100 CCA.

> Cold cranking rating is also called cold cranking amps (CCA).

CRANKING AMPS

Cranking Amps (CA) is an indication of the battery's ability to provide a cranking amperage at 32°F (0°C). This rating uses the same test procedure as the cold cranking rating or CCA discussed earlier, except it uses a higher temperature. To convert CA to CCA, divide the CA by 1.25. For example, a 650-CCA-rated battery is the same as 812 CA. It is important that the technician not misread the rating and think the battery is rated as CCA instead of CA.

Reserve-Capacity Rating

The **reserve-capacity rating** is determined by the length of time, in minutes, that a fully charged battery can be discharged at 25 amps before battery voltage drops below 10.5 volts. This rating gives an indication of how long the vehicle can be driven, with the headlights on, if the charging system should fail.

Battery Size Selection

Some of the aspects that determine the battery rating required for a vehicle include engine size, engine type, climatic conditions, and vehicle options. The requirement for electrical energy to crank the engine increases as the temperature decreases. Battery power drops drastically as temperatures drop below freezing (**Figure 5-27**). The engine also becomes

Temperature	% of Cranking Power
80°F (26.7°C)	100
32°F (0°C)	65
0°F (−17.8°C)	40

Figure 5-27 The effect temperature has on the cranking power of the battery.

harder to crank due to the tendency of oils to thicken when cold, which results in increased friction. As a general rule, it takes 1 amp of cold cranking power per cubic inch of engine displacement. Therefore, a 200-cubic-inch displacement engine should be fitted with a battery of at least 200 CCA. To convert this into metric, it takes 1 amp of cold cranking power for every 16 cm^3 of engine displacement. A 1.6-liter engine should require at least a battery rated at 100 CCA. This rule may not apply to vehicles that have several electrical accessories. The best method of determining the correct battery is to refer to the manufacturer's specifications.

The battery that is selected should fit the battery holding fixture and the holddown must be able to be installed. It is also important that the height of the battery not allow the terminals to short across the vehicle hood when it is shut. BCI group numbers are used to indicate the physical size and other features of the battery. This group number does not indicate the current capacity of the battery.

BATTERY CABLES

Shop Manual
Chapter 5, page 214

Battery cables are high-current conductors that connect the battery to the vehicle's electrical system. Battery cables must be of a sufficient capacity to carry the current required to meet all electrical demands (**Figure 5-28**). Normal 12-volt cable size is usually 4 or 6 gauge. Various forms of clamps and terminals are used to assure a good electrical connection at each end of the cable. Connections must be clean and tight to prevent arcing, corrosion, and high-voltage resistance.

The positive cable is usually red (but not always), and the negative cable is usually black. The positive cable will fasten to the starter solenoid or relay. The negative cable fastens to ground on the engine block or chassis. Some manufacturers use a negative cable with no insulation. Sometimes the negative battery cable may have a body grounding wire to help assure that the vehicle body is properly grounded.

 A BIT OF HISTORY

The storage battery on early automobiles was mounted under the car. It wasn't until 1937 that the battery was located under the hood for better accessibility. Today, with the increased use of maintenance-free batteries, some manufacturers have "buried" the battery again. For example, to access the battery on some vehicles, you must remove the left front wheel and work through the wheel well. Also, some batteries are now located in the trunk area.

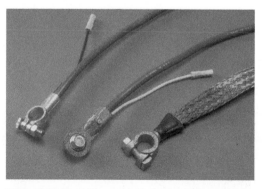

Figure 5-28 The battery cable is designed to carry the high current required to start the engine and supply the vehicle's electrical systems.

AUTHOR'S NOTE It is important to properly identify the positive and negative cables when servicing, charging, or jumping the battery. Do not rely on the color of the cable for this identification; use the markings on the battery case.

AUTHOR'S NOTE Pinch on battery cable clamps is a temporary repair only!

BATTERY HOLDDOWNS

All batteries must be secured in the vehicle to prevent damage and the possibility of shorting across the terminals if the battery tips. Normal vibrations cause the plates to shed their active materials. **Holddowns** reduce the amount of vibration and help increase the life of the battery (**Figure 5-29**).

In addition to holddowns, many vehicles may have a heat shield surrounding the battery (**Figure 5-30**). This heat shield is usually made of plastic and prevents underhood temperatures from damaging the battery.

Shop Manual
Chapter 5,
pages 214, 221

AUTHOR'S NOTE It is important that all holddowns and heat shields be installed to prevent early battery failure.

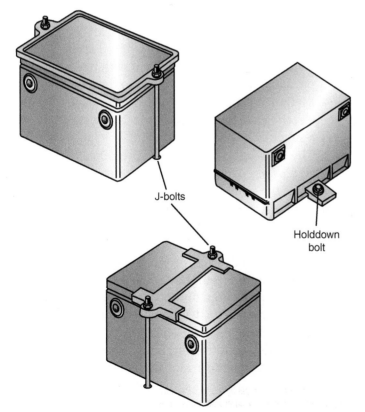

Figure 5-29 Different types of battery holddowns.

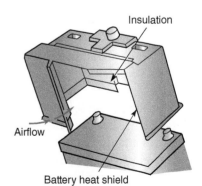

Figure 5-30 Some vehicles are equipped with a heat shield to protect the battery from excessive heat.

SUMMARY

- An automotive battery is an electrochemical device that provides for and stores electrical energy.
- Electrical energy is produced in the battery by the chemical reaction that occurs between two dissimilar plates that are immersed in an electrolyte solution.
- An automotive battery has the following important functions:

    1. It operates the starting motor, ignition system, electronic fuel injection, and other electrical devices for the engine during cranking and starting.
    2. It supplies all the electrical power for the vehicle accessories whenever the engine is not running or is at low idle.
    3. It furnishes current for a limited time whenever electrical demands exceed charging system output.
    4. It acts as a stabilizer of voltage for the entire automotive electrical system.
    5. It stores energy for extended periods of time.

- Electrical loads that are still placed on the battery when the ignition switch is in the OFF position are called key-off or parasitic loads.
- The amount of electrical energy that a battery is capable of producing depends on the size, weight, and active area of the plates and the specific gravity of the electrolyte solution.
- The conventional battery is constructed of seven basic components:

    1. Positive plates.
    2. Negative plates.
    3. Separators.
    4. Case.
    5. Plate straps.
    6. Electrolyte.
    7. Terminals.

- Electrolyte solution used in automotive batteries consists of 64% water and 36% sulfuric acid by weight.
- The electrolyte solution causes chemical actions to take place between the lead dioxide of the positive plates and the sponge lead of the negative plates. The electrolyte is also the carrier that moves electric current between the positive and negative plates through the separators.
- The automotive battery has a fully charged specific gravity of 1.265 corrected to 80°F.

- Grid growth is a condition where the grid grows little metallic fingers that extend through the separators and short out the plates.
- Deep cycling is discharging the battery almost completely before recharging it.
- In a conventional battery, the positive plate is covered with lead dioxide and the negative plate is covered with sponge lead.
- In maintenance-free batteries, the cell plates contain calcium, cadmium, or strontium to reduce gassing and self-discharge.
- The grid construction of the hybrid battery consists of approximately 2.75% antimony alloy on the positive plates and a calcium alloy on the negative plates.
- The recombination battery uses separators that hold a gel-type material in place of liquid electrolyte.
- Absorbed glass mat (AGM) batteries hold their electrolyte in a moistened fiberglass matting that is sandwiched between the battery's high-purity lead plates. Separation of the plates is done by acid-permeated vitreous separators that act as sponges to absorb acid.
- Within a valve-regulated lead-acid (VRLA) battery, the oxygen produced on the positive plates is absorbed by the negative plate, causing a decrease in the amount of hydrogen produced at the negative plate and combining it with the oxygen to produce water that is returned to the electrolyte.
- Methods that are being used and developed for the electrical architecture of the 42-volt system include a single 42-volt system or a dual-voltage system.
- The dual-voltage system may use a dual generator system where one generator operates at 42 volts, while the other operates at 14 volts.
- A dual-stator, dual-voltage system produces dual voltage from a single alternator that has two output voltages.
- A DC/DC converter is configured to provide a 14-volt output from the 42-volt input. The 14-volt output can be used to supply electrical energy to those components that do not require 42 volts.
- Cell construction of the NiCad battery consists of the cathode (positive) electrode made of fiber mesh covered with nickel hydroxide, while the anode (negative) electrode is a fiber mesh that is covered with cadmium. The electrolyte is aqueous potassium hydroxide (KOH).

- During discharge, ions travel from the anode, through the KOH, and on to the cathode. During charging, the opposite occurs.
- The cathode electrode of the NiMH battery is a fiber mesh that contains nickel hydroxide. The anode electrode is made of hydrogen-absorbing metal alloys. The cathode and anode electrodes are separated by a sheet of fine fibers saturated with an aqueous and alkaline electrolyte, KOH.
- Under load, the cell discharges and the hydrogen moves from the anode to the cathode electrode. Since the electrolyte supports only the ion movement from one electrode to the other, it has no active role in the chemical reaction, and the electrolyte level does not change.
- A 300-volt NiMH battery is constructed of 240 cells that produce 1.2 volts each. The cells are made into a module, with each module having six cells. The modules are connected in series to create the total voltage.
- A service disconnect in the HV battery is used to disable the HV system if repairs or service to any part of the system is required. This service connector provides two functions that are used to separate the HV battery pack into two separate batteries, with approximately 150 volts each.
- Contactors are heavy-duty relays that are connected to the positive and negative sides of the HV battery.
- The contactors are normally open and require a 12-volt supply to keep them closed.

- Ultra-capacitors are capacitors constructed to have a large electrode surface area and a very small distance between the electrodes.
- Ultra-capacitors are used in many present-day hybrid vehicles and in some experimental fuel cell electric vehicles because of their ability to quickly discharge high voltages and then be quickly recharged.
- Hybrids that use regenerative braking, a starter/generator with the stop/start feature, and the 42-volt system will use ultra-capacitors to restart the engine.
- The three most common types of battery terminals are:

  1. Post or top terminals: Used on most automotive batteries. The positive post will be larger than the negative post to prevent connecting the battery in reverse polarity.
  2. Side terminals: Positioned in the side of the container near the top. These terminals are threaded and require a special bolt to connect the cables. Polarity identification is by positive and negative symbols.
  3. L terminals: Used on specialty batteries and some imports.

- The most common methods of battery rating are cold cranking, cranking amps, reserve capacity, and ampere-hour.

REVIEW QUESTIONS

Short-Answer Essays

1. Explain the purposes of the battery.
2. Describe how a technician can determine the correct battery to be installed into a vehicle.
3. Describe the methods used to rate batteries.
4. Describe the innovations resulting from the development of the 42-volt system.
5. Explain the effects that temperature has on battery performance.
6. Describe the different loads or demands that are placed on a battery during different operating conditions.
7. List and describe the seven main elements of the conventional battery.
8. What is the purpose of the service disconnect on a HV battery?

9. Describe the process a battery undergoes during charging.
10. Describe the difference in construction of the hybrid battery as compared to the conventional battery.

Fill in the Blanks

1. An automotive battery is an _____ device capable of storing _____ energy that can be converted to electrical energy.

2. When discharging the battery, it changes _____ energy into _____ energy.

3. The assembly of the positive plates, negative plates, and separators is called the _____ _____.

4. The electrolyte solution used in automotive batteries consists of _____ % water and _____ % sulfuric acid.

5. A fully charged automotive battery has a specific gravity of _____ corrected to 80°F (26.7°C).

6. _____ _____ is a condition where the grid grows metallic fingers that extend through the separators and short out the plates.

7. The _____ _____ rating indicates the battery's ability to deliver a specified amount of current to start an engine at low ambient temperatures.

8. The electrolyte solution causes the chemical actions to take place between the lead dioxide of the _____ plates and the sponge lead of the _____ plates.

9. Some of the aspects that determine the battery rating required for a vehicle include engine _____, engine _____, _____ conditions, and vehicle _____.

10. Electrical loads that are still present when the ignition switch is in the OFF position are called _____ loads.

Multiple Choice

1. *Technician A* says the battery provides electricity by releasing free electrons.
   Technician B says the battery stores energy in chemical form.
   Who is correct?
   A. A only
   B. B only
   C. Both A and B
   D. Neither A nor B

2. *Technician A* says the largest demand on the battery is when it must supply current to operate the starter motor.
   Technician B says the current requirements of a starter motor may be over 100 amps.
   Who is correct?
   A. A only
   B. B only
   C. Both A and B
   D. Neither A nor B

3. Which of the following statements about NiMH cells is NOT true?
   A. When the NiMH cell discharges, hydrogen moves from the anode to the cathode electrode.
   B. Nickel-metal hydride batteries have an anode electrode that contains nickel hydroxide.
   C. The alkaline electrolyte has no active role in the chemical reaction.
   D. The plates are separated by a sheet of fine fibers saturated with an aqueous and alkaline electrolyte.

4. The current capacity rating of the battery is being discussed.
   Technician A says the amount of electrical energy that a battery is capable of producing depends on the size, weight, and active area of the plates.
   Technician B says the current capacity rating of the battery depends on the types of materials used in the construction of the battery.
   Who is correct?
   A. A only
   B. B only
   C. Both A and B
   D. Neither A nor B

5. The construction of the battery is being discussed.
   Technician A says the 12-volt battery consists of positive and negative plates connected in parallel.
   Technician B says the 12-volt battery consists of six cells wired in series.
   Who is correct?
   A. A only
   B. B only
   C. Both A and B
   D. Neither A nor B

6. Which of the following statements about battery ratings is true?
   A. The ampere-hour rating is defined as the amount of steady current that a fully charged battery can supply for 1 hour at 80°F (26.7°C) without the cell voltage falling below a predetermined voltage.
   B. The cold cranking amps rating represents the number of amps that a fully charged battery can deliver at 0°F (−17.7°C) for 30 seconds while maintaining a voltage above 9.6 volts for a 12-volt battery.

C. The cranking amp rating expresses the number of amperes a battery can deliver at 32°F (0°C) for 30 seconds and maintain at least 1.2 volts per cell.

D. The reserve-capacity rating expresses the number of amperes a fully charged battery at 80°F can supply before the battery's voltage falls below 10.5 volts.

7. Battery terminology is being discussed.

Technician A says grid growth is a condition where the grid grows little metallic fingers that extend through the separators and short out the plates.

Technician B says deep cycling is discharging the battery almost completely before recharging it.

Who is correct?

A. A only

B. B only

C. Both A and B

D. Neither A nor B

8. Battery rating methods are being discussed.

Technician A says the ampere-hour is determined by the load in amperes a battery is able to deliver for 30 seconds at 0°F (−17.7°C) without terminal voltage falling below 7.2 volts for a 12-volt battery.

Technician B says the cold cranking rating is the amount of steady current that a fully charged battery can supply for 20 hours at 80°F (26.7°C) without battery voltage falling below 10.5 volts.

Who is correct?

A. A only

B. B only

C. Both A and B

D. Neither A nor B

9. The hybrid battery is being discussed.

Technician A says the hybrid battery can withstand at least six deep cycles and still retain 100% of its original reserve capacity.

Technician B says the grid construction of the hybrid battery consists of approximately 2.75% antimony alloy on the positive plates and a calcium alloy on the negative plates.

Who is correct?

A. A only

B. B only

C. Both A and B

D. Neither A nor B

10. *Technician A* says battery polarity must be observed when connecting the battery cables.

Technician B says the battery must be secured in the vehicle to prevent internal damage and the possibility of shorting across the terminals if it tips.

Who is correct?

A. A only

B. B only

C. Both A and B

D. Neither A nor B

CHAPTER 6

STARTING SYSTEMS AND MOTOR DESIGNS

Upon completion and review of this chapter, you should be able to understand and describe:

- The purpose of the starting system.
- The components of the starting system.
- The principle of operation of the DC motor.
- The purpose and operation of the armature.
- The purpose and operation of the field coil.
- The differences between the types of magnetic switches used.
- The differences between starter drive mechanisms.

- The differences between the positive-engagement and solenoid shift starter.
- The operation and features of the permanent magnet starter.
- The principles of operation of the three-phase AC motor.
- The purpose of the inverter module.
- The operating principles of integrated starter generator (ISG) systems.
- The operation of the stop/start system components.

Terms To Know

Amortisseur winding	Field coils	Ratio
Armature	Hold-in windings	Rotating magnetic field
Belt alternator starter (BAS)	Induction motor	Sentry key
Brushes	Integrated starter generator (ISG)	Shunt
Commutator		Slip
Compound motor	Laminated construction	Start clutch interlock switch
Counter electromotive force (CEMF)	Overrunning clutch	Starter drive
	Permanent magnet gear reduction (PMGR)	Static neutral point
Double-start override		Stop/start
Drive coil	Pole shoes	Synchronous motor
Eddy currents	Pull-in windings	Synchronous speed

INTRODUCTION

The internal combustion engine must be rotated before it will run under its own power. The starting system is a combination of mechanical and electrical parts that work together to start the engine. The starting system is designed to change the energy stored in the battery into mechanical energy. To accomplish this conversion, a starter or cranking motor is used. The starting system includes the following components:

Shop Manual
Chapter 6, page 265

1. Battery.
2. Cable and wires.

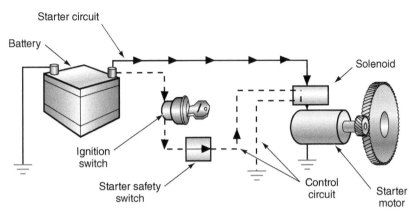

Figure 6-1 Major components of the starting system. The solid line represents the starting (cranking) circuit and the dashed line indicates the starter control circuit.

3. Ignition switch.
4. Starter solenoid or relay.
5. Starter motor.
6. Starter drive and flywheel ring gear.
7. Starter safety switch.

Components in a simplified cranking system circuit are shown in **Figure 6-1**. This chapter examines both this circuit and the fundamentals of electric motor operation.

DIRECT-CURRENT MOTOR PRINCIPLES

DC motors use the interaction of magnetic fields to convert electrical energy into mechanical energy. Magnetic lines of force flow from the north pole to the south pole of a magnet (**Figure 6-2**). If a current-carrying conductor is placed within the magnetic field, two fields will be present (**Figure 6-3**). On the left side of the conductor, the lines of force are in the same direction. This will concentrate the flux density of the lines of force on the left side. This will produce a strong magnetic field because the two fields will reinforce each other. The lines of force oppose each other on the right side of the conductor. This results in a weaker magnetic field. The conductor will tend to move from the strong field to the weak field (**Figure 6-4**). This principle of electromagnetism is used to convert electrical energy into mechanical energy in a starter motor.

Shop Manual
Chapter 6, page 270

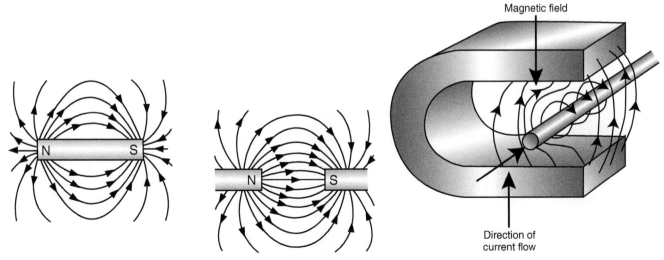

Figure 6-2 Magnetic lines of force flow from the north pole to the south pole.

Figure 6-3 Interaction of two magnetic fields.

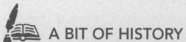

A BIT OF HISTORY

In the early days of the automobile, vehicles did not have a starter motor. The operator had to use a starting crank to turn the engine by hand. Charles F. Kettering invented the first electric self-starter, which was developed and built by the Delco Electrical Plant. The self-starter first appeared on the 1912 Cadillac and was actually a combination starter and generator.

A simple electromagnet-style starter motor is shown in **Figure 6-5**. The inside windings are called the **armature**. The armature is the moveable component of the motor that consists of a conductor wound around a laminated iron core. It is used to create a magnetic field. The armature rotates within the stationary outside windings, called the **field coils**, which has windings coiled around **pole shoes** (**Figure 6-6**). Field coils are heavy copper wire wrapped around an iron core to form an electromagnet. Pole shoes are made of high–magnetic permeability material to help concentrate and direct the lines of force in the field assembly.

When current is applied to the field coils and the armature, both produce magnetic flux lines (**Figure 6-7**). The direction of the windings will place the left pole at a south polarity and the right side at a north polarity. The lines of force move from north to south in the field. In the armature, the flux lines circle in one direction on one side of the loop and in the opposite direction on the other side. Current will now set up a magnetic field

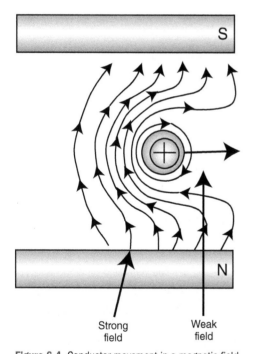

Figure 6-4 Conductor movement in a magnetic field.

Figure 6-5 Simple electromagnetic motor.

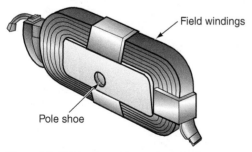

Figure 6-6 Field coil wound around a pole shoe.

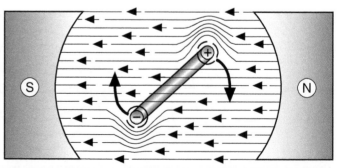

Figure 6-7 Rotation of the conductor is in the direction of the weaker field.

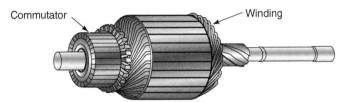

Figure 6-8 Starter armature.

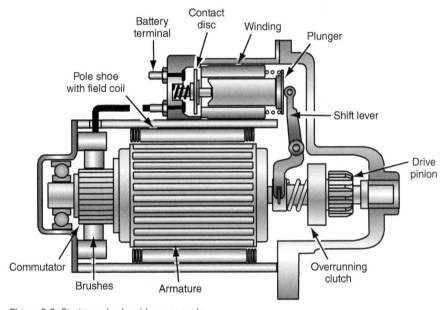

Figure 6-9 Starter and solenoid components.

around the loop of wire, which will interact with the north and south fields and put a turning force on the loop. This force will cause the loop to turn in the direction of the weaker field. However, the armature is limited in how far it is able to turn. When the armature is halfway between the shoe poles, the fields balance one another. The point at which the fields are balanced is referred to as the **static neutral point**.

For the armature to continue rotating, the current flow in the loop must be reversed. To accomplish this, a split-ring **commutator** is in contact with the ends of the armature loops. The commutator is a series of conducting segments located around one end of the armature. Current enters and exits the armature through a set of **brushes** that slide over the commutator's sections. Brushes are electrically conductive sliding contacts, usually made of copper and carbon. As the brushes pass over one section of the commutator to another, the current flow in the armature is reversed. The position of the magnetic fields is the same. However, the direction of current flow through the loop has been reversed. This will continue until the current flow is turned off.

A single-loop motor would not produce enough torque to rotate an engine. Power can be increased by the addition of more loops or pole shoes. An armature with its many windings, with each loop attached to corresponding commutator sections, is shown in **Figure 6-8**. In a typical starter motor (**Figure 6-9**), there are four brushes placed 90° apart from each other that make the electrical connections to the commutator. Two of the brushes are grounded to the starter motor frame and two are insulated from the frame. Also, the armature is supported by bushings or bearings at both ends.

Armature

The armature is constructed with a laminated core made of several thin iron stampings that are placed next to each other (**Figure 6-10**). **Laminated construction** is used because, in a solid iron core, the magnetic fields would generate **eddy currents**.

Shop Manual
Chapter 6, page 288

Eddy currents are closed loops of electrical current that is created within a conductor by a changing magnetic field in the conductor. Eddy currents flow in planes that are perpendicular to the magnetic field, creating a magnetic field that opposes the magnetic field that created it. The eddy current reacts back onto the source of the magnetic field.

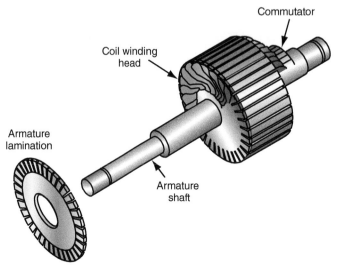

Figure 6-10 Lamination construction of a typical motor armature.

Shop Manual
Chapter 6,
pages 286, 288

These are counter voltages induced in a core. They cause heat to build up in the core and waste energy. By using laminated construction, eddy currents in the core are minimized.

The slots on the outside diameter of the laminations hold the armature windings. The windings loop around the core and are connected to the commutator. Each commutator segment is insulated from the adjacent segments. A typical armature can have more than 30 commutator segments.

A steel shaft is fitted into the center hole of the core laminations. The commutator is insulated from the shaft.

Two basic winding patterns are used in the armature: lap winding and wave winding. In the lap winding, the two ends of the winding are connected to adjacent commutator segments (**Figure 6-11**). In this pattern, the wires passing under a pole field have their current flowing in the same direction.

In the wave-winding pattern, each end of the winding connects to commutator segments that are 90° or 180° apart (**Figure 6-12**). In this pattern design, some windings will have no current flow at certain positions of armature rotation. This occurs because the segment ends of the winding loop are in contact with brushes that have the same polarity. The wave-winding pattern is the most commonly used due to its lower resistance.

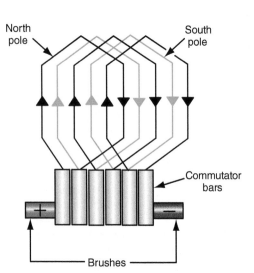

Figure 6-11 Lap winding diagram.

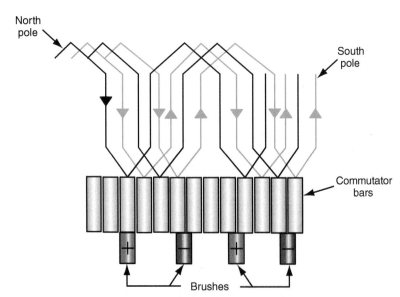

Figure 6-12 Wave-wound armature.

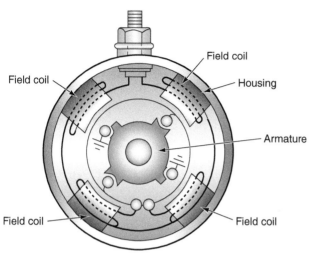

Figure 6-13 Field coils mounted to the inside of starter housing.

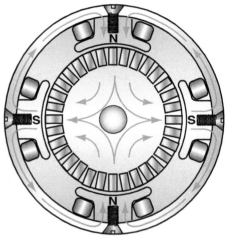

Figure 6-14 Magnetic fields in a four-pole starter motor.

Field Coils

The field coils are electromagnets constructed of wire ribbons or coils wound around a pole shoe. The pole shoes are constructed of heavy iron. The field coils are attached to the inside of the starter housing (**Figure 6-13**). Most starter motors use four field coils. The iron pole shoes and the iron starter housing work together to increase and concentrate the field strength of the field coils (**Figure 6-14**).

When current flows through the field coils, strong stationary electromagnetic fields are created. The fields have a north and south magnetic polarity based on the direction the windings are wound around the pole shoes. The polarity of the field coils alternates to produce opposing magnetic fields.

In any DC motor, there are three methods of connecting the field coils to the armature: series, parallel (shunt), and a compound connection that uses both series and shunt coils.

Shop Manual
Chapter 6, page 286

DC MOTOR FIELD WINDING DESIGNS

The field windings and armature of the DC motor can be wired in various ways. The motor design is referenced by the method these two components are wired together. In addition, many motors are using permanent magnet fields. Also, many newer motors are designed to be brushless.

Series-Wound Motors

Most starter motors are series wound with current flowing first to the field windings, then to the brushes, through the commutator and the armature winding contacting the brushes at that time, and then through the grounded brushes back to the battery source (**Figure 6-15**). This design permits all of the current that passes through the field coils to also pass through the armature.

A series-wound motor will develop its maximum torque output at the time of initial start. As the motor speed increases, the torque output of the motor will decrease. This decrease of torque output is the result of **counter electromotive force (CEMF)** caused by self-induction. Since a starter motor has a wire loop rotating within a magnetic field, it will generate an electrical voltage as it spins. This induced voltage will be opposite the battery voltage that is pushing the current through the starter motor. The faster the armature spins, the greater the amount of induced voltage that is generated. This results in less

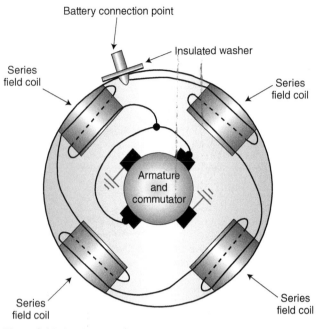

Figure 6-15 A series-wound starter motor.

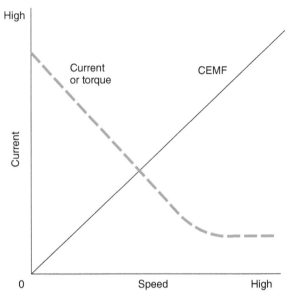

Figure 6-16 Graph illustrating the relationship between CEMF, starter motor speed, and current draw. As speed increases, so does CEMF, reducing current draw and torque.

current flow through the starter from the battery as the armature spins faster. **Figure 6-16** shows the relationship between starter motor speed and CEMF. Notice that at 0 (zero) rpm, CEMF is also at 0 (zero). At this time, maximum current flow from the battery through the starter motor will be possible. As the motor spins faster, CEMF increases and current decreases. Since current decreases, the amount of rotating force (torque) also decreases.

Shunt-Wound Motors

Electric motors, or **shunt** motors, have the field windings wired in parallel across the armature (**Figure 6-17**). Shunt means there is more than one path for current to flow.

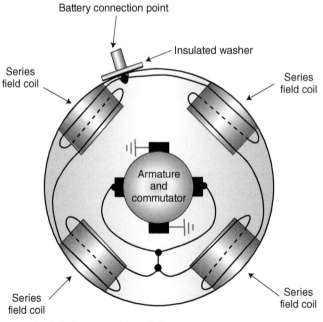

Figure 6-17 A shunt-wound (parallel) starter motor.

A shunt-wound field is used to limit the speed that the motor can turn. A shunt motor does not decrease in its torque output as speeds increase. This is because the CEMF produced in the armature does not decrease the field coil strength. Due to a shunt motor's inability to produce high torque, it is not typically used as a starter motor. However, shunt motors may be found as wiper motors, power window motors, power seat motors, and so on.

Compound Motors

In a **compound motor**, most of the field coils are connected to the armature in series, and one field coil is connected in parallel with the battery and the armature (**Figure 6-18**). This configuration allows the compound motor to develop good starting torque and constant operating speeds. The field coil that is shunt wound is used to limit the speed of the starter motor. Also, on Ford's positive-engagement starters, the shunt coil is used to engage the starter drive. This is possible because the shunt coil is energized as soon as battery voltage is sent to the starter.

Permanent Magnet Motors

Most current vehicles have starter motors that use permanent magnets in place of the field coils (**Figure 6-19**). These motors are also used in many different applications. When a permanent magnet is used instead of coils, there is no field circuit in the motor. By eliminating this circuit, potential electrical problems are also eliminated, such as field-to-housing shorts. Another advantage to using permanent magnets is weight savings; the weight of a typical starter motor is reduced by 50%. Most permanent magnet starters are gear-reduction–type starters.

Multiple permanent magnets are positioned in the housing around the armature. These permanent magnets are an alloy of boron, neodymium, and iron. The field strength of these magnets is much greater than typical permanent magnets. The operation of these motors is the same as other electric motors, except there is no field circuit or windings.

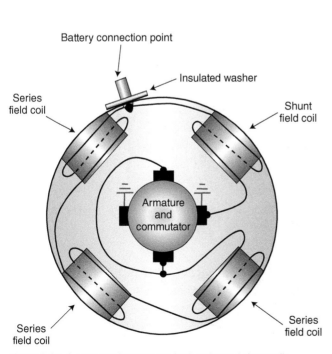

Figure 6-18 A compound motor uses both series and shunt coils.

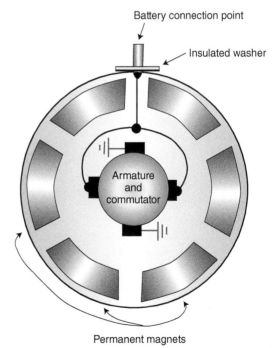

Figure 6-19 A permanent magnet motor has only an armature circuit, as the field is created by strong permanent magnets.

Brushless Motors

The brushless motor uses a permanent magnet rotor and electromagnet field windings (**Figure 6-20**). Since the motor design is brushless, the potential for arcing is decreased and longer service life is expected. In addition, arcing can cause electromagnetic interference that can adversely affect electronic systems. High-output brushless DC motors are used in some hybrid electric vehicle–drive vehicles (**Figure 6-21**).

Control of the stator is by an electronic circuit that switches the current flow as needed to keep the rotor turning. Power transistors that are wired as "H" gates reverse

The field windings of a brushless motor are also called the stator.

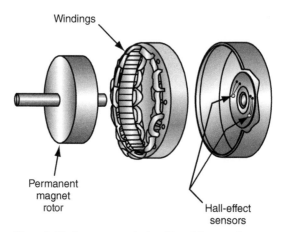

Figure 6-20 Components of a brushless DC motor. The Hall-effect sensor is used to determine rotor position.

Figure 6-21 Brushless motor used by Honda in some of its HEVs.

current flow according to the position of the rotor. Motor speed can be controlled by pulse-width modulation (PWM) of the driver circuits. Rotor position is usually monitored by the use of Hall-effect sensors. However, rotor position can also be determined by monitoring the CEMF that is present in stator windings that are not energized.

STARTER DRIVES

The **starter drive** is the part of the starter motor that engages the armature to the engine flywheel ring gear. A starter drive includes a pinion gear set that meshes with the flywheel ring gear on the engine's crankshaft (**Figure 6-22**). To prevent damage to the pinion gear or the ring gear, the pinion gear must mesh with the ring gear before the starter motor rotates. To help assure smooth engagement, the ends of the pinion gear teeth are tapered (**Figure 6-23**). Also, the action of the armature must always be from the motor to the engine. The engine must not be allowed to spin the armature. The **ratio** of the number of teeth on the ring gear and the starter drive pinion gear is usually between 15:1 and 20:1. This means the starter motor is rotating 15 to 20 times faster than the engine. The ratio of the starter drive is determined by dividing the number of teeth on the drive gear (pinion gear) into the number of teeth on the driven gear (flywheel). Normal cranking speed for the engine is about 200 rpm. If the starter drive had a ratio of 18:1, the starter would be rotating at a speed of 3,600 rpm. If the engine started and was accelerated to 2,000 rpm, the starter speed would increase to 36,000 rpm. This would destroy the starter motor if it was not disengaged from the engine.

The most common type of starter drive is the **overrunning clutch**. The overrunning clutch is a roller-type clutch that transmits torque in one direction and freewheels in the other direction. This allows the starter motor to transmit torque to the ring gear but prevents the ring gear from transferring torque to the starter motor.

In a typical overrunning-type clutch (**Figure 6-24**), the clutch housing is internally splined to the starter armature shaft. The drive pinion turns freely on the armature shaft within the clutch housing. When torque is transmitted through the armature to the clutch housing, the spring-loaded rollers are forced into the small ends of their tapered slots (**Figure 6-25**). They are then wedged tightly against the pinion barrel. The pinion barrel and clutch housing are now locked together; torque is transferred through the starter motor to the ring gear and engine.

Shop Manual
Chapter 6, page 292

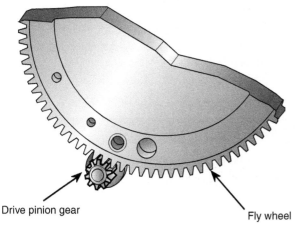

Drive pinion gear

Fly wheel

Figure 6-22 Starter drive pinion gear is used to turn the engine's flywheel.

Figure 6-23 The pinion gear teeth are tapered to allow for smooth engagement.

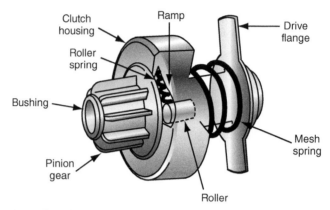

Figure 6-24 Overrunning clutch starter drive.

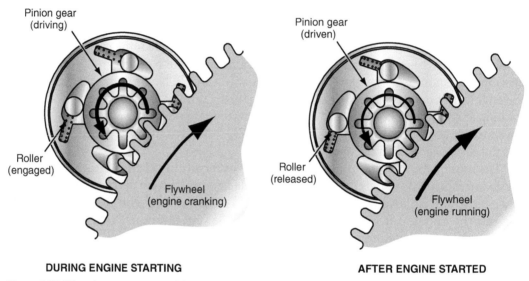

DURING ENGINE STARTING   **AFTER ENGINE STARTED**

Figure 6-25 When the armature turns, it locks the rollers into the tapered notch.

When the engine starts, and is running under its own power, the ring gear attempts to drive the pinion gear faster than the starter motor. This unloads the clutch rollers and releases the pinion gear to rotate freely around the armature shaft.

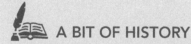 **A BIT OF HISTORY**

The integrated key starter switch was introduced in 1949 by Chrysler. Before this, the key turned the system on and the driver pushed a starter button.

CRANKING MOTOR CIRCUITS

The starting system of the vehicle consists of two circuits: the starter control circuit and the motor feed circuit. These circuits are separate but related. The control circuit consists of the starting portion of the ignition switch, the starting safety switch (if applicable), and the wire conductor to connect these components to the relay or solenoid. The motor feed circuit consists of heavy battery cables from the battery to the relay and the starter or directly to the solenoid if the starter is so equipped.

STARTER CONTROL CIRCUIT COMPONENTS

Magnetic Switches

The starter motor requires large amounts of current [up to 300 amps (A)] to generate the torque needed to turn the engine. The conductors used to carry this amount of current (battery cables) must be large enough to handle the current with very little voltage drop. It would be impractical to place a conductor of this size into the wiring harness to the ignition switch. To provide control of the high current, all starting systems contain some type of magnetic switch. Two basic types of magnetic switches are used: the solenoid and the relay.

Starter-Mounted Solenoids. As discussed in Chapter 3, a solenoid is an electromagnetic device that uses the movement of a plunger to exert a pulling or holding force. In the solenoid-actuated starter system, the solenoid is mounted directly on top of the starter motor (**Figure 6-26**). The solenoid switch on a starter motor performs two functions: it shifts the starter motor pinion gear into mesh with the ring gear. Then it closes the circuit between the battery and the starter motor. This is accomplished by a linkage between the solenoid plunger and the shift lever on the starter motor. In the past, the most common method of energizing the solenoid was directly from the battery through the ignition switch. However, most of today's vehicles use a starter relay in conjunction with a solenoid. The relay is used to reduce the amount of current flow through the ignition switch and is usually controlled by the powertrain control module (PCM). This system will be discussed later in this chapter.

When the circuit is closed and current flows to the solenoid, current from the battery is directed to the **pull-in windings** and **hold-in windings** (**Figure 6-27**). Because it may require up to 50 amps to create a magnetic force large enough to pull the plunger in, both windings are energized to create a combined magnetic field that pulls the plunger. Once the plunger moves, the current required to hold the plunger reduces. This allows the current that was used to pull the plunger in to be used to rotate the starter motor.

Shop Manual
Chapter 6,
pages 277, 279

The two windings of the solenoid are called the **pull-in windings** and the **hold-in windings**. Their names explain their functions.

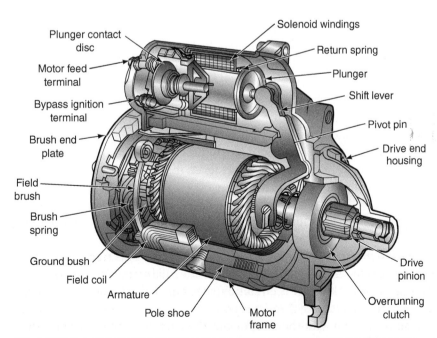

Figure 6-26 Solenoid-operated starter has the solenoid mounted directly on top of the motor.

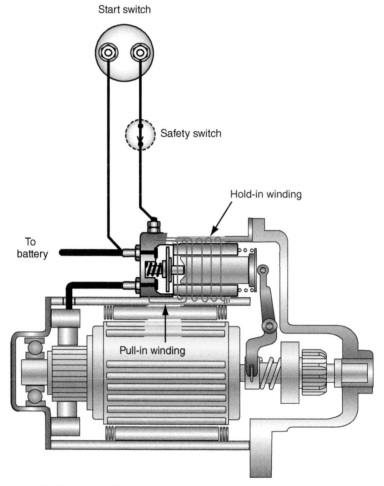

Figure 6-27 The solenoid uses two windings. Both are energized to draw the plunger; then only the hold-in winding is used to hold the plunger in position.

When the ignition switch is placed in the START position, voltage is applied to the S terminal of the solenoid (**Figure 6-28**). The hold-in winding has its own ground to the case of the solenoid. The pull-in winding's ground is through the starter motor. Current will flow through both windings to produce a strong magnetic field. When the plunger is moved into contact with the main battery and motor terminals, the pull-in winding is de-energized. The pull-in winding is not energized because the contact places battery voltage on both sides of the coil (**Figure 6-29**). The current that was directed through the pull-in winding is now sent to the motor.

Because the contact disc does not close the circuit from the battery to the starter motor until the plunger has moved the shift lever, the pinion gear is in full mesh with the flywheel before the armature starts to rotate.

After the engine is started, releasing the key to the RUN position opens the control circuit. Voltage no longer is supplied to the hold-in windings, and the return spring causes the plunger to return to its neutral position.

In Figures 6-28 and 6-29, an R terminal is illustrated. This terminal provides current to the ignition bypass circuit that is used to provide full battery voltage to the ignition coil while the engine is cranking. This circuit bypasses the ballast resistor. The bypass circuit is not used on most ignition systems today.

A common problem with the control circuit is that low system voltage or an open in the hold-in windings will cause an oscillating action to occur. The combination of the

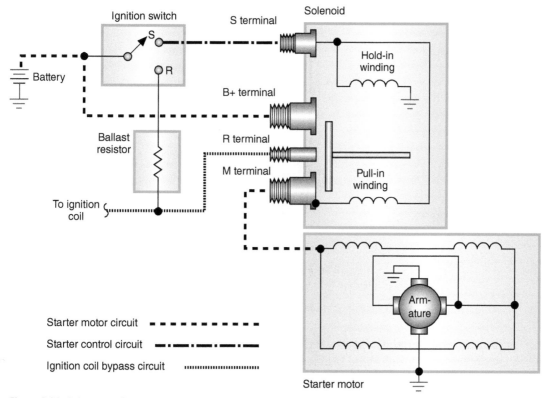

Figure 6-28 Schematic of solenoid-operated starter motor circuit.

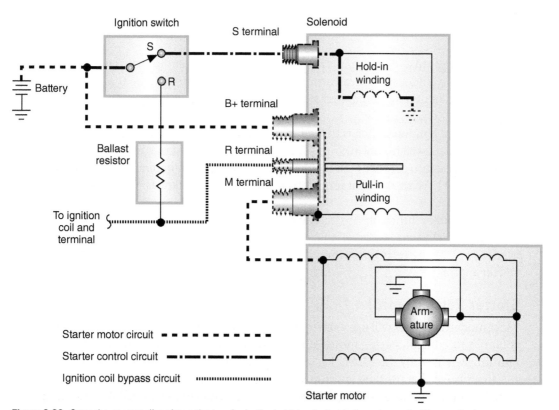

Figure 6-29 Once the contact disc closes the terminals, the hold-in winding is the only one that is energized.

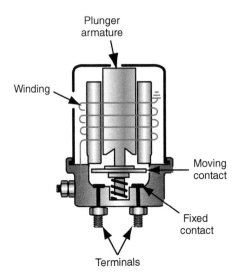

Plunger
armature

Winding

Moving
contact

Fixed
contact

Terminals

Figure 6-30 A remote starter solenoid, often referred to as the starter relay.

pull-in winding and the hold-in winding is sufficient to move the plunger. However, once the contacts are closed, there is insufficient magnetic force to hold the plunger in place. This condition is recognizable by a series of clicks when the ignition switch is turned to the START position. Before replacing the solenoid, check the battery condition; a low battery charge will cause the same symptom.

> **AUTHOR'S NOTE** Some manufacturers use a starter relay in conjunction with a solenoid relay. The relay is used to reduce the amount of current flow through the ignition switch.

Many manufacturers call the remote solenoid the starter relay.

Remote Solenoids. Some manufacturers use a starter solenoid that is mounted near the battery on the fender well or radiator support (**Figure 6-30**). Unlike the starter-mounted solenoid, the remote solenoid does not move the pinion gear into mesh with the flywheel ring gear.

When the ignition switch is turned to the START position, current is supplied through the switch to the solenoid windings. The windings produce a magnetic field that pulls the moveable core into contact with the internal contacts of the battery and starter terminals (**Figure 6-31**). With the contacts closed, full battery current is supplied to the starter motor.

A secondary function of the starter relay is to provide for an alternate path for current to the ignition coil during cranking. This is done by an internal connection that is energized by the relay core when it completes the circuit between the battery and the starter motor.

Starter Relay Controls

Most modern vehicles will use a starter relay in conjunction with a starter motor–mounted solenoid to control starter motor operation. The relay can be controlled through the ignition switch or by the PCM.

In a system that uses the ignition switch to control the relay, the switch will usually be installed on the insulated side of the relay control circuit (**Figure 6-32**). When the ignition switch is turned to the START position, battery voltage is applied to the coil of the relay.

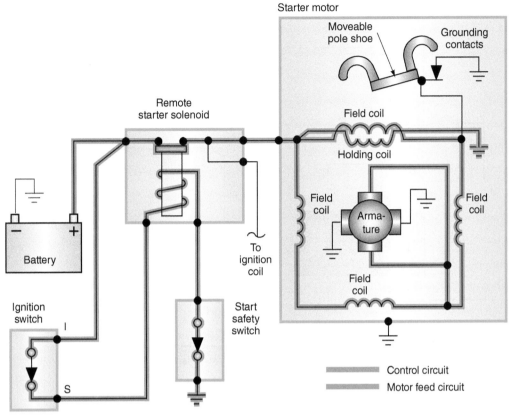

Figure 6-31 Current flow when the remote starter solenoid is energized.

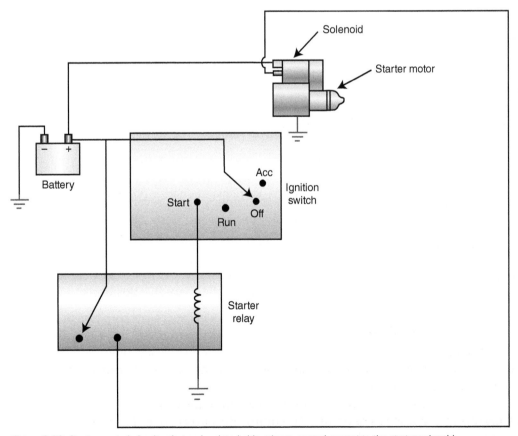

Figure 6-32 Starter control circuit using an insulated side relay to control current to the starter solenoid.

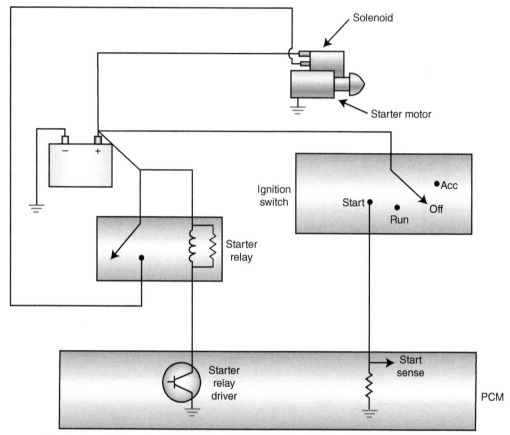

Figure 6-33 Typical PCM starter control circuit.

Sentry key is one of the terms used to describe a sophisticated antitheft system that prevents the engine from starting unless a special key is used.

Since the relay coil is grounded, the coil is energized and pulls the contacts closed. With the contacts closed, battery voltage is applied to the control side of the starter solenoid. The solenoid operates in the same manner as discussed previously.

In this type of system, a very small wire can be used as the ignition switch. This reduces the size of the wiring harness.

In a PCM-controlled system, the PCM will monitor the ignition switch position to determine if the starter motor should be energized. System operation differs among manufacturers. However, in most systems, the PCM will control the starter relay coil ground circuit (**Figure 6-33**). Control by the PCM allows the manufacturer to install software commands such as **double-start override**, which prevents the starter motor from being energized if the engine is already running, and **sentry key** within the PCM.

Ignition Switch

Shop Manual
Chapter 6, page 279

The ignition switch is the power distribution point for most of the vehicle's primary electrical systems (**Figure 6-34**). Most ignition switches have five positions:

1. ACCESSORIES: Supplies current to the vehicle's electrical accessory circuits. It will not supply current to the engine control circuits, starter control circuit, or the ignition system.
2. LOCK: Mechanically locks the steering wheel and transmission gear selector. All electrical contacts in the ignition switch are open. Most ignition switches must be in this position to insert or remove the key from the cylinder.
3. OFF: All circuits controlled by the ignition switch are opened. The steering wheel and transmission gear selector are unlocked.

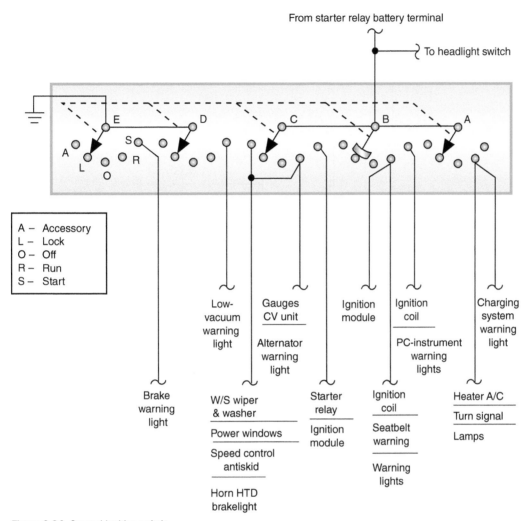

From starter relay battery terminal

To headlight switch

A – Accessory
L – Lock
O – Off
R – Run
S – Start

Low-vacuum warning light

Gauges CV unit

Alternator warning light

Ignition module

Ignition coil

PC-instrument warning lights

Charging system warning light

Brake warning light

W/S wiper & washer

Power windows

Speed control antiskid

Horn HTD brakelight

Starter relay

Ignition module

Ignition coil

Seatbelt warning

Warning lights

Heater A/C

Turn signal

Lamps

Figure 6-34 Ganged ignition switch.

4. ON or RUN: The switch provides current to the ignition, engine controls, and all other circuits controlled by the switch. Some systems will power a chime or light with the key in the ignition switch. Other systems power an antitheft system when the key is removed and turn it off when the key is inserted.

5. START: The switch provides current to the starter control circuit, ignition system, and engine control circuits.

The ignition switch is spring loaded in the START position. This momentary contact automatically moves the contacts to the RUN position when the driver releases the key. All other ignition switch positions are detent positions.

Many manufacturers have moved away from the use of ganged ignition switches with their numerous wires to a multiplexed switch (**Figure 6-35**). The multiplex switch reduces the wires required to determine switch position down to two. The control module then uses a high-side driver to provide other modules their switched battery voltage on the RUN/START circuit. Since the ignition switch is used only as an input, there is no high-current flow through the switch and the wire size can be reduced.

In some instances, the ignition switch is a node on the bus network. In this case, the ignition switch module communicates the switch positions to all other modules that require this information.

Many vehicles are now equipped with starting systems that do not use a conventional ignition switch. These systems use a push button that is an input to a computer. Several

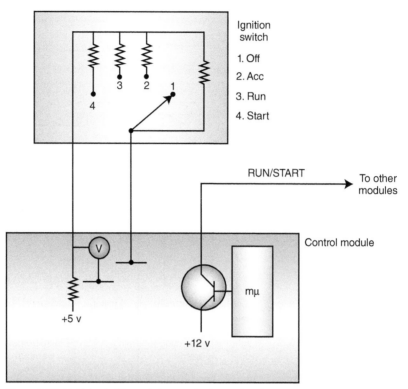

Figure 6-35 Multiplex ignition switches require fewer wires than the ganged switch.

other inputs are involved, so just anyone cannot push the button and take off with the vehicle. This keyless start system, along with remote start, is discussed in Chapter 15 after a thorough discussion of computers, inputs, outputs, and networking is concluded.

Starting Safety Switch

The neutral safety switch is used on vehicles equipped with automatic transmissions. It opens the starter control circuit when the transmission shift selector is in any position except PARK or NEUTRAL. The actual location of the neutral safety switch depends on the kind of transmission and the location of the shift lever. Some manufacturers place the switch in the transmission (**Figure 6-36**).

Vehicles equipped with automatic transmissions require a means of preventing the engine from starting while the transmission is in gear. Without this feature, the vehicle would lunge forward or backward once it is started, causing personal injury or property damage. The normally open neutral safety switch is connected in series in the starting system control

Shop Manual
Chapter 6, page 279

Figure 6-36 The neutral safety switch can be combined with the back-up light switch and installed on the transmission case.

circuit and is usually operated by the shift lever. When in the PARK or NEUTRAL position, the switch is closed, allowing current to flow to the starter circuit. If the transmission is in a gear position, the switch is opened and current cannot flow to the starter circuit.

The neutral safety switch feature can also be a function of the transmission range switch (**Figure 6-37**). The range switch is used by the transmission control module to determine transmission range position. This information is then broadcasted on the network bus to the PCM. If the PCM receives the PARK or NEUTRAL position input, it will allow the starter relay to be energized.

Many vehicles equipped with manual transmissions use a similar type of safety switch. The **start clutch interlock switch** is usually operated by movement of the clutch pedal (**Figure 6-38**). When the clutch pedal is pushed downward, the switch closes and current can flow through the starter circuit. If the clutch pedal is left up, the switch is open and current cannot flow.

Some vehicles use a mechanical linkage that blocks movement of the ignition switch cylinder unless the transmission is in PARK or NEUTRAL (**Figure 6-39**).

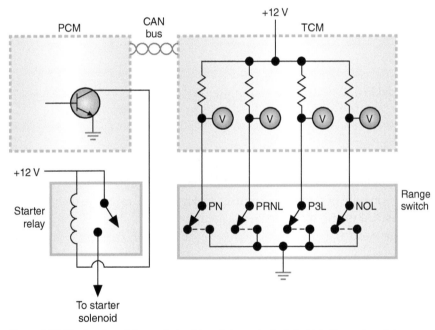

Figure 6-37 The function of the neutral safety switch may be included into the range switch.

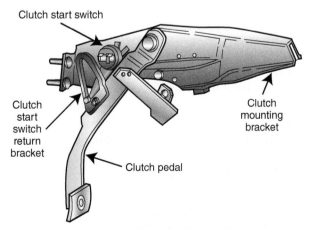

Figure 6-38 Most vehicles with a manual transmission use a clutch start switch to prevent the engine from starting unless the clutch pedal is pressed.

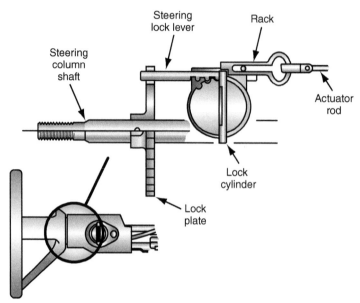

Figure 6-39 Mechanical linkage used to prevent starting the engine while the transmission is in gear.

AUTHOR'S NOTE One-touch and remote starting systems will be discussed in Chapter 14.

CRANKING MOTOR DESIGNS

Shop Manual
Chapter 6, page 284

The most common type of starter motor used today incorporates the overrunning clutch starter drive instead of the old inertia-engagement bendix drive. There are four basic groups of starter motors:

1. Direct drive.
2. Gear reduction.
3. Positive-engagement (moveable pole).
4. Permanent magnet.

Direct Drive Starters

The direct drive starter motor can be either series wound or compound motors.

A common type of starter motor is the solenoid-operated direct drive unit (**Figure 6-40**). Although there are construction differences between applications, the operating principles are the same for all solenoid-shifted starter motors.

When the ignition switch is placed in the START position, the control circuit energizes the pull-in and hold-in windings of the solenoid. The solenoid plunger moves and pivots the shift lever, which in turn locates the drive pinion gear into mesh with the engine flywheel.

When the solenoid plunger is moved all the way, the contact disc closes the circuit from the battery to the starter motor. Current now flows through the field coils and the armature. This develops the magnetic fields that cause the armature to rotate, thus turning the engine.

Gear Reduction Starters

Some gear reduction starter motors are compound motors.

Some manufacturers use a gear reduction starter to provide increased torque (**Figure 6-41**). The gear-reduction starter differs from most other designs, in that the armature does not drive the pinion gear directly. In this design, the armature drives a small gear that is in constant mesh with a larger gear. Depending on the application, the

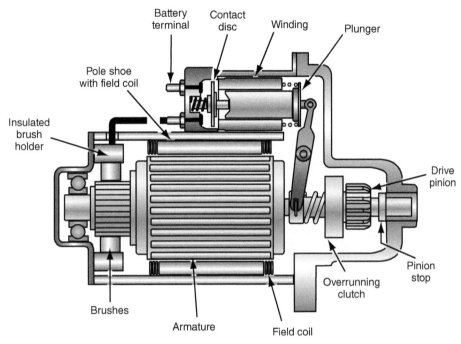

Figure 6-40 Solenoid-operated Delco MT series starter motor.

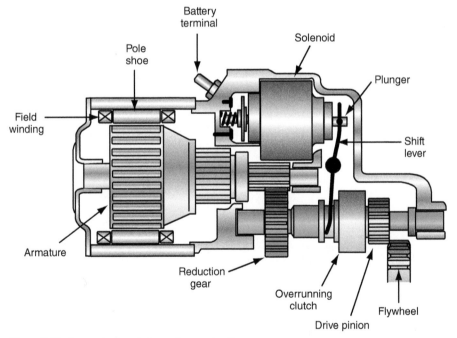

Figure 6-41 Gear reduction starter motor construction.

ratio between these two gears is between 2:1 and 3.5:1. The additional reduction allows for a small motor to turn at a greater torque with less current draw.

The solenoid operation is similar to that of the solenoid-shifted direct drive starter in that the solenoid moves the plunger, which engages the starter drive.

Many gear reduction starters have the commutator and brushes located in the center of the motor.

Positive-Engagement Starters

A commonly used starter on Ford applications in the past was the positive-engagement starter (**Figure 6-42**). Positive-engagement starters use the shunt coil windings of the starter motor to engage the starter drive. The high starting current is controlled by a starter solenoid mounted close to the battery. When the solenoid contacts are closed, current flows

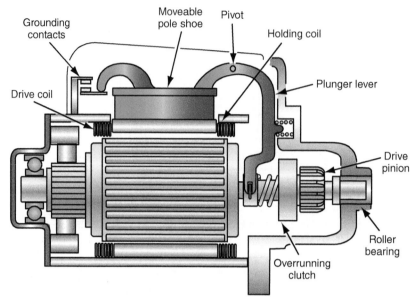

Figure 6-42 Positive-engagement starters use a moveable pole shoe.

Positive-engagement starters are also called moveable-pole shoe starters.

through a drive coil. The drive coil creates an electromagnetic field that attracts a moveable pole shoe. The moveable pole shoe is attached to the starter drive through the plunger lever. When the moveable pole shoe moves, the drive gear engages the engine flywheel.

As soon as the starter drive pinion gear contacts the ring gear, a contact arm on the pole shoe opens a set of normally closed grounding contacts (**Figure 6-43**). With the return to ground circuit opened, all the starter current flows through the remaining three field coils and through the brushes to the armature. The starter motor then begins to rotate. To prevent

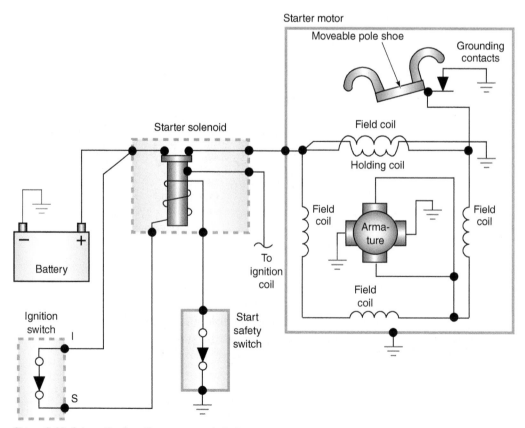

Figure 6-43 Schematic of positive-engagement starter.

the starter drive from disengaging from the ring gear if battery voltage drops while cranking, the moveable pole shoe is held down by a holding coil. The holding coil is a smaller coil inside the main **drive coil** and is strong enough to hold the starter pinion gear engaged.

The **drive coil** is a hollowed field coil that is used to attract the moveable pole shoe.

Permanent Magnet Starters

The **permanent magnet gear reduction (PMGR)** starter design provides for less weight, simpler construction, and less heat generation as compared to conventional field coil starters (**Figure 6-44**). The permanent magnet gear reduction starter uses four or six permanent magnet field assemblies in place of field coils. Because there are no field coils, current is delivered directly to the armature through the commutator and brushes.

Shop Manual
Chapter 6, page 285

The permanent magnet starter also uses gear reduction through a planetary gear set (**Figure 6-45**). The planetary geartrain transmits power between the armature and the pinion shaft. This allows the armature to rotate at increased torque. The planetary gear assembly consists of a sun gear on the end of the armature and three planetary carrier gears inside a ring gear. The ring gear is held stationary. When the armature is rotated, the sun gear causes the carrier gears to rotate about the internal teeth of the ring gear. The planetary carrier is attached to the output shaft. The gear reduction provided for by this gear arrangement is 4.5:1. By providing for this additional gear reduction, the demand for high current is lessened.

The electrical operation between the conventional field coil and PMGR starters remains basically the same (**Figure 6-46**).

> **AUTHOR'S NOTE** The greatest amount of gear reduction from a planetary gear set is accomplished by holding the ring gear, inputting the sun gear, and outputting the carrier.

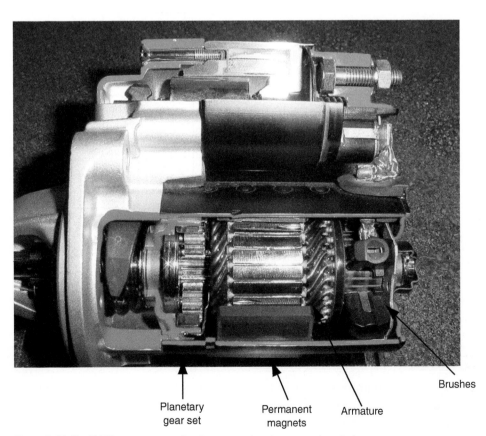

Planetary gear set   Permanent magnets   Armature   Brushes

Figure 6-44 The PMGR motor uses a planetary gear set and permanent magnets.

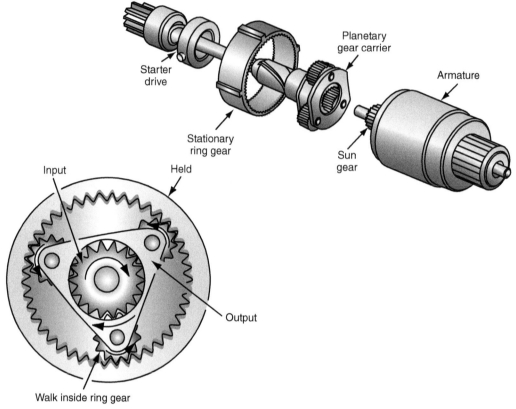

Starter
drive

Planetary
gear carrier

Armature

Stationary
ring gear

Sun
gear

Input

Held

Output

Walk inside ring gear

Figure 6-45 Planetary gear set.

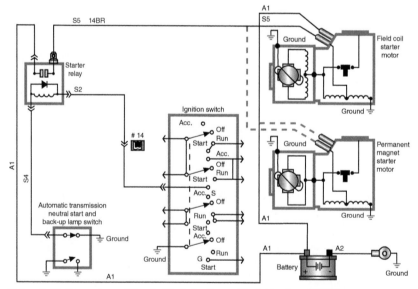

Figure 6-46 Comparison of the electrical circuits used in field coil and PMGR starters.

AC MOTOR PRINCIPLES

A few years ago, the automotive technician did not need to be concerned much about the operating principles of the AC motor. With the increased focus on HEVs and EVs (electric vehicles), this is no longer an option since most of these vehicles use AC motors (**Figure 6-47**).

Figure 6-47 AC three-phase motor used in a HEV.

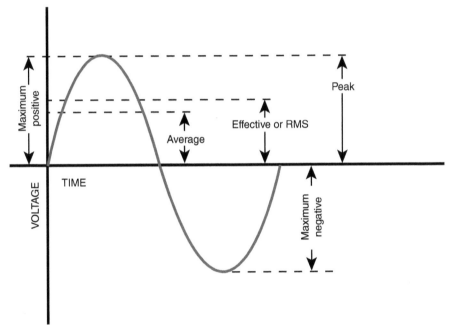

Figure 6-48 AC voltage gradually changes. It is rated at the RMS.

As discussed in Chapter 2, AC voltage has a changing direction of current flow. However, this change does not occur immediately (**Figure 6-48**). Notice that the AC voltage sine wave indicates that in one cycle the voltage will be zero at three different times. Also, notice that as the current changes directions, it gradually builds up or falls in the other direction. The sine wave illustrates that the amount of current in an AC circuit always varies. The current rating is based on the average referred to as a root mean square (RMS) value.

AC Motor Construction

Like the DC motor, the AC motor uses a stator (field winding) and a rotor. Common types of AC motors are the **synchronous motor** and the **induction motor**. In both motor types, the stator comprises individual electromagnets that are either electrically connected to each other or connected in groups. The difference is in the rotor designs. AC motors can use either single-phase or three-phase AC current. Since the three phase is the most common motor used in HEV and EV, we will focus our discussion on these.

A **synchronous motor** operates at a constant speed regardless of load. It generates its own rotor current as the rotor cuts through the magnetic flux lines of the stator field.

AUTHOR'S NOTE Three-phase AC voltage is commonly used in motors because it provides a smoother and more constant supply of power. Three-phase AC voltage is like having three independent AC power sources, which have the same amplitude and frequency but are 120° out of phase with each other.

As in a DC motor, the movement of the rotor is the result of the repulsion and attraction of the magnetic poles. However, the way this works in an AC motor is very different. Because the current is alternating, the polarity in the windings constantly changes. The principle of operation for all three-phase motors is the **rotating magnetic field**. The rotor turns because it is pulled along by a rotating magnetic field in the stator. The stator is stationary and does not physically move. However, the magnetic field does move from pole to pole. There are three factors that cause the magnetic field to rotate (Figure 6-48). The first is the fact that the voltages in a three-phase system are 120° out of phase with each other. The second is the fact that the three voltages change polarity at regular intervals. Finally, the third factor is the arrangement of the stator windings around the inside of the motor.

In **Figure 6-49** the stator is a two-pole, three-phase motor. Two pole means that there are two poles per phase. The motor is wired with three leads: L_1, L_2, and L_3.

AUTHOR'S NOTE The stator of AC motors does not have actual pole pieces, as shown in Figure 6-48. Pole pieces are illustrated to help understand how the rotating magnetic field is created in a three-phase motor.

Each of the poles is wound in such a manner that when current flows through the winding they develop opposite magnetic polarities. All three windings are joined to form a wye connection for the stator. Since each phase reaches its peak at successively later times (**Figure 6-50**), the strongest point of the magnetic field in each winding is also in succession. This succession of the magnetic fields is what creates the effect of the magnetic field continually moving around the stator.

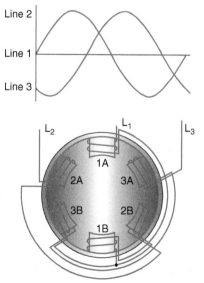

Figure 6-49 The motor stator is energized with the three-phase AC voltage.

TIME

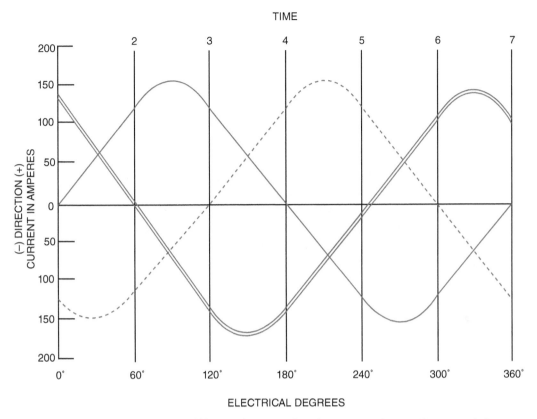

Figure 6-50 The three AC sine waves are 120° apart. At any one time, there are two voltages at the same polarity.

Since the rotating magnetic field will rotate around the stator once for every cycle of the voltage in each phase, the field is rotating at the frequency of the source voltage. Remember that as the magnetic field moves, new magnetic polarities are present. As each polarity change is made, the poles of the rotor are attracted by the opposite poles on the stator. Therefore, as the magnetic field of the stator rotates, the rotor rotates with it. The speed with which the rotor turns depends on the number of windings and poles built into the motor, the frequency of the AC supply voltage, and the load on the rotor's shaft. Frequency modulation (thus motor speed) can be altered by use of controllers.

Synchronous Motors

The speed at which the magnetic field rotates is called the **synchronous speed**. The two main factors determining the synchronous speed of the rotating magnetic field are the number of stator poles (per phase) and the frequency of the applied voltage. A synchronous motor operates at a constant speed, regardless of load. The speed of the rotor is equal to the synchronous speed.

The synchronous motor does not depend on induced current in the rotor to produce torque. The strength of the magnetic field determines the torque output of the rotor, while the speed of the rotor is determined by the frequency of the AC input to the stator.

Synchronous motors cannot be started by applying three-phase AC power to the stator. This is because when the AC voltage is applied to the stator windings, a high-speed rotating magnetic field is present immediately. The rotating magnetic field will pass the rotor so quickly that the rotor does not have time to start turning.

In order to start the motor, the rotor contains a squirrel-cage-type winding made of heavy copper bars connected by copper rings. The squirrel cage is known as the **amortisseur winding**. When voltage is first applied to the stator windings, the resulting rotating magnetic field cuts through the squirrel-cage bars. The cutting action of the field induces a current into the squirrel cage. Since the squirrel cage is shorted, the low voltage that is induced into the squirrel-cage windings results in a relatively large current flow in the cage. This current flow produces a magnetic field within the rotor that is attracted to the rotating magnetic field of the stator. The result is the rotor begins to turn in the direction of rotation of the stator field.

The construction the rotor of a synchronous motor includes wound pole pieces that become electromagnets when DC voltage is applied to them. The excitation current can be applied to the rotor through slip rings or by a brushless exciter. As the rotor is accelerated to a speed of 95% of the speed of the rotating magnetic field, DC voltage is connected to the rotor through the slip rings on the rotor shaft or by a brushless exciter. The application of DC voltage to the rotor windings results in the creation of electromagnets. The electromagnetic field of the rotor is locked in step with the rotating magnetic field of the stator. The rotor will now turn at the same speed as the rotating magnetic field. Since the rotor is turning at the synchronous speed of the field, the cutting action between the stator field and the winding of the squirrel cage has ceased. This stops the induction of current flow in the squirrel cage. The speed of the rotor is locked to the speed of the rotating magnetic field even as different loads are applied.

Induction Motors

An induction motor generates its own rotor current by induced voltage from the rotating magnetic field of the stator. The current is induced in the windings of the rotor as it cuts through the magnetic flux lines of the rotating stator field (**Figure 6-51**). Generally, the rotor windings are in the form of a squirrel cage. However, wound-rotor motors are constructed by winding three separate coils on the rotor 120° apart. The rotor will contain as many poles per phase as the stator winding. These coils are connected to three slip rings located on the rotor shaft so rushes can provide an external connection to the rotor.

> The induction motor is also referred to as an asynchronous motor.

When voltage is first applied to the stator windings, the rotor does not turn. To start the squirrel-cage induction motor, the magnetic field of the stator cuts the rotor bars that induce a voltage into the cage bars. This induced voltage is of the same frequency as the voltage applied to the stator. Since the rotor is stationary, maximum voltage is induced into the squirrel cage and causes current to flow through the cage's bars. The current flow results in the production of a magnetic field around each bar.

The magnetic field of the rotor is attracted to the rotating magnetic field of the stator. The rotor begins to turn in the same direction as the rotating magnetic field. As the rotor increases in speed, the rotating magnetic field cuts the cage bars at a slower rate, resulting

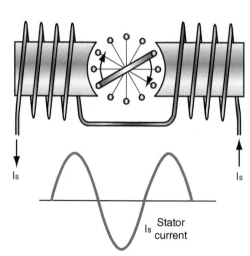

Figure 6-51 Concept of the induction motor.

in less voltage being induced into the rotor. This also results in a reduction of rotor current. With the decrease in rotor current, the stator current also decreases. If the motor is operating without a load, the rotor continues to accelerate until it reaches a speed close to that of the rotating magnetic field. This means that when a squirrel-cage induction motor is first started, it has a current draw several times greater than its normal running current.

If the rotor were to turn at the same speed as the rotating magnetic field, there would be no induced voltage in the rotor and, consequently, no rotor current. This means that an induction motor can never reach synchronous speed. If the motor is operated with no load, the rotor will accelerate until the torque developed is proportional to friction losses. As loads are applied to the motor, additional torque is required to overcome the load. The increase in load causes a reduction in rotor speed. This results in the rotating magnetic field cutting the cage bars at a faster rate. This in turn increases the induced voltage and current in the cage and produces a stronger magnetic field in the rotor; thus, more torque is produced. The increased current flow in the rotor causes increased current flow in the stator. This is why motor current increases as load is added.

The difference between the synchronous speed and actual rotor speed is called **slip**. Slip is directly proportional to the load on the motor. When loads are on the rotor's shaft, the rotor tends to slow and slip increases. The slip then induces more current in the rotor and the rotor turns with more torque, but at a slower speed and therefore produces less CEMF.

In HEVs and EVs, the direction of motor rotation will need to change to meet certain operating requirements. In a three-phase AC motor, the direction of rotation can be changed by simply reversing any two of its stator leads. This causes the direction of the rotating magnetic field to reverse.

An electronic controller is used to manage the flow of electricity from the high-voltage (HV) battery pack to control the speed and direction of rotation of the electric motor(s). The intent of the driver is relayed to the controller by use of an accelerator position sensor. The controller monitors this signal plus other inputs regarding the operating conditions of the vehicle. Based on these inputs, the controller provides a duty cycle control of the voltage levels to the motor(s).

If the HEV or EV uses AC motors, an inverter module is used to convert the DC voltage from the HV battery to a three-phase AC voltage for the motor (**Figure 6-52**). This conversion is done by using sets of power transistors. The transistors modulate the voltage using pulse width while reversing polarity at a fixed frequency (**Figure 6-53**). The inverter module is usually a slave module to the hybrid control processor. Often the inverter module is called the motor control processor since it provides for not only current modification but also motor control.

Figure 6-52 The inverter module controls the speed and direction of the AC motor.

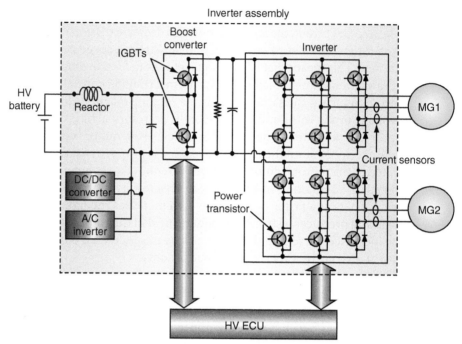

Figure 6-53 Boost and power transistors of the inverter module.

INTEGRATED STARTER GENERATOR

One of the newest technologies to emerge is the **integrated starter generator (ISG)**. Although this system can be used in conventional engine-powered vehicles, one of the key contributors to the hybrid's fuel efficiency is its ability to automatically stop and restart the engine under different operating conditions. A typical hybrid vehicle uses a 14-kW electric induction motor or ISG between the engine and the transmission. The ISG performs many functions such as fast, quiet starting, automatic engine stops/starts to conserve fuel, recharging the vehicle batteries, smoothing driveline surges, and providing regenerative braking.

The ISG is a three-phase AC motor. At low vehicle speeds, the ISG provides power and torque to the vehicle. It also supports the engine when the driver demands more power. During vehicle deceleration, ISG regenerates the power that is used to charge the traction batteries.

The ISG can also convert kinetic energy from AC to DC voltage. When the vehicle is traveling downhill and there is zero load on the engine, the wheels can transfer energy through the transmission and engine to the ISG. The ISG then sends this energy to the HV battery for storage.

An ISG can be mounted externally to the engine and connected to the crankshaft with a drive belt (**Figure 6-54**). This design is called a **belt alternator starter (BAS)**. In these applications, the unit can function as the engine's starter motor as well as a generator driven by the engine.

Both the BAS and the ISG use the same principle to start the engine. Current flows through the stator windings, which generates magnetic fields in the rotor. This will cause the rotor to turn, thus turning the crankshaft and starting the engine. In addition, this same principle is used to assist the engine as needed when the engine is running.

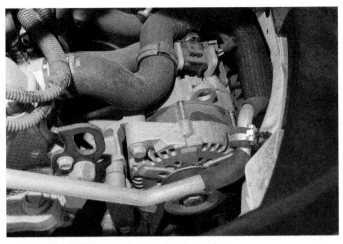

Figure 6-54 A BAS mounted external to the engine.

START/STOP SYSTEMS

Hybrid vehicles utilize the automatic **stop/start** feature to shut off the internal combustion engine (ICE) whenever the vehicle is not moving or when power from the ICE is not required. Conventional vehicles can also benefit from stop/start. Usually, this feature is activated when the vehicle is stopped, no engine power is required, and the driver's foot is on the brake pedal. On manual transmission—equipped vehicles, this feature may be activated when the vehicle is stopped, no ICE power is required, and the clutch pedal is released with the transmission in neutral. Once the driver's foot is removed from the brake pedal (or the clutch pedal is pressed), the starter automatically restarts the ICE in less than one-tenth of a second.

Shop Manual
Chapter 6, page 292

Because conventional vehicles do not have motor generators like an HEV, they may utilize a BAS or an enhanced starter. The enhanced starter is basically a conventional starter motor that has been modified to meet the requirement of multiple restarts. Modifications include dual layer brushes and a unique pinion spring mechanism. This is the least expensive method for adding the stop/start feature.

The enhanced starter is also called an *advanced engagement starter.*

Stop/start systems that use an enhanced starter may incorporate an in-rush current reduction relay (ICR). When the starter motor is initially energized, high current is required to start the rotation. Because voltage drop increases as the amount of current flow increases, during this time there is an increase in the volt drop to the starter motor. The ICR is dual path relay that adds resistance to the starter circuit during initial cranking to reduce the current draw (**Figure 6-55**). Prior to closing of the starter motor solenoid contact, the ICR is energized to open the shorting bar contacts. All current is routed to the starter motor through a resistor bar (approximately 10 ohms). Since the resistance bar is in series with the circuit, the increased resistance reduces the initial in-rush current spike and reduces the voltage drop. The ICR is energized for a very short time (about 185 msec), and it is de-energized. This closes the shorting bar contacts and the current bypasses the resistor bar, allowing full electrical power to the starter motor. Although there is a rebound of the current surge at this time the subsequent voltage drop is not as great as it would be without the ICR.

The tandem solenoid starter uses a co-axial dual solenoid that provides independent control of the starter's pinion gear engagement fork and starter motor rotation (**Figure 6-56**). Solenoid SL1 is used to engage the pinion gear with the flywheel ring

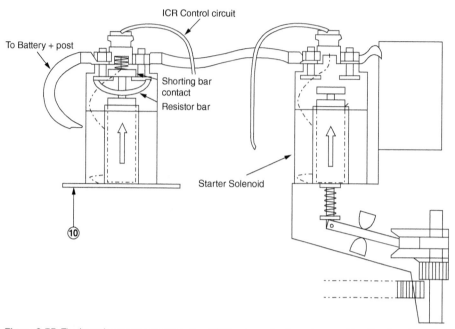

Figure 6-55 The in-rush current reduction relay (ICR) reduces the in-rush current by directing the starter current through a resistor during initial engine cranking.

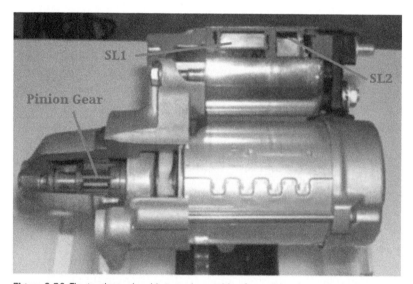

Figure 6-56 The tandem solenoid starter is capable of energizing the starter before engaging the pinion gear.

gear. Solenoid SL2 is used to energize the starter motor. The tandem solenoid allows for engine starting even if the engine has not yet come to a complete stop. This scenario can come into play if the driver has a "change of mind" after the engine has been shut down. When the engine is shutdown at 600 rpms, it takes between 0.5 and 1.5 seconds for the crankshaft to stop rotating. The use of a starter motor that does not allow for engagement to a spinning flywheel requires the engine to come to a complete stop before the starter can be engaged. The tandem solenoid system is capable of first spinning the starter motor, and then engaging the pinion gear. Software controls the timing and synchronization aspects for pinion gear shifting into the spinning flywheel. At higher flywheel speeds the motor is energized first to increase the speed of the pinion gear, and then the pinion gear is shifted forward when the rotation speed of the ring gear and pinion gear match. If the flywheel rpm is slow enough to allow the pinion gear to engage the flywheel, the pinion gear is first moved forward and then the motor is energized.

Another starter design is the permanent engaged (PE) starter. The pinion gear engagement fork is eliminated in the PE starter since the starter motor is mounted to be permanently engaged to the flywheel (**Figure 6-57**). This eliminates the issue of engaging the pinion gear into a rotating flywheel. When the engine is restarted, the motor is simply energized and immediately begins to rotate the crankshaft. The flywheel is fitted with a special clutching mechanism to disconnect it from the engine after the engine starts to prevent continued rotation of the starter motor.

The Mazda i-Stop system is an example of a direct start system that does not use a starter motor for the stop/start function. This system uses direct injection and combustion of the air/fuel mixture to instantly restart the engine. The operating principle of this system is the placement of the pistons into an optimal position during engine shutdown so it can be instantly restarted by injecting fuel into the cylinder (**Figure 6-58**).

The engine is restarted by directly injecting the fuel into the cylinder and then igniting it to create downward piston force. The control module is responsible for identifying and providing precise control over the piston position during engine shutdown. Stopping the pistons when all of them are level with each other provides the correct balance of air volumes in the cylinders and is key to quick restarts.

As the vehicle is coming to a stop, the control module will allow the engine to "pulse" until the cylinder air volumes are balanced. When the engine is stopped, one of the cylinders will be in the combustion stroke. The control module identifies this cylinder and injects fuel directly into it. The atomized fuel is then ignited to allow combustion to take place and forcing the piston to move downward, rotating the crankshaft. At the same time, the starter motor applies a small amount of additional momentum to the crankshaft. As engine speed increases, the cylinders are continuously selected for ignition until the engine reaches its idle speed.

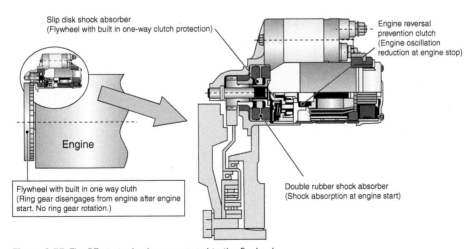

Figure 6-57 The PE starter is always engaged to the flywheel.

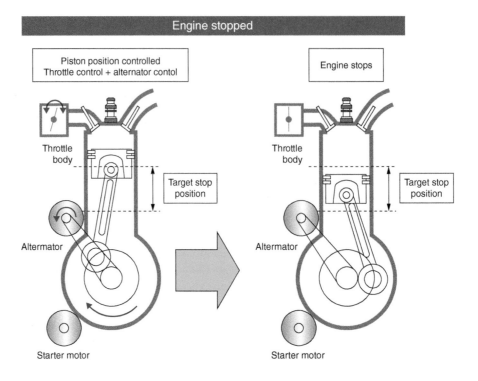

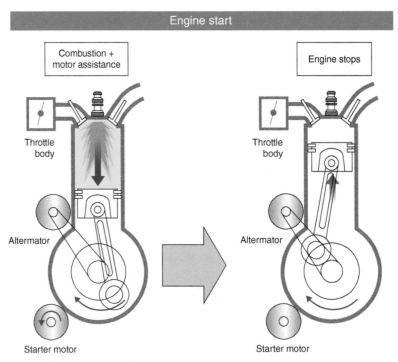

Figure 6-58 Mazda's i-Stop System.

SUMMARY

- The starting system is a combination of mechanical and electrical parts that work together to start the engine.
- The starting system components include the battery, cable and wires, the ignition switch, the starter solenoid or relay, the starter motor, the starter drive and flywheel ring gear, and the starting safety switch.
- The armature is the moveable component of the motor that consists of a conductor wound around a laminated iron core. It is used to create a magnetic field.
- Pole shoes are made of high-magnetic permeability material to help concentrate and direct the lines of force in the field assembly.
- The magnetic forces will cause the armature to turn in the direction of the weaker field.
- Within an electromagnetic style of starter motor, the inside windings are called the armature. The armature rotates within the stationary outside windings, called the field, which has windings coiled around pole shoes.
- The commutator is a series of conducting segments located around one end of the armature.
- A split-ring commutator is in contact with the ends of the armature loops. So, as the brushes pass over one section of the commutator to another, the current flow in the armature is reversed.
- Two basic winding patterns are used in the armature: lap winding and wave winding.
- The field coils are electromagnets constructed of wire coils wound around a pole shoe.
- When current flows through the field coils, strong stationary electromagnetic fields are created.
- In any DC motor, there are three methods of connecting the field coils to the armature: series, parallel (shunt), and a compound connection that uses both series and shunt coils.
- A starter drive includes a pinion gear set that meshes with the engine flywheel ring gear on the engine.
- To prevent damage to the pinion gear or the ring gear, the pinion gear must mesh with the ring gear before the starter motor rotates.
- The bendix drive depends on inertia to provide meshing of the drive pinion with the ring gear.
- The most common type of starter drive is the overrunning clutch. This is a roller-type clutch that transmits torque in one direction and freewheels in the other direction.

- The starting system consists of two circuits called the starter control circuit and the motor feed circuit.
- The components of the control circuit include the starting portion of the ignition switch, the starting safety switch (if applicable), and the wire conductor to connect these components to the relay or solenoid.
- The motor feed circuit consists of heavy battery cables from the battery to the relay and the starter or directly to the solenoid if the starter is so equipped.
- There are four basic groups of starter motors: direct drive, gear reduction, positive engagement (moveable pole), and permanent magnet.
- A synchronous motor operates at a constant speed regardless of load.
- An induction motor generates its own rotor current as the rotor cuts through the magnetic flux lines of the stator field.
- The principle of operation for all three-phase motors is the rotating magnetic field.
- In order to start the synchronous motor, the rotor contains a squirrel-type winding to act as an induction motor.
- Induction motor rotor windings can be in the form of a squirrel cage or constructed by winding three separate coils on the rotor 120° apart.
- The ISG can also convert kinetic energy to storable electric energy. When the vehicle is traveling downhill and there is zero load on the engine, the wheels can transfer energy through the transmission and engine to the ISG. The ISG then sends this energy to the battery for storage and use by the electrical components of the vehicle.
- The belt alternator starter (BAS) is about the same size as a conventional generator and is mounted in the same way.
- The ISG is a three-phase AC motor. At low vehicle speeds, the ISG provides power and torque to the vehicle. It also supports the engine when the driver demands more power.
- Both the BAS and the ISG use the same principle to start the engine. Current flows through the stator windings, which generates magnetic fields in the rotor. This will cause the rotor to turn, thus turning the crankshaft and starting the engine.
- Usually, the stop/start feature is activated when the vehicle is stopped, no engine power is required, and the driver's foot is on the brake pedal. Once the driver's foot is removed from the brake pedal, the starter automatically restarts the ICE.

- The enhanced starter is basically a conventional starter motor that has been modified to meet the requirement of multiple restarts. Modifications include dual layer brushes and a unique pinion spring mechanism.
- Stop/start systems that use an enhanced starter may incorporate an in-rush current reduction relay (ICR).
- The tandem solenoid starter uses a co-axial dual solenoid that provides independent control of the starter's pinion gear engagement fork and starter motor rotation to allow for engine starting even if the engine has not yet come to a complete stop.
- The permanent engaged (PE) starter eliminates the pinion gear engagement fork since the starter motor is mounted to be permanently engaged to the flywheel.
- The Mazda i-Stop system is a direct start system that does not use a starter motor for the stop/start function. This system uses direct injection and combustion of the air/fuel mixture to instantly restart the engine.

REVIEW QUESTIONS

Short-Answer Essays

1. What is the purpose of the starting system?
2. List and describe the purpose of the major components of the starting system.
3. Explain the principle of operation of the DC motor.
4. Describe the types of magnetic switches used in starting systems.
5. Describe the operation of the overrunning clutch drive.
6. Describe the differences between the positive-engagement and solenoid shift starter.
7. Explain the operating principles of the permanent magnet starter.
8. Describe the purpose and operation of the armature.
9. Describe the purpose and operation of the field coil.
10. Describe how the rotor turns in a three-phase AC motor.

Fill in the Blanks

1. DC motors use the interaction of magnetic fields to convert _____ energy into _____ energy.

2. The _____ is the moveable component of the motor, which consists of a conductor wound around a _____ iron core and is used to create a _____ field.

3. Pole shoes are made of high-magnetic _____ material to help concentrate and direct the _____ in the field assembly.

4. The starter motor electrical connection that permits all of the current that passes through the field coils to also pass through the armature is called the _____ motor.

5. _____ _____ _____ is voltage produced in the starter motor itself. This current acts against the supply voltage from the battery.

6. A starter motor that uses the characteristics of a series motor and a shunt motor is called a _____ motor.

7. The _____ _____ is the part of the starter motor that engages the armature to the engine flywheel ring gear.

8. The _____ _____ is a roller-type clutch that transmits torque in one direction and freewheels in the other direction.

9. The two circuits of the starting system are called the _____ _____ circuit and the _____ _____ circuit.

10. There are two basic types of magnetic switches used in starter systems: the _____ and the _____.

Multiple Choice

1. The armature:
   A. Is the stationary component of the starter that creates a magnetic field.
   B. Is the rotating component of the starter that creates a magnetic field.
   C. Carries electrical current to the commutator.
   D. Prevents the starter from engaging if the transmission is in gear.

2. What is the purpose of the commutator?

   A. To prevent the field windings from contacting the armature.

   B. To maintain constant electrical contact with the field windings.

   C. To reverse current flow through the armature.

   D. All of the above.

3. The field coils:

   A. Are made of wire wound around a nonmagnetic pole shoe.

   B. Are always shunt wound to the armature.

   C. Are always series wound with the armature.

   D. None of the above.

4. Which of the following describes the operation of the starter solenoid?

   A. An electromagnetic device that uses movement of a plunger to exert a pulling or holding force.

   B. Both the pull-in and hold-in windings are energized to engage the starter drive.

   C. When the starter drive plunger is moved, the pull-in winding is de-energized.

   D. All of the above.

5. In the ISG, how does current flow to make the system perform as a starter?

   A. Through the rotor to create an electromagnetic field that excites the stator, which causes the rotor to spin.

   B. Through the rotor coils, which cause the magnetic field to collapse around the stator and rotate the crankshaft.

   C. Through the stator windings, which generate magnetic fields in the rotor, causing the rotor to turn the crankshaft.

   D. From the start generator control module to the rotor coils that are connected to the delta wound stator.

6. Permanent magnet starters are being discussed.

   Technician A says the permanent magnet starter uses four or six permanent magnet field assemblies in place of field coils.

   Technician B says the permanent magnet starter uses a planetary gear set.

   Who is correct?

   A. A only   C. Both A and B

   B. B only   D. Neither A nor B

7. Typical components of the control circuit of the starting system include:

   A. Ring gear.   C. Pinion gear.

   B. Magnetic switch. D. All of the above.

8. The characteristic of the series-wound motor is:

   A. Current flows from the armature to the brushes, and then to the field windings.

   B. Current flows from the field windings, to the brushes, and to the armature.

   C. Current flows through shunts to the field windings and the armature.

   D. All of the above.

9. The gear reduction starter uses:

   A. A starter drive that is connected directly to the armature.

   B. A larger gear to drive a smaller gear that is attached to the starter drive.

   C. A smaller gear to drive a larger gear that is attached to the starter drive.

   D. A starter drive that is attached to the commutator ring.

10. A characteristic of permanent magnet starters is:

   A. The use of planetary gears.

   B. Current flows from the field windings to the brushes, and to the armature.

   C. Connection directly to the armature.

   D. All of the above.

CHAPTER 7
CHARGING SYSTEMS

Upon completion and review of this chapter, you should be able to understand and describe:

- The purpose of the charging system.
- The major components of the charging system.
- The function of the major components of the AC generator.
- The two styles of stators.
- How AC current is rectified to DC current in the AC generator.
- The three principal circuits used in the AC generator.
- The relationship between regulator resistance and field current.
- The relationship between field current and AC generator output.

- The differences between A circuit, B circuit, and isolated circuit regulation.
- The operation of charge indicators, including lamps, electronic voltage monitors, ammeters, and voltmeters.
- The use of the ISG and AC motors in an HEV to recharge the HV battery.
- How regenerative braking is used to recharge the HV battery.
- The purpose of the DC/DC converter for charging the HEV auxiliary battery.

Terms To Know

Delta connection	Half-wave rectification	Rotor
Diode rectifier bridge	Heat sink	Sensing voltage
Diode trio	Inductive reactance	Slip rings
Electronic regulator	Pulse-width modulation (PWM)	Stator
Full-wave rectification	Rectification	Wye-wound connection

INTRODUCTION

The automotive storage battery is not capable of supplying the demands of the electrical system for an extended period of time. Every vehicle must be equipped with a means of replacing the current being drawn from the battery. A charging system is used to restore the electrical power to the battery that was used during engine starting. In addition, the charging system must be able to react quickly to high-load demands required of the electrical system. It is the vehicle's charging system that generates the current to operate all of the electrical accessories while the engine is running.

Two basic types of charging systems have been used. The first was a DC (direct current) generator, which was discontinued in the 1960s. Since that time the AC (alternating current) generator has been the predominant charging device. The DC generator and the AC generator use similar operating principles.

In an attempt to standardize terminology in the industry, the term **alternator** is being replaced with **generator**. Often an alternator is referred to as an AC generator.

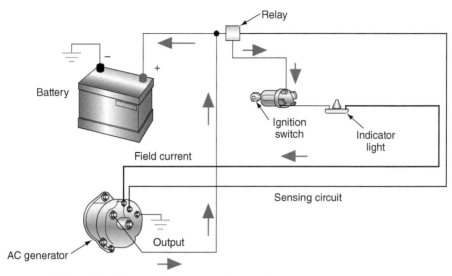

Figure 7-1 Current flow when the charging system is operating.

The purpose of the conventional charging system is to convert the mechanical energy of the engine into electrical energy to recharge the battery and run the electrical accessories. When the engine is first started, the battery supplies all the current required by the starting and ignition systems.

As the battery drain continues, and engine speed increases, the charging system is able to produce more voltage than the battery can deliver. When this occurs, the electrons from the charging device are able to flow in a reverse direction through the battery's positive terminal. The charging device is now supplying the electrical system's load requirements; the reserve electrons build up and recharge the battery.

If there is an increase in the electrical demand, and a drop in the charging system's output equal to the voltage of the battery, the battery and charging system work together to supply the required current.

The entire conventional charging system consists of the following components (**Figure 7-1**):

1. Battery.
2. Generator.
3. Drive belt.
4. Voltage regulator.
5. Charge indicator (lamp or gauge).
6. Ignition switch.
7. Cables and wiring harness.
8. Starter relay (some systems).
9. Fusible link (some systems).

The ignition switch is considered a part of the charging system if it closes the circuit that supplies current to the indicator lamp and stimulates the field coil.

This chapter also covers the operation of the charging systems used on hybrid electric vehicles (HEVs). HEVs can recharge the high-voltage (HV) battery by running the engine and using the ISG or AC motors as generators. They can also use regenerative braking. To charge the auxiliary battery, they may use a DC/DC converter.

PRINCIPLE OF OPERATION

All charging systems use the principle of electromagnetic induction to generate electrical power (**Figure 7-2**). Electromagnetic principle states that a voltage will be

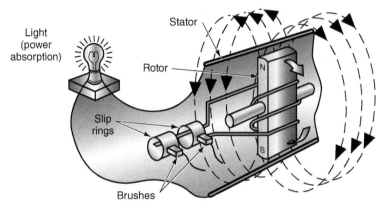

Figure 7-2 Simplified AC generator indicating electromagnetic induction

Shop Manual
Chapter 7,
pages 313, 315

The sine wave produced by a single conductor during one revolution is called single-phase voltage.

produced if motion between a conductor and a magnetic field occurs. The amount of voltage produced is affected by:

1. The speed at which the conductor passes through the magnetic field.
2. The strength of the magnetic field.
3. The number of conductors passing through the magnetic field.

To see how electromagnetic induction produces an AC voltage by rotating a magnetic field inside a fixed conductor (stator), refer to **Figure 7-3**. When the conductor is parallel to the magnetic field, the conductor is not cut by any flux lines (Figure 7-3A). At this point in the revolution, zero voltage and current are being produced.

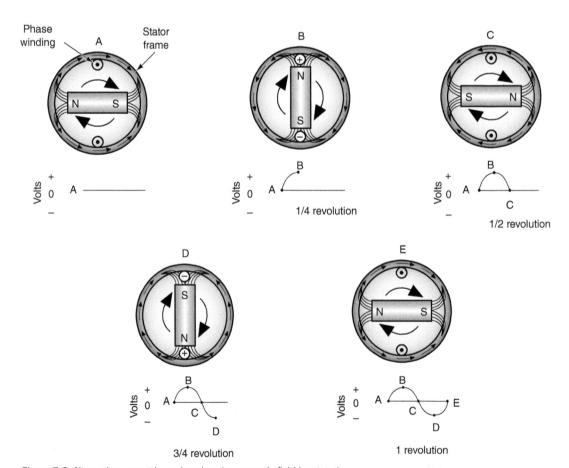

Figure 7-3 Alternating current is produced as the magnetic field is rotated.

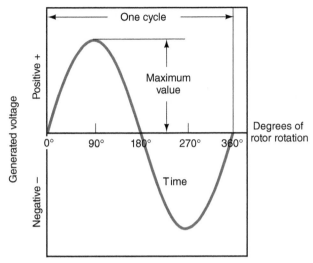

Figure 7-4 Sine wave produced in one revolution of the conductor or magnetic field.

As the magnetic field is rotated 90°, the magnetic field is at a right angle to the conductor (**Figure 7-3B**). At this point in the revolution, the maximum number of flux lines cuts the conductor at the north pole. With the maximum amount of flux lines cutting the conductor, voltage is at its maximum positive value.

When the magnetic field is rotated an additional 90°, the conductor returns to being parallel with the magnetic field (**Figure 7-3C**). Once again, no flux lines cut the conductor and the voltage drops to zero.

An additional 90° revolution of the magnetic field results in the magnetic field being reversed at the top conductor (**Figure 7-3D**). At this point in the revolution, the maximum number of flux lines cuts the conductor at the south pole. Voltage is now at maximum negative value.

When the magnetic field completes one full revolution, it returns to a parallel position with the magnetic field. Voltage returns to zero. The sine wave is determined by the angle between the magnetic field and the conductor. It is based on the trigonometry sine function of angles. The sine wave shown in **Figure 7-4** plots the voltage generated during one revolution.

It is the function of the drive belt to turn the magnetic field. Drive belt tension should be checked periodically to assure proper charging system operation. A loose belt can inhibit charging system efficiency, and a belt that is too tight can cause early bearing failure.

Shop Manual
Chapter 7,
pages 310, 332

AUTHOR'S NOTE The first charging systems used a DC generator that had two field coils which created a magnetic field. Output voltage was generated in the wire loops of the armature as it rotated inside the magnetic field. Current sent to the battery was through the commutator and the generator's brushes.

AC GENERATORS

The DC generator was unable to produce the sufficient amount of current required when the engine was operating at low speeds. With the addition of more electrical accessories and components, the AC generator, or alternator, replaced the DC generator. The main components of the AC generator are (**Figure 7-5**):

1. The rotor.
2. Brushes.

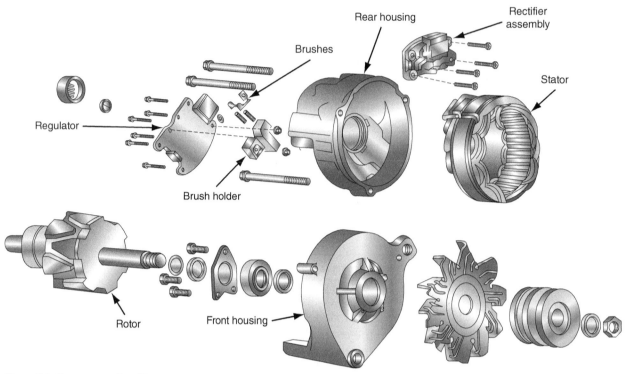

Figure 7-5 Components of an AC generator.

3. The stator.
4. The rectifier bridge.
5. The housing.
6. Cooling fan.
7. Voltage regulator (some models).

Rotors

Shop Manual
Chapter 7, page 341

The current flow through the coil is referred to as field current.

The **rotor** creates the rotating magnetic field of the AC generator. It is the portion of the AC generator that is rotated by the drive belt. The rotor is constructed of many turns of copper wire around an iron core. There are metal plates bent over the windings at both ends of the rotor windings (**Figure 7-6**). The poles (metal plates) do not come into contact with each other, but they are interlaced. When current passes through the coil [1.5 to 3.0 amps (A)], a magnetic field is produced. The strength of the magnetic field is dependent on the amount of current flowing through the coil and the number of windings.

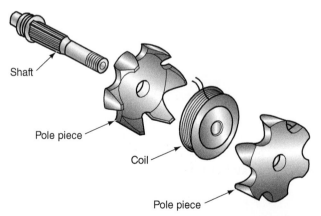

Figure 7-6 Components of a typical AC generator rotor.

The poles will take on the polarity (north or south) of the side of the coil they touch. The right-hand rule will show whether a north or south pole magnet is created. When the rotor is assembled, the poles alternate from north to south around the rotor (**Figure 7-7**). As a result of this alternating arrangement of poles, the magnetic flux lines will move in opposite directions between adjacent poles (**Figure 7-8**). This arrangement provides for several alternating magnetic fields to intersect the stator as the rotor is turning. These individual magnetic fields produce a voltage by induction in the stationary stator windings.

Most rotors have 12 to 14 poles.

The wires from the rotor coil are attached to two **slip rings** that are insulated from the rotor shaft. The slip rings function much like the armature commutator in the starter motor, except they are smooth. The insulated stationary carbon brush passes field current into a slip ring, then through the field coil, and back to the other slip ring. Current then passes through a grounded stationary brush (**Figure 7-9**) or to a voltage regulator.

Brushes

The field winding of the rotor receives current through a pair of brushes that ride against the slip rings. The brushes and slip rings provide a means of maintaining electrical continuity between stationary and rotating components. The brushes (**Figure 7-10**) ride the surface of the slip rings on the rotor and are held tight against the slip rings by spring tension provided by the brush holders. The brushes conduct only the field current (2 to 5 amps). The low current that the brushes must carry contributes to their longer life.

Shop Manual
Chapter 7, page 338

Direct current from the battery is supplied to the rotating field through the field terminal and the insulated brush. The second brush may be the ground brush, which is attached to the AC generator housing or to a voltage regulator.

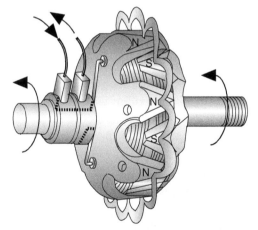

Figure 7-7 The north and south poles of a rotor's field alternate.

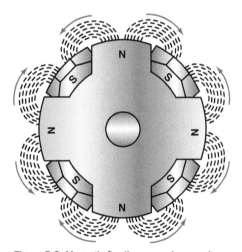

Figure 7-8 Magnetic flux lines move in opposite directions between the rotor poles.

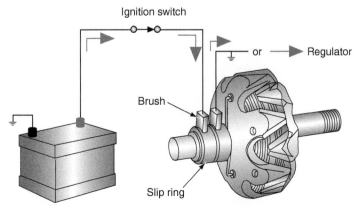

Figure 7-9 The slip rings and brushes provide a current path to the rotor coil.

Figure 7-10 Brushes are the stationary electrical contact to the rotor's slip rings.

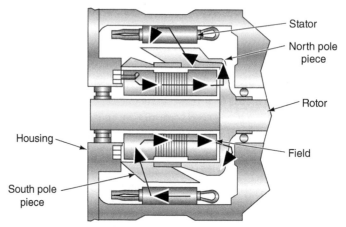

Figure 7-11 Brushless AC generator with stationary field and stator windings.

Some manufacturers developed AC generators that do not require the use of brushes or slip rings. In these AC generators, the field winding and the stator winding are stationary (**Figure 7-11**). A screw terminal is used to make the electrical connection. The rotor contains the pole pieces and is fitted between the field winding and the stator winding.

The magnetic field is produced when current is applied to the field winding. The air gaps in the magnetic path contain a nonmetallic ring to divert the lines of force into the stator winding.

The pole pieces on the rotor concentrate the magnetic field into alternating north and south poles. When the rotor is spinning, the north and south poles alternate as they pass the stator winding. The moving magnetic field produces an electrical current in the stator winding. The alternating current is rectified in the same manner as in conventional AC generators.

Stators

The **stator** contains three main sets of windings wrapped in slots around a laminated, circular iron frame (**Figure 7-12**). The stator is the stationary coil in which electricity is produced. Each of the three windings has the same number of coils as the rotor has pairs of north and south poles. The coils of each winding are evenly spaced around the

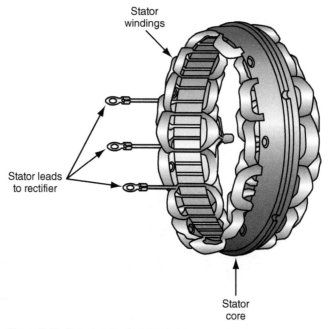

Figure 7-12 Components of a typical stator.

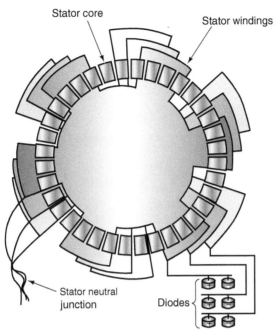

Figure 7-13 Overlapping stator windings produce the required phase angles.

core. The three sets of windings alternate and overlap as they pass through the core (**Figure 7-13**). The overlapping is needed to produce the required phase angles.

The rotor is fitted inside the stator (**Figure 7-14**). A small air gap (approximately 0.015 inch or 0.381 mm) is maintained between the rotor and the stator. This gap allows the rotor's magnetic field to energize all of the windings of the stator at the same time and to maximize the magnetic force.

Each group of windings has two leads. The first lead is for the current entering the winding. The second lead is for current leaving. There are two basic means of connecting the leads. The first method is the **wye-wound connection** (**Figure 7-15**). In the wye connection, one lead from each winding is connected to one common junction. From this junction, the other leads branch out in a Y pattern. A wye-wound AC generator is usually found in applications that do not require high-amperage output.

Shop Manual
Chapter 7, page 341

The common junction in the wye-connected winding is called the stator neutral junction.

Figure 7-14 A small air gap between the rotor and the stator maximizes the magnetic force.

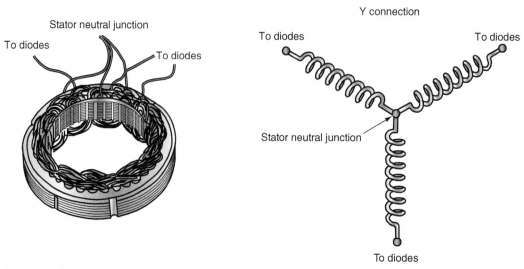

Figure 7-15 Wye-connected stator winding.

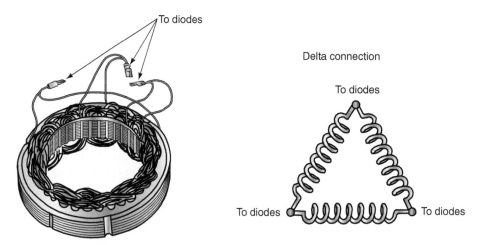

Figure 7-16 Delta-connected stator winding.

The second method of connecting the windings is called the **delta connection** (**Figure 7-16**). The delta connection attaches the lead of one end of the winding to the lead at the other end of the next winding. The delta connection is commonly used in applications that require high-amperage output.

In a wye or delta-wound stator winding, each group of windings occupies one third of the stator, or 120° of the circle. As the rotor revolves in the stator, a voltage is produced in each loop of the stator at different phase angles. The resulting overlap of sine waves that is produced is shown in **Figure 7-17**. Each of the sine waves is at a different phase of its cycle at any given time. As a result, the output from the stator is divided into three phases.

Diode Rectifier Bridge

Shop Manual
Chapter 7,
pages 326, 343

The rectifier bridge is also known as a rectifier stack.

The battery and the electrical system cannot accept or store AC voltage. For the vehicle's electrical system to be able to use the voltage and current generated in the AC generator, the AC current needs to be converted to DC current. This process is called **rectification**. A split-ring commutator cannot be used to rectify AC current to DC current because the stator is stationary in an AC generator. Instead, a **diode rectifier bridge** is used to change the current in an AC generator (**Figure 7-18**). Acting as a one-way check valve, the diodes switch the current flow back and forth so that it flows from the AC generator in only one direction.

Degrees of rotor rotation

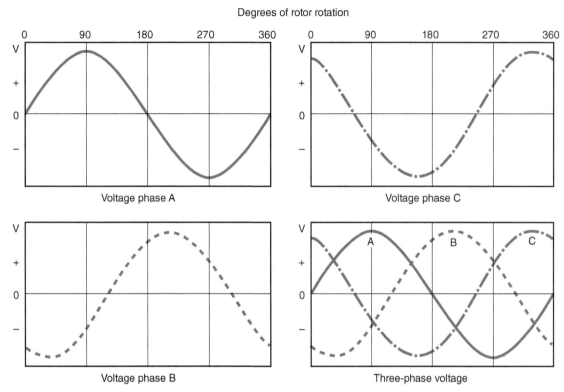

Voltage phase A

Voltage phase C

Voltage phase B

Three-phase voltage

Figure 7-17 The voltage produced in each stator winding is added together to create a three-phase voltage.

Figure 7-18 General Motors' rectifier bridge.

When AC current reverses itself, the diode blocks and no current flows. If AC current passes through a positively biased diode, the diode will block off the negative pulse. The result is the scope pattern shown in **Figure 7-19**. The AC current has been changed to a pulsing DC current. This process is called **half-wave rectification**.

An AC generator usually uses a pair of diodes for each stator winding, for a total of six diodes (**Figure 7-20**). Three of the diodes are positive biased and are mounted in a **heat sink** to dissipate the heat (**Figure 7-21**). The three remaining diodes are negative biased and are attached directly to the frame of the AC generator (**Figure 7-22**). By using a pair of diodes that are reverse-biased to each other, rectification of both sides of the AC sine wave is achieved (**Figure 7-23**). The process of converting both sides of the sine wave to a DC voltage is called **full-wave rectification**. The negative-biased diodes allow the conducting current from the negative side of the AC sine wave to be utilized by the circuit. Diode rectification changes the negative current into positive output.

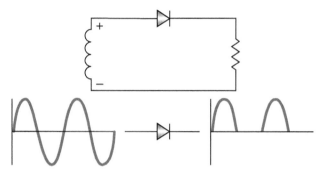

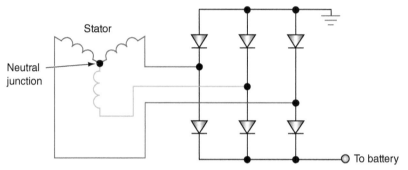

Figure 7-19 AC current rectified to a pulsating DC current after passing through a positive-biased diode. This is called half-wave rectification.

Figure 7-20 A simplified schematic of the AC generator windings connected to the diode rectifier bridge.

Figure 7-21 The positive-biased diodes are mounted into a heat sink to provide protection.

Figure 7-22 Negative-biased diodes pressed into the AC generator housing.

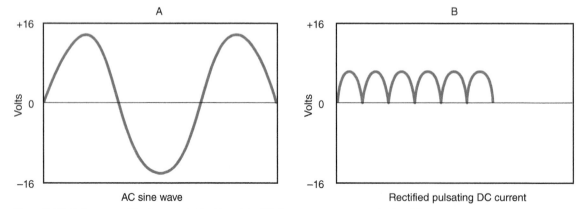

Figure 7-23 Full-wave rectification uses both sides of the AC sine wave to create a pulsating DC current.

With each stator winding connected to a pair of diodes, the resultant waveform of the rectified voltage would be similar to that shown in **Figure 7-24**. With six peaks per revolution, the voltage will vary only slightly during each cycle.

The examples used so far have been for single-pole rotors in a three-winding stator. Most AC generators use either a 12- or a 14-pole rotor. Each pair of poles produces one complete sine wave in each winding per revolution. During one revolution, a 14-pole rotor will produce seven sine waves. The rotor generates three overlapping sine wave voltage cycles in the stator. The total output of a 14-pole rotor per revolution would be 21 sine wave cycles (**Figure 7-25**). With final rectification, the waveform would be similar to the one shown in **Figure 7-26**.

Full-wave rectification is desired because using only half-wave rectification wastes the other half of the AC current. Full-wave rectification of the stator output uses the total potential by redirecting the current from the stator windings so that all current is in one direction.

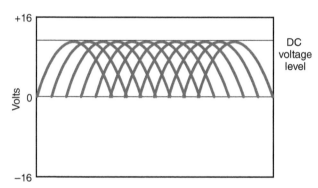

Figure 7-24 With three-phase rectification, the DC voltage level is uniform.

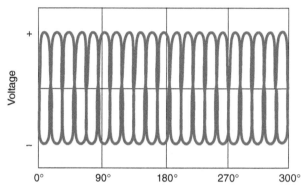

Figure 7-25 Sine wave cycle of a 14-pole rotor and three-phase stator.

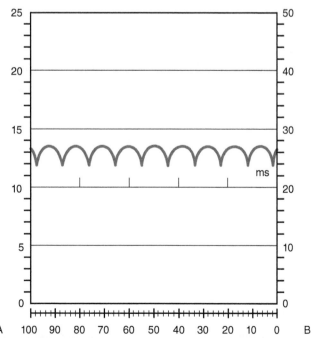

Figure 7-26 Rectified AC output has a ripple pattern that can be shown on an lab scope.

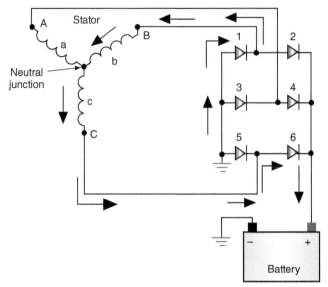

Figure 7-27 Current flow through a wye-wound stator.

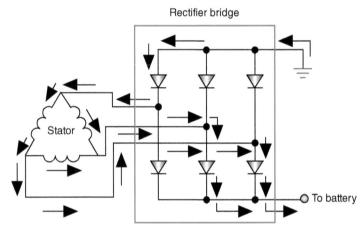

Figure 7-28 Current flow through a delta-wound stator.

A wye-wound stator with each winding connected to a pair of diodes is shown in **Figure 7-27**. Each pair of diodes has one negative and one positive diode. During rotor movement, two stator windings will be in series and the third winding will be neutral. As the rotor revolves, it will energize a different set of windings. Also, current flow through the windings is reversed as the rotor passes. Current in any direction through two windings in series will produce DC current.

The action that occurs when the delta-wound stator is used is shown in **Figure 7-28**. Instead of two windings in series, the three windings of the delta stator are in parallel. This makes more current available because the parallel paths allow more current to flow through the diodes. Since the three outputs of the delta winding are in parallel, current flows from each winding continuously.

AUTHOR'S NOTE Not only do the diodes rectify stator output, but they also block battery drain back when the engine is not running.

AC Generator Housing and Cooling Fan

Shop Manual
Chapter 7, page 338
Most AC generator housings are a two-piece construction, made from cast aluminum (**Figure 7-29**). The two end frames provide support for the rotor and the stator.

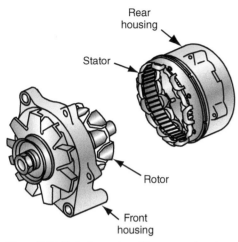

Figure 7-29 Typical two-piece AC generator housing.

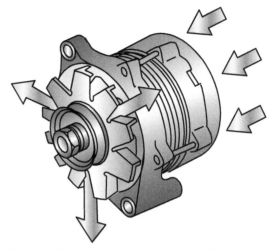

Figure 7-30 The cooling fan draws air in from the rear of the AC generator to keep the diodes cool.

In addition, the end frames contain the diodes, regulator, heat sinks, terminals, and other components of the AC generator. The two end pieces are referred to as:

1. The drive end housing: This housing holds a bearing to support the front of the rotor shaft. The rotor shaft extends through the drive end housing and holds the drive pulley and cooling fan.
2. The slip ring end housing: This housing also holds a rotor shaft that supports a bearing. In addition, it contains the brushes and has all of the electrical terminals. If the AC generator has an integral regulator, it is also contained in this housing.

Because the rotor and stator are made up of coils of wire, heat from the conversion process increases resistance and decreases the AC generator's capabilities. To keep the components of the generator cool, a cooling fan is installed behind the pulley. The spinning fan will draw air in through openings in the slip ring end housing, through the generator, and out openings behind the cooling fan in the drive end housing (**Figure 7-30**).

Liquid-Cooled Generators

High-output generators have a tendency to have higher internal temperatures that can shorten the life of the diodes. To help reduce diode temperatures, some manufacturers are using a liquid-cooled generator (**Figure 7-31**). In addition, since these generators do not use a fan, underhood noises are reduced. The water-cooled generator has water jackets cast into their housing and is connected to the engine's cooling system by hoses.

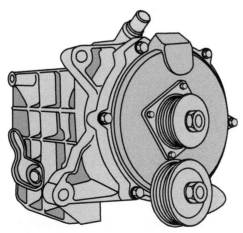

Figure 7-31 Water-cooled generator.

AC GENERATOR CIRCUITS

Shop Manual
Chapter 7, page 313

The **diode trio** is used by some manufacturers to rectify current from the stator so that it can be used to create the magnetic field in the field coil of the rotor. This eliminates extra wiring.

Shop Manual
Chapter 7, page 344

There are three principal circuits used in an AC generator:

1. The charging circuit: Consists of the stator windings and rectifier circuits.
2. The excitation circuit: Consists of the rotor field coil and the electrical connections to the coil.
3. The preexcitation circuit: Supplies the initial current for the field coil that starts the buildup of the magnetic field.

For the AC generator to produce current, the field coil must develop a magnetic field. The AC generator creates its own field current in addition to its output current.

For excitation of the field to occur, the voltage induced in the stator rises to a point that it overcomes the forward voltage drop of at least two of the rectifier diodes. Before the **diode trio** can supply field current, the anode side of the diode must be at least 0.6 volt (V) more positive than the cathode side (**Figure 7-32**). When the ignition switch is turned on, the warning lamp current acts as a small magnetizing current through the field (**Figure 7-33**). This current preexcites the field, reducing the speed required to start its own supply of field current.

AUTHOR'S NOTE If the battery is completely discharged, the vehicle cannot be push started because there is no excitation of the field coil.

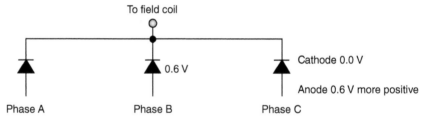

Figure 7-32 The diode trio connects the phase windings to the field. To conduct, there must be 0.6 volt more positive on the anode side of the diodes.

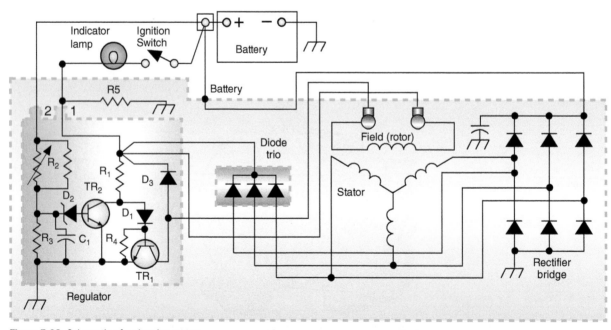

Figure 7-33 Schematic of a charging system.

AC GENERATOR OPERATION OVERVIEW

Shop Manual
Chapter 7, page 313

When the engine is running, the drive belt spins the rotor inside the stator windings. This magnetic field inside the rotor generates a voltage in the windings of the stator. Field current flowing through the slip rings to the rotor creates alternating north and south poles on the rotor.

Shop Manual
Chapter 7, page 321

The induced voltage in the stator is an alternating voltage because the magnetic fields are alternating. As the magnetic field begins to induce voltage in the stator's windings, the induced voltage starts to increase. The amount of voltage will peak when the magnetic field is the strongest. As the magnetic field begins to move away from the stator windings, the amount of voltage will start to decrease. Each of the three windings of the stator generates voltage, so the three combine to form a three-phase voltage output.

In the wye connection (refer to Figure 7-27), output terminals (A, B, and C) apply voltage to the rectifier. Because only two stator windings apply voltage (because the third winding is always connected to diodes that are reverse-biased), the voltages come from points A to B, B to C, and C to A.

To determine the amount of voltage produced in the two stator windings, find the difference between the two points. For example, to find the voltage applied from points A and B, subtract the voltage at point B from the voltage at point A. If the voltage at point A is 8 volts positive and the voltage at point B is 8 volts negative, the difference is 16 volts. This procedure can be performed for each pair of stator windings at any point in time to get the sine wave patterns (**Figure 7-34**). The voltages in the windings are designated as V_a, V_b, and V_c. Designations V_{ab}, V_{bc}, and V_{ca} refer to the voltage difference in the two stator windings. In addition, the numbers refer to the diodes used for the voltages generated in each winding pair.

AUTHOR'S NOTE Alternating current is constantly changing, so this formula would have to be performed at several different times.

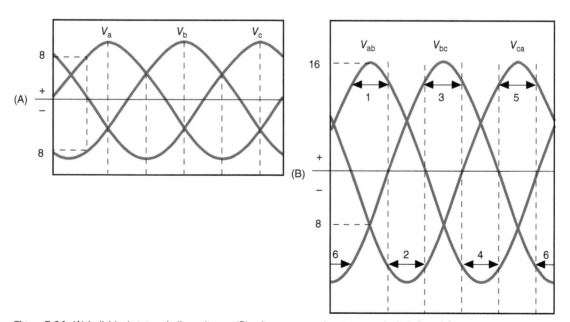

Figure 7-34 (A) Individual stator winding voltages; (B) voltages across the stator terminals A, B, and C.

The current induced in the stator passes through the diode rectifier bridge, consisting of three positive and three negative diodes. At this point, there are six possible paths for the current to follow. The path that is followed depends on the stator terminal voltages. If the voltage from points A and B is positive (point A is positive in respect to point B), current is supplied to the positive terminal of the battery from terminal A through diode 2 (**Figure 7-35**). The negative return path is through diode 3 to terminal B.

Both diodes 2 and 3 are forward-biased. The stator winding labeled C does not produce current because it is connected to diodes that are reverse-biased. The stator current is rectified to DC current to be used for charging the battery and supplying current to the vehicle's electrical system.

When the voltage from terminals C and A is negative (point C is negative in respect to point A), current flow to the battery positive terminal is from terminal A through diode 2 (**Figure 7-36**). The negative return path is through diode 5 to terminal C.

This procedure is repeated through the four other current paths (**Figures 7-37** through **7-40**).

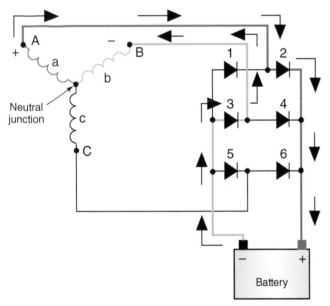

Figure 7-35 Current flow when terminals A and B are positive.

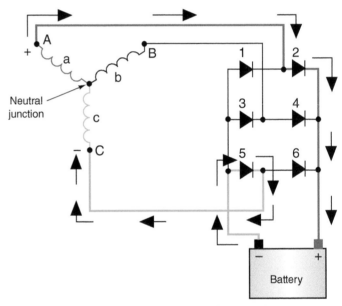

Figure 7-36 Current flow when terminals A and C are negative.

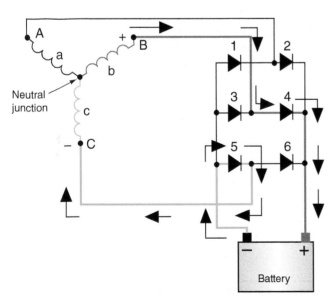

Figure 7-37 Current flow when terminals B and C are positive.

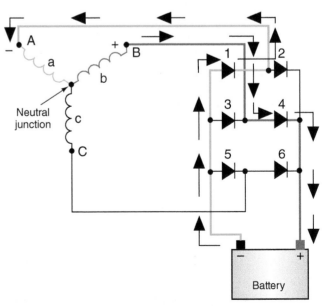

Figure 7-38 Current flow when terminals A and B are negative.

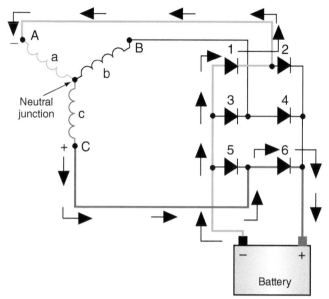

Figure 7-39 Current flow when terminals A and C are positive.

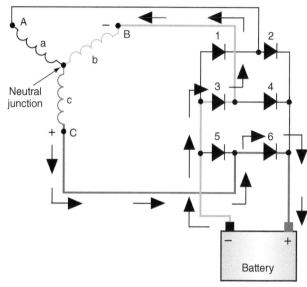

Figure 7-40 Current flow when terminals B and C are negative.

REGULATION

The battery, and the rest of the electrical system, must be protected from excessive voltages. To prevent early battery and electrical system failure, regulation of the charging system voltage is very important. Also, the charging system must supply enough current to run the vehicle's electrical accessories when the engine is running.

AC generators do not require current limiters; because of their design, they limit their own current output. Current limit is the result of the constantly changing magnetic field because of the induced AC current. As the magnetic field changes, an opposing current is induced in the stator windings. This **inductive reactance** in the AC generator limits the maximum current that the AC generator can produce. Even though current (amperage) is limited by its operation, voltage is not. The AC generator is capable of producing as high as 250 volts, if it were not controlled.

Regulation of voltage is done by varying the amount of field current flowing through the rotor. The higher the field current, the higher the output voltage. Control of field current can be done either by regulating the resistance in series with the field coil or by turning the field circuit on and off (**Figure 7-41**). By controlling the amount of current in the field coil, control of the field current and the AC generator output is obtained. To ensure a full battery charge, and operation of accessories, most voltage regulators are set for a system voltage between 13.5 and 14.5 volts.

The regulator must have system voltage as an input in order to regulate the output voltage. The input voltage to the AC generator is called **sensing voltage**. If sensing voltage is below the regulator setting, an increase in charging voltage output results by increasing field current. Higher sensing voltage will result in a decrease in field current and voltage output. A vehicle being driven with no accessories on and a fully charged battery will have a high sensing voltage. The regulator will reduce the charging voltage until it is at a level to run the ignition system while trickle charging the battery. If a heavy load is turned on (such as the headlights), the additional draw will cause a drop in the battery voltage. The regulator will sense this low-system voltage and will increase current to the rotor. This will allow more current to the field windings. With the increase of field current, the magnetic field is stronger and AC generator voltage output is increased. When the load is turned off, the regulator senses the rise in system voltage and cuts back the amount of field current and ultimately AC generator voltage output.

Shop Manual
Chapter 7,
pages 321, 325

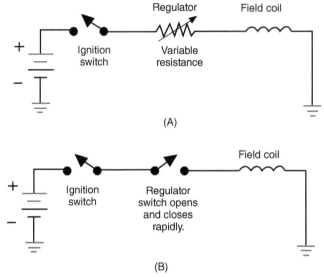

Figure 7-41 The regulator can control the field current by (A) controlling the resistance in series with the coil or (B) by switching the field on and off.

	Volts	
Temperature °F	**Minimum**	**Maximum**
20°F	14.3	15.3
80°F	13.8	14.4
140°F	13.3	14.0
Over 140	Less than 13.3	–

Figure 7-42 Chart indicating relationship between temperature and charge rate.

Another input that affects regulation is temperature. Because ambient temperatures influence the rate of charge that a battery can accept, regulators are temperature compensated (**Figure 7-42**). Temperature compensation is required because the battery is more reluctant to accept a charge at lower ambient temperatures. The regulator will increase the system voltage until it is at a higher level so the battery will accept it.

Field Circuits

Shop Manual
Chapter 7, page 319

The A circuit is called an external grounded field circuit.

Usually the B circuit regulator is mounted externally to the AC generator. The B circuit is an internally grounded circuit.

To properly test and service the charging system, it is important to identify the field circuit being used. Automobile manufacturers use three basic types of field circuits. The first type is called the A circuit. It has the regulator on the ground side of the field coil. The B+ for the field coil is picked up from inside the AC generator (**Figure 7-43**). By placing the regulator on the ground side of the field coil, the regulator will allow the control of field current by varying the current flow to ground.

The second type of field circuit is called the B circuit. In this case, the voltage regulator controls the power side of the field circuit. Also, the field coil is grounded from inside the AC generator.

AUTHOR'S NOTE To remember these circuits: Think of "A" for "After" the field and "B" for "Before" the field.

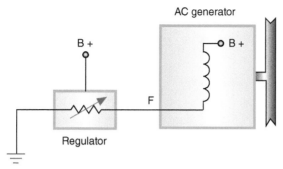

Figure 7-43 Simplified diagram of an A circuit field.

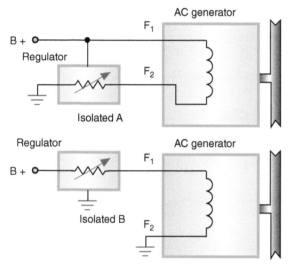

Figure 7-44 In the isolated circuit field AC generator, the regulator can be installed on either side of the field.

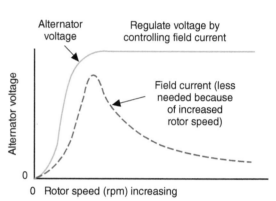

Figure 7-45 Graph showing the relationship between field current, rotor speed, and regulated voltage changes depending on electrical load.

The third type of field circuit is called the isolated field. The AC generator has two field wires attached to the outside of the case. The voltage regulator can be located either on the ground (A circuit) or on the B+ (B circuit) side (**Figure 7-44**). Isolated field AC generators pick up B+ and ground externally.

Regardless of which type of field circuit is used, generator voltage output is regulated by controlling the amount of current through the field windings. The relationship between the field current, rotor speed, and regulated voltage is illustrated in **Figure 7-45**. As rotor speed increases, field current must be decreased to maintain regulated voltage.

Electronic Regulators

The **electronic regulator** uses solid-state circuitry to perform the regulatory functions. Electronic regulators can be mounted either externally or internally to the AC generator. There are no moving parts, so the electronic regulator can cycle between 10 and 7,000 times per second. This quick cycling provides more accurate control of the field current through the rotor.

Pulse-width modulation (PWM) controls AC generator output by varying the amount of time the field coil is energized. For example, assume that a vehicle is equipped with a 100-amp generator. If the electrical demand placed on the charging system requires 50 amps of current, the regulator would energize the field coil for 50% of the time (**Figure 7-46**). If the electrical system's demand was increased to 75 amps, the regulator would energize the field coil 75% of the cycle time.

Shop Manual
Chapter 7, page 325

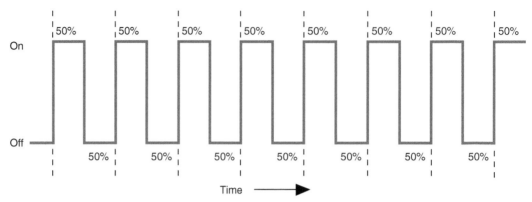

Figure 7-46 Pulse-width modulation with 50% on time.

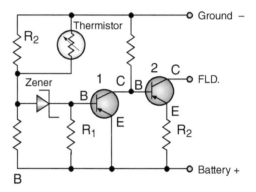

Figure 7-47 A simplified circuit diagram of an electronic regulator utilizing a zener diode.

The electronic regulator uses a zener diode that blocks current flow until a specific voltage is obtained, at which point it allows the current to flow. An electronic regulator is shown (**Figure 7-47**).

Battery voltage is applied to the cathode side of the zener diode as well as to the base of transistor number 1. No current will flow through the zener diode, since the voltage is too low to push through the zener. However, as the AC generator produces voltage, the voltage at the cathode will increase until it reaches the upper limit (14.5 volts) and is able to push through the zener diode. Current will now flow from the battery, through the resistor (R_1), through the zener diode, through the resistor (R_2) and thermistor in parallel, and to ground. Since current is flowing, each resistance in the circuit will drop voltage. As a result, voltage to the base of transistor number 1 will be less than the voltage applied to the emitter. Since transistor number 1 is a PNP transistor and the base voltage is less than the emitter voltage, transistor number 1 is turned on. The base of transistor number 2 will now have battery voltage applied to it. Since the voltage applied to the base of transistor number 2 is greater than that applied to its emitter, transistor number 2 is turned off. Transistor number 2 is in control of the field current and generator output.

The thermistor changes circuit resistance according to temperature. This provides for the temperature-related voltage change necessary to keep the battery charged in cold-weather conditions.

Many manufacturers are installing the voltage regulator inside the AC generator. This eliminates some of the wiring needed for external regulators. The diode trio rectifies AC current from the stator to DC current that is applied to the field windings (**Figure 7-48**).

Current flow with the engine off and the ignition switch in the RUN position is illustrated in **Figure 7-49**. Battery voltage is applied to the field through the common

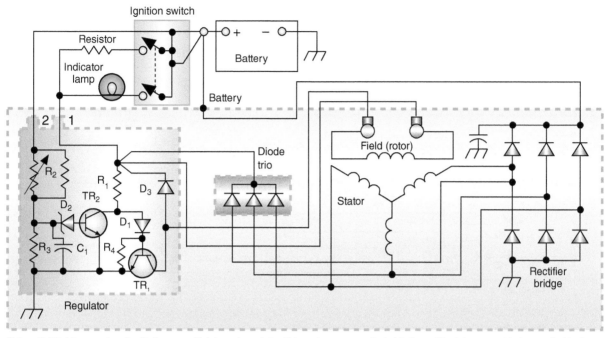

Figure 7-48 AC generator circuit diagram with internal regulator. This system uses a diode trio to rectify stator current to be applied to the field coil. The resistor above the indicator lamp is used to ensure current will flow through terminal 1 if the lamp burns out.

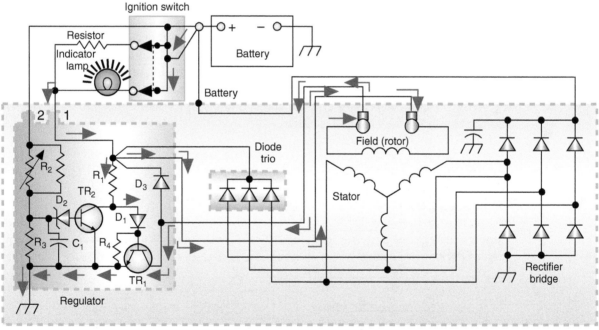

Figure 7-49 Current flow to the rotor with the ignition switch in the RUN position and the engine OFF.

point above R_1. TR_1 conducts the field current coming from the field coil, producing a weak magnetic field. The indicator lamp lights because TR_1 directs current to ground and completes the lamp circuit.

Current flow with the engine running is illustrated in **Figure 7-50**. When the AC generator starts to produce voltage, the diode trio will conduct and battery voltage is available for the field and terminal 1 at the common connection. Placing voltage on both sides of the lamp gives the same voltage potential on each side; therefore, current doesn't flow and the lamp goes out.

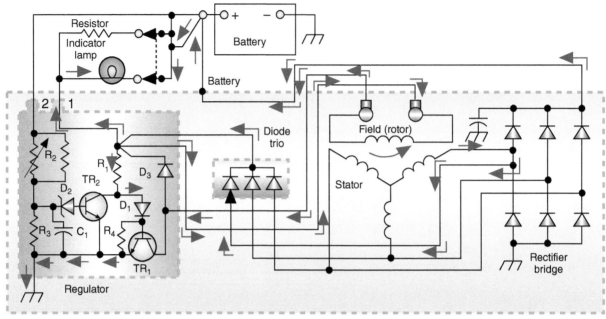

Figure 7-50 Current flow with the engine running and AC generator producing voltage.

Current flow as the voltage output is being regulated is illustrated in **Figure 7-51**. The sensing circuit from terminal 2 passes through a thermistor to the zener diode (D_2). When the system voltage reaches the upper voltage limit of the zener diode, the zener diode conducts current to TR_2. When TR_2 is biased, it opens the field coil circuit and current stops flowing through the field coil. Regulation of this switching on and off is based on the sensing voltage received through terminal 2. With the circuit to the field coil opened, the sensing voltage decreases and the zener diode stops conducting. TR_2 is turned off and the circuit for the field coil is closed.

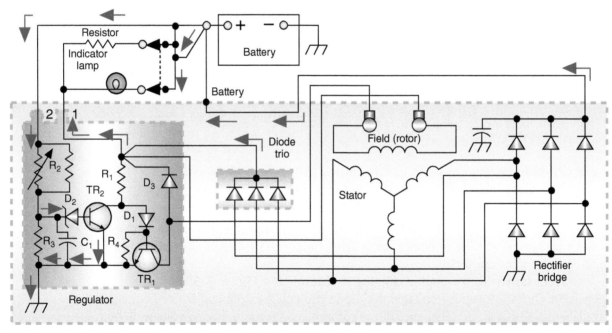

Figure 7-51 When the system voltage is high enough to allow the zener diode to conduct, TR_2 is turned on and TR_1 is shut off, which opens the field circuit.

Computer-Controlled Regulation

Shop Manual
Chapter 7, page 326

On many vehicles after the mid-1980s, the regulator function has been incorporated into the powertrain control module (PCM) (**Figure 7-52**). The computer-controlled regulation system has the ability to precisely maintain and control the changing rate according to the electrical requirements, battery (or ambient) temperature, and several other inputs

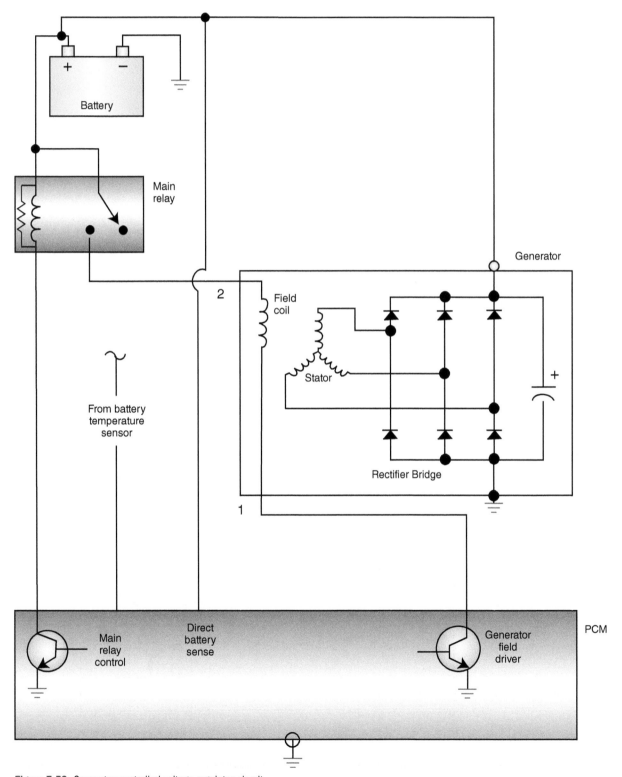

Figure 7-52 Computer-controlled voltage regulator circuit.

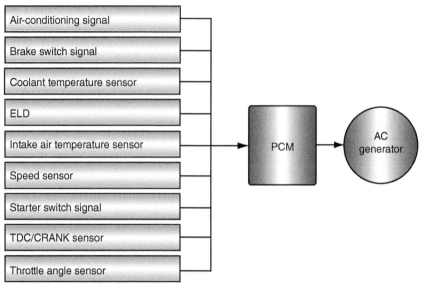

Figure 7-53 The PCM will use various inputs to regulate AC generator output.

(**Figure 7-53**). Regulation of the field circuit is through the ground (A circuit). The PCM will determine the SOC of the battery as inputted by the battery sense circuit. This information is used in conjunction with ambient temperature or battery temperature sensor inputs to determine a target battery voltage. The PCM will control the excitation of the field coil to maintain the target voltage. Typically, the field coil is activated until the voltage on the sense circuit reaches about 0.5 volt above the target voltage, then is turned off. The field coil is reactivated when the sense battery voltages falls to 0.5 volt below the target voltage.

In recent years, there has been an increase in manufacturers that control the field circuit by use of high-side drivers. For example, the later General Motors CS generator systems pulse the voltage output to the field windings (L terminal) from the PCM. This type of generator has a constant field winding ground connection.

Another generator control method is like that used by Mercedes Benz (**Figure 7-54**). This system uses a generator that is a slave module on the local interconnect network (LIN) bus network. The PCM communicates current and voltage requirements over the LIN bus to the internal LIN eight-bit microprocessor. Electronic driver stages control the generator output. This power-on-demand control system reduces fuel consumption since the generator field is turned on only when needed.

When the generator is a slave module on a bus network it is often called an interfaced generator.

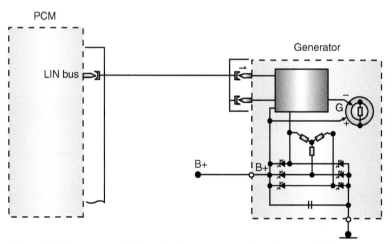

Figure 7-54 A generator with internal microprocessor using a bus network for communications to the PCM.

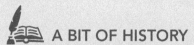

A BIT OF HISTORY

Chrysler equipped its vehicles with AC generators in the late 1950s, making it the first manufacturer to use an AC generator. Chrysler introduced the dual-output AC generator (40 or 90 amps) with computer control in 1985.

In 1985, Chrysler introduced a delta-wound, dual-output, computer-controlled charging system (**Figure 7-55**). This system has some unique capabilities:

1. The system is capable of varying charging system output based on the ambient temperature and the system's voltage needs.
2. The computer monitors the charging system and is capable of self-diagnosis.

When the ignition switch is in the RUN position, the PCM checks the ambient temperature and determines required field current. Based on inputs relating to temperature and system requirements, the PCM determines when current output adjustments are required.

In recent years, Chrysler used a Nippondenso or a Bosch-built AC generator that have a wye-wound stator. This system also uses the PCM to control voltage regulation by varying the field winding ground. Vehicles equipped with a Next Generation Controller (NGC) have high-side control, similar to that of the General Motors' CS series discussed earlier.

Some Mitsubishi AC generators use an internal voltage regulator (**Figure 7-56**). It also has two separate wye-connected stator windings (**Figure 7-57**). Each of the stator windings has its own set of six diodes for rectification.

This AC generator also uses a diode trio to rectify stator voltage to be used in the field winding. The three terminals are connected as follows:

- **B terminal:** Connects the output of both stator windings to the battery, supplying charging voltage.
- **R terminal:** Supplies 12 volts to the regulator.
- **L terminal:** Connects to the output of the diode trio to provide rectified stator voltage to designated circuits.

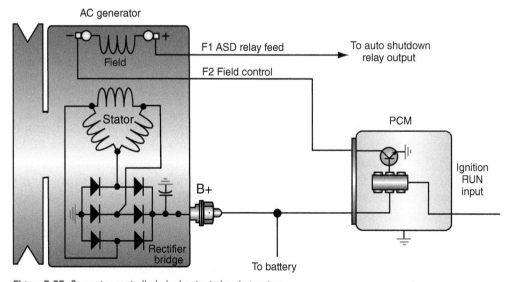

Figure 7-55 Computer-controlled, dual-output charging system.

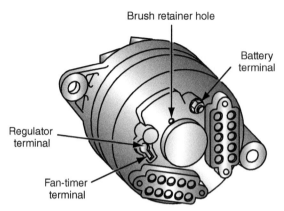

Figure 7-56 Mitsubishi AC generator terminals.

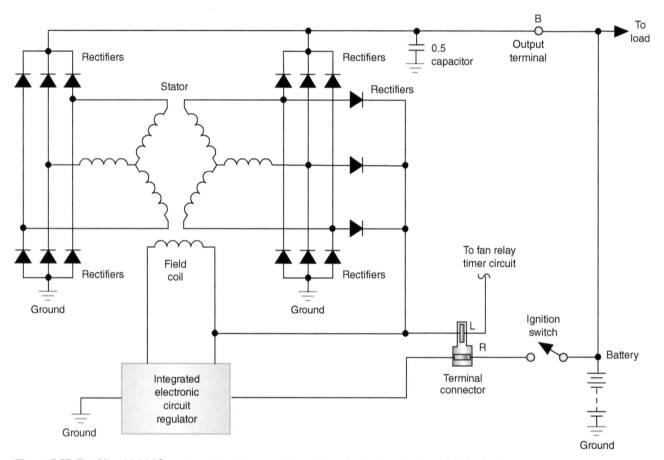

Figure 7-57 The Mitsubishi AC generator uses two separate stator windings and a total of 15 diodes.

Another method that Mitsubishi uses for charging control involves a single wye-wound stator generator with an internal regulator that interfaces with the PCM (**Figure 7-58**). The PCM monitors the state of the field coil through terminal FR of the generator. The PCM sends 5 volts to the FR (field regulation) terminal. As the field coil is turned on and off by the internal regulator, the voltage will cycle between 5 and 0 volts. When the PCM senses 5 volts, it knows the field coil is turned off, and when the voltage drops close to 0 volt, it knows the field coil is turned on. This allows the PCM to control idle speed when the regulator is applying full field. In addition, the PCM can dampen the effects of full fielding during high electrical loads. This will prevent occurrences such as lights flickering bright and dim as the field coil is turned on and off.

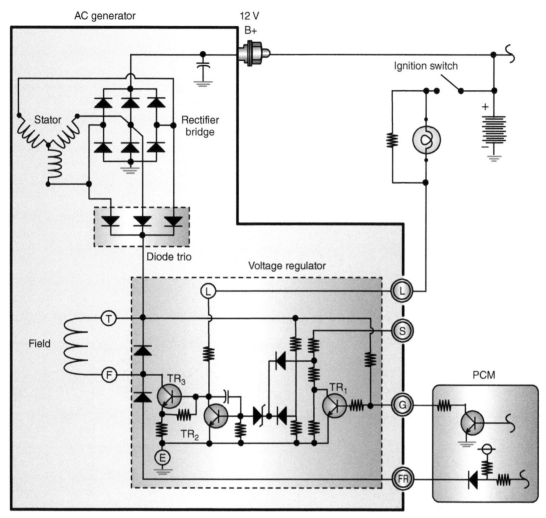

Figure 7-58 Schematic of Mitsubishi charging system using both an internal voltage regulator and a PCM to control output.

When a full field condition is sensed by the PCM, it will modulate its internal transistor. This will then control the power transistor in the internal regulator. The PCM has a maximum authority of 14.4 volts. If the charging system output exceeds this level, the internal regulator turns the power transistor off.

To perform this function, the PCM will duty cycle its internal transistor, which turns the TR_1 in the generator on and off. In Figure 7-58, battery voltage is applied to the S terminal of the generator. This voltage goes through three resistors in series to ground (**Figure 7-59**). Each of the resistors has 2 Ω of resistance. If the PCM internal transistor is turned on, the voltage to the base of TR_1 is pulled low and TR_1 is turned off. With TR_1 turned off, all three of the resistors are involved in the circuit. Each resistor will drop 4 volts. Since R_1 drops 4 volts, 8 volts is applied to the zener diode. This is enough to blow through the diode, applying voltage to the base of TR_2 and turning it on. With TR_2 turned on, base voltage to transistor TR_3 is pulled low and will be turned off. With TR_3 turned off, the field coil is de-energized.

If the internal transistor in the PCM is turned off, battery voltage will be applied to the base of TR_1 and turn it on. TR_1 will now supply an alternate path to ground, bypassing R_3. Now, only R_1 and R_2 are in series and each resistor will drop 6 volts. This means that 6 volts will be applied to the zener diode. This is not enough to blow through the zener; therefore, TR_2 will be off. With TR_2 off, battery voltage is applied to the base of transistor

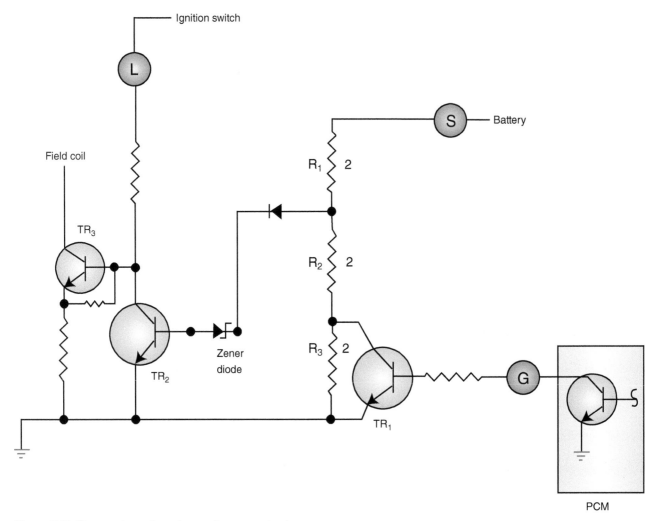

Figure 7-59 Three resistors in series on the sense circuit.

TR_3 and it will be turned on. Since transistor TR_3 is on, the field coil circuit to ground is complete and the coil is energized. The PCM internal transistor switches on and off several times a second to prevent the generator from going to full field too rapidly.

AC GENERATOR OUTPUT RATINGS

Shop Manual
Chapter 7, page 329

The output rating of an AC generator is usually expressed in the amount of amperage that the generator can produce at a specific rotational speed. The International Standards Organization and the Society of Automate Engineers standards (ISO 8854 and SAE J 56) indicate that the rated output is at a rotor shaft speed of 6,000 rpm.

AUTHOR'S NOTE There is a difference between the AC generator's amperage rating and the amount of current it can produce at idle speed. As discussed in the Shop Manual, it is possible to measure the actual output of an alternator under a simulated load. This will allow you to determine what the AC generator is capable of actually producing.

Both ISO and SAE require a format of AC generator ratings that identifies the amperage output at the idle speed (ISO identifies this as the actual engine idle speed and ASE identifies this as 1,500 rpm), the rated amperage output, and the test voltage. This rating is typically stamped or labeled on the housing. For example, a label with "80/200A 13.5V," would indicate the output is rated at 80 amps at idle, 200 amps at 6,000 rpm (shaft speed) at a test voltage of 13.5 volts.

The current output of the alternator is determined by the rotational speed of the rotor, the size of the poles, the size of the stator core, the thickness of the lamination, the number of turns in the coil, and the wire size used for the windings. In addition, high output AC generators use a higher amperage rated diodes rectifier bridge. Standard duty rectifiers use 55 amp diodes while the high output rectifier typically uses 70 amp diodes.

AC Generator Output Supply and Demand

The amperage rating on an AC generator is the amount of current the generator is capable of producing, not the amount it always produces. The output of the AC generator is based on the demands placed on it by the vehicle's electrical system. An AC generator will not produce more current than is required by the demands of the electrical system. For example, an AC generator that is rated at 180 amps will provide 40 amps if that is the total draw of the electrical system.

An AC generator that is under rated for the vehicle can cause problems since it cannot meet the requirements of the electrical system. However, an AC generator that has a higher output than required represents wasted potential. Because the current output is based on demand, the oversized AC generator will not damage the battery or electrical system.

Some conventional vehicles with stop/start use a high output (220 amp) AC generator due to the frequent demands placed on the battery. Because the AC generator is not producing any output while the engine is shut down, a voltage stabilization module (VSM) is used to allow vehicle accessories to continue to function during the engine stop event.

AUTHOR'S NOTE With aftermarket high wattage amplifiers, and other high current components, it is common to install high output AC generators in place of the OEM units. Even if the demands of the added accessories requires 150 amps, and the AC generator produces this amount, the rest of the vehicle's electrical system will not be having to deal with this high amperage. As you know, the current flow in a circuit is determined by the resistance of the circuit; thus, any given electrical component will only draw as much amperage as required to operate. While an aftermarket amplifier may require 150 amps, you do not need to worry about 150 amps going to the headlights and blowing them out.

If you install a high output alternator, replace the ground straps and the cable between the AC generator to the battery with heavier gauge cables.

CHARGING INDICATORS

There are three basic methods of informing the driver of the charging system's condition: indicator lamps, ammeter, and voltmeter.

AUTHOR'S NOTE Greater detail concerning operation of indicator lamps and gauges, including computer-controlled systems, is provided in Chapter 13.

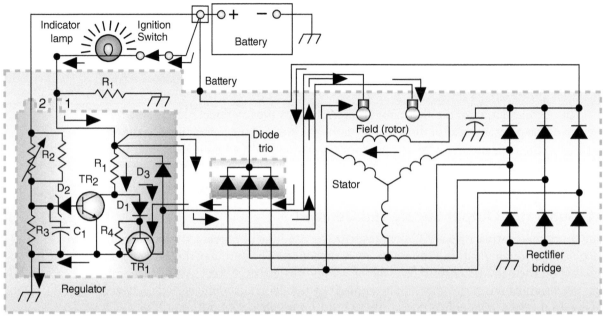

Figure 7-60 Electronic regulator with an indicator light on due to no AC generator output.

Indicator Light Operation

As discussed earlier, the indicator lamp can operate on the basis of opposing voltages. If the AC generator output is less than battery voltage, there is an electrical potential difference in the lamp circuit and the lamp will light. If there is no stator output through the diode trio, then the lamp circuit is completed to ground through the rotor field and TR_1 (**Figure 7-60**).

On most systems, the warning lamp will be "proofed" when the ignition switch is in the RUN position before the engine starts. This indicates that the bulb and indicator circuit are operating properly. Proofing the bulb is accomplished since there is no stator output without the rotor turning.

Ammeter Operation

In place of the indicator light, some manufacturers install an ammeter. The ammeter is wired in series between the AC generator and the battery (**Figure 7-61**). Most ammeters work on the principle of D'Arsonval movement.

The movement of the ammeter needle under different charging conditions is illustrated in **Figure 7-62**. If the charging system is operating properly, the ammeter needle will remain within the normal range. If the charging system is not generating sufficient current, the needle will swing toward the discharge side of the gauge. When the charging system is recharging the battery, or is called on to supply high amounts of current, the needle deflects toward the charge side of the gauge.

It is normal for the gauge to read a high amount of current after initial engine start-up. As the battery is recharged, the needle should move more toward the normal range.

Voltmeter Operation

Because the ammeter is a complicated gauge for most people to understand, many manufacturers use a voltmeter to indicate charging system operation. In early systems, the voltmeter is connected to a voltage output of the ignition switch and ground (**Figure 7-63**).

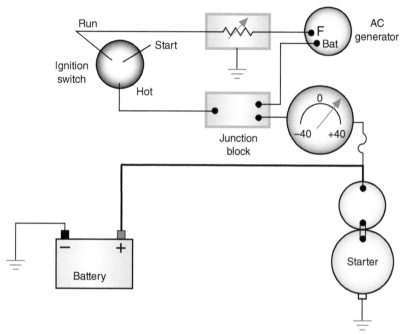

Figure 7-61 Ammeter connected in series to indicate charging system operation.

Ammeter conditions

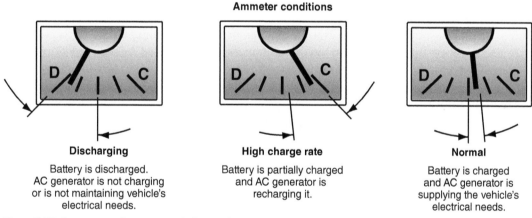

Discharging

Battery is discharged.
AC generator is not charging
or is not maintaining vehicle's
electrical needs.

High charge rate

Battery is partially charged
and AC generator is
recharging it.

Normal

Battery is charged
and AC generator is
supplying the vehicle's
electrical needs.

Figure 7-62 Ammeter needle movement indicates charging conditions.

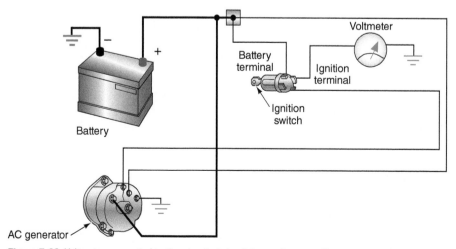

Figure 7-63 Voltmeter connected to the charging circuit to monitor operation.

When the engine starts, it is normal for the voltmeter to indicate a reading between 13.2 and 15.2 volts. If the voltmeter indicates a voltage level that is below 13.2, it may mean that the battery is discharging. If the voltmeter indicates a voltage reading that is above 15.2 volts, the charging system is overcharging the battery. The battery and electrical circuits can be damaged as a result of higher-than-normal charging system voltage output.

In most modern systems, the voltmeter is controlled either directly by the PCM or by information sent to the instrument cluster by the PCM. A dedicated circuit from the battery to the PCM allows the PCM to constantly monitor the battery voltage.

HEV CHARGING SYSTEMS

As discussed in Chapter 6, HEVs utilize the automatic stop/start feature to shut off the engine whenever the vehicle is not moving or when power from the engine is not required. Some HEV systems use a starter/generator unit to perform both functions. The difference between a motor and a generator is that the motor uses two opposing magnetic fields and the generator uses one magnetic field that has rotating conductors. The use of electronics to control the direction of current flow allows the unit to function as both a motor and a generator.

There are two basic designs of the starter/generator. The first design uses a belt alternator starter (BAS) that is about the same size as a conventional generator and is mounted in the same way (**Figure 7-64**). The second design is to mount an integrated starter generator (ISG) at either end of the crankshaft. Most designs have the ISG mounted at the rear of the crankshaft between the engine and transmission (**Figure 7-65**).

The ISG is a three-phase AC motor that can provide power and torque to the vehicle. It also supports the engine when the driver demands more power. As seen in Figure 7-65, the ISG includes a rotor and stator that are located inside the transmission bell housing. The stator is attached to the engine block and is made up of two separate lamination stacks. The rotor is bolted to the engine crankshaft and has both wire-wound and permanent magnet sections. Rectification is accomplished with traditional diodes.

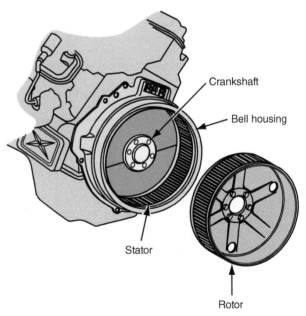

Figure 7-65 The ISG is usually located at the rear of the crankshaft in the bell housing.

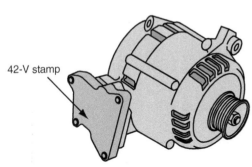

Figure 7-64 A BAS used on some HEVs.

The advantage of this rotor and stator design is that the output at engine idle speed is up to 240 amps. Maximum output can exceed 300 amps.

Generation is done anytime the engine is running. Since the rotor is connected to the crankshaft, it turns at the speed of the engine. Also, during vehicle deceleration the ISG regenerates the power that is used to slow the vehicle to recharge the HV and auxiliary batteries. When the vehicle is traveling downhill and there is zero load on the engine, the wheels can transfer energy through the transmission and engine to the ISG. The energy (AC voltage) is converted to DC voltage and sent to the batteries for storage and use by the electrical components of the vehicle.

Remember that the output of conventional AC generators is dependent upon the intensity of the magnetic field, the number of conductors passing through the magnetic field at any given time, the number of magnets, and the speed at which the lines of flux or the conductors are moving when the intercept occurs. Since the speed at which the magnetic poles move influences the amount of current output, output will be the lowest at idle. Maximum output will not be achieved until the engine speed is increased. It is during the times of low engine speeds that current demand is likely to be at its highest. To increase output, the hybrid rotor has permanent magnets located between the pole pieces of the rotor. The magnetic flux from these permanent magnets goes into the pole piece, through the rotor shaft, and then back through the pole piece on the opposite side of the magnet. The permanent magnet fills the gap between the pole pieces, forcing more of the flux from the rotor into the stator windings. This results in an increase in the alternator's output.

Regulation uses a technique that is referred to as "boost-buck." At low speed and high electrical loads, the wire-wound section is fully energized. This extra magnetic flux then boosts the output of the permanent magnet section. When the engine is operated at a medium speed and with a medium electrical load, the field current is off. During this time, only the permanent magnet section is producing the output. During high-speed, low-electrical-load conditions, the field current is reversed. This bucks the permanent magnet's field and maintains a constant output voltage.

Full hybrid vehicles that are capable of propelling the vehicle in an electric-only mode require HV batteries to power the three-phase AC motors. These batteries may have a capacity of over 300 volts. Many full hybrid vehicles have at least two AC motors located in the transmission or transaxle assembly to operate the planetary gear sets that provide constantly variable gear ratios (**Figure 7-66**). These motors can also be used as generators. If the HV SOC becomes too low, the engine is started and the crankshaft drives motor "A" to generate high-voltage AC current. The current is rectified to DC voltage and sent to recharge the HV battery. Voltage generation can also occur whenever one of the motors slips. In most cases, one of the motors slips at all times. The slipping causes a cutting of the magnetic field and results in AC current. This current is used to supply electrical energy to the other motor (**Figure 7-67**).

Regenerative Braking

About 30% of kinetic energy lost during braking is in heat. When decreasing acceleration, regenerative braking helps minimize energy loss by recovering the energy used to slow the vehicle. This is done by converting rotational energy into electrical energy through the ISG or AC motors. Regenerative braking assumes some of the stopping duties from the conventional friction brakes and uses the electric motor to help slow the vehicle. To do this, the electric motor operates as a generator when the brakes are applied, recovering some of the kinetic energy and converting it into electrical energy. The motor becomes a generator by using the kinetic energy of the vehicle to store power in the battery for later use.

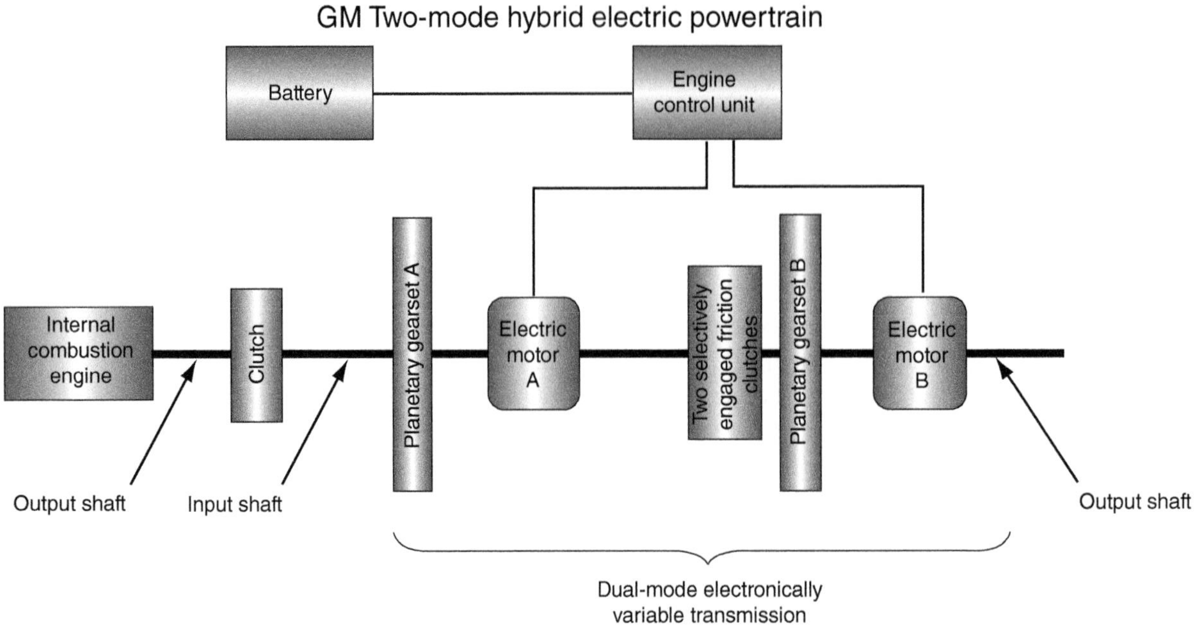

GM Two-mode hybrid electric powertrain

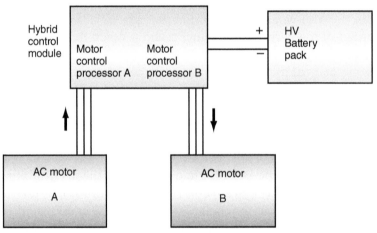

Figure 7-66 Multiple motors are usually used in a full hybrid vehicle.

Figure 7-67 Current generated in one motor can be used to power the other.

When regenerative braking operation takes place, no friction braking occurs. Regenerative braking is mainly a function of light brake pedal application, using inputs from the various sensors such as the pedal angle sensor, the vacuum sensor, and the accelerator pedal position sensor. As soon as the accelerator pedal is released, the hybrid control module will initiate regenerative braking. At this point, the electric motors are being turned by the wheels and act as generators to recharge the HV battery since the rotor is turning within the stator windings. The hydraulic brakes are not used during this phase. If additional vehicle deceleration is required, the hybrid controller can increase the force required to rotate the electric motors. This increases resistance to wheel rotation that helps further slow the vehicle. In cases where more braking power is needed, hydraulic brakes are used.

DC/DC CONVERTER

The AC voltage from the motors during regenerative or charging modes is rectified by the inverter module. Since this module converts the high AC voltage into high DC voltage to recharge the HV battery, it cannot be used to recharge the auxiliary battery. An additional function of the inverter is that of a DC/DC converter. The converter allows the conversion of electrical power between the high-voltage (HV) direct current system and the low-voltage (LV) direct current system. The converter is a bidirectional, solid-state, DC conversion device that charges the 12-volt system from the 300-volt direct current system. The converter replaces the function of the engine-driven generator while maintaining isolation of the HV system.

The conversion of HV to LV is accomplished through magnetic fields instead of physical wired connections. Sets of coils are used to accomplish the voltage conversion (**Figure 7-68**). The coils operate as step-down transformers to reduce the high DC voltage to the low DC voltage. By sequentially inducing and collapsing the magnetic field of the coils, a smooth output on the low voltage side is maintained.

As discussed in Chapter 5, when the HEV is placed into the jump assist mode, the converter is required to charge the 300-volt system by use of the 12-volt battery. In this mode, the coils operate as step-up transformers to increase the 12 volts to the 300 volts required to charge the HV battery.

> **AUTHOR'S NOTE** When there is a high rate of current conversion taking place in the inverter/converter module, a significant amount of heat is produced. Many manufacturers will use a separate cooling system to run coolant through these modules to prevent damage from the heat.

The inverter is PWM controlled by the hybrid control module or by bus communication from the control module. The hybrid control module requests the required amount of voltage to be applied to the LV system. This will normally be a commanded PWM to the converter that is between 33% and 90%. In this range, the converter increases or decreases the charging voltage between 12.5 and 15.5 volts. In the event that the signal is lost from the control module (or the signal is corrupted in any way), the default low-voltage setting is at 13.8 volts.

Figure 7-68 DC/DC converter coils are used as transformers to reduce the 300 volts DC to about 14 volts DC.

SUMMARY

- The most common method of stator connection is called the wye connection. In the wye connection, one lead from each winding is connected to one common junction. From this junction, the other leads branch out in a Y pattern.

- Another method of stator connection is called the delta connection. The delta connection connects the lead of one end of the winding to the lead at the other end of the next winding.

- The diode rectifier bridge provides reasonably constant DC voltage to the vehicle's electrical system and battery. The diode rectifier bridge is used to change the current in an AC generator.

- The conversion of AC current to DC current is called rectification.

- The three principal circuits used in the AC generator are the charging circuit, which consists of the stator windings and rectifier circuits; the excitation circuit, which consists of the rotor field coil and the electrical connections to the coil; and the preexcitation circuit, which supplies the initial current for the field coil that starts the buildup of the magnetic field.

- The voltage regulator controls the output voltage of the AC generator, based on charging system demands, by controlling field current. The higher the field current, the higher the output voltage.

- The regulator must have system voltage as an input in order to regulate the output voltage. The input voltage to the regulator is called sensing voltage.

- Because ambient temperatures influence the rate of charge that a battery can accept, regulators are temperature compensated.

- The A circuit is called an external grounded field circuit and is always an electronic-type regulator. In the A circuit, the regulator is on the ground side of the field coil. The B+ for the field coil is picked up from inside the AC generator.

- Usually the B circuit regulator is mounted externally to the AC generator. The B circuit is an internally grounded circuit. In the B circuit, the voltage regulator controls the power side of the field circuit.

- Isolated field AC generators pick up B+ and ground externally. The AC generator has two field wires attached to the outside of the case. The voltage regulator can be located either on the ground (A circuit) or on the B+ (B circuit) side.

- An electronic regulator uses solid-state circuitry to perform the regulatory functions using a zener diode that blocks current flow until a specific voltage is obtained, at which point it allows the current to flow.

- On most modern vehicles, the regulator function has been incorporated into the vehicle's engine computer. Regulation of the field circuit is through the ground (A circuit).

- There are three basic methods of informing the driver of the charging system's condition: indicator lamps, ammeter, and voltmeter.

- Most indicator lamps operate on the basis of voltage drop. If the charging system output is less than battery voltage, there is an electrical potential difference in the lamp circuit and the lamp will light.

- The ammeter measures charging and discharging current in amperes. The ammeter is wired in series between the AC generator and the battery.

- The voltmeter is usually connected between the battery's positive and negative terminals.

- The hybrid AC generator design consists of a rotor assembly with both wire-wound and permanent magnet sections. The permanent magnets are located between the pole pieces of the rotor.

- The ISG is a three-phase AC motor that can provide power and torque to the vehicle that can also generate voltage whenever the rotor is turning.

- Generation is done anytime the engine is running.

- During vehicle deceleration, the ISG regenerates the power that is used to slow the vehicle to recharge the HV and auxiliary batteries.

- The AC motors used in full hybrid vehicles can be used as generators whenever the engine is running and during vehicle deceleration.

- Generation can also occur whenever one of the motors slips. The slipping causes a cutting of the magnetic field and results in AC current. This current is used to supply electrical energy to the other motor.

- While the vehicle is coasting, the electric motors are being turned by the wheels and act as generators to recharge the HV battery. If additional vehicle deceleration is required, the hybrid controller can increase the force required to rotate the electric motors.

- The AC voltage from the motors during regenerative or charging modes is rectified by the inverter module. This module converts the high AC voltage into high DC voltage to recharge the HV battery.

- The DC/DC converter allows the conversion of electrical power between the HV system and the LV system and replaces the function of the engine-driven generator while maintaining isolation of the HV system.

REVIEW QUESTIONS

Short-Answer Essays

1. List the major components of the charging system.

2. List and explain the function of the major components of the AC generator.

3. What is the relationship between field current and AC generator output?

4. Identify the differences between A, B, and isolated circuits.

5. Explain the operation of non-computer–controlled charge indicator lamps.

6. Describe the two styles of stators.

7. What is the difference between half and full-wave rectifications?

8. Describe how AC voltage is rectified to DC voltage in the AC generator.

9. What is the purpose of the charging system?

10. Explain the meaning of regenerative braking.

Fill in the Blanks

1. The charging system converts the _____ energy of the engine into _____ energy to recharge the battery and run the electrical accessories.

2. All charging systems use the principle of _____ to generate electrical power.

3. The _____ creates the rotating magnetic field of the AC generator.

4. _____ are electrically conductive sliding contacts, usually made of copper and carbon.

5. In the _____ connection stator, one lead from each winding is connected to one common junction.

6. The _____ controls the output voltage of the AC generator, based on charging system demands, by controlling _____ current.

7. In an electronic regulator, _____ controls AC generator output by varying the amount of time the field coil is energized.

8. Full-wave rectification in the AC generator requires _____ pair of diodes.

9. The _____ is the stationary coil that produces current in the AC generator.

10. _____ recovers the heat energy used to brake by converting rotational energy into _____ energy through a system of electric motors and generators.

Multiple Choice

1. The magnetic field current of the AC generator is carried in the:
   - A. Rotor.
   - B. Diode trio.
   - C. Rectifier bridge.
   - D. Stator.

2. The voltage induced in one conductor by one revolution of the rotor is called:
   - A. Three phase
   - B. Half wave
   - C. Single phase
   - D. Full wave

3. Rectification is being discussed.
   Technician A says the AC generator uses a segmented commutator to rectify AC current.
   Technician B says the DC generator uses a pair of diodes to rectify AC current.
   Who is correct?
   - A. A only
   - B. B only
   - C. Both A and B
   - D. Neither A nor B

4. Rotor construction is being discussed.
   Technician A says the poles will take on the polarity of the side of the coil that they touch.
   Technician B says the magnetic flux lines will move in opposite directions between adjacent poles.
   Who is correct?
   - A. A only
   - B. B only
   - C. Both A and B
   - D. Neither A nor B

5. The amount of voltage output of the AC generator is related to:
   - A. Field strength.
   - B. Stator speed.
   - C. Number of rotor segments.
   - D. All of the above.

6. The delta-wound stator:

   A. Shares a common connection point.

   B. Has each winding connected in series.

   C. Does not require rectification.

   D. None of the above.

7. Non-computer–controlled indicator lamp operation is being discussed.

   Technician A says in a system with an electronic regulator, the lamp will light if there is no stator output through the diode trio.

   Technician B says when there is stator output, the lamp circuit has voltage applied to both sides and the lamp will not light.

   Who is correct?

   A. A only

   B. B only

   C. Both A and B

   D. Neither A nor B

8. *Technician A* says only two stator windings apply voltage because the third winding is always connected to diodes that are reverse-biased.

   Technician B says AC generators that use half-wave rectification are the most efficient.

   Who is correct?

   A. A only

   B. B only

   C. Both A and B

   D. Neither A nor B

9. Charging system regulation is being discussed.

   Technician A says the regulation of voltage is done by varying the amount of field current flowing through the rotor.

   Technician B says control of field current can be done either by regulating the resistance in series with the field coil or by turning the field circuit on and off.

   Who is correct?

   A. A only

   B. B only

   C. Both A and B

   D. Neither A nor B

10. In a full hybrid HEV, which component is responsible for recharging the 12-volt battery from the 300-volt direct current system?

    A. The battery control module.

    B. The belt-driven generator.

    C. The DC/DC converter.

    D. The LV control module.

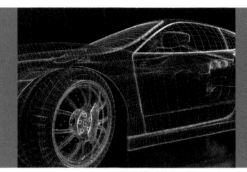

CHAPTER 8

INTRODUCTION TO THE BODY COMPUTER

Upon completion and review of this chapter, you should be able to understand and describe:

- The basic functions of the computer.
- The principle of analog and digital voltage signals.
- The principle of computer communications.
- The basics of logic gate operation.
- The basic function of the microprocessor.
- The basic method by which the microprocessor makes determinations.
- The differences in memory types.
- The operation of low and high-side drivers.
- The operation of common output actuators.

Terms To Know

Actuators	Interface	Program
Adaptive memory	Logic gates	Sequential logic circuits
Adaptive strategy	Low-side drivers	Sequential sampling
Binary code	Microprocessor	Servomotor
Bit	Nonvolatile	Stepper motor
Clock circuit	Output Driver	Volatile
High-side drivers	Output Signal	

INTRODUCTION

A computer is an electronic device that stores and processes data. It is also capable of controlling other devices. This chapter introduces the basic theory and operation of the digital computer used to control most of the vehicle's electrical accessories. The use of computers on automobiles has expanded to include control and operation of several functions, including climate control, lighting circuits, cruise control, antilock braking, electronic suspension systems, and electronic shift transmissions. Some of these are functions of what is known as a body computer module (BCM). Some body computer-controlled systems include direction lights, rear window defoggers, illuminated entry, intermittent wipers, and other systems once thought of as basic.

AUTHOR'S NOTE When computer controls were first installed on the automobile, the aura of mystery surrounding these computers was so great that some technicians were afraid to work on them. Most technicians coming into the field today have grown up around computers and the fear is not as great. Regardless of your comfort level with computers, knowledge is the key in understanding their function. Although it is not necessary to understand all the concepts of computer operation in order to service the systems they control, knowledge of the digital computer will help you feel more comfortable when working on these systems.

COMPUTER FUNCTIONS

A computer processes the physical conditions that represent information (data). The operation of the computer is divided into four basic functions:

1. *Input:* A voltage signal sent from an input device. This device can be a sensor or a switch activated by the driver or technician.
2. *Processing:* The computer uses the input information and compares it to programmed instructions. The logic circuits process the input signals into output demands.
3. *Storage:* The program instructions are stored in an electronic memory. Some of the input signals are also stored for later processing.
4. *Output:* After the computer has processed the inputs and checked its programmed instructions, it will put out control commands to various output devices. These output devices may be the instrument panel display or a system actuator. The output of one computer can also be used as an input to another computer.

 A BIT OF HISTORY

"Computer" was originally a job title, so the first computers were actual people. The job title of computer was used to describe those people who performed the repetitive calculations required to compute navigational tables, tide charts, and planetary positions. Electronic computers were given this name because they performed the work that had previously been assigned to people. People who had the job title of computer relied upon a tool known as the abacus to aid their memory while performing the calculations. Since people who worked as computers would often suffer from boredom that resulted in carelessness and mistakes, inventors started searching for ways to mechanize this task.

During World War II, the United States had battleships capable of firing shells weighing over a ton at targets up to 25 miles away. To determine the trajectory of the shells, physicists provided the equations but solving these equations using human computers proved to be labor intensive. To solve its problems, the U.S. military looked for means of automating these computations. An early success was a computer built from a partnership between Harvard and IBM in 1944 known as the Harvard Mark I. The Harvard Mark I was the first programmable digital computer made in the United States. However, the Harvard Mark I was not exclusively electronic and was constructed out of switches, relays, rotating shafts, and clutches. The computer was 8 feet tall and 51 feet long and weighed 5 tons. It used 500 miles of wire, and required a 50 feet rotating shaft that ran its length that was turned by a 5 horsepower electric motor. The Harvard Mark I contained three quarters of a million components, yet it was capable of storing only 72 numbers. Today's home computers can store 30 million numbers in RAM and another 10 billion numbers on their hard disk. The Harvard Mark I ran nonstop for 15 years. Grace Hopper was one of the primary programmers for the Harvard Mark I and is said to be the founder of the first computer "bug." A dead moth had gotten into the Harvard Mark I and its wings blocked the reading of the holes in the paper tape.

Understanding these four functions will help today's technician organize the trouble-shooting process. When a system is tested, the technician will be attempting to isolate the problem to one of these functions.

ANALOG AND DIGITAL PRINCIPLES

Remembering the basics of electricity, voltage does not flow through a conductor; current flows and voltage is the pressure that "pushes" the current. However, voltage can be used as a signal—for example, difference in voltage levels, frequency of change, or switching from positive to negative values can be used as a signal. The computer uses these voltage signals to perform its functions.

A **program** is a set of instructions the computer must follow to achieve desired results. The program used by the computer is "burned" into integrated circuit (IC) chips using a series of numbers. These numbers represent various combinations of voltages that the computer can understand. The voltage signals to the computer can be either analog or digital. Many of the inputs from the sensors are analog variables. For example, ambient temperature sensors do not change abruptly. The temperature varies in infinite steps from low to high. The same is true for several other inputs such as engine speed, vehicle speed, fuel flow, and so on.

Compared to an analog voltage representation, digital voltage patterns are square-shaped because the transition from one voltage level to another is abrupt (**Figure 8-1**). A digital signal is produced by an on/off or high/low voltage. The simplest generator of a digital signal is a switch (**Figure 8-2**). If 5 volts (V) are applied to the circuit, the voltage sensor will read 5 volts (a high-voltage value) when the switch is open. Closing the switch will result in the voltage sensor reading close to 0 volt. This measuring of voltage drops sends a digital signal to the computer. The voltage values are represented by a series of digits, which create a **binary code**. Binary code is represented by the numbers 1 and 0. Any number and word can be translated into a combination of binary 1s and 0s.

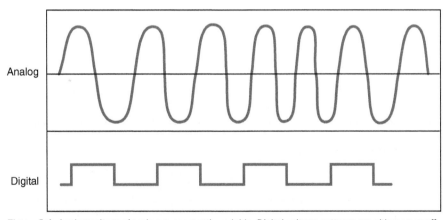

Figure 8-1 Analog voltage signals are constantly variable. Digital voltage patterns are either on or off. Digital signals are referred to as a square sine wave.

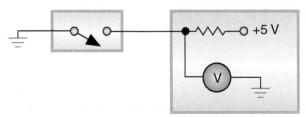

Figure 8-2 Simplified voltage sensing circuit that indicates if the switch is opened or closed.

Binary Numbers

A transistor that operates as a relay is the basis of the digital computer. As the input signal switches from off to on, the transistor output switches from cutoff to saturation. The on and off output signals represent the binary digits 1 and 0.

The computer converts the digital signal into binary code by translating voltages above a given value to 1 and voltages below a given value to 0. As shown in **Figure 8-3**, when the switch is open and 5 volts are sensed, the voltage value is translated into a 1 (high voltage). When the switch is closed, lower voltage is sensed and the voltage value is translated into a 0. Each 1 or 0 represents one bit of information.

In the binary system, whole numbers are grouped from right to left. Because the system uses only two digits, the first portion must equal a 1 or a 0. To write the value of 2, the second position must be used. In binary, the value of 2 would be represented by 10 (one two and zero one). To continue, a 3 would be represented by 11 (one two and one one). **Figure 8-4** illustrates the conversion of binary numbers to digital base

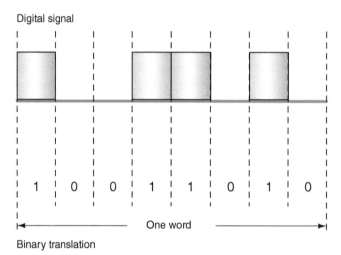

Figure 8-3 Each binary 1 and 0 is one bit of information. Eight bits equal one byte.

Decimal number	Binary number code 8 4 2 1	Binary to decimal conversion
0	0000	= 0 + 0 = 0
1	0001	= 0 + 1 = 1
2	0010	= 2 + 0 = 2
3	0011	= 2 + 1 = 3
4	0100	= 4 + 0 = 4
5	0101	= 4 + 1 = 5
6	0110	= 4 + 2 = 6
7	0111	= 4 + 2 + 1 = 7
8	1000	= 8 + 0 = 8
9	1001	= 8 + 1 = 9

Figure 8-4 Binary number code conversion to base 10 numbers.

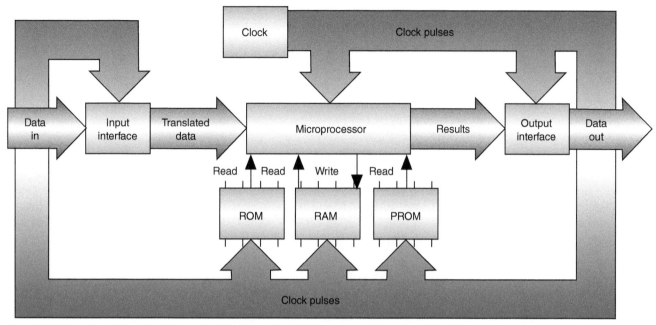

Figure 8-5 Interaction of the main components of the computer. All of the components monitor clock pulses.

ten numbers. Starting from the right, the first position has a place value of 1. Moving to the left each place value doubles. If a thermistor is sensing 150 °F, the binary code would be 10010110. If the temperature increases to 151 °F, the binary code changes to 10010111.

The computer contains a crystal oscillator or **clock circuit** that delivers a constant time pulse. The clock is a crystal that electrically vibrates when subjected to current at certain voltage levels to produces a very regular series of voltage pulses. The clock maintains an orderly flow of information through the computer circuits by transmitting one **bit** of binary code for each pulse (**Figure 8-5**). In this manner, the computer is capable of distinguishing between the binary codes such as 101 and 1001.

A **bit** is a 0 or a 1 of the binary code. Eight bits is called a byte.

Signal Conditioning and Conversion

The input and/or output signals may require conditioning in order to be used. This conditioning may include amplification and/or signal conversion.

Some input sensors produce a very low-voltage signal of less than 1 volt. This signal has an extremely low-current flow. Therefore, this type of signal must be amplified (increased) before it is sent to the microprocessor by an amplification circuit in the input conditioning chip inside the computer (**Figure 8-6**).

For the computer to receive information from the sensor and give commands to actuators, it requires an **interface**. The computer will have two interface circuits: input and output. An interface is used to protect the computer from excessive voltage levels and to translate input and output signals. The digital computer cannot accept analog signals from the sensors and requires an input interface to convert the analog signal to a digital signal. The analog-to-digital (A/D) converter continually scans the analog input signals at regular intervals. For example, if the A/D converter scans the TPS signal and finds the signal at 5 volts, the A/D converter assigns a numeric value to this specific voltage. Then the A/D converter changes this numeric value to a binary code (**Figure 8-7**).

Also, some of the controlled actuators may require an analog signal. In this instance, an output digital to analog (D/A) converter is used.

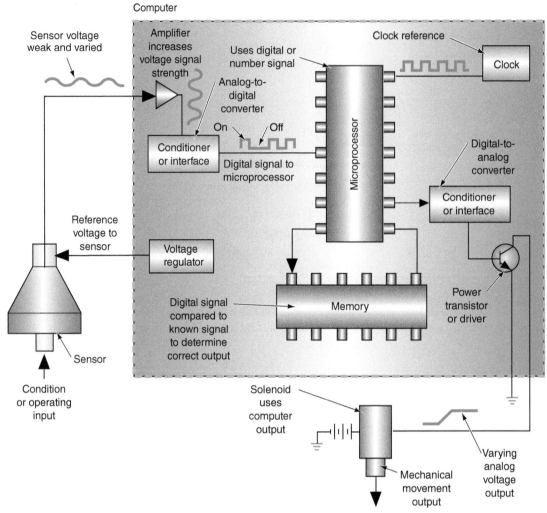

Figure 8-6 Amplification and interface circuits in the computer. The amplification circuit boosts the voltage and conditions it. The interface converts analog inputs into digital signals. The digital-to-analog converter changes the output voltage from digital to analog.

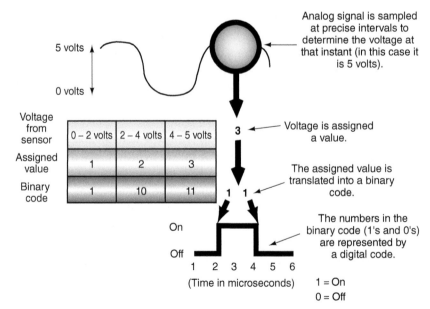

Figure 8-7 The A/D converter assigns a numeric value to input voltages and changes this numeric value to a binary code.

MICROPROCESSOR

The **microprocessor (μP)** is the brain of the computer. The μP is constructed of thousands of transistors that are placed on a small chip. The μP brings information into and out of the computer's memory. The input information is processed in the μP and checked against the program in memory. The μP also checks memory for any other information regarding programmed parameters. The information obtained by the μP can be altered according to the program instructions. The program may have the μP apply logic decisions to the information. Once all calculations are made, the μP will deliver commands to make the required corrections or adjustments to the operation of the controlled system.

The program guides the microprocessor in decision-making. For example, the program may inform the microprocessor when sensor information should be retrieved and then tell the microprocessor how to interpret this information. Finally, the program guides the microprocessor regarding the activation of output control devices such as relays and solenoids. The various memories contain the programs and other vehicle data, which the microprocessor refers to as it performs calculations. As the μP performs calculations and makes decisions, it works with the memories by either reading or writing information to them.

The μP has several main components (**Figure 8-8**). The registers used include the accumulator, the data counter, the program counter, and the instruction register. The control unit implements the instructions located in the instruction register. The arithmetic logic unit (ALU) performs the arithmetic and logic functions.

The terms microprocessor and central processing unit are basically interchangeable.

COMPUTER MEMORY

The computer requires a means of storing both permanent and temporary information. The memories contain many different locations. These locations can be compared to file folders in a filing cabinet, with each location containing one piece of information. An address is assigned to each memory location. This address may be compared to the lettering or numbering arrangement on file folders. Each address is written in a binary code, and these codes are numbered sequentially beginning with 0.

While the engine is running, the engine computer receives a large quantity of information from a number of sensors. The computer may not be able to process all this information immediately. In some instances, the computer may receive sensor inputs, which the computer requires to make a number of decisions. In these cases, the μP writes information into memory by specifying a memory address and sending information to this address.

When stored information is required, the μP specifies the stored information address and requests the information. When stored information is requested from a specific address, the memory sends a copy of this information to the μP. However, the original stored information is still retained in the memory address.

The memories store information regarding the ideal air–fuel ratios for various operating conditions. The sensors inform the computer about the engine and vehicle operating conditions. The μP reads the ideal air–fuel ratio information from memory and compares this information with the sensor inputs. After this comparison, the μP makes the necessary decision and operates the injectors to provide the exact air–fuel ratio the engine requires.

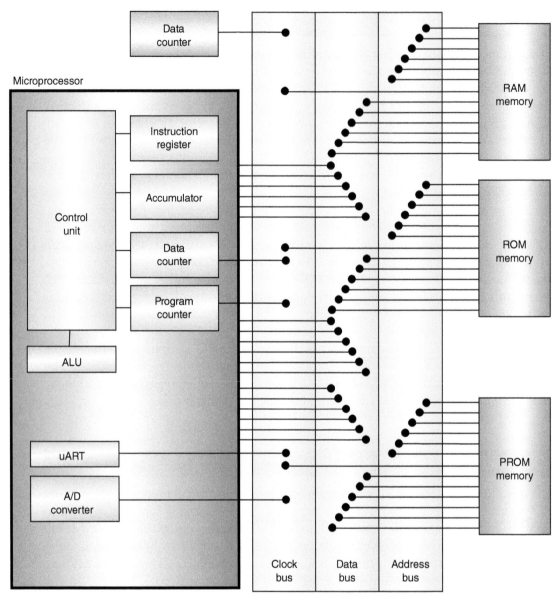

Figure 8-8 Main components of the computer and the μP.

Several types of memory chips may be used in the computer:

The terms *ROM, RAM,* and *PROM* are used fairly consistently in the computer industry. However, the names vary between automobile manufacturers.

1. **Read-only memory (ROM)** contains a fixed pattern of 1s and 0s that represent permanent stored information. This information is used to instruct the μP on what to do in response to input data. The μP reads the information contained in ROM but it cannot write to it or change it. ROM is permanent memory that is programmed in. This memory is not lost when power to the computer is removed. ROM contains formulas and calibrations that identifies the basic purpose and functions of the computer, along with identification data.

2. **Random access memory (RAM)** is constructed from flip-flop circuits formed into the chip. The RAM will store temporary information that can be read from or written to by the μP. RAM stores information that is waiting to be acted upon and the output signals that are waiting to be sent to an output device. RAM can be designed as **volatile** or **nonvolatile**. In volatile RAM, the data will be retained only when current flows through the memory. When the RAM chip is no longer

powered, its data is erased. Typically, RAM that is connected to the battery through the ignition switch will lose its data when the switch is turned off. Nonvolatile RAM retains its data even if the chip is powered down (see below "Nonvolatile RAM").

3. **Keep alive memory (KAM)** is a version of volatile RAM. KAM is connected directly to the battery through circuit protection devices. For example, the µP can read and write information to and from the KAM and erase KAM information. However, the KAM retains information when the ignition switch is turned off. KAM will be lost when the battery is disconnected, the battery drains too low, or the circuit opens.

Shop Manual
Chapter 8, page 365

4. **Programmable read only memory (PROM)** contains specific data that pertains to the exact vehicle in which the computer is installed. This information may be used to inform the µP of the accessories that are equipped on the vehicle. The information stored in the PROM is the basis for all computer logic. For example, this chip installed into a powertrain control module (PCM) will provide the look-up tables for injection pulse width under all engine operating conditions. The information in PROM is used to define or adjust the operating parameters held in ROM.

5. **Erasable PROM (EPROM)** is similar to PROM except that its contents can be erased to allow new programming to be installed. A piece of Mylar tape covers a window. If the tape is removed, the microcircuit is exposed to ultraviolet light that erases its memory (**Figure 8-9**). Once erased, new data can be programmed into the chip.

6. **Electrically erasable PROM (EEPROM)** allows changing the information electrically one bit at a time. Since it is a form of PROM, it contains the specific operating instructions for the system the computer operates. Some manufacturers use this type of memory to store information concerning mileage, vehicle identification number, and options. The flash EEPROM may be reprogrammed through the data link connector (DLC) using special diagnostic equipment. The PROM is erased when a voltage of a specific level is applied to the chip. It then can be loaded with a new set of files. Manufacturers use this type of chip so the computer functions can be updated quickly without replacing the computer itself. For example, if there is a problem with the operation of the memory seat function that is related to the programmed instructions, the use of a EEPROM will allow the new instructions to be loaded into the chip.

Shop Manual
Chapter 8, page 380

7. **Nonvolatile RAM (NVRAM)** is a combination of RAM and EEPROM in the same chip. During normal operation, data is written to and read from the RAM portion of the chip. If the power is removed from the chip, or at programmed timed intervals, the data is transferred from RAM to the EEPROM portion of the chip. When the power is restored to the chip, the EEPROM will write the data back to the RAM. One example of this type of memory would be vehicle mileage tracking. As vehicle mileage is accumulated, the current value is stored in RAM. At set intervals (or during key off) the current vehicle mileage is load into a specified block of the EEPROM chip by first erasing the stored value and then loading the current value.

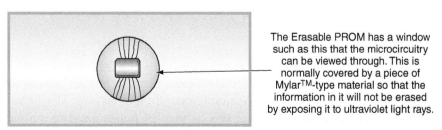

The Erasable PROM has a window such as this that the microcircuitry can be viewed through. This is normally covered by a piece of Mylar™-type material so that the information in it will not be erased by exposing it to ultraviolet light rays.

Figure 8-9 EPROM memory is erased when the ultraviolet rays contact the microcircuitry.

Adaptive Strategy and Memory

If a computer has **adaptive strategy** capabilities, it can actually learn from past experience. For example, the normal voltage input range from an ambient temperature sensor may be 0.6 to 4.5 volts. If the sensor sends a 0.4-volt signal to the computer, the μP interprets this signal as an indication of component wear and stores this altered calibration in the RAM. The μP now refers to this new calibration during calculations, and normal system performance is maintained. If a sensor output is erratic or considerably out of range, the computer may ignore this input. When a computer has adaptive strategy, a short learning period may be necessary under the following conditions:

1. After the battery has been disconnected.
2. When a computer system component has been replaced or disconnected.
3. A new vehicle.
4. Replacement computer is installed.

Adaptive memory is the ability of the computer system to store changing values and to correct operating characteristics. For example, a transmission control module may monitor the transmission's input and output shaft speeds to determine gear ratio. If the input speed sensor indicates a speed of 1,000 rpm and the output speed sensor indicates a speed of 333 rpm, then the controller determines that the ratio is 3:1 (first gear). When the controller determines that it will make the shift to second gear (2:1 ratio), it monitors the sensors to see how long it takes to achieve the ratio change from 3:1 to 2:1. The length of time required represents the amount of fluid needed to stroke the clutch piston and lock up the clutch element. This value is learned so the timing of the shifts can be altered as the clutch elements wear, yet the quality of the shifts will not deteriorate over the life of the transmission.

INFORMATION PROCESSING

The air charge temperature (ACT) sensor input will be used as an example of how the computer processes information. If the air temperature is low, the air is denser and contains more oxygen per cubic foot. Warmer air is less dense and therefore contains less oxygen per cubic foot. The cold, dense air requires more fuel compared to the warmer air that is less dense. The μP must supply the correct amount of fuel in relation to air temperature and density.

An ACT sensor is positioned in the intake manifold where it senses air temperature. This sensor contains a resistive element that has an increased resistance when the sensor is cold. Conversely, the ACT sensor resistance decreases as the sensor temperature increases. When the ACT sensor is cold, it sends a high-analog voltage signal to the computer, and the A/D converter changes this signal to a digital signal.

When the μP receives this ACT signal, it addresses the tables in the ROM. The look-up tables list air density for every air temperature. When the ACT sensor voltage signal is very high, the look-up table indicates very dense air. This dense air information is relayed to the μP, and the μP operates the output drivers and injectors to supply the exact amount of fuel the engine requires (**Figure 8-10**).

Logic Gates

Logic gates are the thousands of field effect transistors (FETs) incorporated into the computer circuitry. These circuits are called logic gates because they act as gates to output voltage signals depending on different combinations of input signals. The FETs use the incoming voltage patterns to determine the pattern of pulses leaving the gate. The

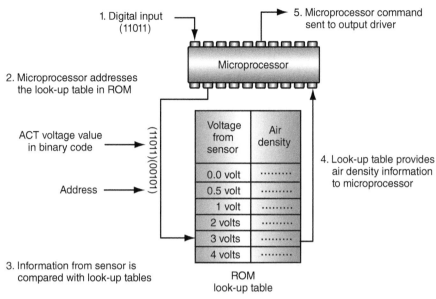

1. Digital input
(11011)

5. Microprocessor command
sent to output driver

Microprocessor

2. Microprocessor addresses
the look-up table in ROM

ACT voltage value
in binary code

(11011)(00101)

Address

Voltage from sensor	Air density
0.0 volt	....
0.5 volt	....
1 volt	....
2 volts	....
3 volts	....
4 volts	....

4. Look-up table provides
air density information
to microprocessor

3. Information from sensor is
compared with look-up tables

ROM
look-up table

Figure 8-10 The microprocessor addresses the look-up tables in the ROM, retrieves air density information, and issues commands to the output devices.

following are some of the most common logic gates and their operations. The symbols represent functions and not electronic construction:

1. **NOT gate:** A NOT gate simply reverses binary 1s to 0s and vice versa (**Figure 8-11**). A high input results in a low output and a low input results in a high output.
2. **AND gate:** The AND gate will have at least two inputs and one output. The operation of the AND gate is similar to two switches in series to a load (**Figure 8-12**). The only way the light will turn on is if switches A *and* B are closed. The output of the gate will be high only if both inputs are high. Before current can be present at the output of the gate, current must be present at the base of both transistors (**Figure 8-13**).

Input   Output

Truth table	
Input	Output
0	1
1	0

Figure 8-11 The NOT gate symbol and truth table. The NOT gate inverts the input signal.

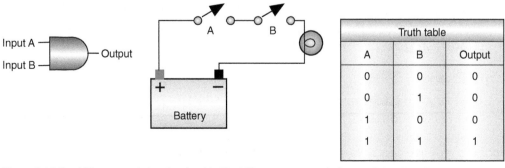

Input A
Input B
Output

A   B

Battery

Truth table		
A	B	Output
0	0	0
0	1	0
1	0	0
1	1	1

Figure 8-12 The AND gate symbol and truth table. The AND gate operates similar to switches in series.

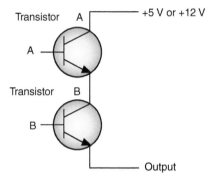

Figure 8-13 The AND gate circuit.

3. **OR gate:** The OR gate operates similarly to two switches that are wired in parallel to a light (**Figure 8-14**). If switch A *or* B is closed, the light will turn on. A high signal to either input will result in a high output.

4. **NAND and NOR gates:** A NOT gate placed behind an OR or AND gate inverts the output signal (**Figure 8-15**).

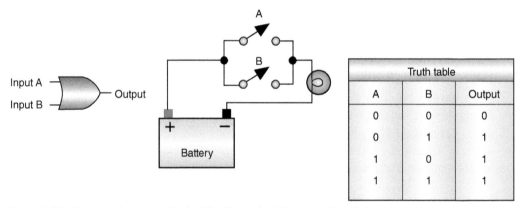

Truth table		
A	B	Output
0	0	0
0	1	1
1	0	1
1	1	1

Figure 8-14 OR gate symbol and truth table. The OR gate is similar to parallel switches.

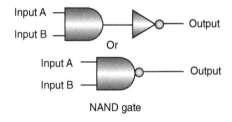

Truth table		
A	B	Output
0	0	1
0	1	1
1	0	1
1	1	0

Truth table		
A	B	Output
0	0	1
0	1	0
1	0	0
1	1	0

Figure 8-15 Symbols and truth tables for NAND and NOR gates. The small circle represents an inverted output on any logic gate symbol.

5. **Exclusive-OR (XOR) gate:** A combination of gates that will produce a high-output signal only if the inputs are different (**Figure 8-16**).

These different gates are combined to perform the processing function. The following are some of the most common combinations:

1. **Decoder circuit:** A combination of AND gates used to provide a certain output based on a given combination of inputs (**Figure 8-17**). When the correct bit pattern is received by the decoder, it will produce the high-voltage signal to activate the relay coil.
2. **Multiplexer (MUX):** The basic computer is not capable of looking at all of the inputs at the same time. A multiplexer is used to examine one of many inputs depending on a programmed priority rating (**Figure 8-18**). This process is called **sequential sampling**. This means the computer will deal with all of the sensors and actuators one at a time.
3. **Demultiplexer (DEMUX):** Operates similar to the MUX except that it controls the order of the outputs (**Figure 8-19**).

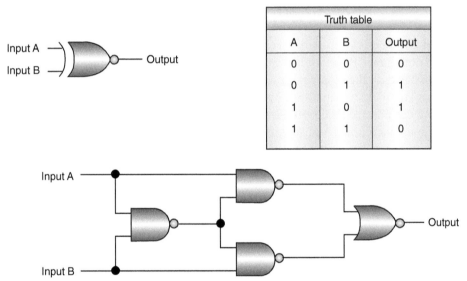

Truth table		
A	B	Output
0	0	0
0	1	1
1	0	1
1	1	0

Figure 8-16 XOR gate symbol and truth table. A XOR gate is a combination of NAND and NOR gates.

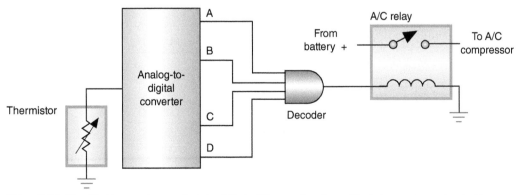

Figure 8-17 Simplified temperature sensing circuit that will turn on the air-conditioning compressor when inside temperatures reach a predetermined value.

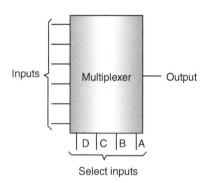

Figure 8-18 Selection at inputs D, C, B, and A will determine which data input will be processed.

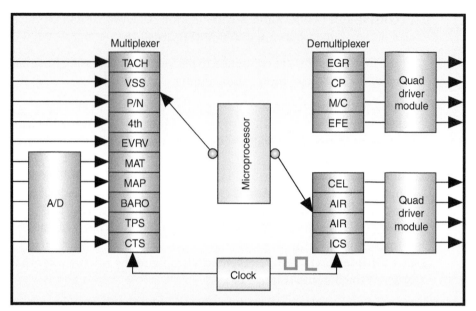

Figure 8-19 Block diagram representation of the MUX and DEMUX circuit.

4. **RS and clocked RS flip-flop circuits:** Logic circuits that remember previous inputs and do not change their outputs until they receive new input signals. **Figure 8-20** shows a basic RS flip-flop circuit. The clocked flip-flop circuit has an inverted clock signal as an input so that circuit operations occur in proper order (**Figure 8-21**). Flip-flop circuits are called **sequential logic circuits** because the sequence of inputs determines the output. A given input affects the output produced by the next input.

5. **Driver circuits:** *Driver* is a term used to describe a transistor device that controls the current in the output circuit. Drivers are controlled by the μP to operate high-current circuits. The high currents handled by a driver are not really that high; they are just more than what is typically handled by a transistor. Several types of driver circuits are used on automobiles, such as Quad, Discrete, Peak and Hold, and Saturated Switch driver circuits.

6. **Registers:** A register is a combination of flip-flops that transfer bits from one to another every time a clock pulse occurs (**Figure 8-22**). It is used in the computer to temporarily store information.

7. **Accumulators:** Registers designed to store the results of logic operations that can become inputs to other modules.

Figure 8-20 (A) RS flip-flop symbol. (B) Truth table. (C) Logic diagram. Variations of the circuit may include NOT gates at the inputs, if used; the truth table outputs would be reversed.

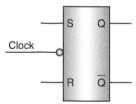

Figure 8-21 Clocked RS flip-flop symbol.

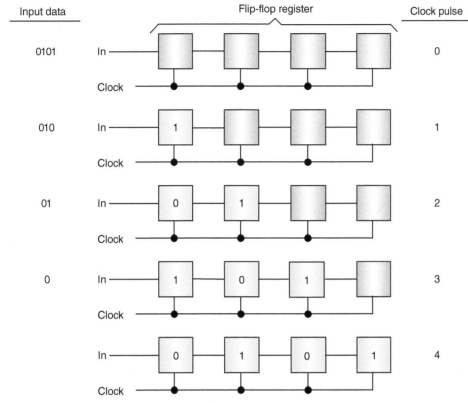

Figure 8-22 It takes four clock pulses to load 4 bits into the register.

HIGH-SIDE AND LOW-SIDE DRIVERS

Usually the computer will control an actuator using **low-side drivers**. These drivers will complete the path to ground through an FET transistor to control the output device. The computer may monitor the voltage on this circuit to determine if the actuator operates when commanded (**Figure 8-23**). Monitoring of the system can be done either by measuring voltage on the circuit or by measuring the current draw of the circuit.

Many manufacturers are now using **high-side drivers**, which control the output device by varying the positive (12-volt) side. High-side drivers consist of a Metal Oxide Field Effect Transistor (MOSFET) that is controlled by a bipolar transistor. The bipolar transistor is controlled by the microprocessor. The advantage of the high-side driver is that it may provide quick-response self-diagnostics for shorts, opens, and thermal conditions. It also reduces vehicle wiring.

High-side driver diagnostic capabilities may include the ability to determine a short- or open-circuit condition. In this case, the high-side driver will take the place of a fuse. In the event of a short-circuit condition, it senses the high-current condition and will turn off the power flow. It may then store a diagnostic trouble code (DTC) in memory. The driver will automatically reset once the short-circuit condition is removed. In addition, the high-side driver monitors its internal temperature. The driver reports the junction temperature to the µP. If a slow-acting resistive short occurs in the circuit, the temperature will begin to climb. Once the temperature reaches 300°F (150°C), the driver will turn off and set a DTC.

The high-side driver is also capable of detecting an open circuit, even if the system is turned off. This is done by reading a feedback voltage to the µP when the driver is off. For example, a 5-volt, 50 µA current can be fed through the circuit, which also has a resistor wired in parallel. Low voltage (less than 2.25 volts) will indicate a normal circuit. If the voltage is high (above 2.25 volts), a high resistance or open circuit is detected. If the open circuit is detected, a DTC is set.

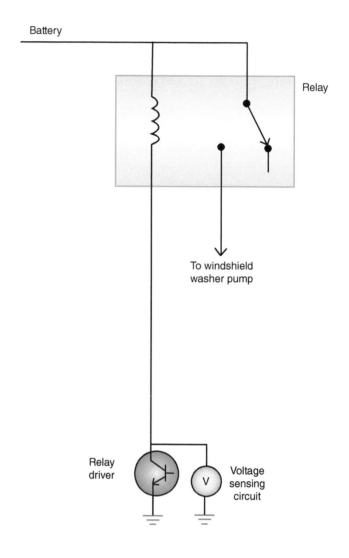

Figure 8-23 Computers using low-side drivers may be able to monitor the circuit for proper operation. When the relay coil is not energized, the sense circuit should see a high voltage (12 V). When the relay is turned on, the voltage should go low (0 V).

OUTPUTS

Once the computer's programming instructs that a correction, adjustment, or activation must be made in the controlled system, an **output signal** is sent to an actuator. This involves translating the electronic signals into mechanical motion.

An **output driver** is used within the computer to control the actuators. The circuit driver usually applies the ground circuit of the actuator. The ground can be applied steadily if the actuator needs to be activated for a selected amount of time. For example, if the BCM inputs indicate that the automatic door locks are to be activated, the actuator is energized steadily until the locks are latched. Then the ground is removed.

Other systems require the actuator to be turned either on and off very rapidly or for a set amount of cycles per second. It is duty cycled if it is turned on and off a set amount of cycles per second. Most duty-cycled actuators cycle ten times per second. To complete a cycle, it must go from off to on to off again. If the cycle rate is ten times per second, one

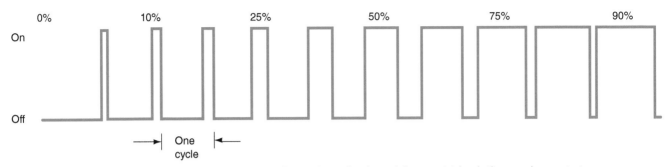

Figure 8-24 Duty cycle is the percentage of on time per cycle. Duty cycle can be changed; however, total cycle time remains constant.

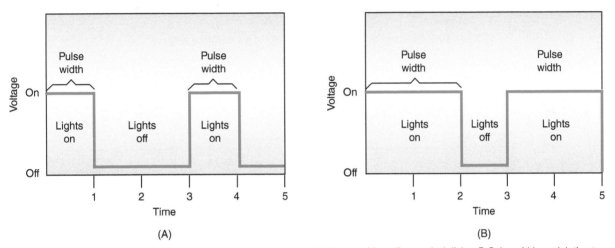

Figure 8-25 Pulse width is the duration of on time. A. Pulse-width modulation to achieve dimmer dash lights. B. Pulse-width modulation to achieve brighter dash illumination.

actuator cycle is completed in one tenth of a second. If the actuator is turned on for 30% of each tenth of a second and off for 70%, it is referred to as a 30% duty cycle (**Figure 8-24**).

If the actuator is cycled on and off very rapidly, the pulse width can be varied to provide the programmed results. For example, the computer program will select an illumination level of the digital instrument panel based on the intensity of the ambient light in the vehicle. The illumination level is achieved through pulse-width modulation of the lights. If the lights need to be bright, the pulse width is increased, which increases the length of on-time. As the light intensity needs to be reduced, the pulse width is decreased (**Figure 8-25**).

Actuators

Most computer-controlled **actuators** are electromechanical devices that convert the output commands from the computer into mechanical action. These actuators are used to open and close switches, control vacuum flow to other components, and operate doors or valves, depending on the requirements of the system.

Shop Manual
Chapter 8, page 375

Although they do not fall into the strict definition of an actuator, the BCM can also control lights, gauges, and driver circuits.

Relays. Relays allow control of a high-current circuit by a very low-current circuit. The computer can control the relay by providing the ground for the relay coil through a low-side driver (**Figure 8-26**). The use of relays protects the computer by keeping the high current from passing through it. For example, the motors used for power door locks require a high current draw to operate them. Instead of having the computer operate the motor directly, it will energize the relay. With the relay energized, a direct circuit from the battery to the motor is completed (**Figure 8-27**). In this illustration, the BCM is controlling the lock relay by a high-side driver.

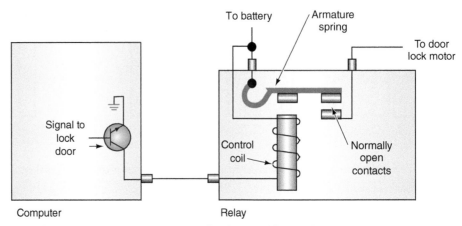

Figure 8-26 The computer's output driver applies the ground for the relay coil.

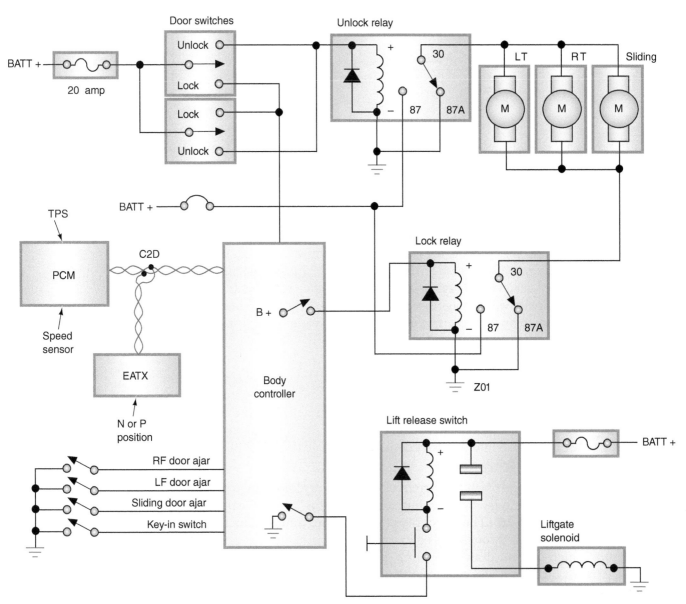

Figure 8-27 The computer controls the operation of the door lock motors by controlling the relays.

Solenoids. Computer control of the solenoid is usually provided by applying the ground through the output driver. A solenoid is commonly used as an actuator because it operates well under duty-cycling conditions.

An example of using a solenoid is to control vacuum to other components. Automatic climate control systems may use vacuum motors to move the blend doors. The computer can control the operation of the doors by controlling the solenoid.

Motors. Many computer-controlled systems use a **stepper motor** to move the controlled device to whatever location is desired. A stepper motor contains a permanent magnet armature with two, four, or more field coils (**Figure 8-28**). By applying voltage pulses to selected coils of the motor, the armature will turn a specific number of degrees. When the same voltage pulses are applied to the opposite coils, the armature will rotate the same number of degrees in the opposite direction.

Some applications require the use of a permanent magnet field **servomotor** (**Figure 8-29**). A servomotor produces rotation of less than a full turn. A feedback mechanism is used to position itself to the exact degree of rotation required. The polarity of the voltage applied to the armature windings determines the direction the motor rotates. The computer can apply a continuous voltage to the armature until the desired result is obtained.

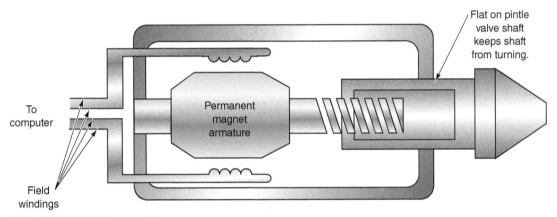

Figure 8-28 Typical stepper motor.

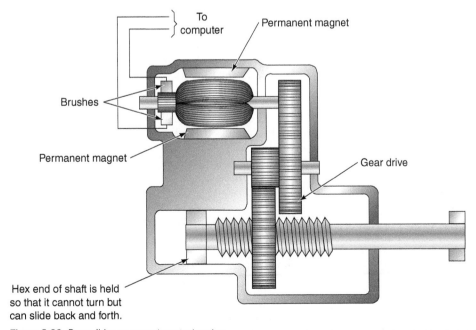

Figure 8-29 Reversible permanent magnet motor.

SUMMARY

- A computer is an electronic device that stores and processes data and is capable of operating other devices.
- The operation of the computer is divided into four basic functions: input, processing, storage, and output.
- Binary numbers are represented by the numbers 1 and 0. A transistor that operates as a relay is the basis of the digital computer. As the input signal switches from off to on, the transistor output switches from cutoff to saturation. The on and off output signals represent the binary digits 1 and 0.
- Logic gates are the thousands of field-effect transistors that are incorporated into the computer circuitry. The FETs use the incoming voltage patterns to determine the pattern of pulses that leave the gate. The most common logic gates are NOT, AND, OR, NAND, NOR, and XOR gates.
- There are several types of memory chips used in the body computer; ROM, RAM, and PROM are the most common types.
- ROM (read-only memory) contains a fixed pattern of 1s and 0s representing permanent stored information used to instruct the computer on what to do in response to input data.

- RAM (random access memory) will store temporary information that can be read from or written to by the μP.
- PROM (programmable read-only memory) contains specific data that pertains to the exact vehicle in which the computer is installed.
- EPROM (erasable PROM) is similar to PROM except its contents can be erased to allow new data to be installed.
- EEPROM (electrically erasable PROM) allows changing the information electrically one bit at a time.
- NVRAM (nonvolatile RAM) is a combination of RAM and EEPROM into the same chip.
- Actuators are devices that perform the actual work commanded by the computer. They can be in the form of a motor, relay, switch, or solenoid.
- A servomotor produces rotation of less than a full turn. A feedback mechanism is used to position itself to the exact degree of rotation required.
- A stepper motor contains a permanent magnet armature with two, four, or more field coils. It is used to move the controlled device to whatever location is desired by applying voltage pulses to selected coils of the motor.

REVIEW QUESTIONS

Short-Answer Essays

1. What is binary code?
2. Describe the basics of NOT, AND, and OR logic gate operation.
3. List and describe the four basic functions of the microprocessor.
4. What is the difference between ROM, RAM, and PROM?
5. Explain the differences between analog and digital signals.
6. What is adaptive strategy?
7. Describe the basic function of a stepper motor.
8. Explain the function of a high-side driver.
9. What is the difference between duty cycle and pulse width?
10. What are the purposes of the interface?

Fill in the Blanks

1. In the binary code, number 4 is represented by _____.
2. The _____ is a crystal that electrically vibrates when subjected to current at certain voltage levels.
3. _____ are registers designed to store the results of logic operations.
4. The _____ is the brain of the computer.
5. _____ contains specific data that pertains to the exact vehicle in which the computer is installed.
6. The _____ gate reverses binary code.
7. _____ drivers complete the actuator control circuit to ground.

8. If a control circuit to an actuator is turned on and off a set amount of cycles per second it is called _____.

9. The _____ function of the computer holds the program instructions.

10. The input _____ converts analog signals to digital signals.

Multiple Choice

1. *Technician A* says during the processing function the computer uses input information and compares it to programmed instructions.
   Technician B says during the output function the computer will put out control commands to various output devices.
   Who is correct?
   A. A only   C. Both A and B
   B. B only   D. Neither A nor B

2. Which of the following is correct?
   A. Analog signals are either high/low, on/off, or yes/no.
   B. Digital signals are infinitely variable within a defined range.
   C. All of the above.
   D. None of the above.

3. Logic gates are being discussed.
   Technician A says NOT gate operation is similar to that of two switches in series to a load.
   Technician B says an AND gate simply reverses binary 1s to 0s and vice versa.
   Who is correct?
   A. A only   C. Both A and B
   B. B only   D. Neither A nor B

4. All of the following statements about computer memory are true, EXCEPT:
   A. RAM stores temporary information that can be written to and read by the CPU.
   B. ROM can be read only by the CPU.
   C. All PROM memory is flashable.
   D. Volatile memory is erased when voltage is removed.

5. Nonvolatile memory is retained if removed from its power source.
   A. True   B. False

6. *Technician A* says the EEPROM can be reprogrammed with new files.
   Technician B says electrostatic discharge can destroy the memory chip.
   Who is correct?
   A. A only   C. Both A and B
   B. B only   D. Neither A nor B

7. *Technician A* says high-side drivers control the ground side of the circuit.
   Technician B says high-side drivers may be capable of determining circuit faults.
   Who is correct?
   A. A only   C. Both A and B
   B. B only   D. Neither A nor B

8. Which of the following would represent the number "255" in binary code?
   A. 00000000   C. 00001111
   B. 11111111   D. 11110000

9. Which of the following is responsible for sequential sampling?
   A. DEMUX   C. MUX
   B. Driver   D. Register

10. *Technician A* says the microprocessor commands actuators by output drivers.
    Technician B says that outputs are never controlled by supplying voltage to the actuator.
    Who is correct?
    A. A only   C. Both A and B
    B. B only   D. Neither A nor B

CHAPTER 9
COMPUTER INPUTS

Upon completion and review of this chapter, you should be able to understand and describe:

- The function of input devices.
- The purpose of the thermistor and how it is used in a circuit.
- The difference between NTC and PTC thermistors.
- The operation and purpose of the Wheatstone bridge.
- The operation and purpose of piezo-electric devices.
- The operation and purpose of piezoresistive devices.
- The function of the potentiometer and how it is used.
- The purpose and operation of magnetic pulse generators.

- The purpose and operation of Hall-effect sensors.
- The function of accelerometers.
- The function of photo cell components such as the photodiode, photoresistor, and the phototransistor.
- The function of the photoelectric sensor.
- The function of the pull-down sense circuit.
- The function of the pull-up sense circuit.
- The purpose and operation of feedback systems.

Terms To Know

Accelerometers
Capacitance discharge sensor
Dual-range circuit
Feedback
Floating
G force
Hall-effect sensor
Infrared temperature sensor
Linearity
Magnetic pulse generators
Magnetically coupled linear sensors
Magnetoresistive effect

Magnetoresistive (MR) sensors
Negative temperature coefficient (NTC) thermistors
Photo cell
Photoconductive mode
Photoresistor
Photovoltaic mode
Pickup coil
Piezoelectric device
Piezoresistive device
Positive temperature coefficient (PTC) thermistors

Potentiometer
Potentiometric pressure sensors
Pull-down circuit
Pull-down resistor
Pull-up circuit
Pull-up resistor
Schmitt trigger
Sensors
Shutter wheel
Strain gauge
Thermistor
Timing disc
Wheatstone bridge

INTRODUCTION

As discussed in Chapter 8, the microprocessor (μP) receives inputs, which it checks with programmed values. Several types of input devices are used to gather information for the computer to use in determining the desired output. Many input devices are also used as a feedback signal to confirm proper positioning of the actuator. Depending on the input, the computer will control the actuator(s) until the programmed results are obtained (**Figure 9-1**). The inputs can come from other computers, the vehicle operator, the technician, or through a variety of sensors.

Driver input signals are usually provided by momentarily applying a ground through a switch. The computer receives this signal and performs the desired function. For example, if the driver wishes to reset the trip odometer on a digital instrument panel, he or she would push the reset switch. This switch will provide a momentary ground that the computer receives as an input and sets the trip odometer to zero.

Switches can be used as an input for any operation that requires only a yes/no, or on/off, condition. Other inputs include those supplied by means of a sensor and those signals returned to the computer in the form of feedback.

This chapter discusses the many different designs of **sensors** and inputs. Sensors convert some measurement of vehicle operation into an electrical signal. Some sensors are nothing more than a switch that completes the circuit. Others are complex chemical reaction devices that generate their own voltage under different conditions. Repeatability, accuracy, operating range, and **linearity** are all requirements of a sensor.

Sensors discussed in this chapter include common forms of electrical and electronic devices used in body and chassis systems. These include the following:

- **Thermistor**—a solid-state variable resistor made from a semiconductor material that changes resistance in relation to temperature changes.
- **Wheatstone bridge**—A series-parallel arrangement of resistors between an input terminal and ground. The sensing circuit will receive a voltage reading that is proportional to the amount of resistance change.

Shop Manual
Chapter 9, page 398

Linearity refers to the sensor signal being as constantly proportional to the measured value as possible. It is an expression of the sensor's accuracy.

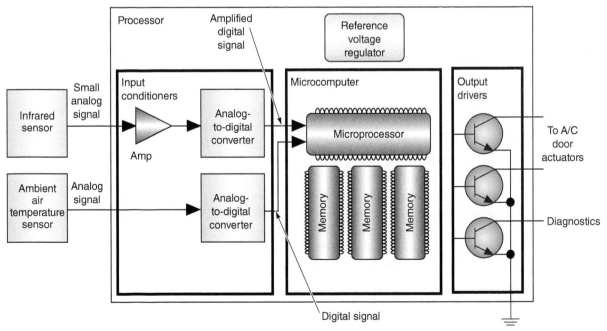

Figure 9-1 The input signals are processed in the microprocessor. The microprocessor directs the output drivers to activate actuators as instructed by the program.

- **Piezoelectric device**—A voltage generator with a resistor connected in series that is used to measure fluid and air pressures.
- **Piezoresistive device**—Similar to a piezoelectric device except they operate like a variable resistor. Its resistance value changes as the pressure applied to the crystal changes.
- **Potentiometer**—A voltage divider that provides a variable direct current (DC) voltage reading to the computer. The potentiometer usually consists of a wire-wound resistor with a movable center wiper.
- **Magnetic pulse generators**—Commonly used to send voltage signals to the computer about the speed of the monitored component. They use the principle of magnetic induction to produce an alternating current (AC) voltage signal that is conditioned by an analog to digital (A/D) converter.
- **Hall-effect sensor**—A sensor that operates on the principle that if a current is allowed to flow through a thin conducting material that is exposed to a magnetic field, another voltage is produced. The sensor contains a permanent magnet, a thin semiconductor layer made of gallium arsenate crystal (Hall layer), and a trigger wheel.

> **AUTHOR'S NOTE** This chapter discusses sensor types that are used in multiple systems. For example, the potentiometer can be used as a throttle position sensor, accelerator position sensor, an A/C mode door position sensor, a seat track position sensor, and so on. Sensors that have a specific function (such as radar, laser, yaw rate, etc.) will be discussed as we explore the system that uses them. Engine management systems use special sensors. These sensors are covered in *Today's Technician: Automotive Engine Performance*. For a more detailed explanation of sensor construction and operation refer to *Today's Technician: Advanced Automotive Electronic Systems*.

THERMISTORS

Shop Manual
Chapter 9, page 400

Thermistors are commonly used to measure the temperature of liquids and air. A thermistor is a solid-state variable resistor made from a semiconductor material, such as metal oxides, that has very reproducible resistance versus temperature properties.

By monitoring the thermistor's resistance value, the computer is capable of observing very small changes in temperature. The computer sends a reference voltage to the thermistor (usually 5 volts) through a fixed resistor. As the current flows through the thermistor resistance to ground, a voltage-sensing circuit measures the voltage after the fixed resistor (**Figure 9-2**). The voltage dropped over the fixed resistor will change as the resistance of the thermistor changes. Using its programmed values, the computer is able to translate the voltage drop into a temperature value.

There are two types of thermistors: **negative temperature coefficient (NTC) thermistors** and **positive temperature coefficient (PTC) thermistors**. NTC thermistors reduce their resistance as the temperature increases, while PTC thermistors increase their resistance as the temperature increases.

Using the circuit shown in Figure 9-2, if the value of the fixed resistor is 10K ohms (Ω) and the value of the thermistor is also 10K ohms, the voltage-sensing circuit will read a voltage value of 2.5 volts (V). If the thermistor is a NTC, as the ambient temperature increases its resistance decreases. If the resistance of the NTC is now 8K ohms, the voltage reading by the voltage-sensing circuit will now be 2.22 volts. As ambient temperature increases and the NTC value continues to decrease, the voltage-sensing circuit will

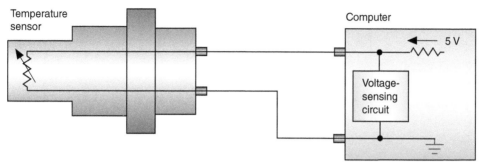

Figure 9-2 A thermistor is used to measure temperature. The sensing unit measures the resistance change and translates the data into temperature values.

measure a voltage decrease (**Figure 9-3**). If the thermistor was a PTC, the opposite would be true and the voltage-sensing circuit would measure an increase in voltage as the ambient temperature increases.

Some temperature-sensing circuits are designed as **dual-range circuits** (**Figure 9-4**). This circuit provides for a switching of the resistance values to allow the microprocessor to measure temperatures more accurately. When the voltage-sensing circuit records a calibrated voltage value (1.25 volts, for example), the microprocessor turns on the transistor, which places the 1K resistor in parallel with the 10K resistor. The circuit will operate the same as described until the 1.25 volts is recorded. With the 1K resistor now involved in the circuit, the resistance of the fixed resistor portion of the circuit is now 909 ohms.

Dual-range circuits are also referred to as dual ramping circuits.

Shop Manual
Chapter 9, page 401

Temperature Sensor			
Voltages versus Temperature Values			
Cold Temperature		Hot Temperature	
Degrees F	Volts	Degrees F	Volts
−20	4.70	110	4.20
−10	4.57	120	4.10
0	4.45	130	4.00
10	4.30	140	3.60
20	4.10	150	3.40
30	3.90	160	3.20
40	3.60	170	3.02
50	3.30	180	2.80
60	3.00	190	2.60
70	2.75	200	2.40
80	2.44	210	2.20
90	2.15	220	2.00
100	1.83	230	1.80
110	1.57	240	1.62
120	1.25	250	1.45

Figure 9-3 Chart of temperature and voltage correlation.

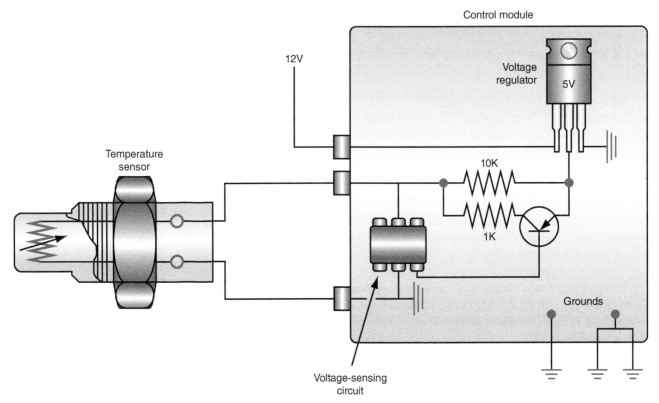

Figure 9-4 Dual-range temperature sensor circuit.

COLD		HOT	
10K-ohm resistor		**909-ohm resistor**	
−20°F	4.7 V	110°F	4.2 V
0°F	4.4 V	130°F	3.7 V
20°F	4.1V	150°F	3.4 V
40°F	3.6 V	170°F	3.0 V
60°F	3.0 V	180°F	2.8 V
80°F	2.4 V	200°F	2.4 V
100°F	1.8 V	220°F	2.0 V
120°F	1.2 V	240°F	1.6 V

Figure 9-5 The same voltage value can represent different temperatures.

At this occurrence, the voltage-sensing circuit will record the sharp increase of voltage and a second range of values can be used. Depending on which side of the switch the voltage is sensed, it can represent two different temperatures (**Figure 9-5**).

Infrared Temperature Sensors

It is becoming more common for manufactures to use an **infrared temperature sensor** instead of in-vehicle temperature sensors. The infrared sensor is capable of measuring surface temperatures without physically touching the surface. Infrared temperature sensors use the principle that all objects emit energy and as the temperature of an object rises, so does the amount of energy it emits. The infrared sensor receives the heat energy radiated from an object and converts the heat to an electrical potential. In climate control systems, this information is used to determine the best control for the desired temperature set by the vehicle occupants.

To determine the temperature of an object, the sensor collects its energy through a lens system. The energy is then focused onto a detector. The detector generates a voltage signal that is sent to the control module. Since the control module has been programmed to know the relationship between the voltage signal and the corresponding temperature, the control module knows the surface temperature of the component.

PRESSURE SENSORS

This section discusses the various types of pressure sensors that are used in automotive applications. In some instances, a simple pressure switch is used. In systems that require monitoring of the exact pressure, electromechanical pressure sensors, piezoresistive sensor, or piezoelectric sensor are used. These sensors convert the applied pressure to an electrical signal. A wide variety of materials and technologies has been used in these devices. These sensors are used to measure the atmospheric air pressure, manifold pressure, pressure of a gas (such as R134a), exhaust pressures, fluid pressures, and so forth. The types of sensors that can be used include: potentiometric, strain gauges using Wheatstone bridges or capacitance discharge, piezoelectric transducers, and pressure differential sensors.

Pressure Switches

Pressure switches will usually use a diaphragm that works against a calibrated spring or other form of tension (**Figure 9-6**). When pressure is applied to the diaphragm that is of a sufficient value to overcome the spring tension, a switch is closed. Current that is supplied to the switch now has a completed path to ground. In a very simple warning light circuit, the closed pressure switch completes the circuit for the bulb and alerts the driver to an unacceptable condition. For example, a simple oil pressure warning lamp circuit will use a pressure switch.

Shop Manual
Chapter 9, page 404

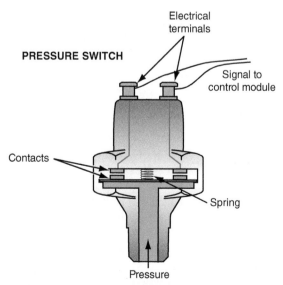

Figure 9-6 Simple pressure switch uses contacts to complete electrical circuit.

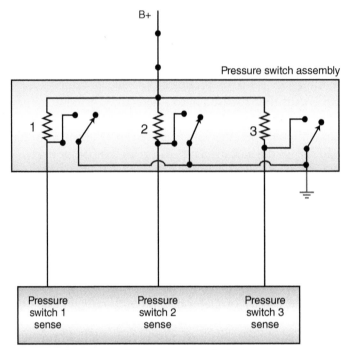

Figure 9-7 Computer-monitored pressure switch circuit.

Computer-monitored pressure switch circuits use the change in voltage as an indication of pressure. When the pressure change (from either low to high or high to low) changes the state of the switch, the voltage change is interpreted by the computer (**Figure 9-7**). Pressure switches are used to monitor the presence of pressure that is above or below a set point; they do not indicate the exact amount of pressure being applied.

Potentiometric Pressure Sensor

One of the basic types of pressure sensor is the **potentiometric pressure sensor.** The potentiometric pressure sensors use a Bourdon tube, a capsule, or bellows to move a wiper arm on a resistive element (**Figure 9-8**). Using the principle of variable resistance, the movement of the wiper across the resistive element will record a different voltage reading to the computer. Although this type of sensor can be used as a computer input, a computer is not always involved. Some early analog instrument panels used this sensor unit with an air core gauge to display engine oil pressure.

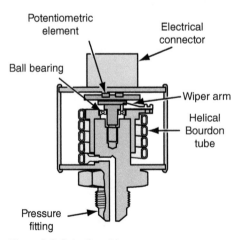

Figure 9-8 Potentiometric pressure sensors use a Bourdon tube, capsule, or bellows to drive a wiper arm on a resistive element.

Piezoresistive Devices

A **strain gauge** sensor determines the amount of applied pressure by measuring the strain a material experiences when subjected to the pressure. In their simplest form, a strain gauge sensor is a piezoresistive device. A piezoresistive sensor behaves like a variable resistor because its resistance value changes as the pressure applied to the sensing material changes. This type of sensor changes resistance values as a function of pressure changes.

A voltage regulator supplies a constant voltage to the sensor. Since the amount of voltage that the sensor drops changes with the change of resistance, the control module can determine the amount of pressure on the sensing material by measuring the voltage drop across the sensor. Piezoresistive sensors are commonly used as a gauge sending unit in standard analog instrument panels (**Figure 9-9**).

> Strain gauges are also called stress gauges.

> The term *piezo* refers to pressure and is derived from the Greek to mean "to be pressed."

Wheatstone Bridges

A Wheatstone bridge is commonly used to measure changes in pressure or strain. Although commonly used for engine control systems, body and chassis control systems that use them include tire pressure monitoring and supplemental restraint systems.

A Wheatstone bridge is nothing more than two simple series circuits connected in parallel across a power supply (**Figure 9-10**). Usually three of the resistors are kept at exactly the same value and the fourth is the sensing resistor. The resistors are placed on a silicon chip that flexes. As the chip flexes, the resistance of the sensing resistor changes. When all four resistors have the same value, the bridge is balanced and the voltage sensor will indicate a value of 0 volt. The output from the amplifier acts as a voltmeter. Remember, since a voltmeter measures electrical pressure between two points, it will display this value. For example, if the reference voltage is 5 volts and the resistors have the same value, then the voltage drop over each resistor is 2.5 volts. Since the voltmeter is measuring the potential on the line between R_s and R_1; and between R_2 and R_3, it will read 0 volt because both of these points have 2.5 volts on them. If there is a change in the resistance value of the sensing resistor, a change will occur in the circuit's balance. The sensing circuit will receive a voltage reading that is proportional to the amount of resistance change. The sensing resistor is a variable resistor that changes resistance as the silicon chip flexes.

> **Shop Manual**
> Chapter 9, page 406

> Wheatstone bridges can have different configurations such as using four variable resistors instead of one sensing resistor and three fixed resistors.

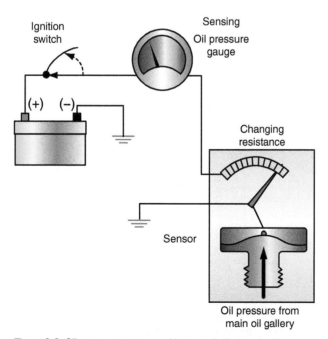

Figure 9-9 Oil pressure sensor used in gauge indicator circuit.

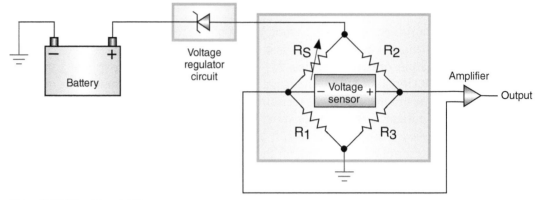

Figure 9-10 Wheatstone bridge.

Piezoelectric Devices

Shop Manual
Chapter 9, page 411

Piezoelectric devices are used to measure pressures by the generation of voltage. Piezoelectric sensors are constructed from alumina ceramics, metalized quartz, single crystals, or ultrasonic transducer materials that make up a bidirectional transducer capable of converting stress into an electric potential (**Figure 9-11**). The piezoelectric materials consist of polarized ions within the crystal. As pressure is applied on the piezoelectric material some mechanical deformation occurs in the polarized crystal, which produces a proportional output charge due to the displacement in the ions. Uses of this type of sensor in the automotive industry include piezoelectric accelerometers, piezoelectric force sensors, and piezoelectric pressure sensors.

The sensor is a voltage generator and has a resistor connected in series with it (**Figure 9-12**). The resistor protects the sensor from excessive current flow in case the circuit becomes shorted. The voltage generator is a thin ceramic disc attached to a metal diaphragm. Pressure that is applied to the diaphragm will transfer pressure onto the piezoelectric crystals in the ceramic disc. The disc generates a voltage that is proportional to the amount of pressure. The voltage generated ranges from zero to one or more volts.

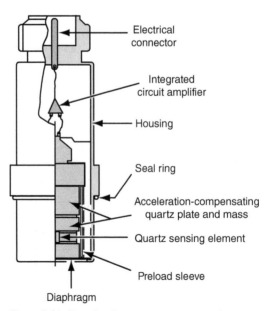

Figure 9-11 Piezoelectric sensors convert stress into an electric potential and vice versa. Sensors based on this technology are used to measure varying pressures.

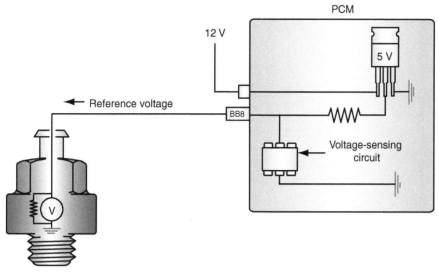

Figure 9-12 Piezoelectric sensor circuit.

Capacitance Discharge Sensors

Another variation of the piezosensor uses capacitance discharge. Instead of using a silicon diaphragm, the **capacitance discharge sensor** uses a variable capacitor. In the capacitor capsule–type sensor, two flexible alumina plates are separated by an insulating washer (**Figure 9-13**). A film electrode is deposited on the inside surface of each plate and a connecting lead is extended for external connections. The result of this construction is a parallel plate capacitor with a vacuum between the plates. This capsule is placed inside a sealed housing that is connected to the sensed pressure. If constructed to measure vacuum, as the pressure increases (goes toward atmospheric), the alumina plates deflect inward, resulting in a decrease in the distance between the electrodes.

Shop Manual
Chapter 9, page 409

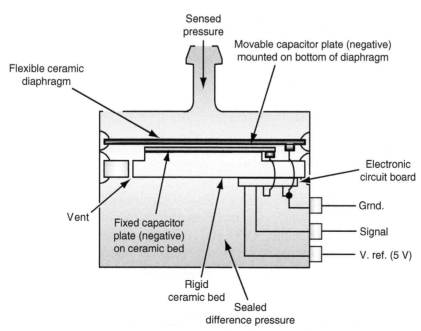

Figure 9-13 Capacitance discharge sensor.

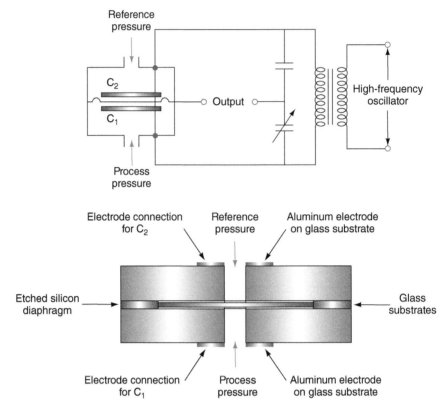

Figure 9-14 Capacitance pressure transducer construction and schematic.

As the distance between the electrodes changes, so does the capacity of the capacitor. A measure of capacitance constitutes a measurement of pressure that is detected by a bridge circuit. The output from the bridge circuit can be either an analog DC voltage or applied to a chip that produces a frequency modulated digital signal.

Capacitance Pressure Transducers. Capacitance pressure transducers are used to measure pressure by changes in capacitance as a e result of the movement of a diaphragm element (**Figure 9-14**). The diaphragm must physically travel a distance that is only a few microns. One side of the diaphragm is exposed to the measured pressure and the other side to the reference pressure. The change in capacitance may be used to control the frequency of an oscillator or to vary the coupling of a voltage signal. Depending on the type of reference pressure, the capacitive transducer can be constructed as either an absolute, gauge, or differential pressure transducer.

POSITION AND MOTION DETECTION SENSORS

Many electronic systems require input data concerning position, motion, and speed. Most motion and speed sensors use a magnet as the sensing element or sensed target to detect rotational or linear speed. The types of magnetic speed sensors include magnetoresistive (MR), inductive, variable reluctance (VR), and Hall-effect. In addition, the potentiometer and commutator pulse counting can be used to detect position.

Some systems require the use of photoelectric sensors that use light-sensitive elements to detect the movement of an object. In addition, solid-state accelerometers, axis rotation sensors, yaw sensors, and roll sensors are becoming common components on many systems. This chapter will explore the operation of common position and motion detection sensors.

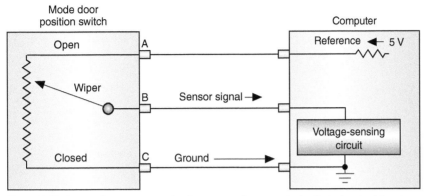

Figure 9-15 A potentiometer sensor circuit measures the amount of voltage drop to determine position.

Potentiometer

A common position sensor used to monitor linear or rotary motion is the potentiometer. A potentiometer is a voltage divider that provides a variable DC voltage reading to the computer. These sensors are typically used to determine the position of a valve, air-conditioning unit door, seat track, and so on.

Shop Manual
Chapter 9, page 413

The potentiometer usually consists of a wire-wound resistor with a movable center wiper (**Figure 9-15**). A constant voltage (usually 5 volts) is applied to terminal A. If the wiper (which is connected to the shaft or movable component of the unit that is being monitored) is located close to this terminal, there will be low voltage drop represented by high-voltage signal back to the computer through terminal B. As the wiper moves toward terminal C, the sensor signal voltage to terminal B decreases. The computer interprets the different voltage value into different shaft positions. The potentiometer can measure linear or rotary movement. As the wiper moves across the resistor, the computer tracks the position of the unit.

Since applied voltage must flow through the entire resistance, temperature and other factors do not create false or inaccurate sensor signals to the computer. A rheostat is not as accurate and its use is limited in computer systems.

Magnetic Pulse Generator

An example of the use of magnetic pulse generators is to determine vehicle and individual wheel speed. The signals from the speed sensors are used for computer-driven instrumentation, cruise control, antilock braking, speed-sensitive steering, and automatic ride control systems.

Shop Manual
Chapter 9, page 416

The timing disc is known as an armature, reluctor, trigger wheel, pulse wheel, toothed wheel, or timing core. It is used to conduct lines of magnetic force.

The components of the pulse generator are as follows (**Figure 9-16**):

1. A **timing disc** that is attached to the rotating shaft or cable. The number of teeth on the timing disc is determined by the manufacturer and depends on application. The teeth will cause a voltage generation that is constant per revolution of the shaft. For example, a vehicle speed sensor may be designed to deliver 4,000 pulses per mile. The number of pulses per mile remains constant regardless of speed. The computer calculates how fast the vehicle is going based on the frequency of the signal.
2. A **pickup coil** consists of a permanent magnet that is wound around by fine wire.

The pickup coil is also known as a stator, sensor, or pole piece.

An air gap is maintained between the timing disc and the pickup coil. As the timing disc rotates in front of the pickup coil, the generator sends an A/C signal (**Figure 9-17**).

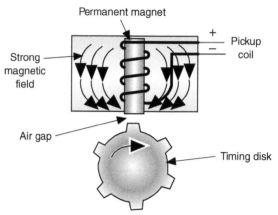

Figure 9-16 Components of the magnetic pulse generator. A strong magnetic field is produced in the pickup coil as the teeth align with the core.

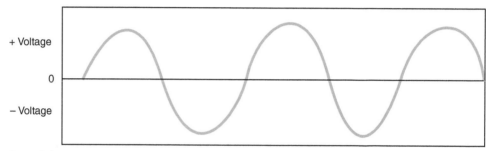

Figure 9-17 Pulse signal sine wave.

As a tooth on the timing disc aligns with the core of the pickup coil, it repels the magnetic field. The magnetic field is forced to flow through the coil and pickup core (refer to Figure 9-16). Since the magnetic field is not expanding, a voltage of zero is induced in the pickup coil. As the tooth passes the core, the magnetic field is able to expand (**Figure 9-18**). The expanding magnetic field cuts across the windings of the pickup coil. This movement of the magnetic field induces a voltage in the windings. This action is repeated every time a tooth passes the core. The moving lines of magnetic force cut across the coil windings and induce a voltage signal.

Magnetic pulse generators are also called magnetic induction sensors.

When a tooth approaches the core, a positive current is produced as the magnetic field begins to concentrate around the coil (**Figure 9-19**). The voltage will continue to climb as long as the magnetic field is expanding. As the tooth approaches the magnet, the magnetic field gets smaller, causing the induced voltage to drop off. When the tooth and core align, there is no more expansion or contraction of the magnetic field (thus no movement) and the voltage drops to zero. When the tooth passes the core, the magnetic field expands and a negative current is produced. The resulting pulse signal is digitalized and sent to the microprocessor.

AUTHOR'S NOTE The magnetic pulse (PM) generator operates on basic magnetic principles. Remember that a voltage can be induced only when a magnetic field is moved across a conductor. The magnetic field is provided by the pickup unit, and the rotating timing disc provides the movement of the magnetic field needed to induce voltage.

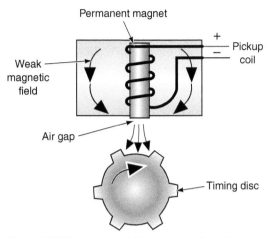

Figure 9-18 The magnetic field expands as the teeth pass the core.

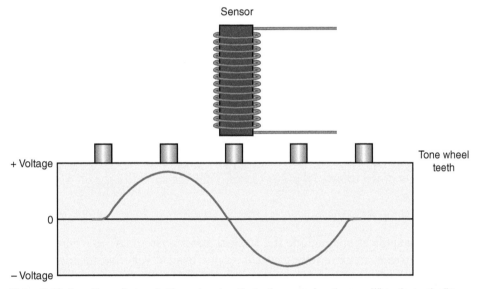

Figure 9-19 A positive voltage swing is produced as the tooth approaches the core. When the tooth aligns with the core, there is no magnetic movement and no voltage. A negative waveform is created as the tooth passes the core.

Magnetoresistive Sensor

Magnetoresistive (MR) sensors consist of the magnetoresistive sensor element, a permanent magnet, and an integrated signal conditioning circuit to make use of the **magnetoresistive effect**. This effect defines that if a current-carrying magnetic material is exposed to an external magnetic field, its resistance characteristics will change. This results in the resistance of the sensing element being a function of the direction and intensity of an applied magnetic field. MMR is the characteristic of some materials to change electrical resistance as a magnetic field is applied. The change in resistance is due to the spin dependence of electron scattering. To understand spin current, consider that the movement of electrons results in a charge current. However, electrons also have the properties of mass, charge, and spin. The spin motion of the electron creates a spin current. Within a semiconductor, spin is a random process.

In the magnetic multilayered structure, two magnetic layers are closely separated by a thin spacer layer. The first magnetic layer will allow electrons in only one spin channel to pass through with little resistance (shown in **Figure 9-20** as the spin-up channel).

Magnetoresistive sensors can determine direction of rotation based on the north and south pole influences of the magnets in the reluctor ring.

Shop Manual
Chapter 9, page 419

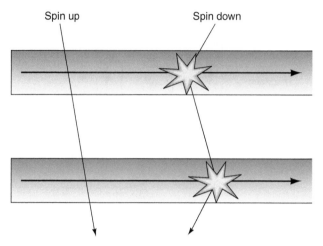

Figure 9-20 The spin-up channel is illustrated as allowing electrons to flow through the magnetic layer with little resistance (i.e., if the second magnetic layer is aligned).

The spin-down channel has resistance as it passes through the first magnetic layer. If the second magnetic layer is aligned, the spin-up channel has low resistance as it passes through the structure. If the second magnetic layer is misaligned, as shown in **Figure 9-21**, then both channels have high resistance. The MR sensor measures the difference in angle between the two magnetic layers. Small angles give a low resistance, while large angles give a higher resistance.

AUTHOR'S NOTE The magnetoresistive principle provides rotational speed measurements down to zero. For this reason, they are sometimes called "zero speed sensors."

Automotive applications include the antilock brake system (ABS) where MR wheel-speed sensors provide an extremely accurate output, even at very low speeds. MR sensors are also used in navigational systems to provide more precise compass readings, in lane change detection systems, and in proximity sensors. MR sensors cannot generate a signal voltage on their own, and must have an external power source. The magnetoresistive bridge changes resistance due to the relationship of the tone wheel and magnetic field surrounding the sensor.

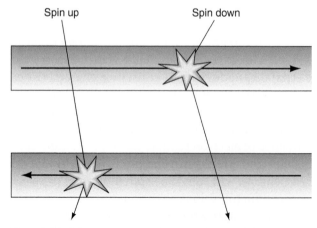

Figure 9-21 If the second magnetic layer is misaligned, then both the spin-up and spin-down channels have high resistance.

The integrated circuit (IC) in the sensor is powered by a 12-volt circuit that is provided by the computer. The IC supplies a constant 7 mA power to the computer. The relationship of the tooth on the tone wheel to the permanent magnet in the sensor signals the IC to enable a second 7-mA power supply. The output of the sensor, sent to the computer, is a DC voltage signal with changing voltage and current levels. When a valley of the tone wheel is aligned with the sensor, the voltage signal is approximately 0.8 volt and a constant 7 mA current is sent to the computer. As the tone wheel rotates, the tooth shifts the magnetic field and the IC enables a second 7-mA current source. The computer senses a voltage signal of approximately 1.6 volts and 14 mA. The computer measures the amperage of the digital signal and interrupts the signal as component speed.

The alternating magnetic poles allows the computer to determine the direction of rotation.

Magnetically Coupled Linear Sensors

Linear sensors are used for such functions as fuel level sending units. The most common type of fuel level sensor is a rheostat style with wire-wound resistor and a movable wiper. The wiper is in constant contact with the winding and may eventually rub through the wire. Many manufacturers are now using **magnetically coupled linear sensors** that are not prone to the wear (**Figure 9-22**).

Magnetically coupled linear sensors used for fuel level sensing have a magnet attached to the end of the float arm. Also, a resistor card and a magnetically sensitive comb are located next to the magnet. When the magnetic field passes the comb, the fingers are pulled against the resistor card contacting resistors that represent the various levels of fuel (**Figure 9-23**). When the tank is full, the float is on top along with the magnet. As the fuel level falls, the float drops and the position of the magnet changes. The magnet is so close to the sensor that it attracts the closest metal fingers. The fingers contact a metal strip. Different contact sites on the strip produce different resistances that are used to determine the fuel level.

Hall-Effect Sensors

Operation of Hall-effect sensors is based on the principle that if a current is allowed to flow through a thin conducting material that is exposed to a magnetic field, another voltage is

Shop Manual
Chapter 9, page 421

Figure 9-22 Magnetically coupled linear sensor used to measure fuel level.

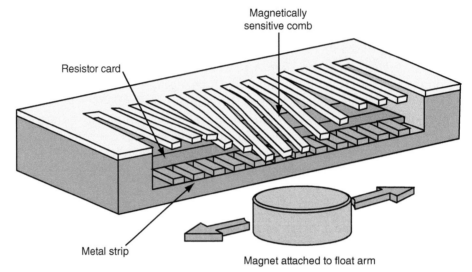

Figure 9-23 Comb design magnetic sensor pulls the fingers against the resistor card.

produced (**Figure 9-24**). The sensor contains a permanent magnet, a thin semiconductor layer made of gallium arsenate crystal (Hall layer), and a **shutter wheel** (**Figure 9-25**). The Hall layer has a negative and a positive terminal connected to it. Two additional terminals located on either side of the Hall layer are used for the output circuit. The shutter wheel consists of a series of alternating windows and vanes. It creates a magnetic shunt that changes the strength of the magnetic field from the permanent magnet.

The permanent magnet is located directly across from the Hall layer so that its lines of flux will bisect at right angles to the current flow. The permanent magnet is mounted so that a small air gap is between it and the Hall layer.

A steady current is applied to the crystal of the Hall layer. This produces a signal voltage that is perpendicular to the direction of current flow and magnetic flux. The signal voltage produced is a result of the effect the magnetic field has on the electrons. When the magnetic field bisects the supply current flow, the electrons are deflected toward the Hall layer negative terminal (**Figure 9-26**). This results in a weak voltage potential being produced in the Hall sensor.

A shutter wheel is attached to a rotational component. As the wheel rotates, the shutters (vanes) pass in this air gap. When a shutter vane enters the gap, it intercepts the magnetic field and shields the Hall layer from its lines of force. The electrons in the supply current are no longer disrupted and return to a normal state. This results in low voltage potential in the signal circuit of the Hall sensor.

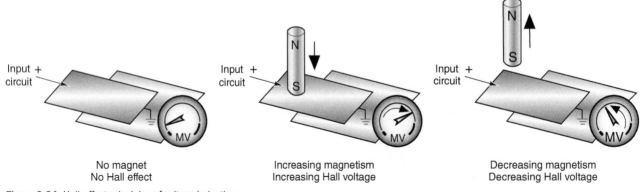

Figure 9-24 Hall-effect principles of voltage induction.

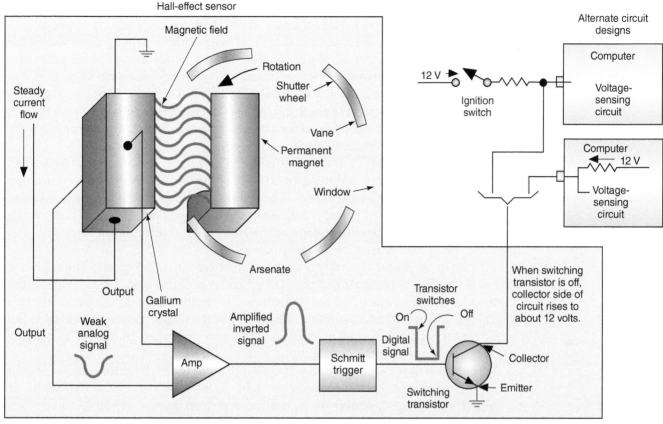

Figure 9.25 Typical circuit of a Hall-effect sensor.

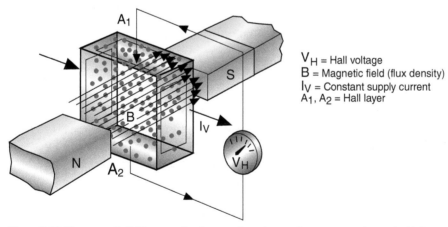

V_H = Hall voltage
B = Magnetic field (flux density)
I_V = Constant supply current
A_1, A_2 = Hall layer

Figure 9-26 The magnetic field causes the electrons from the supply current to gather at the Hall layer negative terminal. This creates a voltage potential.

The signal voltage leaves the Hall layer as a weak analog signal. To be used by the computer, the signal must be conditioned. It is first amplified because it is too weak to produce a desirable result. The signal is also inverted so that a low input signal is converted into a high output signal. It is then sent through a **Schmitt trigger** where it is digitized and conditioned into a clean square wave signal. The signal is finally sent to a switching transistor. The computer senses the turning on and off of the switching transistor to determine the frequency of the signals and calculates the speed.

The Hall effect just discussed describes its usage as a switch to provide a digital signal. It can also be designed as an analog (or linear) sensor that produces an output voltage that is proportional to the applied magnetic field. This makes them useful for determining the position of a component instead of just rotation. For example, this type of sensor can be used to monitor fuel level or to track seat positions in memory seat systems.

A fuel level indication can be accomplished with a Hall-effect sensor by attaching a magnet to the float assembly (**Figure 9-27**). As the float moves up and down with the fuel level, the gap between the magnet and the Hall element will change. The gap changes the Hall effect and thus the output voltage.

As discussed, typical Hall-effect sensors use three wires. However, linear Hall-effect sensors can also be constructed using two-wire circuits (**Figure 9-28**). This is common on systems that use a DC motor drive. The reference voltage to the sensor is supplied through a pull-up resistor. Typically, this reference voltage will be 12 volts. Whenever the motor is operated, the reference voltage will be applied. After the motor is turned off, this reference voltage will remain for a short time.

Internal to the motor assembly is a typical three-terminal Hall sensor. The reference voltage is supplied to terminal 1 of the Hall sensor. A pull-up resistor also connects the reference voltage to terminal 3 of the Hall sensor. This becomes the signal circuit. The two pull-up resistors will be of equal value. Terminal 2 of the Hall sensor is connected to the sensor return circuit. A magnet is attached to the motor armature to provide a changing magnetic field once per motor revolution.

When the Hall sensor is off, the voltage supplied to the Hall sensor will be close to that of the source voltage. Since this is an open circuit condition in the Hall sensor at terminal 3, the voltage drop over the signal circuit pull-up resistor will be 0.

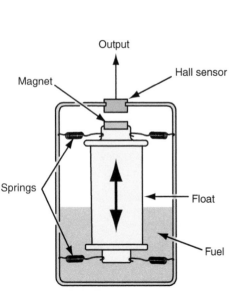

Figure 9-27 Hall-effect sensor used for fuel level indication.

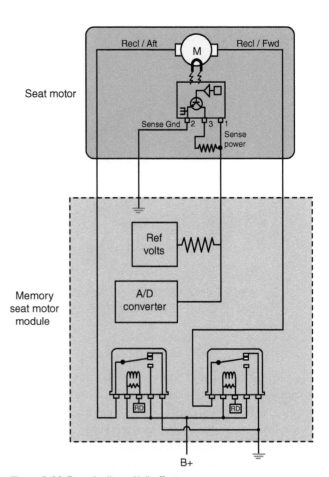

Figure 9-28 Two-wire linear Hall-effect sensor.

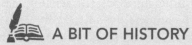

A BIT OF HISTORY

The American physicist, Edwin Herbert Hall (1855–1938), discovered the principle of the Hall-effect sensor in 1879.

When the motor rotates and the influence of the magnetic field turns on the Hall sensor, the signal terminal 3 is connected to ground within the sensor. This pulls the signal voltage low and results in the formation of a series circuit from the reference supply to terminal 3. Since each of the pull-up resistors is equal, the voltage drop will be split between the two. Approximately half the voltage will be dropped across the pull-up resistor in the computer and the other half over the pull-up resistor in the motor assembly. The Hall-effect sensor will remain powered since the reference voltage to terminal 1 is connected between the two resistors and the 6 volts on the circuit is sufficient to operate the sensor.

Accelerometers

Accelerometers are sensors designed to measure the rate of acceleration or deceleration. Common sensors include mass-type, roller-type, and solid-state accelerometers. The first extensive use of the accelerometer was in the air-bag system. The use of accelerometers has expanded greatly in today's vehicles. They are now used on vehicle stability systems, roll over mitigation, hill hold control, electronic steering, and navigational systems. These sensors may perform specific functions other than forward acceleration and deceleration forces. For example, they will operate as a gyro to determine direction change and rotation.

Accelerometers react to the amount of **G force** associated with the rate of acceleration or deceleration. In air-bag systems they are used to determine deceleration forces that indicate the vehicle has been involved in a collision that requires the air bag to be deployed.

Early accelerometers used in air-bag systems were electromechanical designs. The mass-type sensor contains a normally open set of gold-plated switch contacts and a gold-plated ball that acts as a sensing mass (**Figure 9-29**). The gold-plated ball is mounted in a cylinder coated with stainless steel. A magnet holds the ball away from the contacts.

G force is used to describe the measurement of the net effect of the acceleration that an object experiences and the acceleration that gravity is trying to impart to it. Basically, G force is the apparent force that an object experiences due to acceleration.

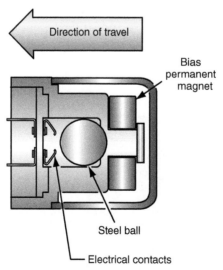

Figure 9-29 Sensing mass held by a magnet will break loose if deceleration forces are severe enough.

When the vehicle is involved in a frontal collision of sufficient force, the sensing mass (ball) moves forward in the sensor and closes the switch contacts.

In many air-bag systems, solid-state accelerometers are used to sense deceleration forces. The piezoelectric accelerometer generates an analog voltage proportional to a G force. The accelerometer contains a piezoelectric element that is distorted during a high G force condition and generates an analog voltage in relation to the force. The analog voltage from the piezoelectric element is sent to a collision-judging circuit in the air-bag computer. If the collision impact is great enough, the computer deploys the air bag.

Accelerometers can also be designed as piezoresistive sensors that use a silicon mass that is suspended from four deflection beams. The deflection beams are the strain sensing elements. The four strain elements are in a Wheatstone bridge circuit. The strain elements on the beam generate a signal that is proportional to the G forces. The resistance changes over the bridge are interpreted by an internal chip that then communicates the status to the control module using a frequency modulated digital pulse.

Shop Manual
Chapter 11, page 512

PHOTO CELLS

AUTHOR'S NOTE Diagnostics of the photo cell is covered in Chapter 11 when these components are introduced as an input for automatic lighting systems.

A **photo cell** describes any component that is capable of measuring or determining light. It is used where settings need to be adjusted to ambient light conditions such as automatic temperature control, automatic headlight operation, night vision assistance, and other convenience and safety systems. It can be used to automatically change display intensity of displays, backlighting of instrument panels and button controls, and to adjust daytime running lights to full power when ambient light levels are low.

There are several different types of semiconductor devices that can be used to measure light intensity. The following are some of the most common.

Photodiode

As discussed in Chapter 3, a photodiode is a light-receiving device that contains a semiconductor PN junction, which is enclosed in a case with a convex lens (**Figure 9-30**). The lens allows ambient light to enter the case and strike the photodiode.

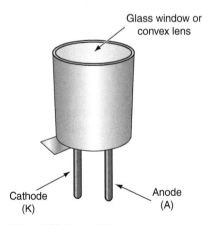

Figure 9-30 Photodiode

Photodiodes can operate in two modes: photovoltaic mode or photoconductive mode.

In the **photovoltaic mode**, the photodiode will generate a voltage in response to light. When light strikes the photodiode, it creates electron-hole pairs. The negatively charged electrons that are generated in the depletion region are attracted to the positively charged ions in the N-type material. The holes are attracted to the negatively charged ions in the P-type material. The result is a separation of charges and a small voltage drop of about 450 mV is developed across the diode. Connecting a load resistor across the voltage source will result in a small current flow from the cathode to the anode.

AUTHOR'S NOTE Operating the photodiode in photovoltaic mode is the principle of the solar cell.

In the **photoconductive mode**, the conductance of the diode changes when light is applied. In this mode, the photodiode is reverse biased when no light is applied. Being reverse biased results in a very wide depletion region and a high resistance across the diode. In this state, there will be only a small reverse current through the diode.

When light is applied to the photodiode, electron-hole pairs are generated. The electrons are attracted to the positive bias voltage, and the holes are attracted to the negative bias voltage. This movement of electrons and holes causes an increase in the reverse current flow. When light is applied, the resistance of the photodiode is very low and decreases as the intensity of light increases. As the resistance continues to decrease, current flow increases.

Photoresistor

A **photoresistor** is a passive light-detecting device composed of a semiconductor material that changes resistance when its surface is exposed to light (**Figure 9-31**). The semiconductor material is shaped into zigzag strip and the ends are attached to the external terminals. Common materials are either cadmium sulfide (CdS) or cadmium selenide (CdSe). A transparent cover allows the ambient light to pass through. The semiconductor material is light sensitive; thus, free electrons are created by light energy instead of heat

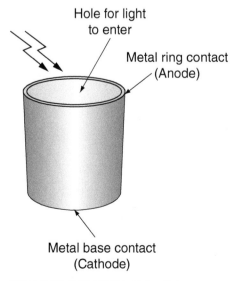

Figure 9-31 Photoresistor construction.

energy as in conventional semiconductor devices. The creation of the free electrons causes the resistance of the semiconductor material to decrease. As the applied light intensity increases, so does the number of free electrons. Thus, as light increases the resistance decreases. This, in turn, means the sense voltage drops as light intensity increases.

Phototransistor

A **phototransistor** is a light-detecting transistor that uses the application of light to generate carriers to supply the base leg current. The intensity of the light controls the collector current of the transistor. Unlike a normal transistor, the phototransistor has only two terminals (**Figure 9-32**). Like a normal transistor, the output current is amplified.

Photoelectric Sensors

Photoelectric sensors use light-sensitive elements to detect rotation (**Figure 9-33**). Photoelectric sensors are commonly made up of an emitter (light source) and a receiver. There are three basic forms of photoelectric sensors:

1. *Direct reflection*—has the emitter and receiver in the same module and uses the light reflected directly off the monitored object for detection.
2. *Reflection with reflector*—also has the emitter and receiver in the same module, but requires a reflector. Motion of an object is detected by its interruptions of the light beam between the sensor and the reflector.
3. *Thru beam*—separates the emitter and receiver and detects the motion of an object when it interrupts the light beam between the emitter and the receiver.

Photoelectric sensors are commonly used to monitor steering wheel rotation.

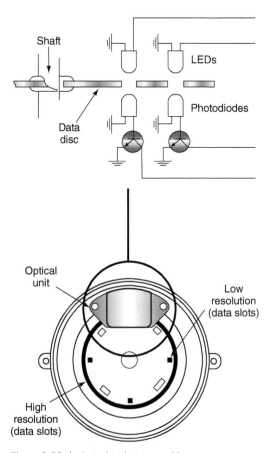

Figure 9-33 A photoelectric-type position sensor.

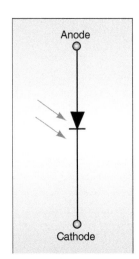

Figure 9.32 Symbol for phototransistor.

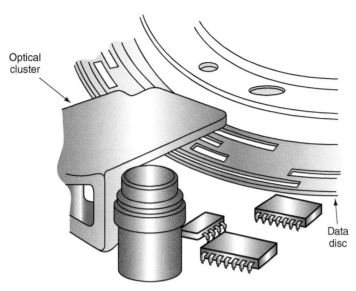

Figure 9-34 The data disc interrupts the light beams from the optical cluster to detect motion.

A thru beam–type photoelectric sensor has a series of LEDs situated across from the same number of photocells. If used as a steering wheel sensor, a code disc (**Figure 9-34**) is attached to the steering shaft and interrupts the infrared light beams from the optics cluster. The optics cluster has three rows of four light detectors that provide a bit code that determines steering wheel position and rotation based on the order of light beam interruption. Based on this input from the photoelectric sensor, the ECU can determine the direction, and how fast, the steering wheel is being turned.

SWITCH INPUTS

Switches are simplest of all input devices. The computer monitors the two states of the switch by measuring the voltage on the sense circuit. There are two types of voltage-sensing circuits used with switches: the **pull-down circuit** and the **pull-up circuit**. Basically, the pull-down circuit closes the switch to ground and the pull-up circuit closes the switch to voltage.

A pull-down voltage-sensing circuit usually uses an internal voltage source within the computer (**Figure 9-35**). It is also possible to use an external voltage source (**Figure 9-36**). The current limiting resistor is used to protect the computer and the circuit. It also prevents input values from **floating**. Floating occurs when the switch is open resulting in the input to the voltage-sensing circuit of the control module being susceptible to electrical noise that may cause the control module to misread the switch state. The current limiting resistor used in the circuit is referred to as a **pull-up resistor** since it assures the proper high voltage reading by connecting the voltage-sensing circuit to an electrical potential that can be removed when the switch is closed.

The pull-up resistor is usually of a very high ohms value to keep amperage to a minimum. This resistor can have a value of 10K to 10M ohms. When the switch is open, there is no current flow through the resistor and no voltage drop over it. This results in the voltage-sensing circuit recording a value equal to the reference voltage. When the switch is closed, current flows through the resistor and results in a voltage drop. Since the switch should provide a clean contact to ground, the voltage-sensing circuit should read a value close to 0 volt.

Shop Manual
Chapter 9, page 404

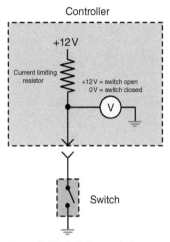

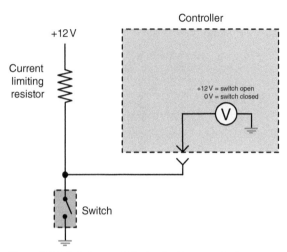

Figure 9-35 Pull-down switch circuit.

Figure 9-36 Pull-down switch circuit with external voltage source.

The pull-up circuit will have a reference voltage through the switch (**Figure 9-37**). Usually the reference voltage will be provided directly from the battery or by the ignition switch. The current limiting resistor performs the same function in this circuit as it does in the pull-down circuit. This resistor is referred to as a **pull-down resistor** since it assures a proper low-voltage reading by preventing float when the switch is open. With the switch in the open position, the voltage-sensing circuit will read 0 volt. With the switch closed, the sense circuit should read close to the reference voltage.

Both of these circuits are limited concerning the ability to determine circuit faults. Since there are only two states for the switch, there are two voltage values that the computer expects to see. An open or short to ground will not produce an unexpected voltage value, but will result in improper system operation. However, the computer may be capable of determining a functionality problem with the input circuit if the seen voltage is implausible for the conditions. For example, if the switch is an operator-activated switch that requests A/C operation and the voltage indicates that the switch may be stuck, the computer can set a stuck switch diagnostic trouble code (DTC) and ignore the input.

To provide continuity diagnostics, the circuit may have a diagnostic resistor wired parallel to the switch (**Figure 9-38**). The computer will be able to recognize three different voltage values. In the example, the current limiting resistor has a value of 10K ohms while the diagnostic resistor has a value of 2K ohms. With the switch in the open state, the voltage-sensing circuit would read 2 volts. With the switch closed, the voltage reading will be close to 0 volt. A reading of 12 volts would indicate an open in the circuit.

Another typical switch that is used is the resistive multiplex switch. This switch is used to provide multiple inputs from a single switch using one circuit (**Figure 9-39**). The control

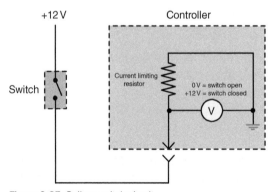

Figure 9-37 Pull-up switch circuit.

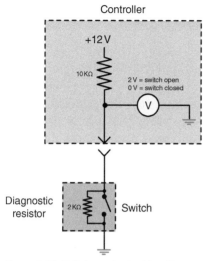

Figure 9-38 Pull-down circuit with a diagnostic resistor.

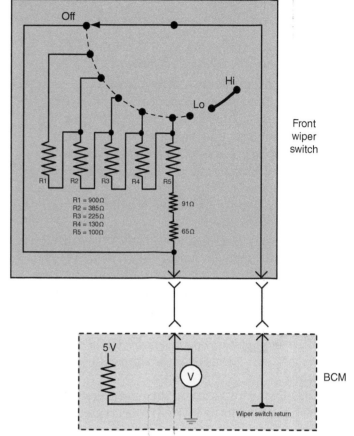

Figure 9-39 Resistive multiplex switch.

module sends a signal voltage to the switch through a fixed resistor. Each switch position has a unique resistance value that will be placed in series with the resistor in the control module. As different switch positions are selected, a different amount of voltage is dropped over the fixed resistor in the control module, and the sensed voltage level changes. Based on the sensed voltage value, the control module interprets what operation the driver is requesting.

FEEDBACK SIGNALS

If the computer sends a command signal to open a blend door in an automatic climate control system, a **feedback** signal may be sent back from the actuator to inform the computer the task was performed. The feedback signal will confirm both the door position and actuator operation (**Figure 9-40**). Another form of feedback is for the computer to monitor voltage as a switch, relay, or other actuator is activated. Changing states of the actuator will result in a predictable change in the computer's voltage-sensing circuit. The computer may set a diagnostic code if it does not receive the correct feedback signal.

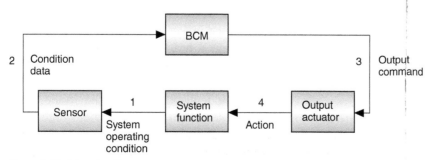

Figure 9-40 Principle of feedback signals.

SUMMARY

- Inputs provide the computer with system operation information or driver requests.
- Driver input signals are usually provided by momentarily applying a ground through a switch.
- Switches can be used as an input for any operation that only requires a yes/no, or on/off, condition.
- Sensors convert some measurement of vehicle operation into an electrical signal. There are many different types of sensors: thermistors, Wheatstone bridge, potentiometers, magnetic pulse generator, and Hall-effect sensors.
- A thermistor is a solid-state variable resistor made from a semiconductor material that changes resistance in relation to temperature changes.
- Negative temperature coefficient (NTC) thermistors reduce their resistance as the temperature increases.
- Positive temperature coefficient (PTC) thermistors increase their resistance as the temperature increases.
- Some temperature-sensing circuits are designed as dual-range circuits to provide more accurate temperature measurements.
- Pressure switches usually use a diaphragm that works against a calibrated spring or other form of tension. When pressure is applied to the diaphragm that is of a sufficient value to overcome the spring tension, a switch is closed.
- A strain gauge sensor determines the amount of applied pressure by measuring the strain a material experiences when subjected to the pressure.
- Piezoresistive devices are similar to variable resistors whose resistance values change as the pressure applied to the crystal changes.
- The Wheatstone bridge is a series-parallel arrangement of resistors between an input terminal and ground.
- Piezoelectric devices are voltage generators with a resistor connected in series that is used to measure fluid and air pressures.
- The capacitance discharge sensor uses a variable capacitor constructed of two flexible alumina plates that are separated by an insulating washer. As the distance between the electrodes changes, so does the capacity of the capacitor. A measure of capacitance constitutes a measurement of pressure that is detected by a bridge circuit.
- Capacitance pressure transducers may be used to measure pressure. The change in capacitance is the result of the movement of a diaphragm element.
- A potentiometer is a variable resistor that usually consists of a wire-wound resistor with a movable center wiper.
- Magnetic pulse generators use the principle of magnetic induction to produce a voltage signal and are commonly used to send data concerning the speed of the monitored component to the computer.
- Magnetoresistive (MR) sensors consist of the magnetoresistive sensor element, a permanent magnet, and an integrated signal conditioning circuit to change resistance due to the relationship of the tone wheel and magnetic field surrounding the sensor.
- Magnetically coupled linear sensors use a movable magnet that is attached to the measured element, a resistor card, and a magnetically sensitive comb. When the magnetic field passes the comb, the fingers are pulled against the resistor card contacting resistors that represent the various positions of the measured element.
- Hall-effect sensors operate on the principle that if a current is allowed to flow through a thin conducting material that is exposed to a magnetic field, another voltage is produced.
- The Hall effect can also be designed as an analog (or linear) sensor that produces an output voltage that is proportional to the applied magnetic field.
- Accelerometers are sensors designed to measure the rate of acceleration or deceleration. Accelerometers react to the amount of G force associated with the rate of acceleration or deceleration.
- The piezoelectric accelerometer generates an analog voltage proportional to a G force.
- Accelerometers can also be designed as piezoresistive sensors.
- A photo cell describes any component that is capable of measuring or determining light. They are used where settings need to be adjusted to ambient light conditions.
- A photodiode is a light-receiving device that contains a semiconductor PN junction, which is enclosed in a case with a convex lens that allows ambient light to enter the case and strike the photodiode.
- Photodiodes can operate in two modes: photovoltaic mode or photoconductive mode.
- A photoresistor is a passive light-detecting device composed of a semiconductor material that changes resistance when its surface is exposed to light.

- A phototransistor is a light-detecting transistor that uses the application of light to generate carriers to supply the base leg current.
- Photoelectric sensors use light-sensitive elements to detect rotation using an emitter (light source) and a receiver.
- The pull-down circuit will close the switch to ground.
- The pull-up circuit will close the switch to voltage.
- The pull-up resistor assures proper high voltage reading by connecting the voltage-sensing circuit to an electrical potential that can be removed when the switch is closed.
- The pull-down resistor assures a proper low-voltage reading by preventing float when the switch is open.
- The resistive multiplex switch is used to provide multiple inputs from a single switch using one circuit.
- Feedback signals are used to confirm position and operation of an actuator.

REVIEW QUESTIONS

Short-Answer Essays

1. What are the functions of input devices?

2. Explain the purpose of the thermistor and how it is used in a circuit.

3. Describe the operation and purpose of the Wheatstone bridge.

4. Explain the operation and purpose of piezoelectric devices.

5. What is the difference between NTC and PTC thermistors?

6. How does the Hall-effect sensor generate a voltage signal?

7. What is the purpose of the multiplex switch?

8. What is meant by feedback as it relates to computer control?

9. Describe the operation of the pull-down sense circuit.

10. Describe the operation of the pull-up sense circuit.

Fill in the Blanks

1. The piezoelectric accelerometer generates an analog voltage proportional to a _____.

2. The _____ resistor assures the proper high-voltage reading by connecting the voltage-sensing circuit to an electrical potential that can be removed when the switch is closed.

3. The resistive multiplex switch is used to provide multiple inputs from a single switch using _____ circuit.

4. Magnetoresistive (MR) sensors consist of the magnetoresistive sensor element, a permanent magnet, and an integrated signal conditioning circuit to change _____ due to the relationship of the tone wheel and magnetic field surrounding the sensor.

5. The capacitance discharge sensor changes its capacitance by the difference in _____ between the electrodes.

6. _____ convert some measurement of vehicle operation into an electrical signal.

7. Negative temperature coefficient (NTC) thermistors _____ their resistance as the temperature increases.

8. _____ sensors operate on the principle that if a current is allowed to flow through a thin conducting material exposed to a magnetic field, another voltage is produced.

9. Magnetic pulse generators use the principle of _____ _____ to produce a voltage signal.

10. _____ means that data concerning the effects of the computer's commands are fed back to the computer as an input signal.

Multiple Choice

1. All of the following can be used to measure movement or position, EXCEPT:
   A. Potentiometer
   B. Magnetic pulse generator
   C. Piezoelectric device
   D. Hall-effect sensor

2. *Technician A* says the piezoresistive sensor changes resistance as a function of temperature.
   Technician B says the piezoresistive sensor outputs its own current based on the pressure it is exposed to.
   Who is correct?
   A. A only   C. Both A and B
   B. B only   D. Neither A nor B

3. The Wheatstone bridge is:
   A. A pressure sensing device that uses a variable capacitor.
   B. A pressure sensing device that uses varying resistances in a series-parallel circuit design.
   C. Used to measure motion by use of magnetic inductance.
   D. None of the above.

4. The piezoelectric sensor operates by:
   A. Altering the resistance values of the bridge circuit located on a ceramic disc.
   B. Dropping voltage over a fixed resistor as pressure is applied to the switching transistor.
   C. Generating a voltage within a thin ceramic disc voltage generator that is attached to a diaphragm, which stresses the crystals in the disc.
   D. None of the above.

5. Capacitance discharge sensors are being discussed.
   Technician A says the size of the electrodes alters as the sensing element is exposed to different pressures.
   Technician B says the distance between the electrodes alters as the sensing element is exposed to different pressures.
   Who is correct?
   A. A only   C. Both A and B
   B. B only   D. Neither A nor B

6. *Technician A* says a potentiometer is a voltage divider circuit used to measure movement of a component.
   Technician B says a magnetoresistive sensor alters current flow through the sense circuit when influenced by the magnetic field.
   Who is correct?
   A. A only   C. Both A and B
   B. B only   D. Neither A nor B

7. *Technician A* says negative temperature coefficient thermistors reduce their resistance as the temperature decreases.
   Technician B says positive temperature coefficient thermistors increase their resistance as the temperature increases.
   Who is correct?
   A. A only   C. Both A and B
   B. B only   D. Neither A nor B

8. *Technician A* says magnetic pulse generators are commonly used to send voltage signals to the computer concerning the speed of the monitored component.
   Technician B says an on–off switch sends a digital signal to the computer.
   Who is correct?
   A. A only   C. Both A and B
   B. B only   D. Neither A nor B

9. Speed sensors are being discussed.
   Technician A says the timing disc is stationary and the pickup coil rotates in front of it.
   Technician B says the number of pulses produced per mile increases as rotational speed increases.
   Who is correct?
   A. A only   C. Both A and B
   B. B only   D. Neither A nor B

10. *Technician A* says a Hall-effect sensor uses a steady supply current to generate a signal.
    Technician B says a Hall-effect sensor consists of a permanent magnet wound with a wire coil.
    Who is correct?
    A. A only   C. Both A and B
    B. B only   D. Neither A nor B

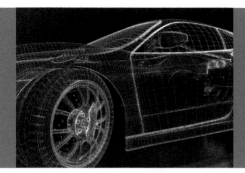

CHAPTER 10
VEHICLE COMMUNICATION NETWORKS

Upon completion and review of this chapter, you should be able to understand and describe:

- The principle of multiplexing.
- The different OBD II multiplexing communication protocols.
- The different classes of communications.
- The operation of a class A multiplexing system.
- The operation of a class B multiplexing system.
- The operation of the controller area network (CAN) bus system.

- The purpose and operation of different supplemental data bus networks.
- The operation of the local interconnect network (LIN) data bus system.
- The operation of the media-oriented system transport (MOST) data bus using fiber optics.
- The operation of wireless networks using Bluetooth technology.

Terms To Know

Asynchronous	ISO 9141-2	Node
Baud rate	ISO-K	Piconets
Bluetooth	J1850	Programmable Communication Interface (PCI)
Bus (–)	K-line	
Bus (+)	L-line	Protocol
Central gateway (CGW)	Local interconnect network (LIN)	Slave modules
Chrysler Collision Detection (CCD)	Master module	Smart sensors
Controller area network (CAN)	Media-oriented system transport (MOST)	Supplemental bus networks
Fiber optics	Multiplexing (MUX)	Termination resistors
ISO 14230-4	Network architecture	Total reflection
		Wireless networks

INTRODUCTION

In the past, if an accessory was added to the vehicle that required input information from sensors, either additional sensors had to be added or the new accessory would have to be spliced into an existing sensor circuit. Either way, the cost of production was increased due to added components and wiring. For example, some early vehicles were equipped with as many as three separate engine coolant temperature sensors. One sensor was used by the powertrain control module (PCM) for fuel and ignition strategies, the other was used by the cooling fan module to operate the radiator fans at the correct speed based on temperature, and the third was used by the instrument cluster for temperature gauge operation.

Today vehicle manufacturers use **multiplexing (MUX)** systems to enable different control modules to share information. Multiplexing provides the ability to use a single circuit to distribute and share data between several control modules throughout the vehicle. Because the data is transmitted through a single circuit, bulky wiring harnesses are eliminated. A MUX wiring system uses bus data links that connect each module and allow for the transporting of data from one module to another. Each module can transmit and receive digital codes over the bus data links. Each computer connected to the data bus is called a **node**. The signal sent from a sensor can go to any of the modules and can be used by the other modules. Before multiplexing, if information from the same sensing device is needed by several controllers, a wire from each controller needs to be connected in parallel to that sensor. If the sensor signal is analog, the controllers need an analog to digital (A/D) converter to read the sensor information. By using multiplexing, the need for separate conductors from the sensor to each module is eliminated and the number of drivers in the controllers is reduced.

Additionally, multiplexing systems have increased system reliability through the reduction of circuits and improved efficiency by less power demands and more accurate control. Multiplexing has increased diagnostic capabilities dramatically by allowing the technician to access diagnostic trouble codes (DTCs), see live data streams, and use bidirectional controls to actuate different components. Government regulations have dictated the use of certain multiplexing systems on the vehicle. Other benefits of the multiplexing system include improved emissions, safety, and fuel economy.

As discussed in Chapter 8, binary code is sent by digital signals to the nodes. The nodes use this code to communicate messages, both internally and with other controllers. A chip is used to prevent the digital codes from overlapping by allowing only one code to be transmitted at a time. Each digital message is preceded by an identification code that establishes its priority. If two modules attempt to send a message at the same time, the message with the higher priority code is transmitted first.

The major difference between a multiplexed system and a nonmultiplexed system is the way data is gathered and processed. In nonmultiplexed systems, the signal from a sensor is sent as an analog signal through a dedicated wire to the computer or computers. At the computer, the signal is changed from an analog to a digital signal. Because each sensor requires its own dedicated signal wire, the number of wires required to feed data from all of the sensors and transmit control signals to all of the output devices is great.

In a MUX system, the signal is sent to a computer where it is converted from analog to digital if needed. Since the computer or control module of any system can process only one input at a time, it calls for input signals as it needs them. By timing the transmission of data from the sensors to the control module, a single data circuit can be used. Between each transmission of data to the control module, the sensor is electronically disconnected from the control module.

MULTIPLEXING COMMUNICATION PROTOCOLS

A **protocol** is a language computers use to communicate with one another over the data bus. Protocols may differ in baud rate and in the method of delivery. For example, some protocols use pulse-width modulation while others use variable pulse width. In addition, there may be differences in the voltage levels that equal a 1 or a 0 bit.

The Society of Automotive Engineers (SAE) has defined different classes of protocols according to their **baud rate** (speed of communication):

- Class A—An **asynchronous** low-speed protocol that has a baud rate of up to 10 kb/s (10,000 bits per second). *Asynchronous protocol* means that the communication between nodes is done only when needed instead of continuously.
- Class B—Medium-speed protocol that has a baud rate between 10 kb/s and 125 kb/s.
- Class C—A high-speed protocol with a baud rate between 125 kb/s and 1,000 kb/s, used for functions that require real-time control.

Since 1996, different automotive diagnostic communication protocols have been required for on-board diagnostics, second-generation (OBD II) compliance. The SAE adopted the OBD II protocols (**Figure 10-1**).

ISO 9141-2 Protocol

The International Standards Organization (ISO) protocol known as the **ISO 9141-2** is a class B system with a baud rate of 10.4 kb/s. ISO 9141-2 is not a network protocol since it can be used only for diagnostic purposes. IS0 9141 standardizes a protocol to be used between the nodes on the data bus and an OBD II standardized scan tool (as per SAE J1978 standards) for diagnostic purposes. This system is a two-wire system (**Figure 10-2**). One wire is called the **K-line** and is used for transmitting data from the module to the scan tool. The scan tool provides the bias voltage onto this circuit and the module pulls the voltage low to transmit its data. The other wire is called the **L-line** and is used by the module to receive data from the scan tool. The module provides the bias onto this circuit and the scan tool pulls the voltage low to communicate.

An adoption of the ISO 9141-2 protocol is the **ISO-K** bus that allows for bidirectional communication on a single wire. Vehicles that use the ISO-K bus require that the scan tool provide the bias voltage to power up the system. The scan tool provides up to 12 volts (V) onto the circuit, and the data is transferred when the voltage is pulled low to create a digital signal (**Figure 10-3**). The actual voltage seen on the circuit can be a little less than 12 volts, based on the number of modules in the circuit.

ISO 14230-4 Protocol

The **ISO 14230-4** protocol uses a single-wire bidirectional data line to communicate between the scan tool and the nodes. Many European manufacturers use this system. This data bus is used only for diagnostics and maintains the ISO 9141 protocol with a baud rate of 10.4 kb/s but uses a different voltage level. The operation of receiving and transmitting data requires a **master module**. The master module controls the transportation of messages by polling all of the **slave modules** and then waiting for the response. The communication occurs as the voltage on the wire is pulled low at a fixed pulse width. When there is no communication, the voltage on the line will be 5 volts.

Shop Manual
Chapter 10, page 450

Biasing refers to the voltage supplied to the bus.

Shop Manual
Chapter 10, page 453

ISO 14230-4 is also called Keyword Protocol 2000 or KWP2000.

ISO 9141-2 (K-line)
ISO 14230-4 [Keyword protocol (KWP) 2000]
J 1850 10.4 kb/s Variable pulse width
J 1850 41.6 kb/s Pulse width modulated
J 2284/ISO 15765-4 Controller area network (CAN)

Figure 10-1 OBD II communication protocols.

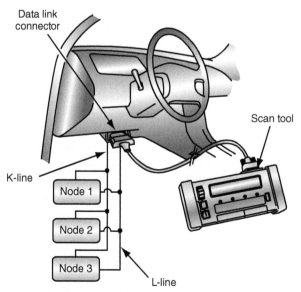

Figure 10-2 The two-wire ISO 9141-2 data bus used for diagnostic purposes.

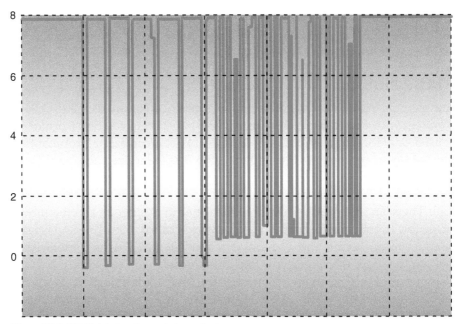

Figure 10-3 ISO-K data bus transmission voltages.

J1850 Protocol

Shop Manual
Chapter 10, page 457

The **J1850** bus system is the class B standard for OBD II for model year vehicles from 1996 to 2007. The J1850 standard allows for two different versions based on baud rate. The first supports a baud rate of 41.6 kb/s. Ford uses this protocol and calls it the Standard Corporate Protocol (SCP). This system uses a pulse-width modulated (PWM) signal that is transferred using a twisted pair of wires. The second protocol supports a baud rate of 10.4 kb/s average. General Motors and Chrysler have adopted this protocol. General Motors calls its system Class 2 and Chrysler calls its system Programmable Communication Interface (PCI). These systems use a variable pulse width (VPW) data bus with a single wire. The J1850 protocol was required for OBD II compliance until the 2008 model year.

J2284/ISO 15765-4 Protocol

Shop Manual
Chapter 10, page 460

The principal SAE class C protocol is J2284. This system is referred to as **controller area network (CAN)**. The CAN network system can support baud rates up to 1 Mb/s and is designed for real-time control of specific systems. Robert Bosch developed CAN in the early 1980s, and it has been a very popular bus system in Europe. Until recently most CAN bus–equipped vehicles used the CAN bus for communications only between modules and not for diagnostics with a scan tool. Scan tool diagnostics are usually performed over the ISO-K bus.

New U.S. regulations are requiring the use of the CAN bus system under industry standard J2284. This is the protocol for communications with a scan tool on U.S.-sold vehicles. Although the CAN bus system has been used since the 1980s, J2284 makes it unique, in that the CAN bus system will be used for diagnostics. Manufacturers were required to use CAN for diagnostic communications beginning in the 2008 model year.

> **YOU SHOULD KNOW** As a vehicle bus system, the CAN bus is not a legislated system. The SAE has mandated the CAN C network as the protocol for scan tool communications to the PCM. Thus, manufacturers can continue to use any system they wish for vehicle communications.

MULTIPLEXING SYSTEMS

The following are some examples of how data bus messages are transmitted based on the different classes and protocols. Although protocols are in place, manufacturers have some freedom to design the system they wish to use. The following examples will explain the common methods that are employed.

Class A Data Bus Network

One of the earliest multiplexing systems was developed by Chrysler in 1988 and used through the 2003 model year. This system is called **Chrysler Collision Detection (CCD)**. The term *collision* refers to the collision of data occurring simultaneously. This bus circuit uses two wires. The advantages of the CCD system include:

1. Reduction of wires.
2. Reduction of drivers required in the computers.
3. Reduced load across the sensors.
4. Enhanced diagnostics.

The CCD system uses a twisted pair of wires to transmit the data in digital form. One of the wires is called the **bus (+)** and the other is **bus (−)**. Negative voltages are not used. The (+) and (−) indicate that one wire is more positive than the other when the bus is sending the dominant bit "0." All modules that are connected to the CCD bus system have a special CCD chip installed (**Figure 10-4**). In most vehicles (but not all), the body control module (BCM) provides the bias voltage to power the bus circuits. Since the BCM powers the system, its internal components are illustrated (**Figure 10-5**). The other modules will operate the same as the BCM to send messages.

The bias voltage on the bus (+) and bus (−) circuits is approximately 2.50 volts when the system is idle (no data transmission). This is accomplished through a regulated 5-volt circuit and a series of resistors. The regulated 5 volts sends current through a 13-kΩ resistor to the bus (−) circuit (**Figure 10-6**). The current is then sent through the

The **Chrysler Collision Detection (CCD)** system is also referred to as C^{2}D (C squared D).

Shop Manual
Chapter 10, page 454

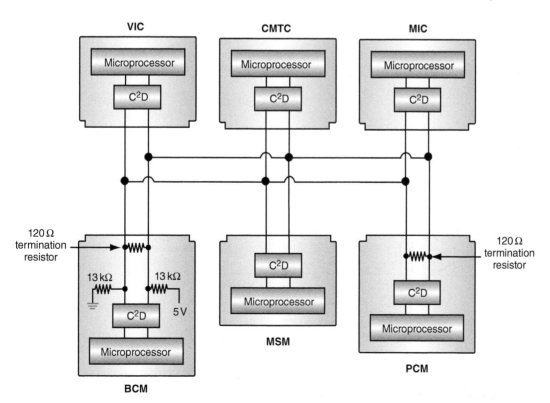

Figure 10-4 Each module on the CCD bus system has a CCD (C^{2}D) chip.

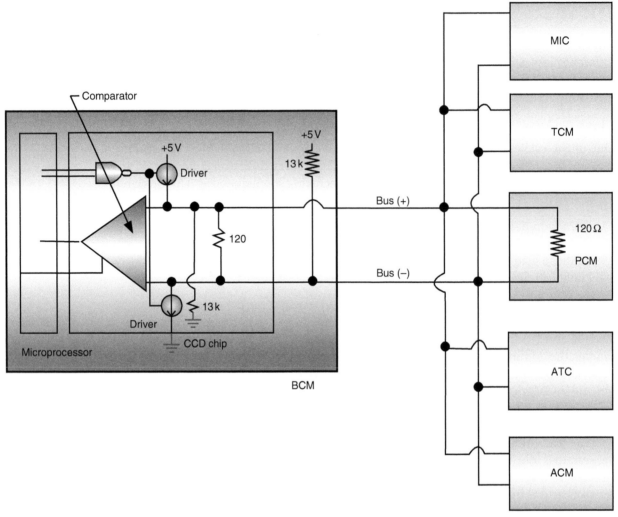

Figure 10-5 CCD bus circuit.

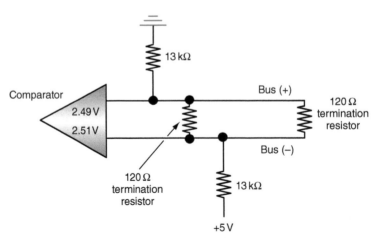

Figure 10-6 The bus is supplied 2.5 volts through the use of pull-up and pull-down resistors.

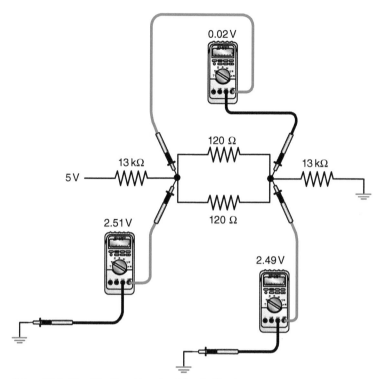

Figure 10-7 Simplified bus bias circuit for clarification.

two 120-Ω resistors that are wired in parallel and to the bus (+) circuit. Finally, the current is sent to ground through a second 13-kΩ resistor. A simplified schematic of this biasing circuit is shown along with the normal voltage drops that occur as a result of the resistors (**Figure 10-7**). The two 120-Ω resistors are referred to as **termination resistors**. Termination resistors are used to control induced voltages. Since voltage is dropped across resistors, the induced voltage is terminated. One termination resistor is internal to the BCM, while the other is located in the PCM.

With the bus circuits at the proper voltage levels, communication can occur. The comparator in the CCD chip acts as a voltmeter. If the positive lead of the voltmeter is connected to the bus (+) circuit and the negative lead is connected to the bus (−) circuit, the voltmeter will read the voltage difference between the circuits. At idle, the difference is 0.02 volt. When a module needs to send a message, the microprocessor will use the NAND gate to turn on and off the two drivers at the same time (**Figure 10-8**). The driver to the bus (+) circuit provides alternate 5 volts to the bus (+) circuit. The comparator will measure this voltage. At the same time, the driver to the bus (−) circuit provides an alternate ground path for the original 5 volts. This alternate ground bypasses the termination resistors and the second 13-kΩ resistor. Since the first 13-kΩ resistor is now the only one in the circuit, all of the voltage is dropped across it and the comparator will see low voltage on the bus (−) circuit.

Since the drivers are turned on and off at a rate of 7812.5 times per second, the voltage will not go to a full 5 volts on bus (+) nor to 0 volt on bus (−). However, bus (+) voltage is pulled higher *toward* 5 volts and bus (−) is pulled lower *toward* 0 volt (**Figure 10-9**). Once the comparator measures a voltage difference greater than 0.060 volt, the computers will recognize a bit value change. When the bus circuit is idle (0.02 volt difference), the bit value is 1. Once the voltage difference increases, the bit value changes to 0.

Class B Data Bus Network

The J1850 protocol is an example of a class B bus network. We will look at a VPW 10.4 kb/s system that Chrysler uses. This system is similar to General Motors, class 2 bus.

Shop Manual
Chapter 10, page 457

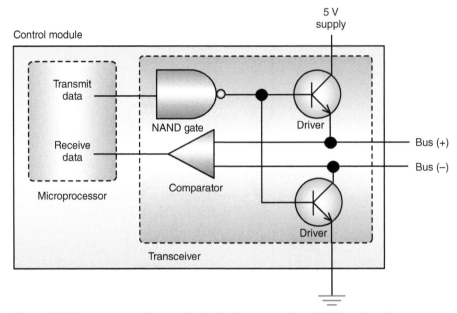

Figure 10-8 For a message to be transmitted, the drivers are activated, which pulls up bias on bus (+) and pulls down bias on bus (−).

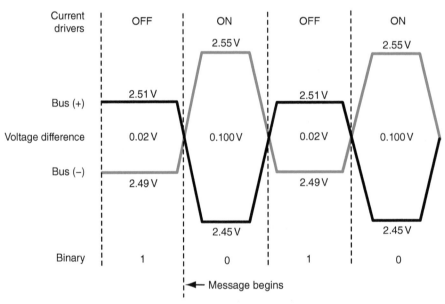

Figure 10-9 As the drivers are activated, the voltage difference between bus (+) and bus (−) increases over their voltage values at idle. The difference in voltage determines if a binary 1 or 0 is being transmitted.

Beginning in the 1998 model year, Chrysler began to phase out the CCD bus system and replace it with a new **Programmable Communication Interface (PCI)** bus system. Since this system is similar to that of other manufacturers, it will be used for discussion purposes.

The PCI system is a single-wire, bidirectional communication bus. Each module on the bus system supplies its own bias voltage and has its own termination resistors (**Figure 10-10**). Like the CCD system, the modules of the PCI system are connected in parallel. As a message is sent, a variable pulse-width modulation (VPWM) voltage between 0 and 7.75 volts is used to represent the 1 and 0 bits (**Figure 10-11**).

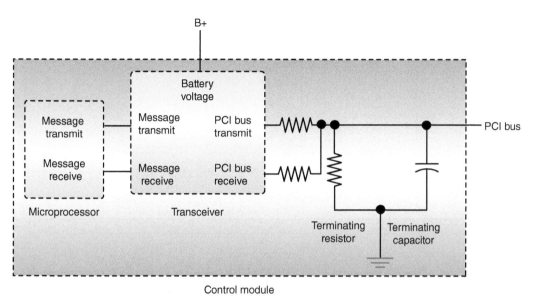

Figure 10-10 Bias and termination are supplied by each module on the PCI bus system.

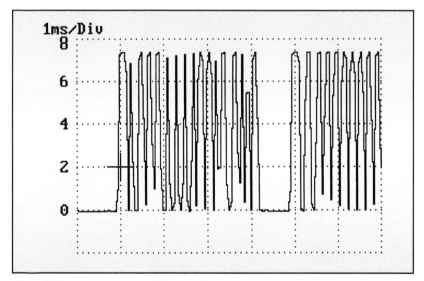

Figure 10-11 Lab scope trace of PCI bus voltages.

The voltage signal is not a clean digital signal. Rather, the voltage traces appear to be trapezoidal in shape because the voltage is slowly ramped up and down to prevent magnetic induction.

> **AUTHOR'S NOTE** The reference to "slowly ramped up and down" is relative. In the PCI bus, an average of 10,400 bits is transmitted per second. So, relatively speaking, the voltage is slow to go to 7.75 volts and slow to return to 0 volt.

The length of time the voltage is high or low determines if the bit value is 1 or 0 (**Figure 10-12**). The typical PCI bus message will have the following elements (**Figure 10-13**):

- SOF—Start of frame pulse used to notify other modules that a message is going to be transmitted.

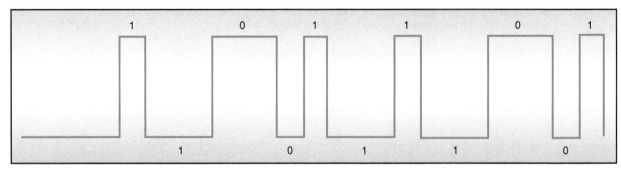

Figure 10-12 The VPWM determines the bit value.

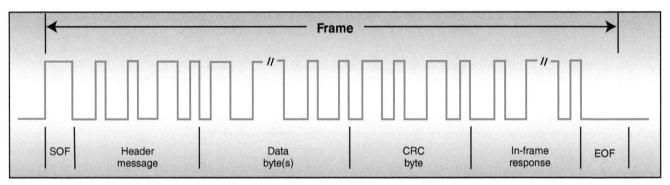

Figure 10-13 Components of a typical PCI bus message.

- Header—One to three bytes of information concerning the type, length, priority, target module, and sending module.
- Data byte(s)—The message that is being sent. This can be up to 8 bytes in length.
- Cyclic redundancy check (CRC) byte—Detects if the message has been corrupted or if there are any other errors.
- In-frame response (IFR) byte(s)—If the sending module requires an acknowledgment or an immediate response from the target module, this request will be received with the message. The IFR is the target module sending the requested information to the original sending module.
- EOF—A period of time with no voltage that identifies the module is finished communicating its message.

Figure 10-14 illustrates the type of information that is sent over the PCI bus system.

Controller Area Network

Shop Manual
Chapter 10, page 460

AUTHOR'S NOTE The following is an example of the CAN bus system. Again, baud rates and design vary between manufacturers. This is provided as a common method used. By understanding this system, you should be able to grasp any system design.

Most vehicles that follow the J2284 protocol integrate either two or three CAN bus networks that operate at different baud rates. The lower-speed bus is typically used for vehicle body functions such as seat, window, radio, and instrumentation. A highspeed bus is used for real-time functions, such as engine management and antilock brake operation. The third CAN bus would be used for diagnostics.

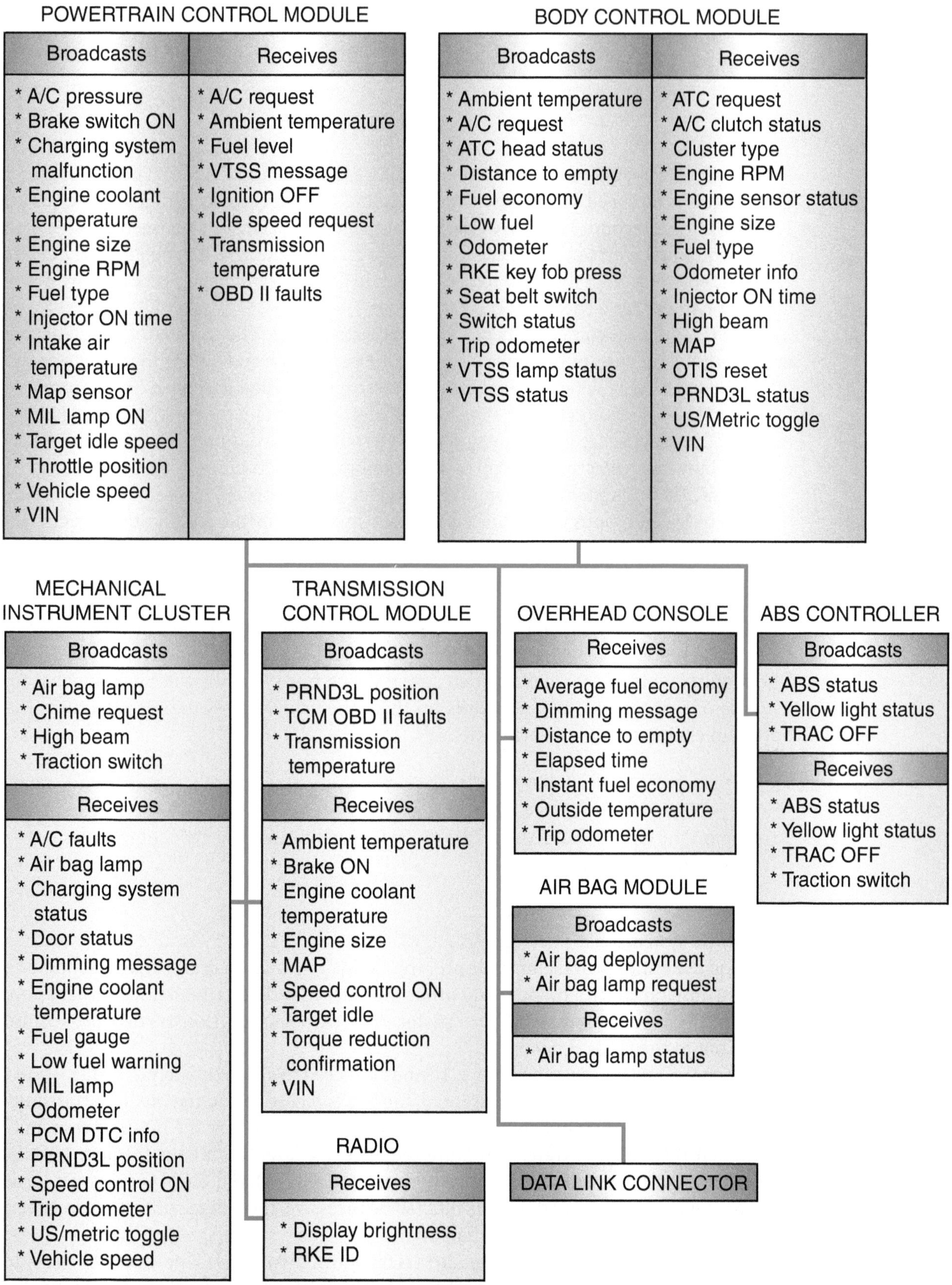

POWERTRAIN CONTROL MODULE

Broadcasts	Receives

* A/C pressure
* Brake switch ON
* Charging system
  malfunction
* Engine coolant
  temperature
* Engine size
* Engine RPM
* Fuel type
* Injector ON time
* Intake air
  temperature
* Map sensor
* MIL lamp ON
* Target idle speed
* Throttle position
* Vehicle speed
* VIN
	* A/C request
	* Ambient temperature
	* Fuel level
	* VTSS message
	* Ignition OFF
	* Idle speed request
	* Transmission
	  temperature
	* OBD II faults

BODY CONTROL MODULE

Broadcasts	Receives
* Ambient temperature * A/C request * ATC head status * Distance to empty * Fuel economy * Low fuel * Odometer * RKE key fob press * Seat belt switch * Switch status * Trip odometer * VTSS lamp status * VTSS status	* ATC request * A/C clutch status * Cluster type * Engine RPM * Engine sensor status * Engine size * Fuel type * Odometer info * Injector ON time * High beam * MAP * OTIS reset * PRND3L status * US/Metric toggle * VIN

MECHANICAL INSTRUMENT CLUSTER

Broadcasts
* Air bag lamp * Chime request * High beam * Traction switch

Receives

* A/C faults
* Air bag lamp
* Charging system
  status
* Door status
* Dimming message
* Engine coolant
  temperature
* Fuel gauge
* Low fuel warning
* MIL lamp
* Odometer
* PCM DTC info
* PRND3L position
* Speed control ON
* Trip odometer
* US/metric toggle
* Vehicle speed

TRANSMISSION CONTROL MODULE

Broadcasts
* PRND3L position * TCM OBD II faults * Transmission   temperature

Receives

* Ambient temperature
* Brake ON
* Engine coolant
  temperature
* Engine size
* MAP
* Speed control ON
* Target idle
* Torque reduction
  confirmation
* VIN

OVERHEAD CONSOLE

Receives
* Average fuel economy * Dimming message * Distance to empty * Elapsed time * Instant fuel economy * Outside temperature * Trip odometer

AIR BAG MODULE

Broadcasts
* Air bag deployment * Air bag lamp request

Receives
* Air bag lamp status

ABS CONTROLLER

Broadcasts
* ABS status * Yellow light status * TRAC OFF

Receives
* ABS status * Yellow light status * TRAC OFF * Traction switch

RADIO

Receives
* Display brightness * RKE ID

DATA LINK CONNECTOR

Figure 10-14 Chart of messages received and broadcasted by each module on the PCI bus.

✎📖 **A BIT OF HISTORY**
The first production car to use a CAN network was the 1991 Mercedes S-Class.

The CAN bus system uses terminology such as CAN B and CAN C. The letters *B* and *C* distinguish the speed of the bus. CAN B is a medium-speed bus with a speed of up to 125,000 bits per second. The CAN C bus has a speed of 500,000 bits per second. A vehicle can be equipped with both of these bus networks. In addition, a diagnostic CAN C bus can be used to connect the scan tool. Diagnostic CAN C (which can be called by many different names) has a speed of 500,000 bits per second.

The CAN bus circuit consists of a pair of twisted wires. The transfer of digital data is done by simultaneously pulling the voltage on one circuit high and pulling the voltage on the other circuit low. The wires for the CAN bus system are twisted to reduce electromagnetic interference. This requires 33 to 50 twists per meter. To maintain the twist, the bus wire pair is in adjacent cavities at connectors. Wires are routed to avoid parallel paths with high-current sources, such as ignition coil drivers, motors, and high-current PWM circuits.

On a CAN bus system, each module provides its own bias. Because of this, communication between groups of modules is still possible if an open occurs in the bus circuit. The CAN bus transceiver has drivers internal to the transceiver chip to supply the voltage and ground to the bus circuit.

Each CAN bus system has its advantages and limitations. For example, the high-speed CAN C bus may be functional only when the ignition is on. On the other hand, the CAN B bus can remain active when the ignition is turned off, if a module requires it to be active. The requirements of each module determine which bus system it will be connected to. The use of two separate bus networks on the same vehicle gives the manufacturer the optimum characteristics of each system.

AUTHOR'S NOTE Some CAN C bus networks do become active based on an event on the CAN B bus. Also, if the CAN C bus is used for vehicle interior systems, it is event driven.

Vehicle systems that exchange data at real time use CAN C. Typically, these modules would be the antilock brake module and the PCM. The manufacturer may also include the transmission control module and other modules that require real-time information. Other modules that may need to transfer data with the ignition turned off will be connected to the CAN B bus.

When CAN C bus becomes active, the bus is biased to approximately 2.5 volts. When both CAN C (+) and CAN C (−) are equal, the bus is recessive and the bit "1" is transmitted. When CAN C (+) is pulled high and CAN C (−) is pulled low, the bit "0" is transmitted. When the bit "0" is transmitted, the bus is considered dominant (**Figure 10-15**). To be dominant, the voltage difference between CAN C (+) and CAN C (−) must be at least 1.5 volts and not more than 3.0 volts. To be recessive, the voltage difference between the two circuits must not be more than 50 mV.

The optimum CAN C bus termination is 60 ohms. Two CAN C modules will provide 120 ohms of termination each. Since the modules are wired in parallel, total resistance is 60 ohms. The two modules that provide termination are typically located the farthest apart from each other. The terminating modules have two 60-ohm resistors that are connected in series to equal the 120 ohms. Common to both resistors are the connections to the center tap and ultimately through a capacitor to ground. This center tap may also

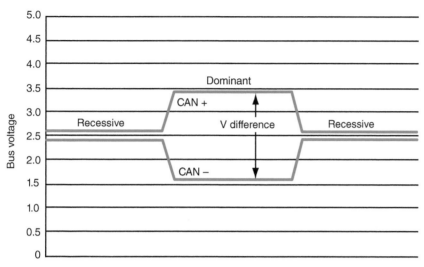

Figure 10-15 Voltages on the CAN C bus.

be connected to the transceiver (**Figure 10-16**). The other ends of the resistors are connected to CAN (+) and to CAN (−).

AUTHOR'S NOTE Some CAN bus networks will have additional termination resistance in other modules. Due to this, the total resistance of the bus may be lower than 60 ohms.

When CAN B (+) is approximately 0 to 0.2 volt and CAN B (−) is 4.8 to 5 volts, the bus is idle or recessive. In this state, the logic is "1." When CAN B (+) is pulled between 3.6 and 5 volts and CAN B (−) is pulled low between 1.4 and 0 volts, the bus is considered dominant and the logic is "0" (**Figure 10-17**). When CAN B (+) is approximately 0 volt and CAN B (−) is near battery voltage, the bus is asleep.

Each module on the CAN B bus supplies its own termination resistance. Total bus termination resistance is determined by the number of modules installed on the vehicle. Internal to CAN B modules are two termination resistors. The resistors connect CAN B (+) and CAN B (−) to their respective transceiver termination pins (**Figure 10-18**). To provide termination and bias, the transceiver internally connects the CAN B (+) resistor to ground and the CAN B (−) resistor to a 5-volt source. When the CAN B bus goes into sleep mode, the termination pin connected to CAN B (−) switches from 5 volts to battery voltage by the transceiver.

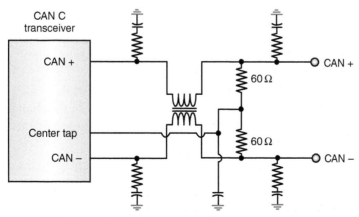

Figure 10-16 Termination resistance of a CAN C module.

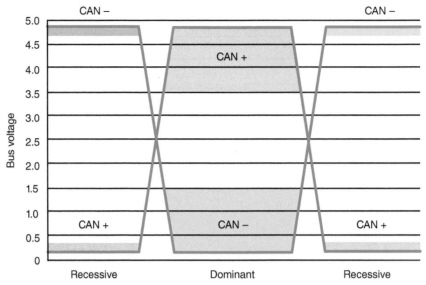

Figure 10-17 Typical CAN B bus voltages.

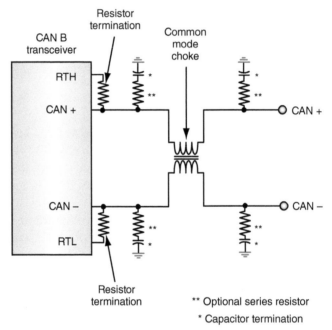

Figure 10-18 CAN B bus module termination resistance.

Some of the network messages are defined by the following:

■ Cyclic: A message launched on a periodic schedule. An example is the ignition on status broadcasts on the CAN B bus every 100 milliseconds.
■ Spontaneous: An application-driven message.
■ Cyclic and change: A message launched on a periodic schedule as long as the signal D is not changing. The message is relaunched whenever the signal changes.
■ By active function: A message that is transmitted only at a specific rate when the message does not equal a default value.

Most CAN B bus systems are fault tolerant and can operate with one of their conductors shorted to ground or with both of their conductors shorted together. Provided there is an electrical potential between one of the CAN B circuits and chassis ground, communication may still be possible. Due to its high speed, CAN C is not fault tolerant.

YOU SHOULD KNOW Since the CAN B bus can be designed to have a baud rate between 10.4 kb/s and 125 kb/s, fault tolerance diminishes as the baud rate increases.

A **central gateway (CGW)** module is used where all three CAN bus networks (CAN B, CAN C, and diagnostic CAN C) connect together. Similar to a router in a computer network, this module allows data exchange between the different busses. The CGW can take a message from one bus and transfer that message to the other bus without changing the message. If several messages are being sent simultaneously, some of the messages will be captured in a buffer and sent out based on priority. The CGW also monitors the CAN network for failures and logs a network DTC (U code) if it detects a malfunction.

The CGW is also the gateway to the CAN network for the scan tool. The scan tool is connected to the gateway using its own CAN bus circuit known as diagnostic CAN C. Because CAN C is used for diagnostics, data can be exchanged with the scan tool at a real-time rate (500 kb/s). As a result, a scan tool that is compatible with the CAN bus system is required for vehicle diagnosis.

Since a variety of generic scan tools may be connected to the vehicle, mandated regulations prohibit scan tools from containing any termination resistance. Due to the speed of the CAN C bus, termination resistors that are used need to be closely matched. If termination resistance resides in the tool, the variance in the tool termination could affect the operation of the diagnostic CAN C bus. For this reason, entire termination for the diagnostic CAN C bus resides in the CGW. The configuration of the resistors is similar to a dominant CAN C module, except that two 30-ohm resistors are connected in series.

AUTHOR'S NOTE The most aggressive company advancing CAN bus networking into the automobile design is Intel.

NETWORK ARCHITECTURE

We have discussed the functions and operations of the bus circuits and have mentioned that there are differences between manufacturers. In the CAN bus networks, these differences are not only in circuit operation but also in **network architecture**. *Network architecture* refers to how the modules are connected on the network. Keep in mind that the network architecture can include different bus classes. Typical networks include linear, stub, ring, and star architectures.

Network Architecture is also called *topology*.

The linear network system connects the modules in series (**Figure 10-19**). In this system data is transmitted from one module to the next. All modules between the transmitting module and the receiving module are involved in the transmission of the message. In this type of system, a short to ground, short to power, circuits shorted together, and opens will affect only the modules downstream of the fault. However, a module that does not power on will also cause messages not to be transmitted if it is between the transmitting and the receiving modules.

The linear network is also known as a daisy chain.

Figure 10-19 Linear network architecture configuration.

The stub network is also called a backbone network since it uses a single main line to connect the modules. It can also be referred to as simply "bus network" because it was the first type used.

Shop Manual
Chapter 10, page 463

The stub network is the most common. The modules are connected through a parallel circuit layout (**Figure 10-20**). As mentioned, fault tolerance of this system depends on the speed of the network.

The ring (or loop) network has the modules wired in series. Unlike the linear system, the ring does not have a beginning or an end (**Figure 10-21**). Messages can be transmitted in either direction based on the proximity the transmitting and receiving modules are to each other in the network.

The star (or hub) network has all of the modules connected to a central point (**Figure 10-22**). The modules are electrically connected in a parallel circuit. In this system, the use of termination resistors that reside in the modules may be eliminated since the termination resides in the star connector(s) (**Figure 10-23**).

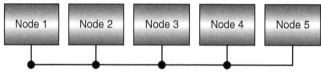

Figure 10-20 The stub network architecture is a common configuration.

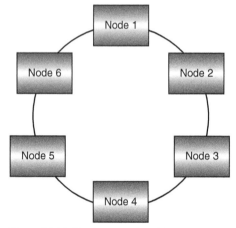

Figure 10-21 The ring network architecture configuration.

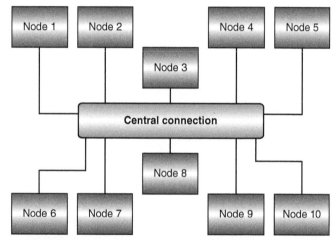

Figure 10-22 The star network architecture uses a common connection point.

Figure 10-23 The star connector is the central location point and may have the termination resistors reside within them.

SUPPLEMENTAL DATA BUS NETWORKS

Since no one data bus can handle all of the requirements of computer-controlled operations on today's vehicles, **supplemental bus networks** are also used. These are bus networks that are on the vehicle in addition to the main bus network. For example, a vehicle may have the CAN bus network and supplemental bus network to handle specified requirements. This section discusses the common bus networks that may also be on the vehicle.

Local Interconnect Network Data Bus

The **local interconnect network (LIN)** bus is a single-master module, multiple-slave module, low-speed network. The term *local interconnect* refers to all of the modules in the LIN network being located within a limited area. The LIN master module is connected to the CAN bus and controls the data transfer speed. The master module translates data between the slave module and the CAN bus (**Figure 10-24**). Diagnosis of the salve modules is performed through the master module. The termination resistance of the master module is 1 kΩ.

Shop Manual
Chapter 10, page 465

The LIN bus system can support up to 15 slave modules. Slave modules use 30-kΩ termination resistors. The slave modules can be actual control modules or sensors and actuators. **Smart sensors** are capable of sending digital messages on the LIN bus. The intelligent actuators receive commands in digital signals on the LIN bus from the master module. Only one pin of the master module is required to monitor several sensors and actuators.

The data transmission is variable between 1 kb/s and 20 kb/s over a single wire. The specific baud rate is programmed into each module. When a steady 12 volts is applied to the circuit, there is no message and the recessive bit is sent (bit 1). To transmit a dominant bit (bit "0"), the circuit is pulled low by a transceiver in the module that is transmitting the message (**Figure 10-25**). The master module or the slave module can send messages.

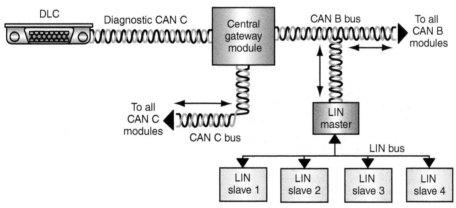

Figure 10-24 The LIN master communicates messages from the slaves onto the CAN bus.

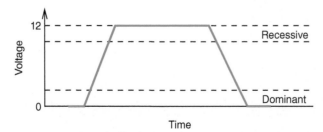

Figure 10-25 Voltages of the LIN bus.

Media-Oriented System Transport Data Bus

The **media-oriented system transport (MOST)** cooperation data bus system is based on standards established by cooperative efforts between automobile manufacturers, suppliers, and software programmers. The result is a data system that is specifically designed for the transmission of audio and video data. The MOST bus uses **fiber optics** to transmit data at a speed of up to 25 Mb/s. The fiber optics use light waves to transmit the data without the effects of electromagnetic interference or radio frequency interference. In the past, video and audio were transmitted as analog signals. With the MOST system using a fiber optics data bus, the data communications are digital.

Modules on the MOST data bus use an LED (light-emitting diode), photodiode, and a MOST transceiver to communicate with light signals (**Figure 10-26**). The LED and photodiode are part of the fiber-optic transceiver. The photodiode changes light signals into voltage that is then transmitted to the MOST transceiver. The LED is used to convert voltage signals from the MOST transceiver into light signals.

The conversion of light to voltage signals in the photodiode is by subjecting the PN junction of the photodiode with light. When light penetrates the junction, the energy converts to free electrons and holes. The electrons and holes pass through the junction in direct proportion to the amount of light. The photodiode is connected in series with a resistor (**Figure 10-27**). As the voltage through the photodiode increases, the voltage drop across the resistor also increases. Since the voltage drop changes with light intensity, the light signals change to voltage signals.

The microprocessor commands the MOST transceiver to send messages to the fiber-optic transceiver as voltage signals. Also, the MOST transceiver sends voltage signals from the fiber-optic transceiver to the microprocessor.

The modules are connected in a ring fashion by fiber-optic cable (**Figure 10-28**). Messages are sent in one direction only. The master module usually starts the message, but not always. The master module sends the message onto the data bus with a duty cycle

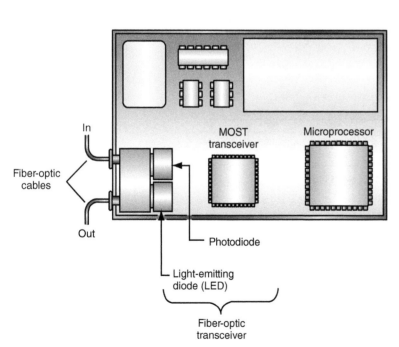

Figure 10-26 Typical MOST data system controller components.

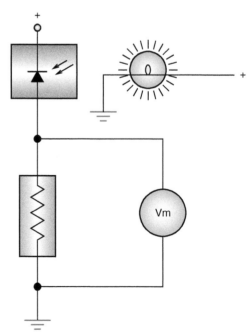

Figure 10-27 The voltage drop across the resistor changes in relation to the amount of light applied to the photodiode.

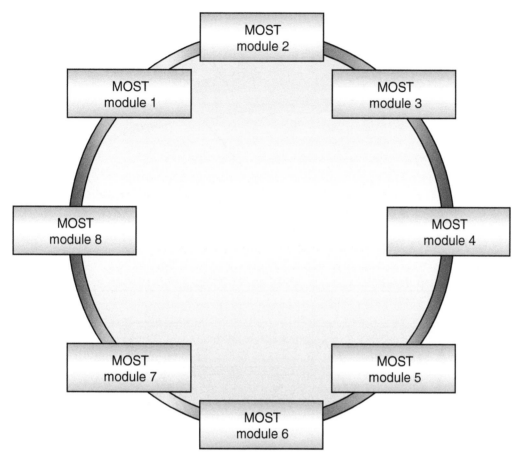

Figure 10-28 The MOST data system transfers data in a single direction through the use of a ring configuration.

frequency of 44.1 kHz. This frequency corresponds to the frequency of digital audio and video equipment. A message that originates from a module is sent to the next module in the ring. That module then sends it to the next module, and this continues until the originating module receives its own message. At this time the ring is closed and the message is no longer passed on. If a module receives a message that it does not need, the message is sent through the MOST transceiver and back to the fiber-optic transceiver without being transmitted to the microprocessor. If the MOST bus is powered down (asleep), it can be awakened by the ignition switch input or an input from a module. When an input is received, the master module will send a wake-up message to all of the modules in the ring.

The fiber-optic cable is constructed with several layers (**Figure 10-29**). The core consists of polymethyl methacrylate. Light travels through the core of the cable based on the principle of **total reflection**. Total reflection is achieved when a light wave strikes a layer that is between a dense and a thin material. The core of the cable is coated with an optically transparent reflective coating. The core makes up the dense material, and the coating is the thin material. The casing of the cable is made from polyamide, which protects the core from outside light. An outer cover is colored so the cable can easily be identified, but it also protects the cable from damage and high temperatures.

If the cable is laid out straight, some of the light waves travel through the core in a straight line. However, most of the light waves travel in a zigzag pattern (**Figure 10-30**). The zigzag pattern is a result of the total reflection. If the fiber-optic cable is bent, the light waves are reflected by total reflection at the borderline of the core coating and are guided through the bend (**Figure 10-31**).

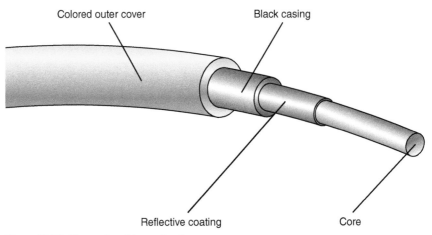

Figure 10-29 Fiber-optic cable construction.

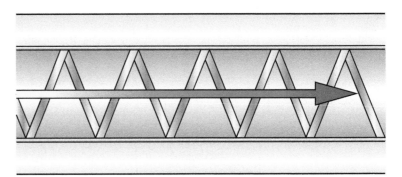

Figure 10-30 Light waves traveling through a straight section of the fiber-optic cable.

Figure 10-31 Light waves traveling through a curve in the fiber-optic cable.

Wireless Bus Networks

Wireless networks can connect modules together to transmit information without the use of physical connections by wires. For example, tire pressure information can be transmitted from a sensor in the tire to a module on the vehicle without wires. Although there are different technologies used for wireless communications, a popular one is called **Bluetooth**. Bluetooth technology allows several modules from different manufacturers to be connected using a standardized radio transmission. Laptop computers, notepads, and hands-free cell phones are all examples of devices that may be connected in the vehicle through Bluetooth.

 A BIT OF HISTORY

The Bluetooth Special Interest Group consists of more than 2,000 companies. The name *Bluetooth* comes from the Viking King Harald Blåtand who was nicknamed Bluetooth. King Harald ruled Denmark between AD 940 and 985. During his reign King Harald united Denmark and Norway. The name *Bluetooth* was adopted for a particular wireless communications technology because it shares the same philosophy as the king, multinational unity. Bluetooth unifies multinational companies. Bluetooth was initially a code name for the project; however, it has now become the trademark name.

Radio transmission uses the 2.40-GHz to 2.48-GHz frequency range. Transmitting on this band does not require a license or a fee. The transmission rate is up to 1 Mb/s, and the normal operating range for transmissions is about 33 feet (10 meters). The short transmitting range makes it possible to integrate the antenna, control, encryption, and transmission/receiver technology into a single module.

Since the frequency range Bluetooth uses is the same as that of other wireless devices such as garage door openers, microwaves, and many types of medical equipment, the technology uses special measures to protect against interference. These measures include:

- Dividing the data into short message packages using a duration of about 625 milliseconds.
- Using a check sum of 16 bits to confirm that the messages are not corrupted.
- Automatically repeating the transmission of faulty data.
- Using language coding that is converted into digital signals.
- Changing the transmitting/receiving frequencies at random, 1,600 times per second.

For security, the data is encrypted using a key that is 128 bits long. The receiver is checked for authenticity with the key. In addition, a secret password is used so devices can connect to each other. The key uses rolling code and is different for each connection.

To connect devices, each device is first adapted through the use of a personal identification number (PIN). When the PIN is entered, **piconets** are formed. These are small transmission cells that assist in the organization of the data. Each piconet will allow for up to eight devices to be operated at the same time. One device in each piconet will be assigned as the master. The master is responsible for establishing the connection and synchronizing the other devices. Once the PIN is entered, two Bluetooth-compatible devices will automatically establish a connection.

AUTHOR'S NOTE Each device has an address that is 48 bits long and is unique worldwide. This means over 281 trillion devices can be identified worldwide.

SUMMARY

- Multiplexing (MUX) is a system in which electrical signals are transmitted by a peripheral serial bus instead of by conventional wires. This allows several devices to share signals on a common conductor.

- A MUX wiring system uses bus data links that connect each module and allow for the transporting of data from one module to another.

- Each computer connected to the data bus is called a node.

- The Society of Automotive Engineers (SAE) has defined different classes of protocols according to their baud rate (speed of communication).

- ISO 9141-2 is not a network protocol since it can be used only for diagnostic purposes. It is a class B system that has a baud rate of 10.4 kb/s.

- An adoption of the ISO 9141-2 protocol is the ISO-K bus that allows for bidirectional communication on a single wire.

- The ISO 14230-4 protocol uses a single-wire, bidirectional data line to communicate between the scan tool and the nodes.

- J1850 protocol is the class B standard for OBD II. The J1850 standard allows for two different versions based on baud rate.

- The controller area network (CAN) system can support baud rates up to 1 Mb/s and is designed for real-time control of specific systems.

- One of the earliest multiplexing systems was developed by Chrysler in 1988 and used through the 2003 model year. This system is called Chrysler Collision Detection (CCD).

- The CCD system uses a twisted pair of wires to transmit the data in digital form. One of the wires is called the bus (+) and the other is called the bus (−).

- The CCD bias voltage on the bus (+) and bus (−) circuits is approximately 2.50 volts when the system is idle (no data transmission). This is accomplished through a regulated 5-volt circuit and a series of resistors.

- The Programmable Communication Interface (PCI) system is a single-wire, bidirectional communication bus. Each module on the bus system supplies its own bias voltage and has its own termination resistors.

- The modules of the PCI system send messages by a variable pulse-width modulation (VPWM) voltage; between 0 and 7.75 volts is used to represent the 1 and 0 bits.

- Most vehicles that are following the J2284 protocol integrate either two or three CAN bus networks that operate at different baud rates.

- The lower-speed bus is typically used for vehicle body functions such as seat, window, radio, and instrumentation control. A high-speed bus is used for real-time functions such as engine management and antilock brake operation.

- The circuitry of the CAN bus usually consists of a pair of twisted wires. For digital data to transfer, voltage is simultaneously pulled high on one circuit and pulled low on the other.

- A central gateway (CGW) module is used where all three CAN bus networks (CAN B, CAN C, and diagnostic CAN C) connect together. Similar to a router in a computer network, this module allows data exchange between the different busses. The CGW is also the gateway to the CAN network for the scan tool.

- The local interconnect network (LIN) was developed to supplement the CAN bus system. The term *local interconnect* refers to all of the modules in the LIN network being located within a limited area.

- The LIN bus is a single-master, multiple-slave, low-speed network.

- The media-oriented system transport (MOST) cooperation data bus system is based on standards established by a cooperative effort between automobile manufacturers, suppliers, and software programmers that resulted in a data system specifically designed for the data transmission of media-oriented data. MOST uses fiber optics to transmit data at an extremely fast rate, up to 25 Mb/s.

- Modules on the MOST data bus use an LED, a photodiode, and a MOST transceiver to communicate with light signals.

- Wireless networks can connect modules together to transmit information without the use of physical connection by wires.

- Bluetooth technology allows several modules from different manufacturers to be connected using a standardized radio transmission.

REVIEW QUESTIONS

Short-Answer Essays

1. Explain the principle of multiplexing.

2. Briefly describe the different multiplexing communication protocols.

3. Explain the different classes of communications.

4. Briefly explain the principle of operation of the CCD bus system as an example of class A multiplexing.

5. Briefly explain the principle of operation of the PCI bus system as an example of class B multiplexing.

6. Briefly explain the principle of operation of the controller area network (CAN) bus system.

7. Describe the purpose of the supplemental data bus networks.

8. Explain the operation of the local interconnect network data bus system.

9. Describe the operation of the media-oriented system transport data bus using fiber optics.

10. Explain the operation of wireless networks using Bluetooth technology.

Fill in the Blanks

1. Multiplexing provides the ability to use a _____ circuit to distribute and share data between several control modules.

2. Each computer connected to the data bus is called a _____.

3. The _____ protocol is the class B standard for OBD II.

4. The CCD system uses a twisted pair of wires to transmit the data in _____ form.

5. _____ are used to control induced voltages.

6. The PCI system uses a _____ pulse-width modulation voltage between 0 and 7.75 volts to represent the 1 and 0 bits.

7. A _____ module is used where all three CAN bus networks (CAN B, CAN C, and diagnostic CAN C) connect together.

8. In the MOST data bus system, the _____ changes light signals into voltage that is then transmitted to the MOST transceiver.

9. The modules of the MOST data bus system are connected in a _____ fashion by fiber-optic cable.

10. _____ technology allows several modules from different manufacturers to be connected using a standardized radio transmission.

Multiple Choice

1. In the CAN bus system:
   A. CAN B is high speed and not fault tolerant.
   B. CAN C is fault tolerant.
   C. Can C is low speed and used for many body control functions.
   D. None of the above.

2. All of the following statements concerning multiplexing are true **EXCEPT:**
   A. Hard wiring and vehicle weight are reduced.
   B. A single computer controls all of the vehicle functions.
   C. Enhanced diagnostics are possible.
   D. Reduces driver requirements in the computer.

3. The purpose of the central gateway module in a CAN bus system is:
   A. To provide a means for the modules on the different CAN bus networks to communicate with each other.
   B. To provide a method for the scan tool to communicate with the modules.
   C. Both A and B.
   D. Neither A nor B.

4. *Technician A* says some multiplexing systems use a data bus that consists of a twisted pair of wires.
   Technician B says some multiplexing systems use a data bus that consists of a single wire.
   Who is correct?
   A. A only   C. Both A and B
   B. B only   D. Neither A nor B

5. On a data bus that uses two wires, why are the wires twisted?

A. To increase the physical strength of the wires.

B. For identification of the circuit.

C. To reduce the effect of the high-current flow through the bus wires on other circuits.

D. To minimize the effects of an induced voltage on the data bus.

6. In the local interconnect network (LIN) bus:

A. The master controller is connected to the CAN bus and controls the data transfer speed.

B. The master controller translates data between the slave modules and the CAN bus.

C. Supporting up to 15 slave modules is possible.

D. All of the above.

7. Protocol is defined as:

A. A common communication method.

B. A method of reducing electromagnetic interference.

C. Data transmission through a single circuit.

D. All of the above.

8. *Technician A* says multiplexed circuits are used to communicate multiple messages over a single circuit.

Technician B says multiplexed circuits communicate by transmitting serial data.

Who is correct?

A. A only   C. Both A and B

B. B only   D. Neither A nor B

9. In a single-wire bus network, electromagnetic interference is controlled by:

A. Using a shielded wire.

B. Slowly ramping up and down the voltage levels.

C. Locating the wire outside of the normal wiring harness.

D. None of the above.

10. A computer that can communicate on a data bus is known as:

A. Node.   C. Protocol.

B. Byte.   D. Transceiver.

CHAPTER 11
LIGHTING CIRCUITS

Upon completion and review of this chapter, you should be able to understand and describe:

- The operation and construction of automotive lamps.
- The operation and construction of various headlights.
- The function of headlight system, including the computer-controlled headlight system.
- The function of automatic headlight on/off and time delay features.
- The operation of most common types of automatic headlight dimming systems.
- The operation of the SmartBeam headlight system as an example of today's sophisticated headlight systems.
- The function of automatic headlight leveling systems.
- The operation of adaptive headlight systems.

- The purpose and function of daytime running lamps.
- The operation of the concealed headlight system.
- The operation of the various exterior light systems, including parking, tail, brake, turn, side, clearance, and hazard warning lights.
- The operating principles of the turn signal and hazard light flashers.
- The operation of adaptive brake light systems.
- The operation of the various interior light systems, including illuminated entry systems and instrument panel lights.
- The use and function of fiber optics.
- The purpose and operation of lamp outage indicators.

Terms To Know

Adaptive brake light

Adaptive headlight system (AHS)

Automatic headlight dimming

Automatic on/off with time delay

Ballast

Bi-xenon headlamps

Composite headlight

Courtesy lights

Daytime running lamps (DRL)

Dimmer switch

Double-filament lamp

Flasher

Halogen

Headlight leveling system (HLS)

High-intensity discharge (HID)

Ignitor

Illuminated entry systems

Incandescence

Instrument panel dimming

Lamp

Lamp outage module

LUX

Night vision

Prisms

Projector headlight

Sealed-beam headlight

Sensitivity control

Timer control

Vaporized aluminum

Wake-up signal

INTRODUCTION

Today's technician is required to understand the operation and purpose of the various lighting circuits on the vehicle. If a lighting circuit is not operating properly, the safety of the driver, passengers, people in other vehicles, and pedestrians is in jeopardy. When today's technician performs repairs on the lighting systems, the repairs must meet at least two requirements: they must assure vehicle safety and meet all applicable laws.

The lighting circuits of today's vehicles can consist of more than 50 light bulbs and hundreds of feet of wiring. Incorporated within these circuits are circuit protectors, relays, switches, lamps, and connectors. In addition, most of today's vehicles have very sophisticated lighting systems that use computers and sensors. The lighting circuits consist of an array of interior and exterior lights, including headlights, taillights, parking lights, stop lights, marker lights, dash instrument lights, and courtesy lights.

With the addition of solid-state circuitry in the automobile, manufacturers have been able to incorporate several different lighting circuits or modify the existing ones. Some of the refinements that were made to the lighting system include automatic headlight washers, automatic headlight dimming, automatic on/off with timed-delay headlights, and illuminated entry systems. Some of these systems use sophisticated body computer-controlled circuitry and fiber optics.

Some manufacturers have included such basic circuits as turn signals into their body computer to provide for pulse-width dimming in place of a flasher unit. The body computer can also be used to control instrument panel lighting based on inputs, including if the side marker lights are on or off. By using the body computer to control many of the lighting circuits, the amount of wiring has been reduced. In addition, the use of computer control has provided a means of self-diagnosis in some applications.

This chapter discusses the types of lamps used, describes the headlight circuit, discusses the concealed headlight systems, and explores the various exterior and interior light circuits individually.

LAMPS

A **lamp** generates light through a process of changing energy forms called **incandescence**. The lamp produces light as a result of current flow through a filament. The filament is enclosed within a glass envelope and is a type of resistance wire that is generally made from tungsten (**Figure 11-1**). As current flows through the tungsten filament, it gets very hot. The conversion of electrical energy to heat energy in the resistive wire filament is so intense that the filament starts to glow and emits light. The lamp must have a vacuum surrounding the filament to prevent it from burning so hot that the filament burns in two. The glass envelope that encloses the filament maintains the presence of vacuum. When the lamp is manufactured, all the air is removed and the glass envelope seals out the air. If air is allowed to enter the lamp, the oxygen would cause the filament to oxidize and burn.

Many lamps are designed to execute more than one function. A **double-filament lamp** has two filaments, so the bulb can perform more than one function (**Figure 11-2**). It can be used in the stop light circuit, taillight circuit, and turn signal circuit combined.

It is important that any burned-out lamp be replaced with the correct lamp. The technician can determine what lamp to use by checking the lamp's standard trade number (**Table 11-1**).

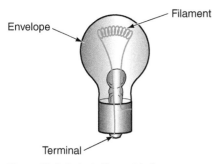

Figure 11-1 A single-filament bulb.

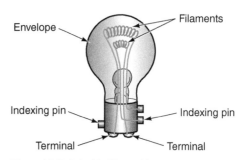

Figure 11-2 A double-filament lamp.

Table 11-1   A table of some typical automotive light bulb examples

Typical Automotive Light Bulbs			
Trade Number	Design Volts	Design Amperes	Watts: $P = A \times V$
168	14.0	0.35	4.9
192	13.0	0.33	4.3
194	14.0	0.27	3.8
194E-1	14.0	0.27	3.8
194NA	14.0	0.27	3.8
912	12.8	1.00	12.8
921	12.8	1.40	17.92
1141	12.8	1.44	18.4
1142	12.8	1.44	18.4
1156	12.8	2.10	26.9
1157	12.8	2.10/0.59	26.9/7.6
1157A	12.8	2.10/0.59	26.9/7.6
1157NA	12.8	2.10/0.59	26.9/7.6
2057	12.8	2.10/0.48	26.9/6.1
2057NA	12.8	2.10/0.48	26.9/6.1
3057	12.8–14.0	2.10/0.48	26.9/6.72
3156	12.8	2.10	26.9
3157	12.8–14.0	2.10/0.59	26.9/8.26
3457	12.8–14.0	2.23/0.59	28.5/8.26
4157	12.8–14.0	2.23/0.59	28.5/8.26
6411	12.0	0.833	10.0
6418	12.0	0.417	5.0
7440	12.0	1.75	21.0
7443	12.0	1.75/0.417	21.0/5.0
7507	12.0	1.75	21.0

HEADLIGHTS

Types of automotive headlights include standard sealed beam, halogen sealed beam, composite, high-intensity discharge (HID), projector, bi-xenon, and LED.

Sealed-Beam Headlights

Shop Manual
Chapter 11, page 503

 A BIT OF HISTORY

In 1940, Federal Motor Vehicle Safety Standard 108 required that all cars sold in the United States use two 7-inch round sealed-beam headlights. The headlight was technically known as the Par 56 for parabolic aluminized reflector and 56/8-inch (7-inch) diameter. The headlight incorporated both the low and high beams. In 1957, the rules were updated to allow two pairs of 6-inch headlamps, one low beam and one high beam on each side. In 1975, rectangular sealed-beam headlights were approved. The last vehicles to continue the use of round 7-inch headlights were the Jeep Wrangler, Toyota FJ Cruiser, and the Mercedes-Benz G-Wagon. The Jeep Wrangler used actual sealed-beam headlights until 2006.

The **sealed-beam headlight** is a self-contained glass unit made up of a filament, an inner reflector, and an outer glass lens (**Figure 11-3**). The standard sealed-beam headlight does not surround the filament with its own glass envelope (bulb). The glass lens is fused to the parabolic reflector, which is sprayed with **vaporized aluminum** that gives a reflecting surface that is comparable to silver. The reflector intensifies the light that the filament produces, and the lens directs the light to form the required light beam pattern. The inside of the lamp is filled with argon gas. All oxygen must be removed from the standard sealed-beam headlight to prevent the filament from oxidizing.

The lens is designed to produce a broad, flat beam. The light from the reflector is passed through concave **prisms** in the glass lens (**Figure 11-4**). Lens prisms redirect the light beam and create a broad, flat beam. **Figure 11-5** shows the horizontal spreading and the vertical control of the light beam to prevent upward glaring.

By placing the filament in different locations on the reflector, the direction of the light beam is controlled (**Figure 11-6**). In a dual-filament lamp, the lower filament is used for the high beam and the upper filament is used for the low beam.

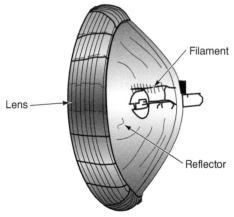

Figure 11-3 Sealed-beam headlight construction.

Figure 11-4 The lens uses prisms to redirect the light.

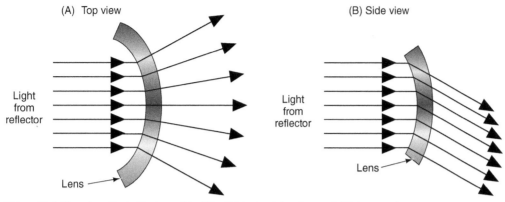

Figure 11-5 The prism directs the beam into (A) a flat horizontal pattern and (B) downward.

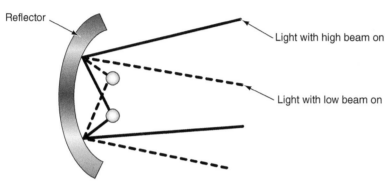

Figure 11-6 Filament placement controls the projection of the light beam.

Halogen Headlights

The sealed-beam **halogen** lamp most commonly used in automotive applications consists of a small bulb filled with iodine vapor. The bulb is made of a high-temperature-resistant quartz that surrounds a tungsten filament. This inner bulb is installed in a sealed glass housing (**Figure 11-7**). Adding halogen to the bulb allows the tungsten filament to withstand higher temperatures than that of conventional sealed-beam lamps. The higher temperatures mean the bulb is able to burn brighter.

In a conventional sealed-beam headlight, the heating of the filament causes atoms of tungsten to be released from the surface of the filament. These released atoms deposit on

Halogen is the term used to identify a group of chemically related nonmetallic elements. These elements include chlorine, fluorine, and iodine.

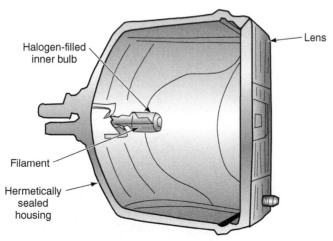

Figure 11-7 A sealed-beam halogen headlight with iodine vapor bulb.

the glass envelope and create black spots that affect the light output of the lamp. In a halogen lamp, the iodine vapor causes the released tungsten atoms to be redeposited onto the filament. This virtually eliminates any black spots. It also allows for increased high-beam output of 25% over conventional lamps and for longer bulb life.

> **AUTHOR'S NOTE** Because the filament is contained in its own bulb, cracking or breaking of the lens does not prevent halogen headlight operation. As long as the filament envelope is not broken, the filament will continue to operate. However, a broken lens will result in poor light quality and should be replaced.

Composite Headlights

Composite headlights are units that have a lens and housing assembly with a replaceable headlight bulb. The bulb can be halogen or HID.

Shop Manual
Chapter 11, page 505

Using the **composite headlight** system offers vehicle manufacturers more flexibility to produce almost any style of headlight lens they desire (**Figure 11-8**). This improves the aerodynamics, fuel economy, and styling of the vehicle. Composite headlight systems use a replaceable halogen bulb placed into a formed headlight assembly.

Many manufacturers vent the composite headlight housing because of the increased amount of heat that the bulb develops. However, since the housings are vented, condensation may develop inside the lens assembly. This condensation is not harmful to the bulb and does not affect the headlight operation. When the headlights are turned on, the heat generated from the halogen bulbs will dissipate the condensation quickly. Ford uses integrated non-vented composite headlights. On these vehicles, condensation is not considered normal. The assembly should be replaced.

HID Headlights

Shop Manual
Chapter 11, page 505

High-intensity discharge (HID) headlamps use an inert gas to amplify the light produced by arcing across two electrodes. These headlamps (**Figure 11-9**) put out three times more light and twice the light spread on the road than conventional halogen headlamps (**Figure 11-10**). They also use about two-thirds less power to operate and will last two to three times longer. HID lamps produce light in both ultraviolet and visible wavelengths. This advantage allows highway signs and other reflective materials to glow. This type of lamp first appeared on select models from BMW in 1993, Ford in 1995, and Porsche in 1996.

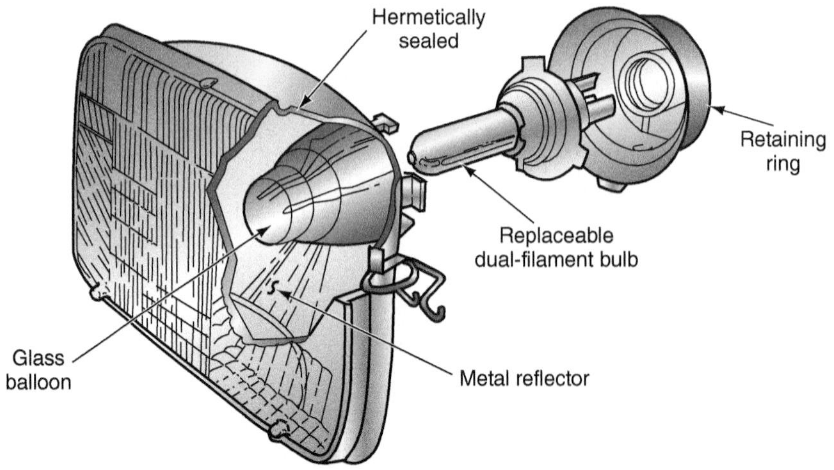

Figure 11-8 A composite headlight system with a replaceable halogen bulb.

Figure 11-9 HID headlamps; note the reduced size of the headlamp assemblies within the lens.

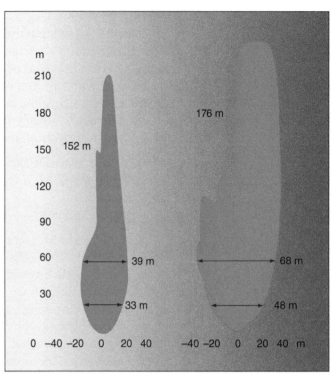

Figure 11-10 Comparison between light intensity and pattern. Halogen lamp is shown on the left and the xenon (HID) lamp on the right.

The HID lamp (**Figure 11-11**) consists of an outer bulb made of cerium-doped quartz that houses the inner bulb (arc tube). The inner bulb is made of fused quartz and contains two tungsten electrodes. It also is filled with xenon gas, mercury, and metal halides (salts).

The HID lamp does not rely on a glowing filament for light. Instead, it uses a high-voltage arcing bridge across the air gap between the electrodes. The xenon gas amplifies the light intensity given off by the arcing. The HID system requires the use of an **ignitor** and a **ballast** to provide the electrical energy required to arc the electrodes (**Figure 11-12**).

Figure 11-11 HID bulb element.

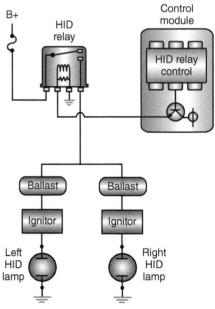

Figure 11-12 HID headlight schematic showing the use of a ballast and an ignitor.

The ignitor is usually built into the base of the HID bulb and will provide the initial 10,000 to 25,000 volts (V) to jump the gap. A cold lamp requires about 10 kV, while a hot lamp will require around 25 kV to ignite. Once the gap has been jumped and the gas warms, the ballast will provide the required voltage to maintain the current flow across the gap. *Ballast* refers to a device that limits the current delivered through an electrical circuit. The ballast must deliver 35 watts to the lamp when the voltage across the lamp is between 70 and 110 volts. The ballast usually consists of a transformer and a capacitor (**Figure 11-13**). The HID system consists of four components:

- High-frequency DC/DC converter.
- Low-frequency DC/AC inverter.
- Ignition circuit.
- Digital signal controller.

The DC/DC converter increases the 12 volts from the battery to the level required for the ignition circuit. After ignition, the output voltage from the converter drops to about 85 volts. The DC/AC inverter converts the voltage output of the converter to an AC square wave current that is used to put an equal charge on the electrodes. The converter operates on different frequencies depending on the stage the lamp is in. For the turn-on stage, the converter frequency is 1 kHz and increases to 20 Hz during warm-up. After warm-up, the frequency is set at 200 Hz. The ignitor generates high-voltage pulses that are sent to the electrodes to start the lamp.

The HID headlight has six stages of start-up (**Figure 11-14**). Prior to the turn-on phase, the gas is an insulator and the resistance is infinite. During the turn-on phase, the

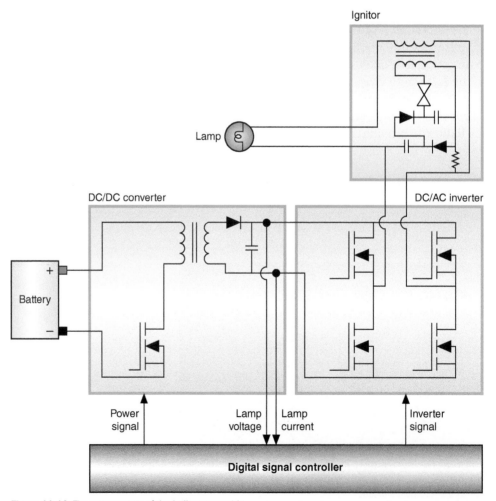

Figure 11-13 The components of the ballast assembly.

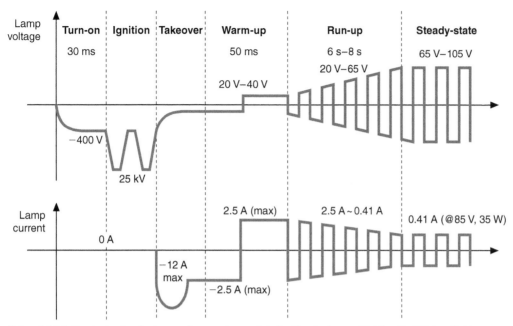

Figure 11-14 Graph showing the lamp voltage and current during the six stages of lighting the HID headlight.

voltage generated in the ballast is sent to the ignitor. The ignitor then produces a high-voltage pulse that is sent to the HID lamp electrodes. The gas that was an insulator becomes conductive under the presence of the high voltage. An arc jumps across the electrodes and light is produced. Once the gas has been ignited, a large amount of current is required to maintain the arc. This takeover high current is initially provided by the ballast's capacitor. Once the ballast has taken over the current, the converter provides the current required to warm up the gas. The warm-up of the gas takes about 50 milliseconds; then the run-up stage starts. During this stage, the voltage is increased while the current is decreased. Once steady state is reached, the voltage output is held to about 85 volts while the current is held at 410 mA.

The greater light output of these lamps allows the headlamp assembly to be smaller and lighter. These advantages allow designers more flexibility in body designs as they attempt to make their vehicles more aerodynamic and efficient. For example, the Infiniti Q45 models are equipped with a seven-lens HID system (**Figure 11-15**) to provide stylish looks and high lamp output.

Figure 11-15 Seven-lens HID headlamp.

Figure 11-16 Projector headlight.

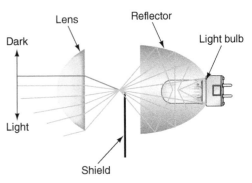

Figure 11-17 Construction of the projector headlight.

Projector Headlights

The projector is a type of headlight; it can use an HID or halogen bulb. Like a light bulb used in a projector, the light is focused in front of the vehicle.

A **projector headlight** is designed to produce a stronger beam to light at a more focused area (**Figure 11-16**). Projector headlights provide very accurate light dispersion with less scatter. They offer a crisp line of light that is focused on the ground and illuminates most everything under the line.

The bulb is positioned at the rear of an elliptical reflector and a condenser lens is located in front of the bulb (**Figure 11-17**). Between the bulb and the lens a shield is placed that provides the light beam cutoff. The light from the reflector is not lost in this configuration; it is harnessed and focused.

Bi-Xenon Headlamps

Due to the increased amount of light that the HID and projector headlamps produce, it is not used in a quad headlight system as high-beam lamps. Instead, it is used only as low-beam lamps and a halogen bulb for the high beam. Because the quad headlamp system uses all four bulbs for high-beam operation, using a quad lamp system with HID or projector lamps would blind any oncoming drivers due to the excessive amount of light output. Also, it is not possible to reduce the light intensity of an HID or projector headlight by pulse-width modulation of the current to the element.

To overcome these issues, and to still be able to use the HID or projector headlight, many manufacturers use **bi-xenon headlamps** in their dual headlamp systems. *Bi-xenon* refers to the use of a single xenon lamp to provide both the high-beam and the low-beam operations. The full light output is used to produce the high beam. Low beam is formed by either moving the xenon bulb within the lens or moving a shutter between the bulb and the lens.

 A BIT OF HISTORY

The Model T Ford used a headlight system that had a replaceable bulb. The owner's manual warned against touching the reflector except with a soft cloth.

Systems that use the shutter will have a motor within the headlamp assembly that raises and lowers the shutter (**Figure 11-18**). The position of the shutter dictates the amount of projected light and its pattern.

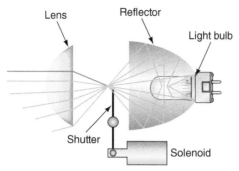

Figure 11-18 A shutter controlled by an actuator motor is used to control the amount of light being projected ahead of the vehicle.

AUTHOR'S NOTE The actual direction that the shutter travels, to block or unblock, varies based on application.

In some systems, a motor is used to change the position of the bulb. The bulb is physically raised in the reflector housing to produce the high-beam output. In low-beam mode, the bulb is lowered in the reflector housing. The amount of reflection dictates the light intensity and pattern.

Using the same lamp for both low- and high-beam operations permits both modes to have the same light color. This produces less visual contrast when switching between modes and is less stressful to the eyes of the driver.

HEADLIGHT SWITCHES

Shop Manual
Chapter 11, page 534

The headlight switch may be located either on the dash by the instrument panel or on the steering column multifunction switch (**Figure 11-19**). Depending on application, the headlight switch is one of two types. The first is a switch that carries current to the lamps directly or to a control device, such as a relay. The second type is a switch that is only an input to the BCM and does not directly activate the lights. Either switch will control a majority of the vehicle's lighting systems.

Figure 11-19 (A) Instrument panel-mounted headlight switch. (B) Steering column-mounted headlight switch.

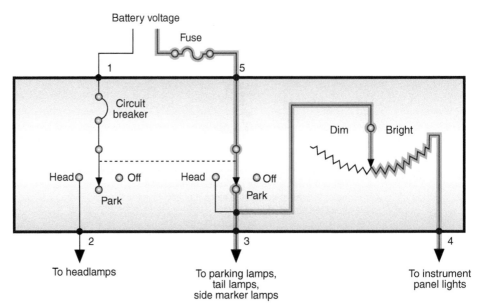

Figure 11-20 Operation with the headlight switch in the PARK position.

Shop Manual
Chapter 11, page 532

The most common style of current-carrying headlight switch is the three-position type with OFF, PARK, and HEADLIGHT positions. The headlight switch will generally receive direct battery voltage to two terminals of the switch. This allows the light circuits to be operated without having the ignition switch in the RUN or ACC (accessory) position.

When the headlight switch is in the OFF position, the open contacts prevent battery voltage from continuing to the lamps. When the switch is in the PARK (parking lights) position, battery voltage that is present at terminal 5 is applied through the closed contacts to the side marker, taillight, license plate, and instrument cluster lights (**Figure 11-20**). These circuits are usually protected by a 15- to 20-amp (A) fuse that is separate from the headlight circuit.

When the switch is located in the HEADLIGHT position, battery voltage that is present at terminal 1 is applied through the circuit breaker and the closed contacts to light the headlights. Battery voltage from terminal 5 continues to light the lights that were on in the PARK position (**Figure 11-21**). The circuit breaker is used to prevent temporary overloads to the system from totally disabling the headlights.

The headlight circuits just discussed are designed with insulated side switches and grounded bulbs. In this system, battery voltage is applied to the headlight switch. The switch must be closed in order for current to flow through the filaments and to ground. The circuit is complete because the headlights are grounded to the vehicle body or chassis. Many import manufacturers use a system design that has insulated bulbs and ground side switches. In this system, when the headlight switch is located in the HEADLIGHT position, the contacts are closed to complete the circuit path to ground. The headlight switch is located after the headlight lamps in the circuit. Battery voltage is applied directly to the headlights when the relays are closed. But the headlights will not light until the switch completes the ground side of the relay circuits. In this system, both the headlight and dimmer switches complete the circuits to ground.

The headlight switch rheostat is a variable resistor that the driver uses to control the instrument cluster illumination lamp brightness. As the driver turns the light switch knob, the resistance in the rheostat changes. The greater the resistance, the dimmer the

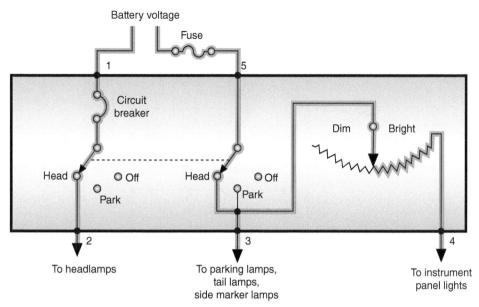

Figure 11-21 Operation with the headlight switch in the HEADLIGHT position.

instrument panel illumination lights glow. In vehicles that have the headlight switch located in the steering column, the rheostat may be a separate unit located on the dash near the instrument panel.

In computer-controlled headlight systems, the headlight switch is usually a resistive multiplex switch that provides multiple inputs over a single circuit.

Shop Manual
Chapter 11, page 508

AUTHOR'S NOTE   The resistive multiplex switch was discussed in Chapter 9.

The BCM will control the relays associated with the exterior lighting. If input from the switch indicates that the headlights are requested, the BCM will energize the low-beam headlight relay (**Figure 11-22**). With this relay energized, current flows to the low-beam lamps. Voltage is also applied to the multifunction switch (dimmer switch function). If the multifunction switch is placed in the high-beam position, it will complete the path for the flow of current to the coil of the high-beam headlight relay. Since this relay is connected to ground, the coil is energized and the contacts move to supply current to the high-beam lamps. The low-beam relay must be energized to turn on the high beams unless the flash-to-pass function is being performed. The flash-to-pass function of the multifunction switch bypasses the BCM control of the circuit. The flash-to-pass circuit illuminates the high-beam headlights even with the headlight switch in the OFF or PARK position (**Figure 11-23**). In this illustration, battery voltage is supplied to terminal B1 of the headlight switch and on to the dimmer switch. Battery voltage is available to the dimmer switch through this wire in both the OFF and PARK positions of the headlight switch. When the driver activates the flash-to-pass feature, the contacts in the dimmer switch complete the circuit to the high-beam filaments.

The **dimmer switch** provides the means for the driver to select high- or low-beam operation and to switch between the two. In systems that do not utilize computer control,

Shop Manual
Chapter 11, page 536

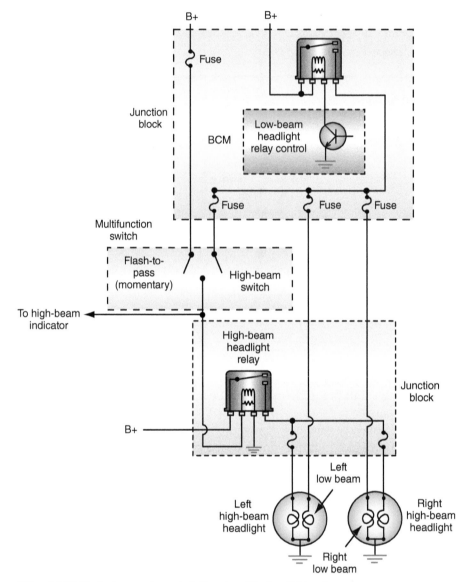

Figure 11-22 The body computer controls the relay of the headlight system.

the dimmer switch is connected in series within the headlight circuit and controls the current path for high and low beams. Typically, the dimmer switch is located on the steering column as part of the multifunction switch. It may be located on the steering wheel to prevent early failure and to increase driver accessibility (**Figure 11-24**). This switch is activated by the driver pulling on the stock switch (turn signal lever).

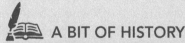 A BIT OF HISTORY

Foot-operated dimmer switches became standard equipment in 1923. The drivers operate this switch by pressing on it with their foot. Positioning the switch on the floor board made the switch subject to damage because of rust, dirt, and so on.

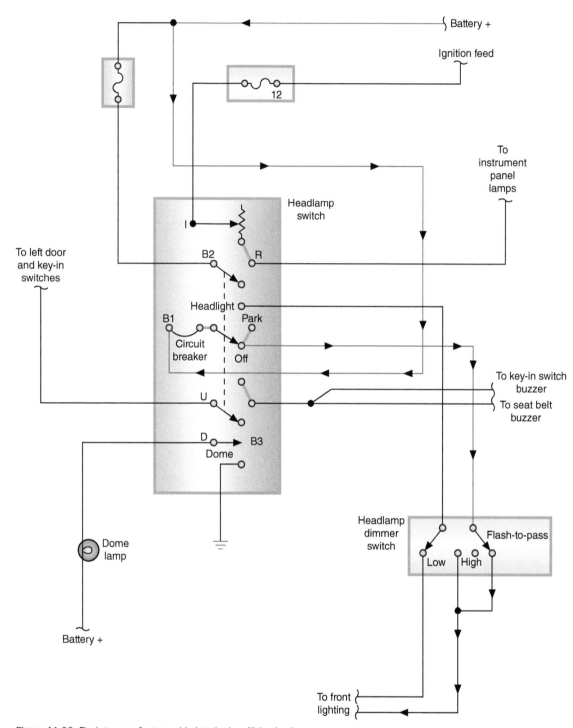

Figure 11-23 Flash-to-pass feature added to the headlight circuit.

In most modern systems, the dimmer switch is an input to the BCM. In this case, the BCM will activate the high-beam relay. The flash-to-pass feature may also be controlled by the BCM.

Some computer-controlled headlight systems use high-side drivers (HSD) (**Figure 11-25**). In the example shown, the lamp and fog lamp relays are controlled using low-side drivers (relay drivers). The difference is that the headlight switch input is sent to the BCM, and then the BCM sends the request to the integrated power

Shop Manual
Chapter 11, page 510

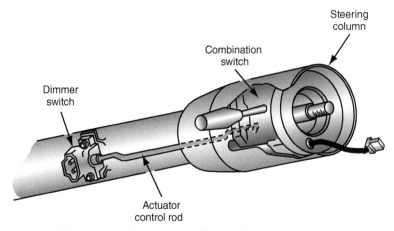

Figure 11-24 A steering column–mounted dimmer switch.

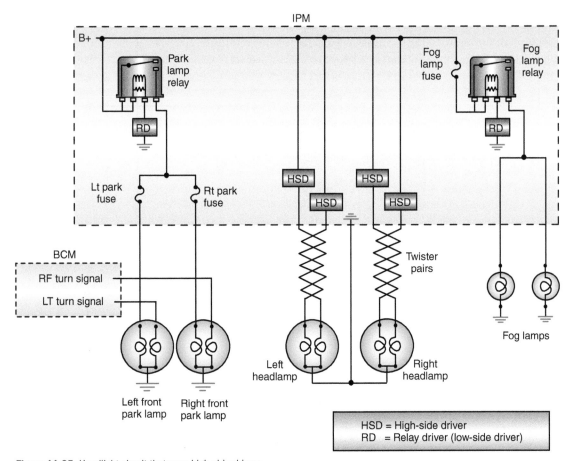

Figure 11-25 Headlight circuit that uses high-side drivers.

module (IPM). The IPM then turns on and off the relays as needed. The operation of the headlights is unique. Based on the input from the headlight switch, the BCM sends the requested state to the IPM. The IPM then turns on the headlights by supplying power from the HSDs. This headlamp system does not require relays or fuses because the HSDs perform these functions. If a high-current condition occurs, the HSDs sense this and turn off the circuit until the cause of the high current is no longer present. The output of the HSDs is pulse-width modulated signal at a frequency of 90 Hz in order to maintain a constant 13.5 volts to the headlight bulbs, relative to the battery voltage. This is done to increase the life of the headlight bulbs. An additional benefit of using HSDs is their ability to perform diagnostics of the system, set diagnostic trouble codes, and turn on indicator lights to notify the driver of a malfunction.

AUTOMATIC ON/OFF WITH TIME DELAY

AUTHOR'S NOTE This system is given several different names by the various manufacturers. Some of the more common names include Twilight Sentinel, Auto-lamp/Delayed Exit, and Safeguard Sentinel.

Shop Manual
Chapter 11, page 511

The **automatic on/off with time delay** has two functions: to turn on the headlights automatically when ambient light decreases to a predetermined level and to allow the headlights to remain on for a certain amount of time after the vehicle has been turned off. The common components of the automatic on/off with time delay include the following:

1. Photo cell and amplifier.
2. Power relay.
3. Timer control.

The photo cell is a variable resistor that uses light to change resistance.

In a typical automatic on/off with time delay headlight system, a photo cell is located inside the vehicle's dash to sense outside light (**Figure 11-26**). In most systems, the headlight switch must be in the AUTO position to activate the automatic mode. The driver can override the automatic on/off feature by placing the headlight switch in the HEADLIGHT or OFF position.

Shop Manual
Chapter 11, page 512, 514

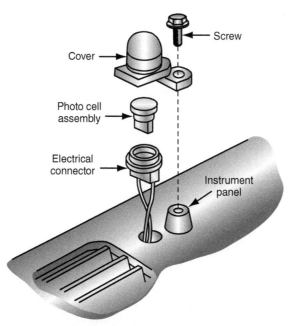

Figure 11-26 Most automatic on/off headlight systems have the photo cell located in the dash to sense incoming light levels.

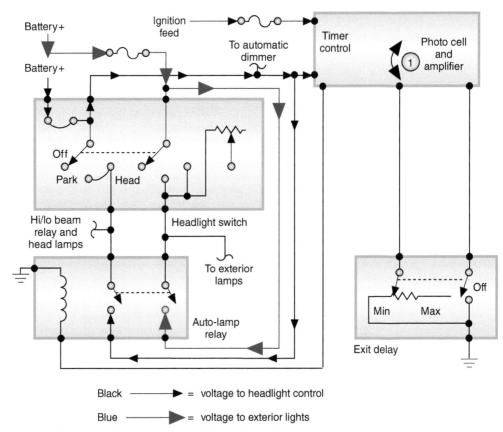

Battery+

Battery+

Ignition feed

To automatic dimmer

Timer control

Photo cell and amplifier

Off

Park   Head

Hi/lo beam relay and head lamps

Headlight switch

To exterior lamps

Auto-lamp relay

Min   Max

Off

Exit delay

Black ———————▶ = voltage to headlight control

Blue ———————▶ = voltage to exterior lights

Figure 11-27 Schematic of automatic headlight on/off with time delay system.

Early systems utilize a **timer control** that works along with a potentiometer in the headlight switch (**Figure 11-27**). The timer control unit controls the automatic operation of the system and the length of time the headlights stay on after the ignition switch is turned off. The timer control signals the sensor-amplifier module to energize the relay for the requested length of time.

To activate the automatic on/off feature, the photo cell and amplifier must receive voltage from the ignition switch. As the ambient light level decreases, the internal resistance of the photo cell increases. When the resistance value reaches a predetermined level, the photo cell and amplifier trigger the sensor-amplifier module. The sensor-amplifier module energizes the relay, turning on the headlights and exterior parking lights.

General Motors's (GM) Twilight Sentinel System is an example of early use of the BCM to control system operation (**Figure 11-28**). The BCM senses the voltage drop across the photo cell and the delay control switch. If the ambient light level drops below a specific value, the BCM grounds the headlamp and park lamp relay coils. The BCM also keeps the headlights on for a specific length of time after the ignition switch is turned off.

Most current systems continue to use the same principle. The difference is that the time control is a function of the module (usually the BCM). An interface module, such as the instrument panel or overhead console, provides a means for the driver to select the desired amount of time delay after the ignition switch is turned off. The BCM keeps the headlight relay energized for the length of time programmed. When the ignition switch is turned off, all other relays controlling exterior lighting are de-energized except the headlight relay.

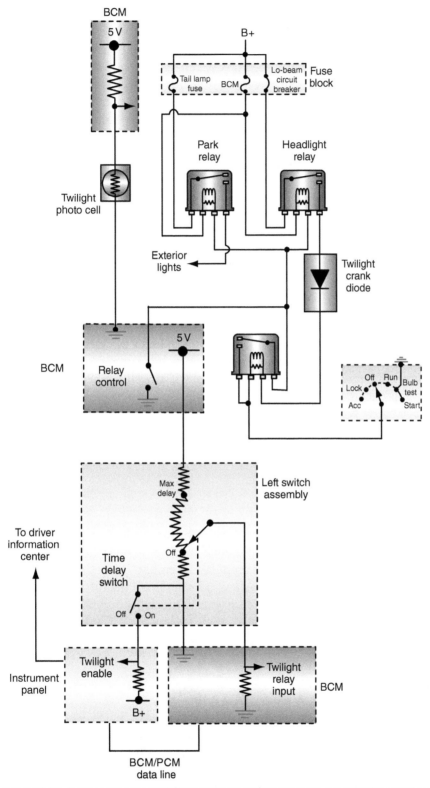

Figure 11-28 Some systems use the BCM to sense inputs from the photo cell and delay control switch.

AUTHOR' NOTE Some systems that do not have the automatic headlight on/off feature can activate the time delay feature when the headlight switch is moved from the HEADLIGHT position to OFF after the ignition is turned off.

AUTOMATIC HEADLIGHT DIMMING

Modern **automatic headlight dimming** systems use solid-state circuitry and electro-magnetic relays or HSDs to control beam switching. Automatic headlight dimming auto-matically switches the headlights from high beams to low beams under two different conditions: when light from oncoming vehicles strikes the photo cell–amplifier or light from the taillights of a passing vehicle strikes the photo cell–amplifier.

Most systems consist of the following major components:

1. Light-sensitive photo cell and amplifier unit.
2. High–low beam relay.
3. Sensitivity control.
4. Dimmer switch.
5. Flash-to-pass relay.
6. Wiring harness.

The photo cell–amplifier is usually mounted behind the front grill, but ahead of the radiator. Some systems use a **sensitivity control** that sets the intensity level at which the photo cell–amplifier will energize. This control is set by the driver and is located next to, or is a part of, the headlight switch assembly (**Figure 11-29**). The driver can adjust the sensitivity level of the system by rotating the control knob. An increase in the sensitivity level will make the headlights switch to the low beams sooner (approaching vehicle is farther away). A decrease in the sensitivity level will switch the headlights to low beams when the approaching vehicle is closer. Rotating the knob to the full counterclockwise position places the system into manual override.

> **AUTHOR'S NOTE** Many vehicle manufacturers install the sensor-amplifier in the rearview mirror support.

The high–low relay is a single-pole, double-throw unit that provides the switching of the headlight beams. The relay also contains a clamping diode for electrical transient damping to protect the photo cell and amplifier assembly.

The dimmer switch is usually a flash-to-pass design. If the turn signal lever is pulled partway up, the flash-to-pass relay is energized. The high beams will stay on as long as the

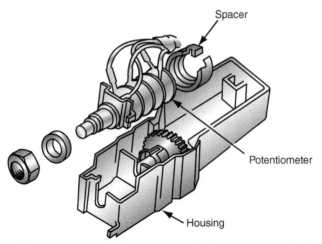

Figure 11-29 The driver sets the sensitivity of the automatic headlight dimmer system by rotating the potentiometer to change resistance values.

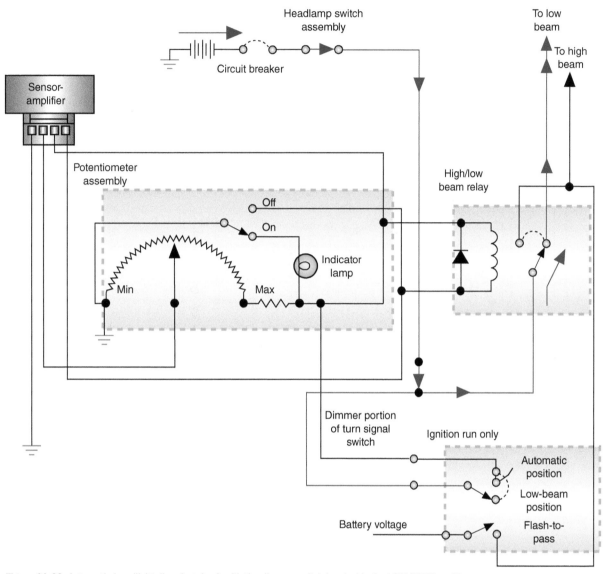

Figure 11-30 Automatic headlight dimming circuit with the dimmer switch located in the LOW-BEAM position.

lever is held in this position, even if the headlights are off. In addition, the driver can select either low beams or automatic operation through the dimmer switch.

Although the components are similar in most systems, there are differences in system operations. Systems differ in how the manufacturer uses the relay to do the switching from high beams to low beams. The system can use an energized relay to activate either the high beams or the low beams. If the system uses an energized relay to activate the high beams, the relay control circuit is opened when the dimmer switch is placed in the low-beam position or the driver manually overrides the system (**Figure 11-30**). With the headlight switch in the ON position and the dimmer switch in the low-beam position, battery voltage is not applied to the relay coil and the relay coil is not energized. Since the dimmer switch is in the low-beam position, it opens the battery feed circuit to the relay coil and the automatic feature is bypassed.

With the dimmer switch in the automatic position, battery feed is provided for the relay coil (**Figure 11-31**). Ground for the relay coil is through the sensor-amplifier. The energized coil closes the relay contacts to the high beams and battery voltage is applied to the headlamps. When the photo cell sensor receives enough light to overcome the

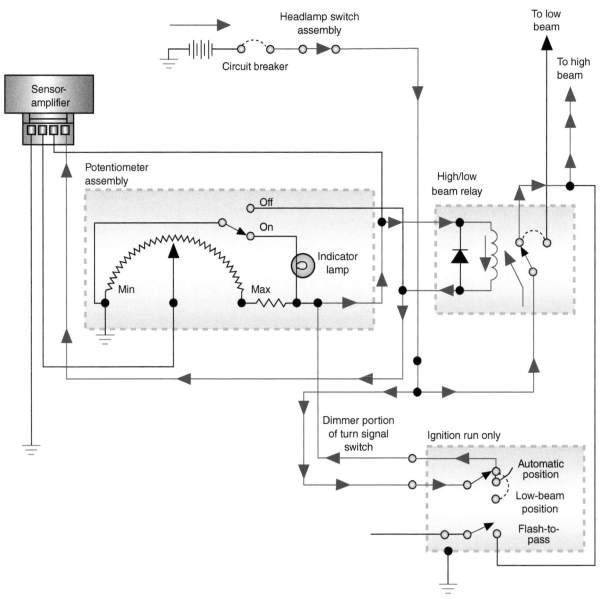

Figure 11-31 Automatic headlight dimming circuit with the dimmer switch located in the automatic position and no oncoming light sensed.

sensitivity setting, the amplifier opens the relay's circuit to ground. This de-energizes the relay coil and switches battery voltage from the high-beam to the low-beam position.

If the system uses an energized relay to switch to low beams, placing the dimmer switch in the low-beam position will energize the relay. With the headlights turned on and the dimmer switch in the automatic position, battery voltage is applied to the photo cell–amplifier, one terminal of the high–low control, and through the relay contacts to the high beams. The voltage drop through the high–low control is an input to the photo cell–amplifier. When enough light strikes the photo cell–amplifier to overcome the sensitivity setting, the amplifier allows battery current to flow through the high–low relay, closing the contact points to the low beams. Once the light has passed, the photo cell–amplifier opens battery voltage to the relay coil and the contacts close to the high beams.

When flash-to-pass is activated, the switch closes to ground. This bypasses the sensitivity control and de-energizes the relay to switch from low beams to high beams.

Today the headlight system can be very sophisticated. The following is an example of how the SmartBeam system performs auto headlamp and auto high-beam operation. This system provides lighting levels based on certain conditions and operate the high beams by sensing light levels.

The SmartBeam system uses a forward-facing, 5,000-pixel, digital imager camera that is attached to the rearview mirror mount (**Figure 11-32**). The camera's field of vision is in front of the vehicle within 2° of the vehicle's centerline and 10° horizontally.

Shop Manual
Chapter 11, page 515

The operation of SmartBeam requires interaction with several vehicle modules. **Figure 11-33** shows how one system interacts between modules. Ambient light levels for automatic headlight operation are provided by the light rain sensor module (LRSM), if equipped. If the LRSM is not used on the vehicle, then the system will use a photo cell

Figure 11-32 The SmartBeam auto headlight system uses a digital camera to determine oncoming light intensity.

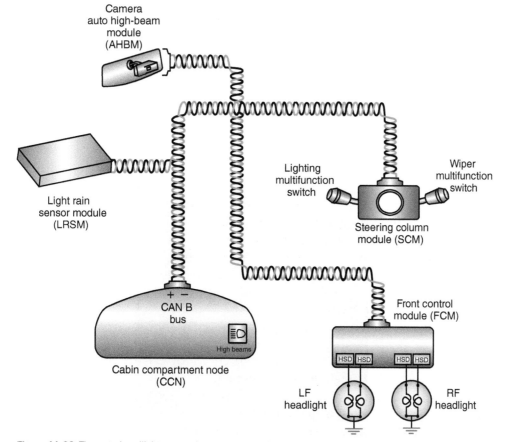

Figure 11-33 The auto headlight system integrates several modules.

located on top of the dash. The headlamp switch position is signaled by the lighting multifunction switch and is an input to the steering column module (SCM). The SCM sends switch position status over the data bus. The front control module (FCM) uses HSDs to provide power to both the low- and high-beam bulbs. The cabin compartment node (CCN) controls the operation of the high-beam indicator.

The decisions made for headlight intensity are based on the sensed intensity of light, the light's location, and the light's movement. The system is capable of distinguishing between light types, such as mercury vapor used for street lighting. In addition, it can distinguish colors, so it is possible to identify the red lights used on tail lamps and sign colors. Distinguishing light types and colors is done by identifying the wavelength of the light source. If it is determined that the approaching light source is another vehicle, a data bus message is sent from the auto high-beam module (AHBM) to the module that controls headlight operation (FCM in this case) to deactivate the high beams.

For this system to be operational, the headlight switch must be placed in the auto headlamp position and the "LOW/HIGH BEAM" option must be selected from the configurable display (**Figure 11-34**). The engine must also be running for the auto headlight function to operate. For the SmartBeam function to operate, the vehicle speed must be over 20 mph.

The auto headlight function will use either a photo cell in the dash or a part of the mirror assembly to sense ambient light intensity. When the engine is running and the ambient light levels are less than 1,000 **LUX**, the auto headlamp low beam becomes operational. LUX is the International System unit of measurement of the intensity of light. It is equal to the illumination of a surface 1 meter away from a single candle (one lumen per square meter).

Once vehicle speed exceeds 20 mph, if the ambient light level sensed at the SmartBeam camera is 5 LUX or less, a PWM voltage is applied to the high-beam circuit by the controlling module (FCM). Within 2½ to 5 seconds, the high beams will be at full intensity. By using PWM, the high beams are ramped up and down; this eliminates the usual flash that occurs as the high beams are turned on and off. Drivers in oncoming vehicles do not see any indication of the beam change since it is gradual and based on distance.

When another vehicle approaches, the camera determines the light intensity from its headlights. Once the light intensity reaches a predetermined level, a bus message is sent to the control module (FCM) to deactivate the high beams. The voltage to the high beams is ramped down using PWM until the beams are turned off.

Figure 11-34 For the auto headlight system to operate, the drive must activate it.

AUTHOR'S NOTE Federal law specifies that the high-beam indicator is turned on at the initial start of the ramping up to high-beam operation and that it remains on during all high-beam operation. High beam is considered in operation throughout the ramp-down phase; thus, the indicator light stays on until the high beams are totally turned off.

If the driver uses the high-beam switch to manually turn on the high beams, SmartBeam operation is defeated. Also, the system is momentarily defeated if the driver uses the flash-to-pass function. If the headlamp switch is placed in any other position other than AUTO, both the auto headlamp and the SmartBeam functions are defeated.

Initial camera calibration and verification are performed at the factory as the vehicle is near completion. During this time the camera is precisely aligned. Once the camera is properly aligned, logic used by the AHBM will make adjustments to fine-tune the alignment based on sensed lighting inputs while the vehicle is driven. These adjustments occur as the processing logic looks for a light source that represents oncoming headlights. This would include light sources that start as low intensity and then gradually increase in brightness and have movement, indicating a gradual rate of approach. In addition, the light source must be coming from just to the left of center. These conditions indicate a vehicle is approaching from a distance. The computer logic of the AHBM will apply a weighting factor to its calibration to correct the aim. If the system is out of calibration, the light-emitting diode (LED) in the mirror will flash.

Headlight systems that utilize a single high-intensity discharge (HID) bulb can use a shutter to control light intensity between low- and high-beam operation. A motor or solenoid is used to alter the position of the shutter (**Figure 11-35**). As in the other systems discussed, a photo cell or camera is used to determine the light intensity of the headlights from oncoming vehicles. Usually the shutter is spring loaded to the low-beam position. In this position the shutter blocks some of the HID's light from being directed at the lens. When the high beams are selected, the shutter is moved so the HID is totally uncovered and its full light intensity is directed through the lens. On solenoid-controlled systems, the solenoid is de-energized to allow the shutter to return to its default position of low-beam operation. The solenoid is energized to open the shutter for high-beam operation. When the oncoming vehicle is close enough that the headlights must change from high to low beams, the solenoid is de-energized until the oncoming light has passed.

Figure 11-35 Single HID headlight systems can use a shutter blade operated by a solenoid or motor to control high- and low-beam operations.

HEADLIGHT LEVELING SYSTEM

Shop Manual
Chapter 11, page 523

Some European and Asian markets require a **headlight leveling system (HLS)**. This system is now being installed on domestic vehicles. The HLS uses front lighting assemblies with a leveling actuator motor. Some systems use a leveling switch that the driver controls. The switch will allow the headlights to be adjusted into different vertical positions (usually four). This allows the driver to compensate for headlight position that can occur when the vehicle is loaded.

Electric motors use a pushrod assembly to change the position of the headlight reflector (**Figure 11-36**). When different voltage levels are input to leveling motors from the multiplexed switch, the motors move the reflector to the selected position.

The automatic headlight level system (AHLS) places the headlight reflectors in the correct position based on inputs from a vehicle height sensor and wheel speed sensors. This system will maintain a constant level without input from the driver and can continuously alter the level to different driving conditions. The AHLS can be a stand-alone system with its own control module, or be part of the adaptive headlight system. Regardless, the operation is very similar.

The height sensor is attached at the rear of the vehicle to measure the distance between the rear suspension and the vehicle body (**Figure 11-37**). This information is monitored whenever the ignition switch is in the RUN position (**Figure 11-38**). Speed sensor information is provided over the bus network by the ABS module. The inputs from the height and speed sensors indicate the posture of the vehicle.

When the ignition is placed in the RUN position, the ECU will command the leveling actuators to lower the beams to its lowest travel and then command the beams to their proper position. This provides an initial set point. The actuators use a stepper motor and the ECU counts the steps to determine the position of the reflectors. If the vehicle posture changes due to additional loads or different driving conditions, the ECU commands the actuator to move the reflector up or down as needed to match the posture of the vehicle.

Figure 11-36 Beam level actuator with a push rod connection to the headlight reflector.

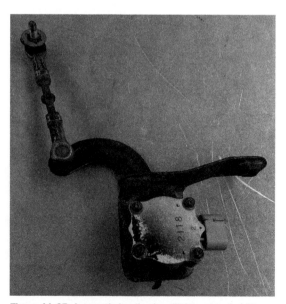

Figure 11-37 Automatic leveling headlight system height sensor.

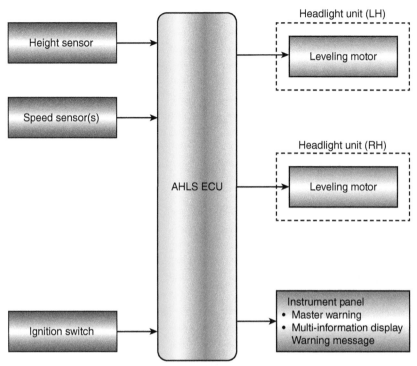

Figure 11-38 Schematic of automatic headlight leveling system.

ADAPTIVE HEADLIGHTS

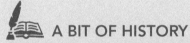 **A BIT OF HISTORY**

The 1928 Wills-Knight 70A Touring car had a third headlight mounted on the centerline of the vehicle above the front bumper. This headlight was a directional light that was attached to the steering system so it would rotate left and right as the steering wheel was turned.

The **adaptive headlight system (AHS)** is designed to enhance nighttime safety by turning the headlight beams to follow the direction of the road as the vehicle enters a turn (**Figure 11-39**). This provides the driver additional reaction time if approaching any road hazards. Conventional fixed headlights illuminate the road straight in front of the vehicle. On turns, the light beam can shine into oncoming traffic or leave the road beyond the turn in darkness. The AHS uses headlight assemblies that swivel into the direction the vehicle is steering. This provides a 90% improvement in illumination of the road so obstacles become visible sooner.

The AHS control module monitors vehicle speed, steering angle, and yaw (degree of rotation around the vertical axis) (**Figure 11-40**). As these sensor inputs indicate the vehicle is entering a turn, electric motors turn the headlights into the direction of the turn. The input from the steering angle sensor is used to determine how much the headlight beams need to rotate. As the headlight beams rotate, they follow the road and guide the

Shop Manual
Chapter 11, page 526

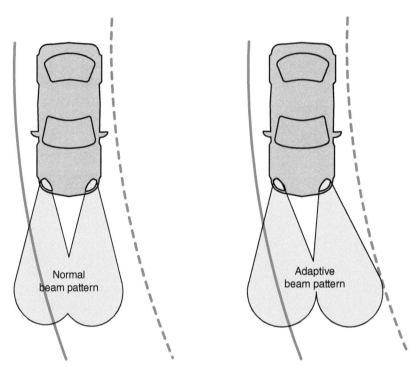

Figure 11-39 The adaptive headlight system will illuminate the road beyond a turn for added safety.

driver into the turn (**Figure 11-41**). The headlights can follow the turns of the road through an arc of 15°. The headlight located on the side of the vehicle opposite the direction of turn will rotate about 7°.

AUTHOR'S NOTE The AHS is activated whenever the vehicle is traveling forward. If the vehicle is in reverse, or if the steering wheel is turned when the vehicle is not moving, then the AHS is deactivated.

Some systems are capable of predicting the arrival of the vehicle at a turn. These systems use not only steering angle, vehicle speed, and yaw rate inputs, but also GPS-fed road data. They adjust not only the rotation of the headlamp motors but also the pattern of illumination of the bi-xenon lamps.

AUTHOR'S NOTE While only 28% of driving is conducted at night, more than 62% of pedestrian fatalities occur at night. When driving at 60 mph, the standard automobile headlights allow only 3.5 seconds reaction time.

Night vision systems use military-style thermal imaging to allow drivers to see things that they may not be able to see with the naked eye (**Figure 11-42**). The system allows the driver to see up to five times more of the road than with just headlights.

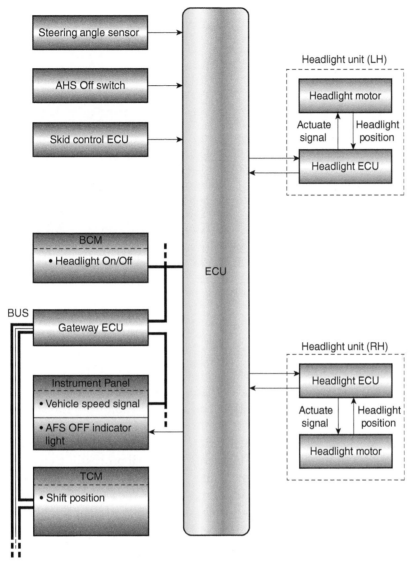

Figure 11-40 Schematic of adaptive headlight system.

Figure 11-41 The adaptive headlight assembly uses motors to rotate the beam.

A fixed-lens, high-resolution, "far infrared" camera is located behind the front grill and generates an electronically processed video image that is displayed in real time either in the head-up display (HUD) or on a thin-film transistor (TFT) monitor in the instrument panel.

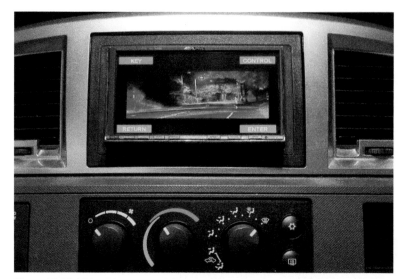

Figure 11-42 Night vision uses an infrared camera to project objects onto a screen. The objects may be out of the range of the headlights.

The lens of the camera is an infrared sensor designed to operate at room temperature. To achieve this, the lens has its own heating and cooling system. The special camera creates a temperature pattern (thermogram) that is refreshed 30 times a second. The heat from a person or an animal is greater than the heat radiation from the surroundings of the camera.

The system works by registering small differences in temperature and displaying them in 16 different shades of gray on the HUD screen. A signal processor translates the thermogram data and sends it to the display. Using a wavelength of 6 to 12 micrometers, the lens of the camera detects the infrared heat emitted from the vehicle's surroundings. This heat radiation is displayed as a negative image. The display of people, animals, or other heat-generating objects is based on the temperature of the object. Cold objects appear as a dark image, while warmer objects are seen as white or a light color.

Another variation of the night vision system uses near-infrared. This system has two infrared emitters that are integrated into the headlight assemblies. The infrared reflection from the objects in front of the vehicle is captured by a camera located behind the rear-view mirror. The camera converts the reflection into a digital signal by a charge-coupled device (CCD). The signal from the CCD is sent to an image processor that changes the signal into a format that can be viewed on the display.

DAYTIME RUNNING LAMPS

Shop Manual
Chapter 11, page 526

All late-model Canadian vehicles and many domestic vehicles are equipped with **daytime running lamps (DRL)**. The basic idea behind DRLs is to dimly light the headlamps during the day. This allows other drivers and pedestrians to see the vehicle from a distance. Manufacturers have taken many different approaches to achieve this lighting. Most have a control module or relay (**Figure 11-43**) that turns the lights on when the engine is running and allows normal headlamp operation when the driver turns on the headlights. Daytime running lamps generally use the high-beam or low-beam headlight system at a reduced intensity.

The dimmer headlights can result from headlight current being passed through a resistor when the daytime running lamps are activated. The resistor reduces the voltage available, and the current flowing through the circuit, to the headlights. The resistor is bypassed during normal headlamp operation.

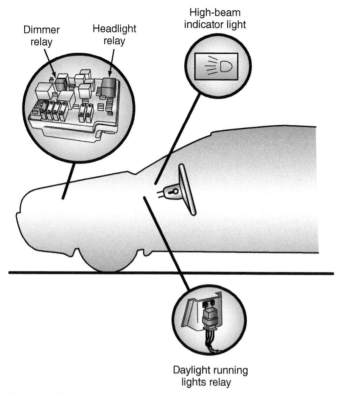

Dimmer relay   Headlight relay   High-beam indicator light

Daylight running lights relay

Figure 11-43 A daytime running light relay.

Other systems use a control module that sends a duty-cycled output to either the high-beam or the low-beam headlights (depending on the manufacturer). The duty cycle reduces the output of the headlights by 50% to 75%. Most systems that use the high-beam headlights have a method of turning off the high-beam indicator lamp in the instrument cluster if DRLs are activated. However, on some systems, it is normal for the high-beam indicator to be dimly lit during DRL operation.

GM's DRL system includes a solid-state control module assembly, a relay, and an ambient light sensor assembly. The system lights the low-beam headlights at a reduced intensity when the ignition switch is in the RUN position during daylight. The DRL system is designed to light the low-beam headlamps at full intensity when low-light conditions exist.

As the intensity of the light reaching the ambient light sensor increases, the electrical resistance of the sensor assembly decreases. When the DRL control module assembly senses the low resistance, the module allows voltage to be applied to the DRL diode assembly and then to the low-beam headlamps. Because of the voltage drop across the diode assembly, the low-beam headlamps are on with a low intensity.

As the intensity of the light reaching the ambient light sensor decreases, the electrical resistance of the sensors increases. When the DRL module assembly senses high resistance in the sensor, the module closes an internal relay, which allows the low-beam headlamps to illuminate with full intensity.

If the normal headlights are controlled by PWM from a high-side driver (refer to Figure 11-25), the DRLs are controlled by the same drivers. The headlight switch inputs to the computer if the switch position is in the OFF or AUTO modes of operation. In these modes, DRL is activated by PWM of the voltage to the lamps at a reduced rate. The low PWM results in low-luminous output of the lamps.

AUTHOR'S NOTE BMW uses a DRL system that is integrated into the corona rings of the HID headlights.

Most DRL systems also use the parking brake switch as an input. If the parking brake is applied while the engine is running, the daytime running light feature is turned off.

EXTERIOR LIGHTS

The parking light system can be activated with the ignition switch in the OFF position.

The headlight switch controls parking lights and taillights. They can be turned on without having to turn on the headlights. Usually, the first detent on the headlight switch is provided for this function. When the headlight switch is placed in the PARK or HEADLIGHT position, the front parking lights, taillights, side marker lights, and rear license plate light(s) are all turned on. The front parking lights may use dual-filament bulbs. The other filament is used for the turn signals and hazard lights.

Shop Manual
Chapter 11, page 541

Taillight Assemblies

Most taillight assemblies include the brake, parking, rear turn signal, and rear hazard lights. The center high-mounted stop light (CHMSL), back-up lights, and license plate lights can be included as part of the taillight circuit design. Depending on the manufacturer, the taillight assembly can be wired to use single-filament or dual-filament bulbs. When single-filament bulbs are used, the taillight assembly is wired as a three-bulb circuit. A three-bulb circuit uses one bulb each for the tail, brake, and turn signal lights on each side of the vehicle. When dual-filament bulbs are used, the system is wired as a two-bulb circuit. Each bulb can perform more than one function.

In a three-bulb taillight system, the brake lights are controlled directly by the brake light switch. In most applications, the brake light switch is attached to the brake pedal. When the brakes are applied, the pedal moves down and the switch plunger closes the contact points and lights the brake lights (**Figure 11-44**). On some vehicles, the brake

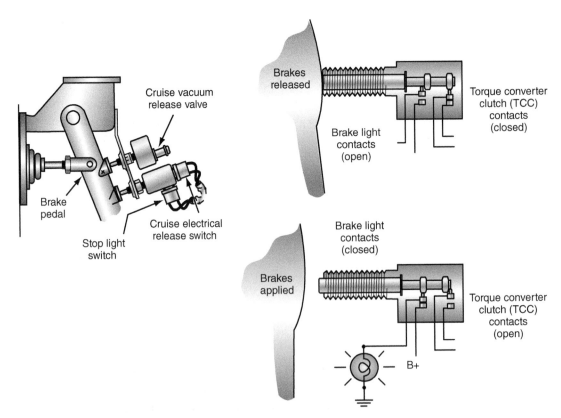

Figure 11-44 Operation of a brake light switch.

light switch may be a pressure-sensitive switch located in the brake master cylinder. When the brakes are applied, the pressure developed in the master cylinder closes the switch to light the lamps.

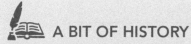

A BIT OF HISTORY

Taillights on both sides of the car didn't appear until 1929.

The brake light switch receives direct battery voltage through a fuse, which allows the brake lights to operate when the ignition switch is in the OFF position. Once the brake light switch is closed, voltage is applied to the brake lights. The brake lights on both sides of the vehicle are wired in parallel. The bulb is grounded to complete the circuit.

Shop Manual
Chapter 11, page 541

Many brake light systems use dual-filament bulbs that perform multiple functions. Usually, the filament of the dual-filament bulb, which is also used by the turn signal and hazard lights, is used for the brake lights (the high-intensity filament). In this type of circuit, the brake lights are wired through the turn signal and hazard switches. If neither turn signal is on, the current is sent to both brake lights (**Figure 11-45**). If the left turn signal is on, current for the right brake light is sent to the lamp through the turn signal switch terminal 5. The left brake light does not receive any voltage from the brake switch because the turn signal switch opens that circuit (**Figure 11-46**). The left-rear lamp will flash as the turn signal flasher provides pulsed voltage into switch terminal 3 and out terminal 8.

A BIT OF HISTORY

In 1921, turn signals were made standard equipment by Leland Lincoln. This marquee later joined Ford Motor Company. Leland Lincolns were built by Henry Leland, who was the originator of Cadillac. Early turn signals were not like those used today; many were steel arms with reflective material on them. These arms pivoted out on the side of the car as it was turning. This style continued for many years until Buick introduced electric turn signals to the public in 1939.

AUTHOR'S NOTE Because the turn signal switches used in a two-bulb system also control a portion of the operation of the brake lights, they have a complex system of contact points. The technician must remember that many brake light problems are caused by worn contact points in the turn signal switch.

All brake lights must be red and, starting in 1986, the vehicle must have a CHMSL. This light must be located on the centerline of the vehicle and no lower than 3 inches below the bottom of the rear window (6 inches on convertibles). In a three-bulb system, wiring for the CHMSL is in parallel to the brake lights (**Figure 11-47**).

In a two-bulb circuit, the CHMSL can be wired in one of two common methods. The first method is to connect to the brake light circuit between the brake light switch and the

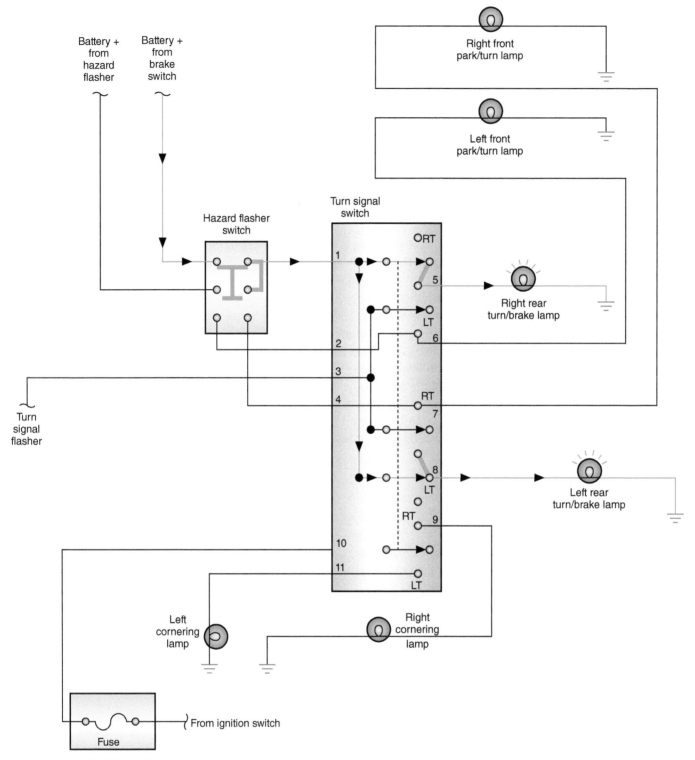

Figure 11-45 Brake light operation with the turn signals in the neutral position.

turn signal switch (**Figure 11-48**). This method is simple to perform. However, it increases the number of conductors needed in the harness.

A common method used by manufacturers is to install diodes in the conductors that are connected between the left- and right-side bulbs (**Figure 11-49**). If the brakes are applied with the turn signal switch in the neutral position, the diodes will allow voltage to flow to the CHMSL. If the turn signal switch is placed in the left-turn position, the left light must receive a pulsating voltage from the flasher. However, the steady voltage being

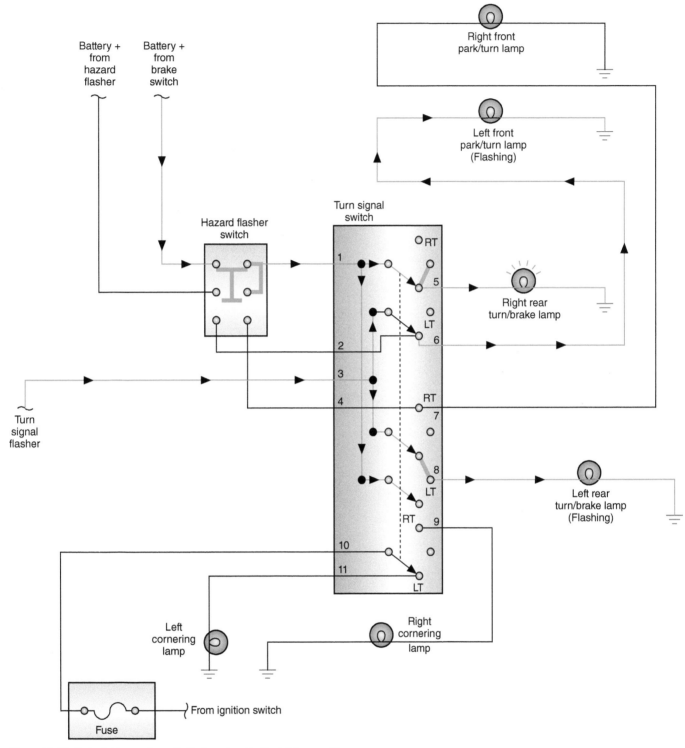

Figure 11-46 Brake light operation with the turn signal in the left-turn position.

applied to the right brake light would cause the left light to burn steady if the diode was not used. Diode 1 will block the voltage from the right lamp, preventing it from reaching the left light. Diode 2 will allow this voltage from the right brake light circuit to be applied to the CHMSL.

Computer control of the parking light and taillight systems has greatly simplified the operation of these systems. Because the switches are used as inputs, this eliminates the complexity and contacts associated with the steering wheel switch. However, not all inputs

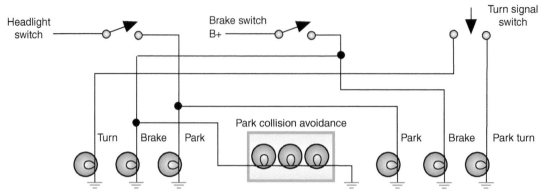

Figure 11-47 Wiring of a CHMSL in a three-bulb circuit.

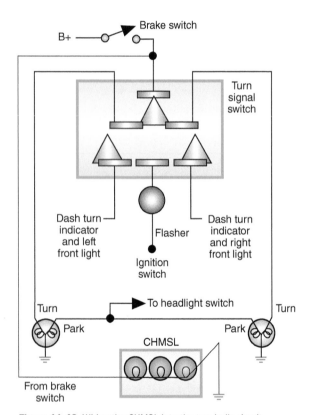

Figure 11-48 Wiring the CHMSL into the two-bulb circuit between the brake light switch and the turn signal switch.

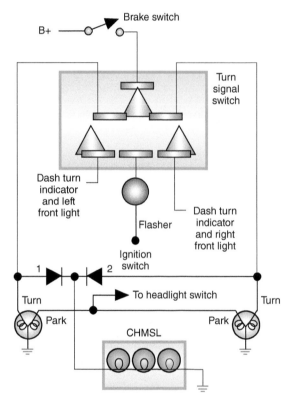

Figure 11-49 Two-bulb taillight circuit incorporating a CHMSL into the brake light system.

go directly to the BCM. For example, the brake light switch input may be an input to the antilock brake system (ABS) module. This module will then send a data bus message to the BCM to illuminate the brake lights. In addition, several bus networks may be used in the operation of the exterior lighting system.

Computer-controlled parking light relays are controlled by the BCM in the same manner as discussed for headlight control. Because the BCM controls the operation of the relays, it can turn off the exterior lights if the driver forgets to do so. Once the ignition switch is turned off, a timer is stated. After a programmed length of time, the BCM will turn off all the relays associated with the exterior lighting system, even if the switch is still in the headlight position.

Turn Signals

Turn signals are used to indicate the driver's intention to change direction or lanes. The driver will actuate a turn signal switch that is located on the right side of the steering column (**Figure 11-50**). In the neutral position, the contacts open, preventing current flow. When the driver moves the turn signal lever to indicate a left turn, the turn signal switch closes the contacts to direct voltage to the front and rear lights on the left side of the vehicle (**Figure 11-51**). When the turn signal switch is moved to indicate a right turn, the contacts are moved to direct voltage to the front and rear turn signal lights on the right side of the vehicle.

Shop Manual
Chapter 11, page 541

A **flasher** is used to open and close the turn signal circuit at a set rate. With the contacts closed, power flows from the flasher through the turn signal switch to the lamps. The flasher consists of a set of normally closed contacts, a bimetallic strip, and a coil heating element (**Figure 11-52**). These three components are wired in series. As current flows through the heater element, it increases in temperature, which heats the bimetallic strip. The strip then bends and opens the contact points. Once the points are open, current flow stops. The bimetallic strip cools and the contacts close again. With current flowing again, the process is repeated. Because the flasher is in series with the turn signal switch, this action causes the turn signal lights to turn on and off.

Shop Manual
Chapter 11, page 545

The hazard warning system is part of the turn signal system. It has been included on all vehicles sold in North America since 1967. All four turn signal lamps flash when the hazard warning switch is turned on. Depending on the manufacturer, a separate flasher can be used for the hazard lights than the one used for the turn signal lights. The operation of the hazard flasher is identical to that of the turn signal. The only difference is that the hazard flasher is capable of carrying the additional current drawn by all four turn signals. Also, it receives its power source directly from the battery. **Figure 11-53** shows the current flow through the hazard warning system.

Computer-controlled turn and hazard lighting systems may not use a flasher. Instead, the function of the flasher is incorporated within the BCM. Also, the BCM will use a noise generator to make the audible clicking sound to indicate that the turn signals or hazards are activated.

Adaptive Brake Lights

Since normal brake lights behave the same regardless of how hard the brakes are applied, drivers behind the vehicle that stops have no indication if the vehicle stops under normal or emergency conditions. The BMW X6 and other vehicles are using an **adaptive brake light** system that selects different illumination levels or methods of display for the rear

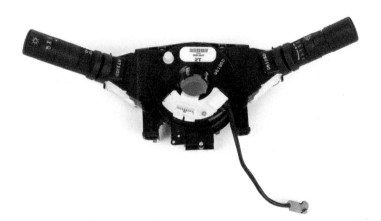

Figure 11-50 Typical turn signal switch.

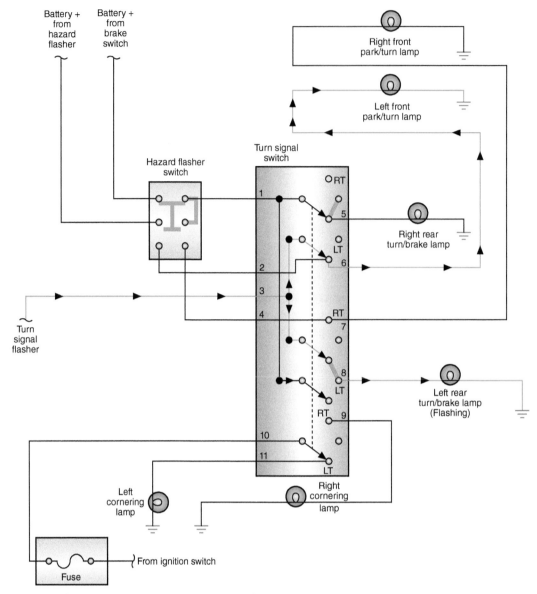

Figure 11-51 Turn signal circuit with the switch in the left-turn position.

brake lights depending on conditions. This increases the visibility of the vehicle to other drivers and provides them with greater information concerning the events ahead of them.

One method is to use the normal brake lights (including the center high-mounted stop light) during moderate braking events. However, under intense braking or ABS braking intervention, additional lights in the tail lamp assembly are used (**Figure 11-54**). This increases the intensity of the brake light assembly. Another method is to use the normal brake lights during moderate braking. However, during intense braking or ABS intervention, the rear brake lights will blink several times per second.

Regardless of the method of display, the adaptive brake light system uses information gathered from sensors used by the ABS module and vehicle speed to determine the rate of deceleration and to calculate the intensity of the braking event.

Fog Lights

For increased safety when driving in snow, sleet, heavy rains, and heavy fog conditions, some vehicles are equipped with fog lights. Fog lights can also be installed as an after-market accessory. Headlights will reflect off heavy, dense fog and cause a white haze that

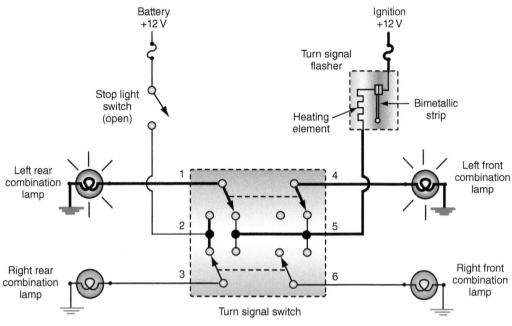

Figure 11-52 The flasher uses a bimetallic strip and a heating coil to flash the turn signal lights.

can reduce visibility. Fog lights emit a specialized beam to penetrate through the snow, rain, or fog, providing the driver with a better and safer field of vision.

Fog lights are installed on each side of the vehicle, generally low on the front fascia. Due to their mounting location, fog lights illuminate below the normal line of sight. This minimizes the amount of reflected light, to help the driver see better.

Common fog light circuits use a relay (**Figure 11-55**). Also, most fog light circuits are wired, so the fog lights will come on only if the headlight switch is in the PARK or low-beam position.

Back-Up Lights

All vehicles sold in North America after 1971 are required to have back-up lights. Back-up lights illuminate the road behind the vehicle and warn other drivers and pedestrians of the driver's intention to back up. **Figure 11-56** illustrates a back-up light circuit. Power is supplied through the ignition switch when it is in the RUN position. When the driver shifts the transmission into reverse, the back-up light switch contacts are closed and the circuit is completed.

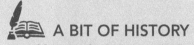 **A BIT OF HISTORY**

The 1921 Wills-St. Claire was the first car to feature a back-up lamp.

Many vehicles equipped with automatic transmissions incorporate the back-up light switch into the neutral safety switch. Most manual transmissions are equipped with a separate switch. Either style of switch can be located on the steering column, on the floor console, or on the transmission (**Figure 11-57**). Depending on the type of switch used, there may be a means of adjusting the switch to assure that the lights are not on when the vehicle is in a forward gear selection.

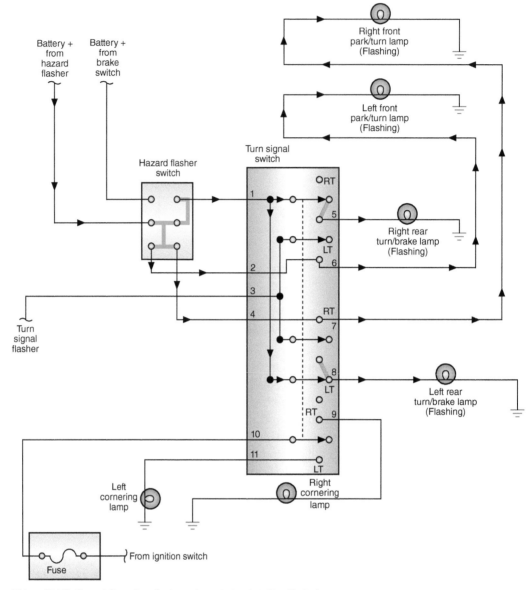

Figure 11-53 Current flow when the hazard warning system is activated.

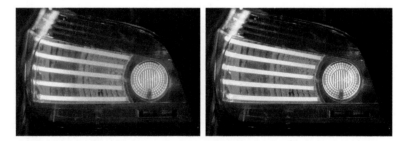

Figure 11-54 Adaptive brake lights can use two illumination levels. Left is normal braking, while the right photo is during hard braking.

Side Marker Lights

Side marker lights are installed on all vehicles sold in North America since 1969. These lamps permit the vehicle to be seen when entering a roadway from the side. This also provides a means for other drivers to determine vehicle length. The front side marker light

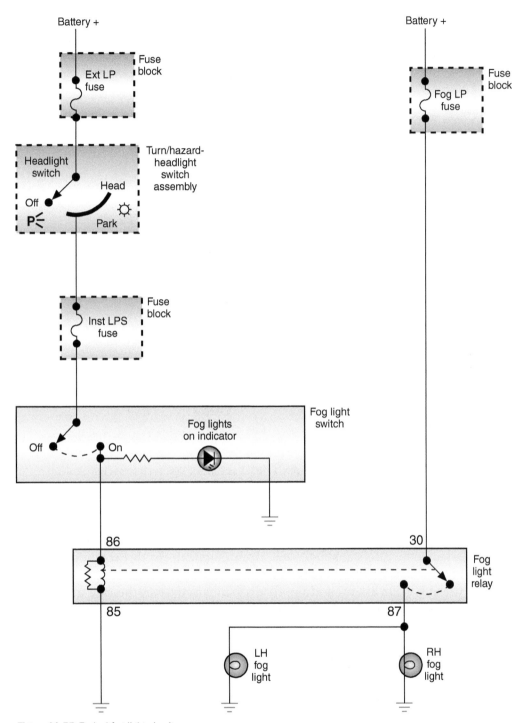

Figure 11-55 Typical fog light circuit.

lens must be amber and the rear lens must be red. Vehicles that use wrap-around headlight and taillight assemblies also use this lens for the side marker lights (**Figure 11-58**). Vehicles that surpass certain length and height limits are also required to have clearance lights that face both the front and rear of the vehicle.

The common method of wiring the side marker lights is in parallel with the parking lights. Wired in this manner, the side marker lights will illuminate only when the headlight switch is in the PARK or HEADLIGHT position.

Many vehicle manufacturers use a method of wiring the side marker lights so that they flash when the turn signals are activated. The side marker light is wired across the parking light and turn signal light (**Figure 11-59**). If the parking lights are on, voltage is applied

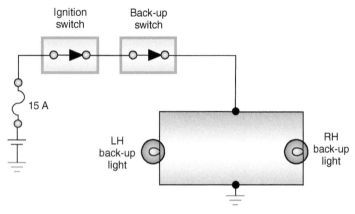

Figure 11-56 Back-up light circuit.

Figure 11-57 A combination back-up and neutral safety switch installed on an automatic transmission.

Figure 11-58 Wrap-around headlights serve as side marker lights.

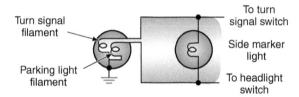

Figure 11-59 A side marker light wired across two circuits.

to the side marker light from the parking light circuit. Ground for the side marker light is provided through the turn signal filament. Because of the large voltage drop across the side marker lamp, the turn signal bulb will barely illuminate. In this condition, the side marker light stays on constantly (**Figure 11-60**).

If the parking lights are off and the turn signal is activated, the side marker light receives its voltage source from the turn signal circuit. Ground for the side marker light is provided through the parking light filament. The voltage drop over the side marker light is so high that the parking light will not illuminate. The side marker light will flash with the turn signal light (**Figure 11-61**).

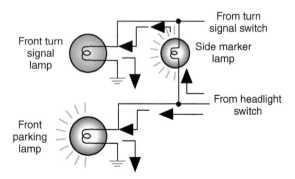

Figure 11-60 Current flow to the side marker light with the parking light on.

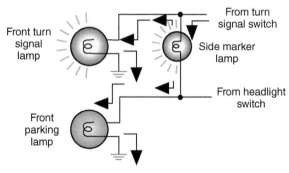

Figure 11-61 Side marker operation, with the turn signal switch activated.

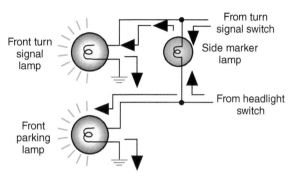

Figure 11-62 Side marker light operation with the turn signal and parking light switches activated.

If the turn signal is activated while the parking lights are illuminated, the side marker light will flash alternately with the turn signal light. When both the turn signal light and the parking light are on, there is equal voltage applied to both sides of the side marker light. There is no voltage potential across the bulb, so the light does not illuminate (**Figure 11-62**). The turn signal light turns off as a result of the flasher opening. Then the turn signal light filament provides a ground path and the side marker light comes on. The side marker light will stay on until the flasher contacts close to turn on the turn signal light again.

LED Exterior Lighting

Many car manufacturers use LED lighting technology for several exterior lighting functions, including the CHMSL, taillight assemblies (**Figure 11-63**), side marker lights, and turn-indicating outside mirrors. Using LEDs in rear-lighting applications (especially the CHMSL) provide one means of increasing traffic safety. The reaction time of the following driver in response to brake light illumination is shorter for CHMSLs equipped with LEDs than for those equipped with conventional incandescent bulbs. This is due to the shorter LED illumination time of less than 100 milliseconds compared to 300 milliseconds of an incandescent bulb. At 75 mph (120.7 km/hour), the faster illumination time means an additional 21 feet (6.4 meters) of braking distance for the following drivers. Another advantage of LED lighting is the extended life compared to a bulb.

Figure 11-63 LED taillight assembly.

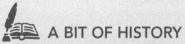

A BIT OF HISTORY

The 1984 Corvette was the first automobile to use LEDs. They were used for the CHMSL.

An example of the use of LED technology in rear-lighting systems is the Cadillac STS. Each tail lamp assembly has 30 points of illumination by using two vertical boards, each board consisting of 15 LEDs. The CHMSL is approximately 1/2 inch (12 mm) thick and consists of 78 points of illumination.

When using LEDs, the assemblies use a larger number of points of illumination due to LEDs not having the same light intensity as compared to an incandescent bulb and the light emitted by an LED is in a narrow angle of view. The increased number of LEDs assures that the light will be viewed from several angles.

Figure 11-64 illustrates the diagram for an LED taillight assembly. The brightness of the assembly is dependent upon the resistance in the circuit. Because resistance controls the amount of current in the circuit, changing circuit resistance will also change circuit current. When the park light switch is closed, the emitter for TR_1 is pulled low through the 33-ohm (Ω) resistance of R_2. This turns on TR_1 and TR_2 and the LEDs are illuminated.

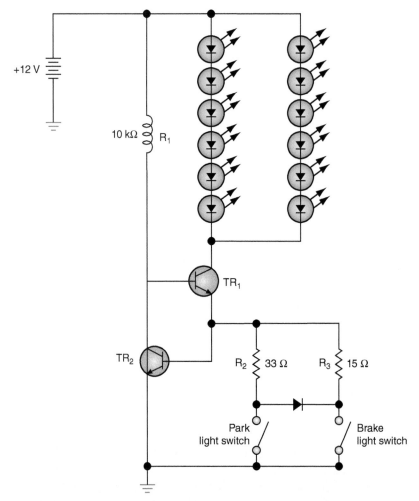

Figure 11-64 Schematic of an LED taillight assembly used for park light and brake light circuits.

Figure 11-65 LED headlight.

When the brake light switch is closed, the emitter for TR_1 is pulled low through the 15 ohms of resistance of R_3. Now the circuit has less resistance than when the park switch was closed. The current increases and the LEDs are illuminated at a brighter level.

Red, green, blue (RGB) LEDs can emit any color light from the same LED unit. A unit can be made up of three different LEDs that emit the red, green, and blue colors that make up the color spectrum. A single LED can be used if it is doped with three different substances to create the colors. With either system, different colors can be produced by varying the voltage to the LEDs. If the red, blue, and green are mixed when producing equal brightness, a white light is produced. Varying the voltages will vary the brightness of the LED, and mixing the red, green, and blue will result in different colors and shades. This allows a single LED strip to be used for multiple functions, such as taillight assemblies.

Manufacturers have been developing LED headlights (**Figure 11-65**). The 2009 Cadillac Escalade Platinum was the first vehicle in the U.S. market to offer an all-LED headlight system. In the European market, the 2007 Lexus LS600 hybrid and the Audi R8 offered forms of the LED headlight system.

LED headlights match the output of HID headlights, without the generation of heat. Heat is controlled by heat sinks, so the bulbs remain cool. LED headlights can increase the light output pattern, providing for safer nighttime driving. In addition, they can last for up to 40,000 hours of operation.

AUTHOR'S NOTE Aftermarket LED headlight kits are available for many vehicles. These kits typically will utilize the existing headlamp housing and are "plug and play."

INTERIOR LIGHTS

Interior lighting includes courtesy lights, map lights, and instrument panel lights.

Shop Manual
Chapter 11, page 546

Courtesy Lights

Courtesy lights illuminate the vehicle's interior when the doors are open. Courtesy lights operate from the headlight and door switches and receive their power source directly from a fused battery connection. The switches can be either ground switch circuit (**Figure 11-66**) or insulated switch circuit design (**Figure 11-67**). In the insulated switch circuit, the switch is used as the power relay to the lights. In the grounded switch circuit,

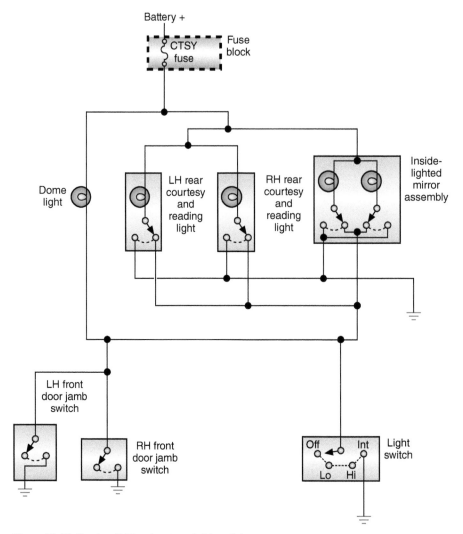

Figure 11-66 Courtesy lights using ground side switches.

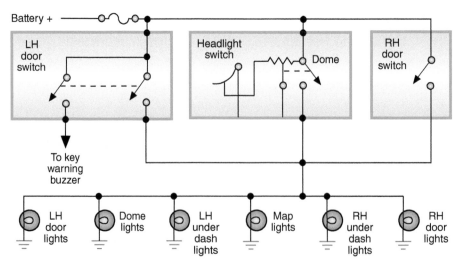

Figure 11-67 Courtesy lights using insulated side switches.

the switch controls the grounding portion of the circuit for the lights. The courtesy lights may also be activated by the headlight switch. When the headlight switch knob is turned to the extreme counterclockwise position, the contacts in the switch close and complete the circuit.

Reading and Map Lights

A BIT OF HISTORY
In 1913, the Spaulding touring car had such luxuries as four seats with folding backs, air mattresses, and electric reading lamps.

Individual switches and controls to allow passengers in the vehicle to turn on individual lights are incorporated within most courtesy light systems (refer to Figure 11-67). The system shown has individual two-position switches that allow the passenger to turn on a light. When the switch is pressed, it completes the circuit to ground for that light only.

Illuminated Entry Systems

Illuminated entry systems turn on the courtesy lights before the doors are opened. Most modern illuminated entry systems incorporate solid-state circuitry that includes an illuminated entry actuator and side door switches in the door handles. Illumination of the door lock tumblers can be provided by the use of fiber optics or LEDs.

Shop Manual
Chapter 11, page 549

When either of the front door handles is lifted, a switch in the handle will close the ground path from the actuator. This signals the logic module to energize the relay (**Figure 11-68**). With the relay energized, the contacts close and the interior and door lock lights come on. A timer circuit that will turn off the lights after 25 to 30 seconds is incorporated. If the ignition switch is placed in the RUN position before the timer circuit turns off the interior lights, the timer sequence is interrupted and the interior lights turn off.

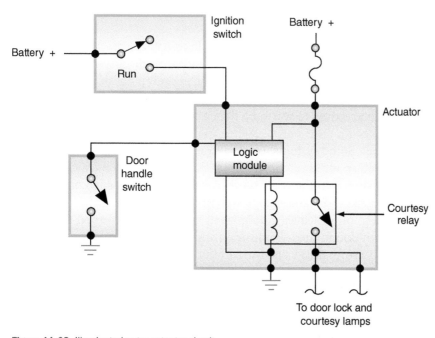

Figure 11-68 Illuminated entry actuator circuit.

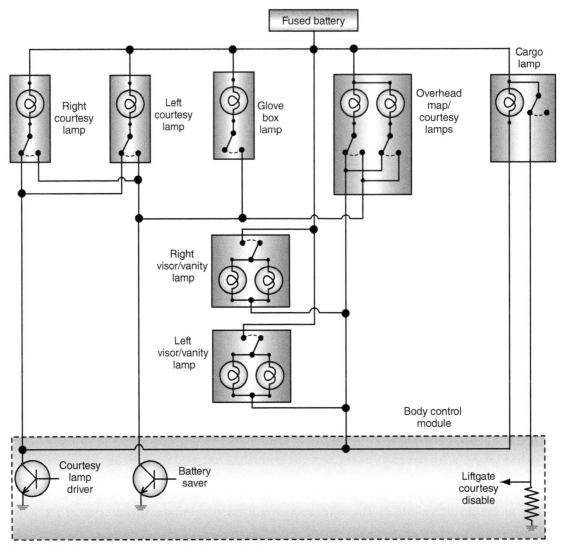

Figure 11-69 Body computer control of the illuminated entry system.

Some manufacturers have incorporated the illuminated entry actuator into their BCM (**Figure 11-69**). Activation of the system is identical as discussed. The signal from the door handle switch can also be used as a **wake-up signal** to the BCM. A wake-up signal is used to notify the BCM that an engine start and operation of accessories is going to be initiated soon. This signal is used to warm up the circuits that will be processing information.

The signal from the door handle switch informs the body computer to activate the courtesy light driver. Some systems use a pair of door jamb switches that signal the body computer to keep the courtesy lights on when the door is open. When the door is closed and the ignition switch is in the RUN position, the lights are turned off.

Referring to Figure 11-69, this type of system can turn off the interior lights if the driver forgets to turn them off or leaves a door open. After a programmed period of time has elapsed with the ignition in the OFF position and no other inputs received, the BCM will turn off the courtesy lamp driver and turn off the lamps. The battery saver driver is turned on whenever the BCM receives its wake-up signal and provides ground for the lamps through the switches. This allows the vehicle occupants to turn on and off individual reading lights. Once the ignition key is turned off and a programmed length of time has elapsed, the BCM will turn off the battery saver driver. This will assure that all lights are turned off while the vehicle is not being used even though a switch was left in the ON position.

Another feature of this system is fade-to-off. Whenever the BCM determines it will turn off the courtesy lamps (by either the ignition switch being placed in the RUN position or the timer expiring), the driver circuit will gradually change the duty cycle, resulting in the lamps dimming as they go off.

Some manufacturers use the twilight photo cell to inform the body computer of ambient light conditions. If the ambient light is bright, the photo cell signals the BCM that courtesy lights are not required.

INSTRUMENT CLUSTER AND PANEL LAMPS

Consider the following three types of lighting circuits within the instrument cluster:

1. *Warning lights* alert the driver to potentially dangerous conditions such as brake failure or low oil pressure.
2. *Indicator lights* include turn signal indicators.
3. *Illumination lights* provide indirect lighting to illuminate the instrument gauges, speedometer, heater controls, clock, ashtray, radio, and other controls.

Shop Manual
Chapter 11, page 553

The power source for the instrument panel lights is provided through the headlight switch. The contacts are closed when the headlight switch is located in the PARK or HEADLIGHT position. The current must flow through a variable resistor (rheostat) that is either a part of the headlight switch or a separate dial on the dash. The resistance of the rheostat is varied by turning the knob. By varying the resistance, changes in the current flow to the lamps control the brightness of the lights (**Figure 11-70**).

The BCM can also be used to control the **instrument panel dimming** feature. The body computer uses inputs from the panel dimming control and photo cell to determine the illumination level of the instrument panel lights (**Figure 11-71**). With the ignition switch in the RUN position, a 5-volt signal is supplied to the panel dimming control potentiometer. The wiper of the potentiometer returns the signal to the BCM.

When the dimmer control moves toward the dimmer positions, the increased resistance results in a decreased voltage signal to the BCM. By measuring the voltage that is returned, the BCM is able to determine the resistance value of the potentiometer. The BCM controls the intensity level of the illumination lamps by pulse-width modulation.

Some digital instrument panel modules use an ambient light sensor in addition to the rheostat. The ambient sensor will control the display brightness over a 35 to 1 range and the rheostat will control over a 30 to 1 range. When the headlights are on, the module

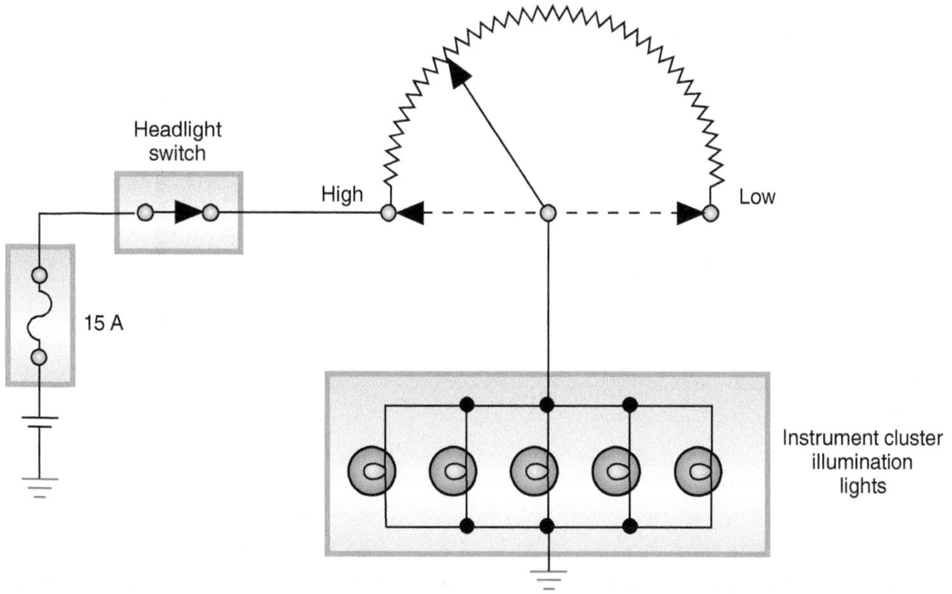

Figure 11-70 A rheostat controls the brightness of the instrument panel lights.

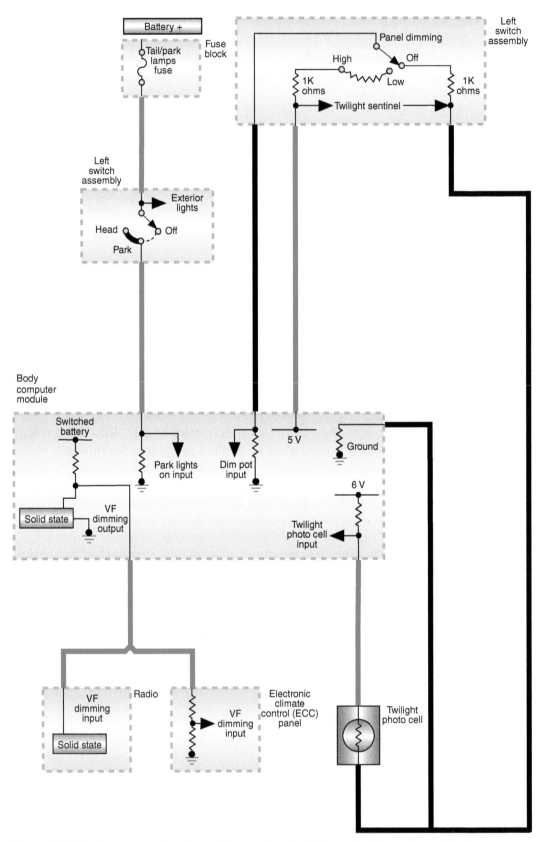

Figure 11-71 The dimming control and photo cell are inputs to the BCM to control instrument panel dimming.

compares the values from both inputs and determines the illumination level. When the headlights are off, the module uses only the ambient light sensor for its input.

A variation of this operation is that the BCM will receive the light intensity level request from the headlight rheostat as it did before, but the requested level is then sent to the instrument cluster by the multiplexing circuit. The instrument cluster uses this input information to control lamp intensity through its own microprocessor.

FIBER OPTICS

Fiber optics is the transmission of light through polymethyl methacrylate plastic that keeps the light rays parallel even if extreme bends are in the plastic. The invention of fiber optics has made it possible to provide illumination of several objects by a single light source (**Figure 11-72**). Plastic fiber-optic strands are used to transmit light from the source to the object to be illuminated. The strands of plastic are sheathed by a polymer that insulates the light rays as they travel within the strands. The light rays travel through the strands by means of internal reflections.

Shop Manual
Chapter 11, page 558

 A BIT OF HISTORY

Optical communication systems date back to the optical telegraph that French engineer Claude Chappe invented in the 1790s. Alexander Graham Bell patented an optical telephone system, which he called the Photophone, in 1880. During the 1920s, John Logie Baird of England and Clarence W. Hansell of the United States patented the idea of using arrays of hollow pipes or transparent rods to transmit images for television or facsimile systems. However, the first person known to have demonstrated image transmission through a bundle of optical fibers was Heinrich Lamm, a medical student in Munich.

His goal was to look inside inaccessible parts of the body and, in a 1930 paper, he reported transmitting the image of a light bulb filament through a short bundle.

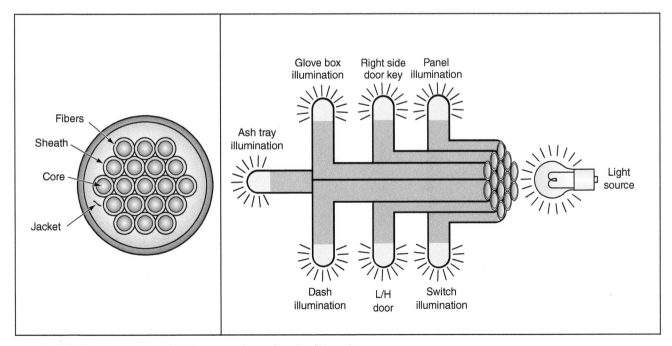

Figure 11-72 One light source can illuminate several areas by using fiber optics.

Fiber optics is commonly used in the operation of indicator lights to show the driver that certain lights are functioning. Many vehicles with fender-mounted turn signal indicators use fiber optics from the turn signal light to the indicator (**Figure 11-73**). The indicator will show light only if the turn signal light is on and working properly.

Some manufacturers use fiber optics to provide illumination of the lock cylinder halo during illuminated entry operation (**Figure 11-74**). When the illuminated entry system is activated, the light collector provides the source light to the fiber optics and the halo lens receives the light from the fiber-optic cable.

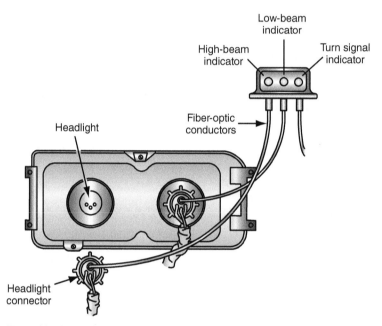

Figure 11-73 Fiber optics can be used to indicate the operation of exterior lights to the driver.

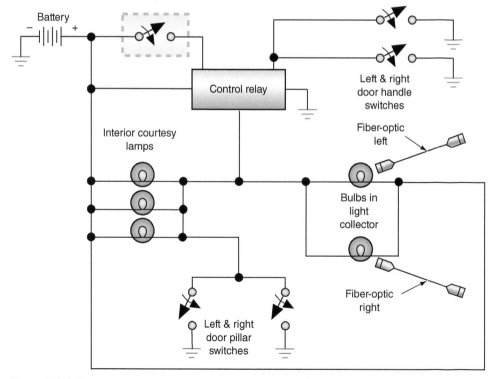

Figure 11-74 Fiber optics used in an illuminated entry system.

The advantage of fiber optics is that it can be used to provide light in areas where bulbs would be inaccessible for service. Other uses of fiber optics include the following:

- Lighting ash trays.
- Illuminating instrument panels.
- Dash lighting over switches.

LAMP OUTAGE INDICATORS

A common lamp outage indicator uses a translucent drawing of the vehicle (**Figure 11-75**). If one of the monitored systems fails or is in need of driver attention, the graphic display illuminates a light to indicate the location of the problem.

The basic lamp outage indicator system is used to monitor the stop light circuit. This system consists of a reed switch and opposing electromagnetic coils (**Figure 11-76**). When the ignition switch is turned to the RUN position, battery voltage is applied to the normally open reed switch. When the brake light switch is closed, current flows through the coils on the way to the brake light bulbs. If both bulbs are operating properly, the coils create opposing magnetic fields that keep the reed switch in the open position. If one of the brake light bulbs burns out, current will flow through only one of the coils, which attracts the reed switch contacts and closes them. This completes the brake light warning circuit and illuminates the warning light on the dash. The warning light will remain on as long as the brake light switch is closed.

AUTHOR'S NOTE Opposing magnetic fields are created because the coils are wound in opposite directions.

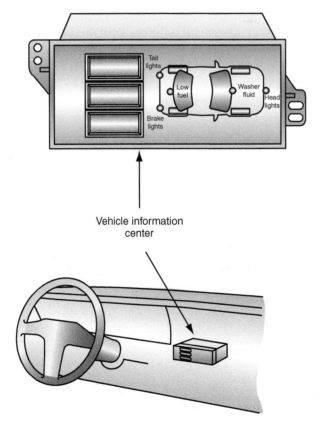

Figure 11-75 Many vehicle information systems use a graphic display to indicate warning areas to the driver.

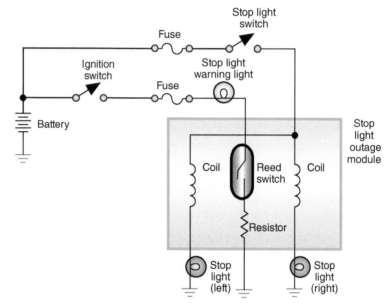

Figure 11-76 Stop light lamp outage indicator circuit.

Some manufacturers use a **lamp outage module** either as a stand-alone module or in conjunction with the BCM. A lamp outage module is a current-measuring sensor that contains a set of resistors, wired in series with the power supply to the headlights, tail-lights, and stop lights. If the module is a stand-alone unit, it will operate the warning light directly. The module monitors the voltage drop of the resistors. If the circuits are operating properly, there is a 500-mV input signal to the module. If one of the monitored bulbs burns out, the voltage input signal drops to about 250 mV. The module completes the ground circuit to the warning light to alert the driver that a bulb has burned out. The module is capable of monitoring several different light circuits.

Many vehicles today use a computer-driven information center to keep the driver informed of the condition of monitored circuits (**Figure 11-77**). The vehicle information center usually receives its signals from the BCM (**Figure 11-78**). In this system, the lamp outage module is used to send signals to the BCM. The BCM will illuminate a warning light, give a digital message, or activate a voice warning device to alert the driver that a light bulb is burned out.

A burned-out light bulb means there is a loss of current flow in one of the resistors of the lamp outage module. A monitoring chip in the module compares the voltage drop

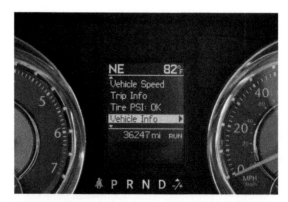

Figure 11-77 The computer-driven vehicle information center keeps the driver aware of the condition of monitored systems.

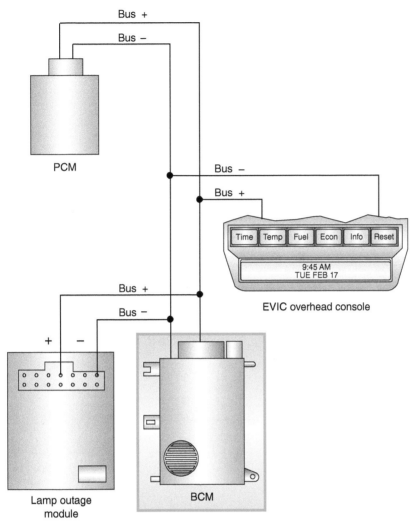

Bus +

Bus −

PCM

Bus −

Bus +

| Time | Temp | Fuel | Econ | Info | Reset |

9:45 AM
TUE FEB 17

EVIC overhead console

Bus +

Bus −

+   −

Lamp outage
module

BCM

Figure 11-78 The body computer can be used to receive signals from various inputs and to give signals to control the information center.

across the resistor. If there is no voltage drop across the resistor, there is an open in the circuit (burned-out light bulb). When the chip measures no voltage drop across the resistor, it signals the BCM, which then gives the necessary message to the vehicle information center.

AUTHOR'S NOTE In these systems the bulbs are monitored only when current is supplied to them.

General Motors uses the lamp monitor module to connect the light circuits to ground (**Figure 11-79**). When the circuits are operating properly, the ground connection in the module causes a low-circuit voltage. Input from the lamp circuits is through two equal-resistance wires. The module output to the bulbs is from the same module terminals as the inputs.

If a bulb burns out, the voltage at the lamp monitor module terminal will increase. The module will open the appropriate circuit from the BCM, signaling the BCM to send

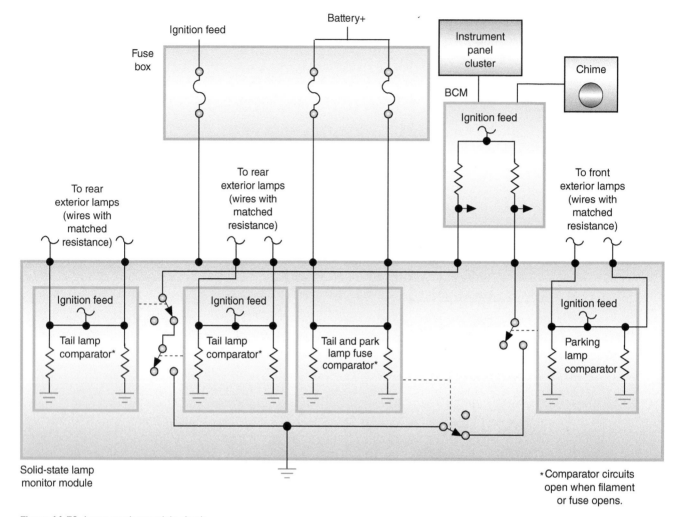

Figure 11-79 Lamp monitor module circuit.

a communication to the instrument panel cluster computer, which displays the message in the information center.

Vehicles that use HSDs to control lamp illumination are also capable of detecting lamp outage. Often these systems can determine if a lamp circuit is open without the lamp system being activated. This is done by sending a small diagnostic current through the circuit. If the circuit is complete, an expected voltage drop will occur. If the circuit is open, the voltage reading will remain high and the module will request the lamp outage indicator to come on.

Shop Manual
Chapter 11, page 510

SUMMARY

- The most commonly thought of light circuit is the headlights, but there are many lighting systems in the vehicle.
- Different types of lamps are used to provide illumination for the systems. The lamp may be either a single-filament bulb that performs a single function or a double-filament bulb that performs several functions.

- The headlight lamps can be one of four designs: standard sealed beam, halogen sealed beam, composite, or high-intensity discharge (HID).
- The headlight filament is located on a reflector that intensifies the light, which is then directed through the lens. The lens is designed to change the circular light pattern into a broad, flat light beam. Placement of the filament in the reflector provides for low- and high-beam light patterns.

- In addition to the headlight system, the lighting systems include the following:

  Stop lights.
  Turn signals.
  Hazard lights.
  Parking lights.
  Taillights.
  Back-up lights.
  Side marker lights.
  Courtesy lights.
  Instrument panel lights.

- The headlight switch can be used as the control of many of these lighting systems. Most headlight switches have a circuit breaker that is an integral part of the switch. The circuit breaker provides protection of the headlight system without totally disabling the headlight operation if a circuit overload is present.

- A rheostat is used in conjunction with the headlight switch to control the brightness of the instrument panel illumination lights.

- Many computer-controlled headlight systems use resistive multiple headlight switches as an input.

- Computer-controlled headlight systems can use relays operated by the control module or high-side drivers (HSDs) to illuminate the lamps.

- The automatic on/off with time delay has two functions: to turn on the headlights automatically when ambient light decreases to a predetermined level and to allow the headlights to remain on for a certain amount of time after the vehicle has been turned off.

- Most automatic headlight dimming systems consist of a light-sensitive photo cell and amplifier unit, high–low beam relay, sensitivity control, dimmer switch, flash-to-pass relay, and a wiring harness.

- The SmartBeam system uses a forward-facing, 5,000-pixel, digital imager camera.

- The operation of SmartBeam requires interaction with several vehicle modules, including the light rain sensor module (LRSM), the steering column module (SCM), the front control module (FCM), and the cabin compartment node (CCN).

- Decisions about headlight intensity are based on the sensed intensity of light, the light's location, and the light's movement.

- A headlight leveling system (HLS) uses front lighting assemblies with a leveling actuator motor.

- The adaptive headlight system (AHS) is designed to enhance nighttime safety by turning the headlight beams to follow the direction of the road as the vehicle enters a turn.

- The AHS uses sensors that measure vehicle speed, steering angle, and yaw (degree of rotation around the vertical axis). Based on this information, small electric motors turn the headlights so the beam falls on the road ahead, guiding the driver into the turn.

- The night vision system uses an infrared camera to transfer images to a display panel to enable the driver to identify and react to obstacles outside of the headlight range.

- Daytime running lamps can use a relay or module to illuminate the low- or high-beam lamps at a reduced output.

- The adaptive brake light system selects different illumination levels or methods of display for the rear brake lights depending on conditions.

- The illuminated entry system turns on the interior lights prior to the door being opened. The system may also be capable of turning off the lights if the driver fails to shut a door when exiting.

- Instrument panel dimming is usually done by the BCM providing a pulse-width modulation to the illumination lamps or LEDs.

- Fiber optics is the transmission of light through polymethyl methacrylate plastic that keeps the light rays parallel even if there are extreme bends in the plastic.

- The lamp outage indicator alerts the driver, through an information center on the dash or console, that a light bulb has burned out.

REVIEW QUESTIONS

Short-Answer Essays

1. List the common components of the automatic headlight dimming system.

2. Explain the operation of body computer–controlled instrument panel illumination dimming.

3. Explain the operation of the SmartBeam headlight system.

4. Describe the function of automatic headlight leveling systems.

5. Describe the purpose and function of daytime running lamps.

6. Explain the use and function of fiber optics.

7. What is meant by pulse-width dimming?

8. What three lighting circuits are incorporated within the instrument cluster?

9. Describe the operation of HID headlights.

10. Describe the purpose of bi-xenon headlights.

Fill in the Blanks

1. The photo cell will have _____ resistance as the ambient light level increases.

2. In some illuminated entry systems, the _____ signals the body computer that the courtesy lights are not required.

3. The body computer uses inputs from the _____ and _____ to determine the illumination level of the instrument panel lights.

4. The body computer dims the illumination lamps by using a _____ signal to the panel lights.

5. Most computer-controlled headlight systems use a _____ _____ switch for an input.

6. Fiber optics are commonly used as _____ lights.

7. Lamp outage modules detect _____ _____ in a normally operating circuit.

8. HSDs supply _____ to the lamps.

9. When today's technician performs repairs on the lighting systems, the repairs must meet at least two requirements: they must assure vehicle _____ and meet all _____ _____.

10. A _____ is a device that produces light as a result of current flow through a _____.

Multiple Choice

1. All of the following statements about the SmartBeam system are true, **EXCEPT:**

   A. The system uses a digital camera to determine light intensity.

   B. The system is capable of detecting movement of oncoming light.

   C. The system is capable of distinguishing colors.

   D. The AHBM turns off the high-beam relay when oncoming light intensity is 10 LUX or more.

2. Computer-controlled instrument panel dimming is being discussed.

   Technician A says the body computer dims the illumination lamps by varying resistance through a rheostat that is wired in series to the lights.

   Technician B says the body computer can use inputs from the panel dimming control and photo cell to determine the illumination level of the instrument panel lights on certain systems.

   Who is correct?

   A. A only   C. Both A and B

   B. B only   D. Neither A nor B

3. Which statement about fiber optics is correct?

   A. Fiber optics is the transmission of light through several plastic strands that are sheathed by a polymer.

   B. Fiber optics is used only for exterior lighting.

   C. Fiber optics can be used only in applications where the conduit can be laid straight.

   D. All of the above.

4. The purpose of the headlight leveling system is to:

   A. Reduce the need to align the light beams.

   B. To allow the driver to raise or lower the light beams as vehicle loads change.

   C. To allow the driver to raise or lower the light beams as the vehicle ascends and descends hills.

   D. All of the above.

5. *Technician A* says computer-controlled headlight systems can use relays that the BCM operates.

   Technician B says computer-controlled headlight systems can use high-side drivers to operate the lamps.

   Who is correct?

   A. A only   C. Both A and B

   B. B only   D. Neither A nor B

6. In the SmartBeam system, the headlight intensity is based on which of the following:

   A. Movement of the light.

   B. Intensity of the light.

   C. Location of the light.

   D. All of the above.

7. Which statement is the most correct?

    A. Lamp outage modules can use voltage drop to determine circuit operation.

    B. HSDs can be used only to detect opens in activated circuits.

    C. Low-side, driver-controlled lamps cannot determine lamp outage conditions.

    D. All of the above.

8. In a composite headlight:

    A. The bulb is replaceable.

    B. A cracked lens will prevent lamp operation.

    C. All of the above.

    D. None of the above.

9. The CHMSL circuit is being discussed.

    Technician A says the diodes are used to assure proper turn signal operation.

    Technician B says the diodes are used to prevent radio static when the brake light is activated.
    Who is correct?

    A. A only

    B. B only

    C. Both A and B

    D. Neither A nor B

10. Which of the following best describes the function of the bi-xenon headlight system?

    A. Uses double-filament headlights to provide the high-beam output.

    B. Uses a chamber filled with multiple xenon gases that are ignited at different temperatures to provide high-beam and low-beam operations.

    C. Uses multiple chambers of xenon gas that are ignited based on dimmer switch position. One chamber is for low beam, and both chambers are ignited for high beam.

    D. None of the above.

CHAPTER 12

INSTRUMENTATION AND WARNING LAMPS

Upon completion and review of this chapter, you should be able to:

- Describe the operation of electromagnetic gauges, including d'Arsonval, three coil, two coil, and air core.
- Describe the operation of electronic fuel, temperature, oil, and voltmeter gauges.
- Describe the operation of quartz analog instrumentation.
- Explain the function and operation of the various gauge sending units, including thermistors, piezoresistive, and mechanical variable resistors.
- Describe the purpose of speedometers and odometers.

- Describe the purpose of the tachometer.
- Describe the operating principles of the digital speedometer.
- Explain the operation of integrated circuit (IC) chip and stepper motor odometers.
- Explain the operation of various warning lamp circuits.
- Explain the operation of various audible warning systems.
- Explain the operation of body computer–controlled instrument panel illumination light dimming.

Terms To Know

Air-core gauge	Head-up display (HUD)	Prove-out circuit
Bucking coil	H-gate	Sending unit
Buffer circuit	High-reading coil	Speedometer
d'Arsonval gauge	International Standards Organization (ISO)	Tachometer
Digital instrument clusters	Low-reading coil	Three-coil gauge
Electromagnetic gauges	Odometer	Two-coil gauge
Electromechanical gauge	Oscillator	Warning lamp
Gauge	Pinion factor	Watchdog circuit

INTRODUCTION

Instrument gauges and indicator lights monitor the various vehicle operating systems. They provide information to the driver of their current operation (**Figure 12-1**). Warning devices also provide information to the driver; however, they are usually associated with an audible signal. Some vehicles use a voice module to alert the driver to certain conditions.

Early instrument cluster gauges were analog or swing needle type. Although many modern vehicles still use the analog gauge, they are now computer driven. These instruments provide far more accurate readings than their conventional analog counterparts.

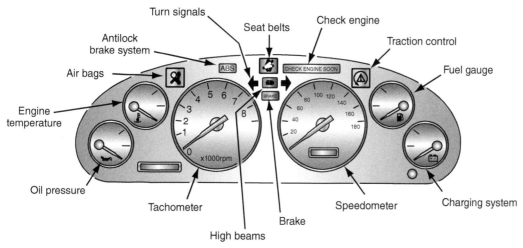

Figure 12-1 Typical gauge and warning indicator layout of an instrument panel.

This chapter introduces you to the most commonly used computer-driven instrumentation systems. These systems include the speedometer, odometer, fuel, oil, tachometer, and temperature gauges.

The computer-driven instrument panel uses a microprocessor (μP) to process information from various sensors and to control the gauge display. Depending on the manufacturer, the microprocessor can be a separate computer that receives direct information from the sensors and makes the calculations, or can use the body control module (BCM) to perform all functions.

In addition, there are many types of information systems used today. These systems keep the driver informed of a variety of monitored conditions, including vehicle maintenance, trip information, and navigation.

ELECTROMECHANICAL GAUGES

A **gauge** is a device that displays the measurement of a monitored system by the use of a needle or pointer that moves along a calibrated scale. The **electromechanical gauge** acts as an ammeter since the gauge reading changes with variations in resistance. The gauge is called an electromechanical device because it is operated electrically, but its movement is mechanical. There are two basic types of electromechanical gauges: the bimetallic gauge and the electromagnetic gauge. Conventional analog instrument clusters that used these types of gauges had a direct connection to the sending unit. The resistance of the sending unit determined the location of the needle on the gauge face. A short study of the different types of gauges is provided to give a foundation to the study of computer-driven gauges.

Bimetallic gauges (or thermoelectric gauges) are not used in today's automobiles. These gauges are simple dial and needle indicators that transformed the heating effect of electricity into mechanical movement. The construction of the bimetallic gauge featured an indicating needle linked to the free arm of a U-shaped bimetallic strip (**Figure 12-2**). The free arm had a heater coil connected to the gauge terminal posts. When current flowed through the heater coil, it heated the bimetallic arm and caused the arm to bend and move the needle across the gauge dial. The amount the bimetallic strip bent is proportional to the heat produced in the heater coil; the greater the heat created, the more the needle would move. Current flow was controlled by the changing resistance of a sending unit.

Electromagnetic gauges produce needle movement by magnetic forces instead of heat. There are four types of electromagnetic gauges: the d'Arsonval, the three coil, the two coil, and the air core.

Shop Manual
Chapter 12, pages 604

Bimetallic and some electromagnetic gauges require a consistent voltage. The instrument panel printed circuit has a instrument voltage regulator (IVR) to stabilize the battery voltage to the gauges.

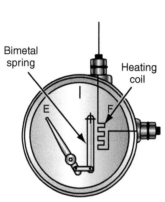

Figure 12-2 Bimetallic gauge construction.

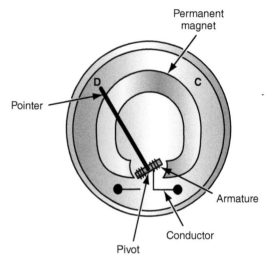

Figure 12-3 d'Arsonval gauge needle movement.

The **d'Arsonval gauge** uses the interaction of a permanent magnet and an electromagnet, and the total field effect to cause needle movement. The d'Arsonval gauge consists of a permanent horseshoe–type magnet that surrounds a movable electromagnet (armature) that is attached to a needle (**Figure 12-3**). When current flows through the armature, it becomes an electromagnet and is repelled by the permanent magnet. When current flow through the armature is low, the strength of the electromagnet is weak and needle movement is small. When the current flow is increased, the magnetic field created in the armature is increased and needle movement is greater. The armature has a small spring attached to it to return the needle to zero when current is not applied to the armature.

The **three-coil gauge** uses the interaction of three electromagnets and the total field effect upon a permanent magnet to cause needle movement. The three-coil gauge consists of a permanent magnet with a needle attached to it. The permanent magnet is surrounded by three electromagnets. There may also be a quantity of silicone dampening fluid to restrict needle movement due to vehicle movement. The current flow controlled by the resistance from the variable resistor–type sending unit determines the magnetic strength of the coils.

The **three-coil gauge** is also known as a magnetic bobbin gauge.

The three coils of fine wire are wound on a square plastic frame. The needle shaft is supported by a bearing sleeve extending from the frame. The needle shaft connects the pointer.

The three coils are the **bucking coil**, the **low-reading coil**, and the **high-reading coil** (**Figure 12-4**). The bucking coil produces a magnetic field that bucks or opposes the low-reading coil. The low-reading coil and the bucking coil are connected but are wound in opposite directions. The high-reading coil is positioned at a 90° angle to the low-reading and bucking coils. To compensate for production tolerances in the coils, a selective shunt resistor is attached to the back of the gauge housing. This selective resistor bypasses a certain amount of current past the coils.

A majority of the current flows through the low-reading coil since its resistance is less than that of the shunt resistor.

When voltage is applied to the gauge, there are two paths in which the current can flow. One path is through the low-reading coil and through the sending unit to ground (**Figure 12-5**). The second path is through the low-reading coil to the bucking coil and the high-reading coil to ground (**Figure 12-6**). The amount of resistance that the sending unit has determines the path. If there is less resistance to ground through the sending unit than through the bucking and high-reading coils, most of the current will take the path through the sending unit. If the resistance through the sending unit is greater than the resistance through the coils, very little current will flow through the sending unit.

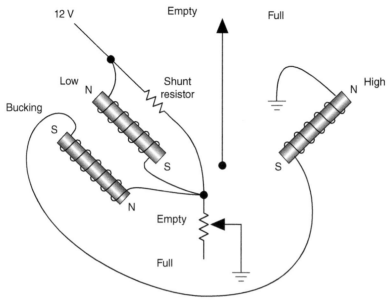

Figure 12-4 Three-coil gauge circuit.

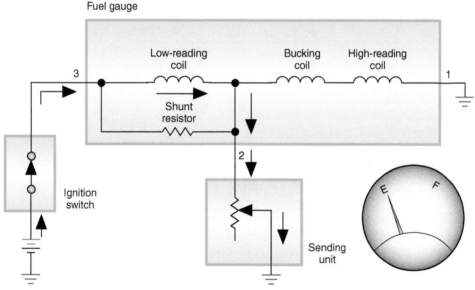

Figure 12-5 With sending unit resistance low, the needle is attracted to the low-reading coil.

When sending unit resistance is low, more current flows through the low-reading coil than through the bucking and high-reading coils. This causes the needle to be attracted to the left, and the gauge reads toward zero. When sending unit resistance is high, very little current will flow through the sending unit. The current now flows through the three coils. The magnetic field created by the bucking coil cancels the magnetic field of the low-reading coil. The high-reading coil's magnetic field then attracts the needle and it swings toward maximum.

At an intermediate sending unit resistance value, the current can flow through both paths. If resistance was equal in the two paths, the needle would point at the midrange. As the magnetic field(s) changes—the result of resistance changes in the sending unit—the needle will swing toward the more powerful magnetic field.

The **two-coil gauge** uses the interaction of two electromagnets and the total field effect on an armature to cause needle movement. There are different designs of the

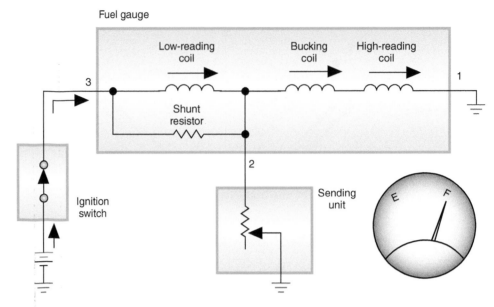

Figure 12-6 With sending unit resistance high, the needle is attracted to the high-reading coil.

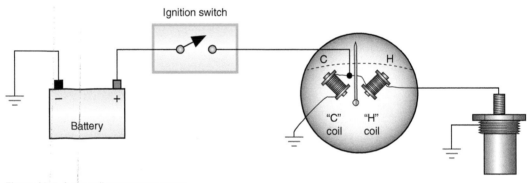

Figure 12-7 A two-coil temperature gauge.

two-coil gauge, depending upon the gauge application. For example, in a coolant tempera-ture gauge (**Figure 12-7**), both coils receive battery voltage. One of the coils is grounded directly, while the other is grounded through the sender unit. As the resistance in the sender unit varies—as a result of temperature changes—the current flow through that coil changes. The two magnetic fields have different strengths, depending upon the amount of current flow through the sender unit. A two-coil gauge that is constructed to be used as a fuel gauge will have the E coil receiving battery voltage (**Figure 12-8**). At the end of the coil, the voltage is divided. One path is through the F coil to ground, while the other is through the sender unit to ground. The stronger the current flow through a coil, the more the needle will move toward that coil.

The most common style of gauge used today is the **air-core gauge**. The air-core gauge works on the same principle as the two coil by using the interaction of two electromagnets and the total field effect upon a permanent magnet to cause needle movement. This gauge has the pointer connected to a permanent magnet (**Figure 12-9**). Two windings are placed at different angles, one wound around the other. There is no core inside of the windings. Instead, the permanent magnet is placed inside the windings. The magnet aligns itself to a resultant field, in the field windings, according to the resistance of the sender unit. The sender unit resistance varies the strength of the field winding, which opposes the strength of the reference winding. The strength of the electromagnetic field depends upon the resistance in the sending unit.

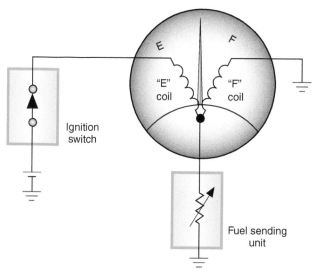

Figure 12-8 A two-coil fuel gauge.

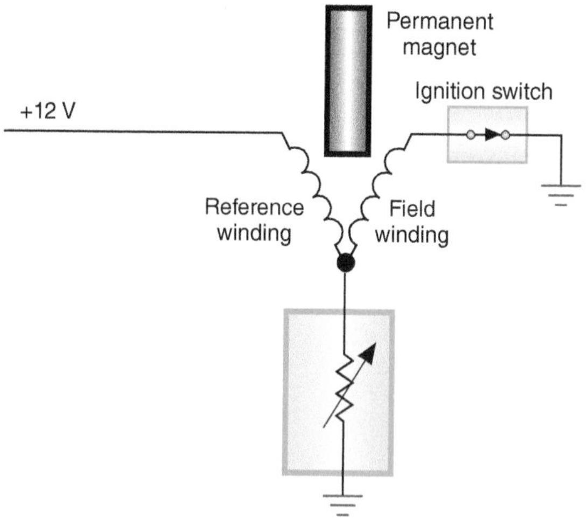

Figure 12-9 Air-core fuel gauge circuit.

QUARTZ ANALOG INSTRUMENTATION

Computer-driven quartz swing needle displays are similar in design to the air-core electromagnetic gauges used in conventional analog instrument panels (**Figure 12-10**). Any, or all, of the gauges in the instrument cluster may be of this type. We will look at the **speedometer** as an example of operation. The speedometer is used to indicate the speed of the vehicle.

Conventional speedometers used a cable that was connected to the output shaft of the transmission (or transfer case if four-wheel drive). The rotation of the output shaft caused the speedometer cable to rotate within its housing. The cable transferred the rotation to the speedometer assembly. This system relied on a rotating permanent magnet that produced a rotating magnetic field around a drum. The rotating magnetic field generated circulating eddy currents in the drum that produced a small magnetic field that interacted with the field of the rotating magnet. This interaction of the two magnetic fields pulled the drum and needle around with the rotating magnet. It is not hard to see that this system would not be extremely accurate. Today's vehicles use sensors and computer logic to display vehicle speed.

In many quartz analog speedometer gauge systems, a permanent magnet generator sensor is installed in the transaxle, transmission, or differential. As the permanent magnet

Shop Manual
Chapter 12,
pages 599, 607

Shop Manual
Chapter 12, page 598

Figure 12-10 Electronically controlled swing needle instrumentation.

Shop Manual
Chapter 12, page 601

(PM) generator is rotated, it causes a small alternating current (AC) voltage to be induced in its coil. This AC voltage signal is sent to a **buffer circuit** that changes the AC voltage from the PM generator into a digitalized signal (**Figure 12-11**). The signal is then sent to the processing unit (**Figure 12-12**). The signal is passed to a quartz clock circuit, a gain

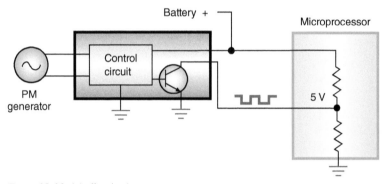

Figure 12-11 A buffer circuit.

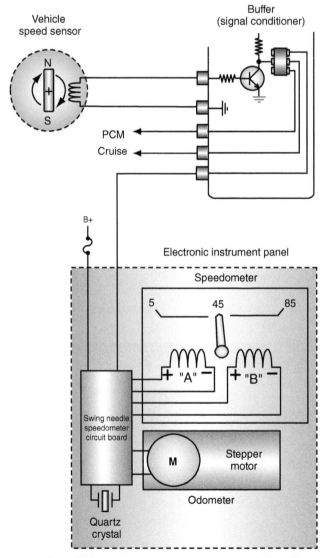

Figure 12-12 Quartz swing needle speedometer schematic. The "A" coil is connected to system voltage and the "B" coil receives a voltage that is proportional to input frequency. The magnetic armature reacts to the changing magnetic fields.

selector circuit, and a driver circuit. The driver circuit sends voltage pulses to the coils of the gauge; the coils operate like conventional air-core gauges to move the needle.

 A BIT OF HISTORY

One of the early styles of speedometers used a regulated amount of air pressure to turn a speed dial. The air pressure was generated in a chamber containing two intermeshing gears. The gears were driven by a flexible shaft that was connected to a front wheel or the driveshaft. The air was applied against a vane inside the speed dial. The amount of air applied was proportional to the speed of the vehicle.

Often the sensor used to determine vehicle speed has multiple purposes. In this case, the signal being generated may not be an accurate representation of vehicle speed because it comes before the final drive unit. The control module may need to do additional calculations to make the speedometer accurate. For example, Chrysler vehicles equipped with the 41TE or 42LE electronic shift transaxles use an output speed sensor that generates an AC signal from a 24-tooth tone wheel on the rear planetary unit. This signal is sent to the transmission control module (TCM). The TCM applies **pinion factor** to the hertz count of the speed sensor signal to calculate vehicle speed. Pinion factor is a calculation using the final drive ratio and the tire circumference to obtain accurate vehicle speed signals. The TCM then transmits this information to the powertrain control module (PCM) at a set rate of 8,000 pulses per mile by pulsing the dedicated circuit. The PCM then sends the vehicle speed signal over the data bus circuit to all modules that require it.

In some applications, the mechanical instrument cluster (MIC) receives the vehicle speed message from the BCM instead of receiving it directly from the PCM. The MIC then sets the needle to read the vehicle speed. Even though the MIC is on the bus system, it does not respond to the vehicle speed signal the PCM sends. It is programmed to accept messages only from the BCM.

Shop Manual
Chapter 12, page 593

Pinion factor information is set into the TCM at the factory. If the TCM is replaced in the field, the scan tool must be used to program the tire size used on the vehicle. In some systems, the gear ratio also has to be programmed. If the pinion factor is not programmed into the TCM, the speedometer and cruise control systems will not function.

Some manufacturers will use the wheel speed sensors from the antilock brake system (ABS) to determine vehicle speed. Usually the two front or the two rear sensor inputs are averaged. The ABS module then determines vehicle speed and broadcasts the information on the data bus.

The other air-core gauges (temperature, fuel level, etc.) work as described earlier with conventional instrument clusters. The difference is that the sending unit input goes to a module. The current flow through the gauge coils is controlled by the module, based on the sending unit resistance.

GAUGE SENDING UNITS

The **sending unit** is the sensor for the gauge. It is a variable resistor that changes resistance values with changes in the monitored conditions. There are three types of sending units that are associated with the gauges just described: (1) a thermistor, (2) a piezoresistive sensor, and (3) a mechanical variable resistor. These same types of sending units can also be used for computer-driven instrument clusters.

Shop Manual
Chapter 12, page 508

In the conventional coolant temperature sensing circuit, current is sent from the gauge unit into the top terminal of the sending unit, through the variable resistor (thermistor), and to the engine block (ground). The resistance value of the thermistor changes in proportion to coolant temperature (**Figure 12-13**). As the temperature rises, the

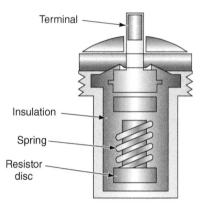

Figure 12-13 A thermistor used to sense engine temperature.

resistance decreases and the current flow through the gauge increases. As the coolant temperature lowers, the resistance value increases and the current flow decreases.

In a computer-driven gauge or digital display, the control module will send a 5-volt reference voltage through a pull-up resistor and then to the temperature sensor. This type of circuit was discussed in Chapter 9. As the resistance changes, the voltage dropped over the pull-up resistor changes and the voltmeter reading indicates the engine temperature (**Figure 12-14**). The PCM sends the temperature information over the data bus to the instrument cluster (or BCM). The module will then send a current to the gauge coils to move the pointer to the correct temperature reading.

The piezoresistive sensor sending unit is threaded into the oil delivery passage of the engine, and the pressure that is exerted by the oil causes the flexible diaphragm to move (**Figure 12-15**). The diaphragm movement is transferred to a contact arm that slides along

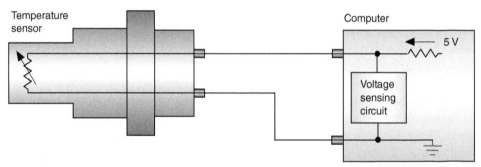

Figure 12-14 A typical temperature sensing circuit.

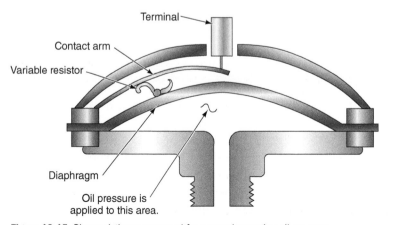

Figure 12-15 Piezoresistive sensor used for measuring engine oil pressure.

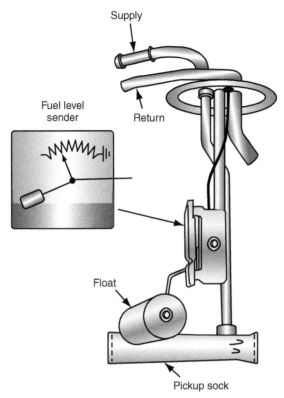

Supply

Fuel level
sender

Return

Float

Pickup sock

Figure 12-16 Fuel gauge sending unit.

the resistor. The position of the sliding contacts on the arm in relation to the resistance coil determines the resistance value, and the amount of current flow through the gauge to ground.

Another style is a transducer that operates much like a Wheatstone bridge MAP sensor. The function of the gauge is the same as that just discussed for the computer-driven temperature gauge, based on data bus messages from the PCM.

Some computer-driven instrument clusters have an oil gauge but do not use a sensor. These systems use an oil pressure switch. When oil pressure is greater than 6 psi (41 kPa), the switch opens the sense circuit. This pulls the sense voltage high. The gauge displays an oil pressure that is based on a calculated value determined by engine run time, engine temperature, engine load value, and ambient temperature. If the oil pressure drops below 6 psi (41 kPa), the switch closes the circuit to ground, pulling the sense voltage low. The instrument cluster gauge now reads 0. As long as the switch is open, the gauge will indicate normal oil pressure.

A fuel level sending unit is an example of a mechanical variable resistor (**Figure 12-16**). The sending unit is located in the fuel tank and has a float that is connected to the wiper of a variable resistor. The floating arm rises and falls with the difference in fluid level. This movement of the float is transferred to the sliding contacts. The position of the sliding contacts on the resistor determines the resistor value.

DIGITAL INSTRUMENTATION

Digital instrumentation is far more precise than conventional analog gauges. Analog gauges display an average of the readings received from the sensor; a digital display will present exact readings. In some systems, the information to the gauge is updated as often as 16 times per second.

Shop Manual
Chapter 12, page 607

Digital instrument clusters use digital and linear displays to notify the driver of monitored system conditions (**Figure 12-17**). Most digital instrument clusters provide for display in English or metric values. Also, many gauges are a part of a multigauge system. Drivers select which gauges they wish to have displayed. Most of these systems will automatically display the gauge to indicate a potentially dangerous situation. For example, if the driver has chosen the oil pressure gauge to be displayed and the engine temperature increases above set limits, the temperature gauge will automatically be displayed to warn the driver. A warning light and/or a chime will also activate to get the driver's attention.

Most electronic instrument panels have self-diagnostic capabilities. The tests are initiated through a scan tool or by pushing selected buttons on the instrument panel. The instrument panel cluster also initiates a self-test every time the ignition switch is turned to ACC or RUN. Usually, the entire dash is illuminated and every segment of the display is lighted. **International Standards Organization (ISO)** symbols are used to represent the gauge function (**Figure 12-18**). These symbols may flash during this test. At the completion of the test, all gauges display current readings. A code is displayed to alert the driver if a fault is found.

Speedometers

Shop Manual
Chapter 12, page 502

Ford, General Motors (GM), and Toyota have used optical vehicle speed sensors (VSSs). The Ford and Toyota optical sensors are operated from the conventional speedometer cable. The cable rotates a slotted wheel between a light-emitting diode (LED) and a phototransistor (**Figure 12-19**). As the slots in the wheel break the light, the transistor

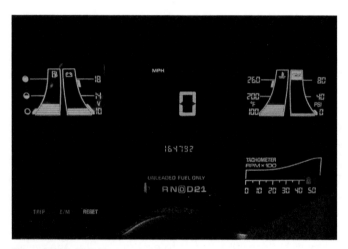

Figure 12-17 Digital instrument cluster.

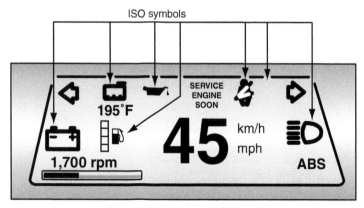

Electronic instrument cluster

Figure 12-18 A few of the ISO symbols used to identify the gauge.

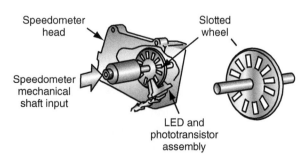

Figure 12-19 Optical speed sensor.

conducts an electronic pulse signal to the speedometer. An integrated circuit rectifies the analog input signal from the optical sensor and counts the pulses per second. The value is calculated into miles per hour and displayed in the digital readout. The display is updated every 1/2 second. If the driver selects the readout to be in kilometers per hour, the computer makes an additional calculation to convert the readout. These systems may use a conventional gear-driven odometer.

The early style of GM speed sensor also operated from the conventional speedometer cable. The LED directs its light onto the back of the speedometer cup. The cup is painted black, and the drive magnet has a reflective surface applied to it. As the drive magnet rotates in front of the LED, its light is reflected back to a phototransistor. A small voltage is created every time the phototransistor is hit with the reflective light.

Figure 12-20 is a schematic of an instrument panel cluster that uses a PM generator for the VSS. As the PM generator is rotated, it causes a small AC voltage to be induced in its coil. This AC voltage signal is sent to the PCM and is shared with the BCM. The signal is rectified into a digital signal that is used to control the output to the instrument panel cluster (IPC) module. The BCM calculates the vehicle speed and provides this information to the IPC module through the serial data link. The IPC module turns on the proper display.

The electronic speedometer receives voltage signals from the vehicle speed sensor (VSS). This sensor can be a PM generator, Hall-effect sensor, or an optical sensor.

Shop Manual
Chapter 12, page 593

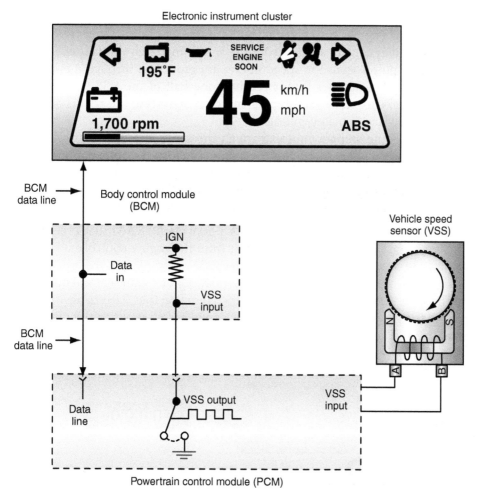

Figure 12-20 The instrument panel cluster module receives its instructions from the BCM, which shares the signals from the VSS with the ECM.

The microprocessor initiates a self-check of the electronic instrument cluster any time the ignition switch is placed in the ACC or RUN position. The self-check usually runs for about 3 seconds. The most common sequence for the self-check is as follows:

1. All display segments are illuminated.
2. All displays go blank.
3. 0 mph or 0 km/h is displayed.

Shop Manual
Chapter 12, page 602

In addition to the methods of sensing speed mentioned earlier, Hall-effect sensors are also used. The sensor is attached to a gear-driven wheel that rotates a trigger wheel. The gear is determined by tire size and the final drive gear ratio of the vehicle. As the trigger wheel rotates, it causes the Hall-effect sensor to change voltages to high and low at a set amount each revolution. The amount of switches per revolution remains constant regardless of vehicle speed. Once the control module receives a programmed number of switches (e.g., 8,000), it knows it has traveled 1 mile.

Odometers

Shop Manual
Chapter 12, page 599

The **odometer** is a counter that uses the speedometer inputs to indicate the total miles accumulated on the vehicle. Many vehicles also have a second odometer that can be reset to zero; this is referred to as a trip odometer.

Early odometers were driven by the speedometer cable through a worm gear. If the speedometer uses an optical sensor, the odometer may be of conventional design. Two other types of odometer are used with electronic displays: the electromechanical type with a stepper motor and the electronic design using an IC chip.

Stepper Motor. The electromechanical odometer uses a direct current (DC) stepper motor that receives control signals from the speedometer circuit (**Figure 12-21**). The digital signal impulses from the speedometer are processed through a circuit that will halve the signal. The stepper motor receives one-half of the VSS signals sent to the instrument panel cluster. As the stepper motor is activated, the rollers are rotated to accurately display accumulated mileage.

GM controls the stepper motor through the same impulses that are sent to the speedometer. The stepper motor uses these signals to turn the odometer drive IC on and off. An **H-gate** arrangement of four transistors is used to drive the stepper motor by alternately activating a pair of its coils (**Figure 12-22**). The H-gate is constantly reversing system polarity, causing the permanent magnet poles to rotate in the same direction.

IC Chip. The IC chip–type odometer uses a nonvolatile RAM that receives distance information from the speedometer circuit or from the engine controller. The controller can update the odometer display every 1/2 second.

In many systems, distance is updated to the RAM every 10 miles and whenever the ignition switch is turned off.

Many instrument panel clusters cannot display both trip mileage and odometer readings at the same time. Drivers must select which function they wish to have displayed. By depressing the trip reset button, a ground is applied as an input to the microprocessor (**Figure 12-23**). The microprocessor clears the trip odometer readings from memory and

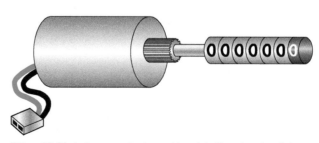

Figure 12-21 A stepper motor is used to rotate the odometer dial.

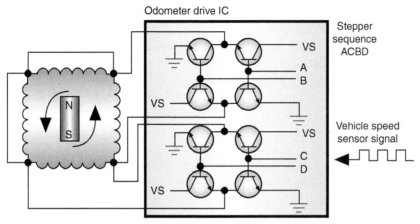

Figure 12-22 The H-gate energizes two coils at a time and constantly reverses system polarity.

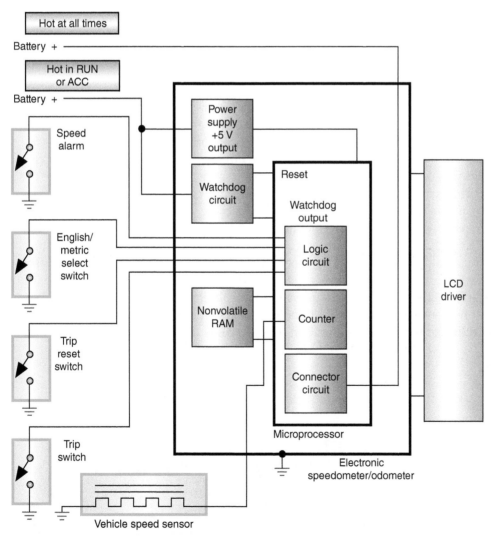

Figure 12-23 The trip reset button provides a ground signal to the logic circuit, which is programmed to erase the trip odometer memory while retaining total accumulated mileage in the odometer.

returns the display to zero. The trip odometer continues to store trip mileage even if this function is not selected for display.

If the IC chip fails, some manufacturers provide for replacement of the chip. Depending on the manufacturer, the new chip may be programmed to display the last odometer reading. Most replacement chips display an X, S, or * to indicate the odometer has been changed. If the odometer IC chip cannot be programmed to display correct accumulated mileage, a door sticker must be installed to indicate that the odometer has been replaced.

> **AUTHOR'S NOTE** Federal Motor Vehicle Safety Standards require the odometer be capable of storing up to 500,000 miles in nonvolatile memory. Most odometer readouts are up to 199,999.9 miles.

If an error occurs in the odometer circuit, the display changes to notify the driver. The form of error message differs among manufacturers. In some systems, the word *ERROR* is displayed, while others may use dashed lines.

Federal and state laws prohibit tampering with the correct mileage as indicated on the odometer. If the odometer needs to be replaced, it must be set to the reading of the original odometer, or a door sticker must be installed indicating the reading of the odometer when it was replaced.

Tachometers

A **tachometer** is an instrument that measures the speed of the engine in revolutions per minute (rpm). The electric tachometer receives voltage pulses from the ignition system, usually the ignition coil (**Figure 12-24**). The tachometer signal is picked up from the negative (−) side of the coil as the switching unit opens the primary circuit. Each of the voltage pulses represents the generation of one spark at the spark plug. The rate of spark plug firing is in direct relationship to the speed of the engine. A circuit within the tachometer converts the ignition pulse signal into a varying voltage. The voltage is applied to a voltmeter that serves as the engine speed indicator.

Shop Manual
Chapter 12, page 603

The digital tachometer can be a separate function that is displayed at all times, or a part of a multigauge display. The digital tachometer receives its voltage signals from the ignition module or PCM via the bus network and displays the readout in a bar graph (**Figure 12-25**). The multigauge system has a built-in power supply that provides a 5-volt reference signal to the other monitored systems for the gauge. Also, the gauge has a **watchdog circuit** incorporated in it. The power on/off watchdog circuit supplies a

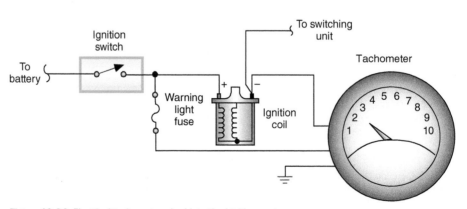

Figure 12-24 Electrical tachometer wired into the ignition system.

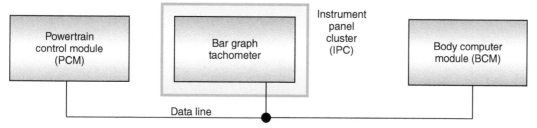

Figure 12-25 The IPC "listens in" on the communications between the PCM and the BCM to gather information on engine speed.

reset voltage to the microprocessor in the event that pulsating output signals from the microprocessor are interrupted.

Electronic Fuel Gauges

Most digital fuel gauges use a fuel level sender that decreases resistance value as the fuel level decreases. This resistance value is converted to voltage values by the microprocessor. A voltage-controlled **oscillator** changes the signal into a frequency signal. The microprocessor counts the cycles and sends the appropriate signal to operate the digital display (**Figure 12-26**).

An F is displayed when the tank is full and an E is displayed when less than 1 gallon is remaining in the tank. Other warning signals include incandescent lamps, a symbol on the dash, or flashing of the fuel ISO symbol. If the warning is displayed by a bulb, usually a switch is located in the sending unit that closes the circuit. The microprocessor usually controls flashing digital displays.

The bar graph–style gauge uses segments that represent the amount of fuel remaining in the tank (**Figure 12-27**). The segments divide the tank into equal levels. The display also includes the F, 1/2, and E symbols along with the ISO fuel symbol. A warning to the driver is displayed when only one bar is lit. The gauge also alerts the driver to problems in the circuit. A common method of indicating an open or short is to flash the F, 1/2, and E symbols while the gauge reads empty.

Shop Manual
Chapter 12, page 608

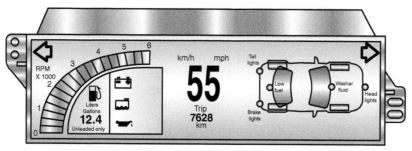

Figure 12-26 The digital fuel gauge displays remaining fuel in gallons or liters.

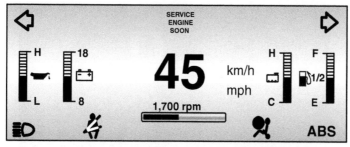

Figure 12-27 Bar graph style of electronic instrumentation. Each segment represents a different value.

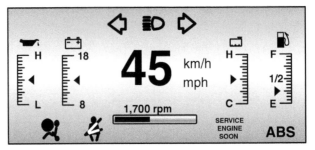

Figure 12-28 Floating pointer indicates the value received from the sensor.

Other Digital Gauges

Most of the gauges used to display temperature, oil pressure, and charging voltage are of bar graph design. Another popular method is to use a floating pointer (**Figure 12-28**).

The temperature gauge usually receives its input from a negative temperature coefficient (NTC) thermistor. When the engine is cold, the resistance value of the thermistor is high, resulting in a high-voltage input to the microprocessor. This input signal is translated into a low-temperature reading on the gauge. As the engine coolant warms, the resistance value drops. At a predetermined resistance level, the microprocessor will activate an alert function to warn the driver of excessive engine temperature.

The voltmeter calculates charging voltage by comparing the voltage supplied to the instrument panel module to a reference voltage signal. The oil pressure gauge uses a piezoresistive sensor that operates like those used for conventional analog gauges.

Digital gauges perform self-tests. If a fault is found, a warning signal will be displayed to the driver. A "CO" indicates the circuit is open, and a "CS" indicates the circuit is shorted. The gauge will continue to display these messages until the problem is corrected.

LCD MONITORS

Thin-film liquid crystal display (LCD) monitors are popular for displaying instrument panel gauge readings (**Figure 12-29**). In addition, it is commonly used for radio, audio, and navigational information. LCD monitors provide a high-resolution graphic display.

The thin-film transistor (TFT) is the most common switching device. TFT is an arrangement of tiny transistors and capacitors in a matrix on the glass of the display. TFT LCD monitors have a sandwich-type structure with liquid crystal filled between two glass plates (**Figure 12-30**). The TFT glass has as many TFTs as the number of pixels displayed.

Figure 12-29 Monitor display of instrument panel gauges.

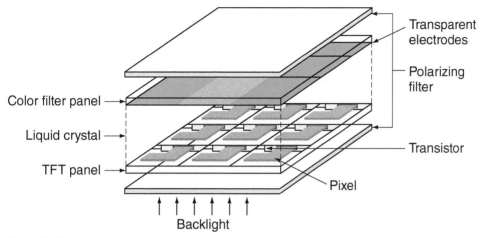

Figure 12-30 Sandwich structure of the TFT LCD monitor.

A TFT LCD module consists of a TFT panel, driving-circuit unit, backlight system, and assembly unit (**Figure 12-31**). The monitor is divided into millions of pixel units that are formed by liquid crystal cells. The cells change the polarization direction of light passing through them in response to an electrical voltage. As the polarization direction changes, the amount of light allowed to pass through a polarizing layer changes. The LCD monitor uses the matrix driving method that displays characters and pictures in sets of dots.

The TFT substrate contains the TFTs, storage capacitors, pixel electrodes, and circuit wiring (**Figure 12-32**). The color filter generates the colors of the display. The movement

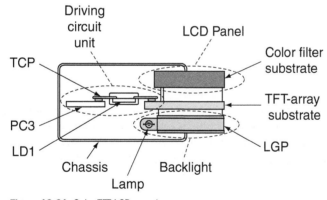

Figure 12-31 Color TFT LCD panel.

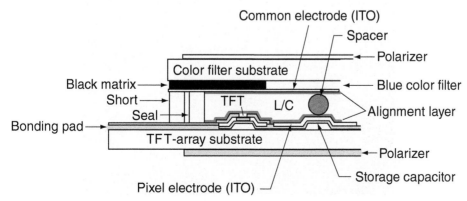

Figure 12-32 Cross section of the TFT substrate.

of the liquid crystals is based on the difference in voltage potential between the TFT glass and the color filter glass. The amount of backlight supplied is determined by the amount of movement of the liquid crystals in such a way as to generate color. The color filter contains a black matrix and a resin film. The resin film includes the three primary color pigments of red, green, and blue (RGB).

HEAD-UP DISPLAY

Shop Manual
Chapter 12, page 612

Some manufacturers have equipped select models with a **head-up display (HUD)** feature. This system displays visual images onto the inside of the windshield in the driver's field of vision (**Figure 12-33**). With the display located in this area, drivers do not need to remove their eyes from the road to check the instrument panel. The images are projected onto the windshield from a vacuum fluorescent light source, much like a movie projector.

The head-up control module is mounted in the top of the instrument panel. This module contains a computer and an optical system that projects images to a holographic combiner integrated into the windshield above the module. The holographic combiner projects the images in the driver's view just above the front end of the hood. The HUD may contain the following displays and warnings:

1. Speedometer reading with US/metric indicator.
2. Turn signal indicators.
3. High-beam indicator.
4. Low-fuel indicator.
5. Check gauges indicator.

The head-up control switch contains a head-up display on/off switch, USC/metric switch, and a head-up dimming switch. The head-up dimming switch is a rheostat that sends an input signal to the head-up module. The vertical position of the head-up display may be moved with one of the switches in the control switch assembly that is connected through a mechanical cable-drive system to the head-up module. Moving the vertical position switch moves the position of the head-up module.

The VSS signal information is sent from the PCM to the head-up module for the speedometer display. The check gauges and low-fuel signals are sent from the instrument cluster to the head-up module. A high-beam indicator input signal is sent from the dimmer switch to the head-up module. This module also receives inputs from the signal light switch to operate the signal light indicators.

Figure 12-33 The HUD displays various information inside the windshield.

TRAVEL INFORMATION SYSTEMS

The travel information system can be a simple calculator that computes fuel economy, distance to empty, and remaining fuel (**Figure 12-34**). Other systems provide a much larger range of functions.

Shop Manual
Chapter 12, page 611

Fuel data centers display the amount of fuel remaining in the tank and provide additional information for the driver (**Figure 12-35**). By depressing the RANGE button, the BCM calculates the distance until the tank is empty by using the amount of fuel remaining and the average fuel economy. When the INST button is depressed, the fuel data center displays instantaneous fuel economy. The display is updated every 1/2 second and is computed by the BCM.

Depressing the AVG button displays average fuel economy for the total distance traveled since the reset button was last pushed. FUEL USED displays the amount of fuel that has been used since the last time this function was reset. The RESET button resets the average fuel economy and fuel-used calculations. The function to be reset must be displayed on the fuel data center.

Deluxe systems may incorporate additional features such as outside temperature, compass, elapsed time, estimated time of arrival, distance to destination, day of the week, time, and average speed. **Figure 12-36** shows the inputs that are used to determine many of these functions. The sensors shown in **Figure 12-37** determine fuel system calculations. Injector on time and vehicle speed pulses determine the amount of fuel flow. Some manufacturers use a fuel flow sensor that provides pulse information to the microprocessor concerning fuel consumption (**Figure 12-38**).

Figure 12-34 Fuel data display panel.

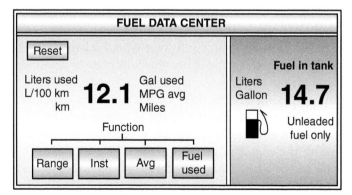

Figure 12-35 Fuel data center.

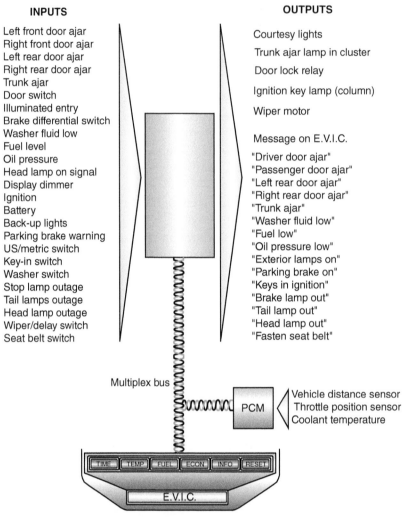

INPUTS

Left front door ajar
Right front door ajar
Left rear door ajar
Right rear door ajar
Trunk ajar
Door switch
Illuminated entry
Brake differential switch
Washer fluid low
Fuel level
Oil pressure
Head lamp on signal
Display dimmer
Ignition
Battery
Back-up lights
Parking brake warning
US/metric switch
Key-in switch
Washer switch
Stop lamp outage
Tail lamps outage
Head lamp outage
Wiper/delay switch
Seat belt switch

OUTPUTS

Courtesy lights
Trunk ajar lamp in cluster
Door lock relay
Ignition key lamp (column)
Wiper motor

Message on E.V.I.C.

"Driver door ajar"
"Passenger door ajar"
"Left rear door ajar"
"Right rear door ajar"
"Trunk ajar"
"Washer fluid low"
"Fuel low"
"Oil pressure low"
"Exterior lamps on"
"Parking brake on"
"Keys in ignition"
"Brake lamp out"
"Tail lamp out"
"Head lamp out"
"Fasten seat belt"

Multiplex bus

PCM ← Vehicle distance sensor
      Throttle position sensor
      Coolant temperature

TIME TEMP FUEL ECON INFO RESET

E.V.I.C.

Figure 12-36 Inputs used for the electronic vehicle information center.

WARNING LAMPS

Shop Manual
Chapter 12, page 609

A **warning lamp** is a lamp that is illuminated to warn the driver of a possible problem or hazardous condition. A warning lamp may be used to warn of low oil pressure, high coolant temperature, defective charging system, or a brake failure. A warning lamp can be operated by two methods: a sending unit circuit hardwired to the instrument cluster or computer-controlled lamp drivers.

Sending Unit–Controlled Lamps

Shop Manual
Chapter 12, page 609

Unlike gauge sending units, the sending unit for a warning lamp is nothing more than a simple switch. The style of switch can be either normally open or normally closed, depending on the monitored system.

Most oil pressure warning circuits use a normally closed switch (**Figure 12-39**). A diaphragm in the sending unit is exposed to the oil pressure. The switch contacts are controlled by the movement of the diaphragm. When the ignition switch is turned to the RUN position with the engine not running, the oil warning lamp turns on. Because there is no pressure to the diaphragm, the contacts remain closed and the circuit is complete to ground. When the engine is started, oil pressure builds and the diaphragm moves the contacts apart. This opens the circuit and the warning lamp goes off. The

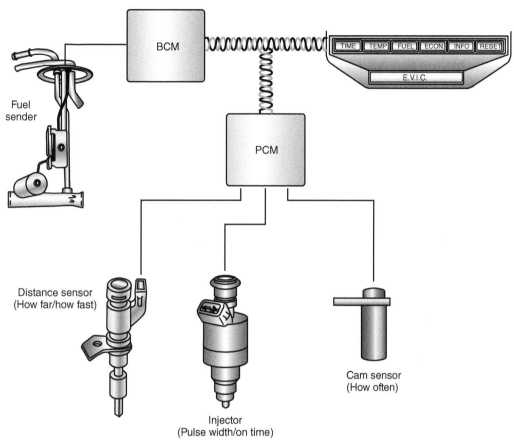

Figure 12-37 Fuel data system inputs. The injector on time is used to calculate the rate of fuel flow.

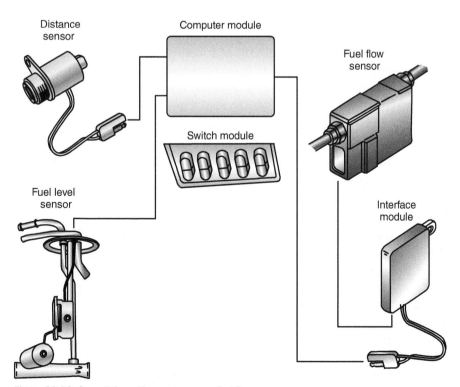

Figure 12-38 Some information centers use a fuel flow sensor.

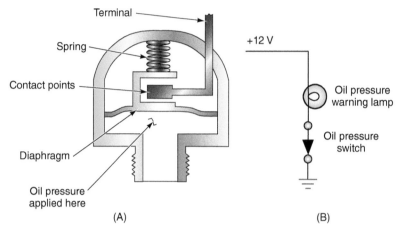

Figure 12-39 (A) Oil pressure light sending unit. (B) Oil pressure warning lamp circuit.

amount of oil pressure required to move the diaphragm is about 3 psi. If the oil warning lamp comes on while the engine is running, it indicates that the oil pressure has dropped below the 3 psi limit.

Most coolant temperature warning lamp circuits use a normally open switch (**Figure 12-40**). The temperature sending unit consists of a fixed contact and a contact on a bimetallic strip. As the coolant temperature increases, the bimetallic strip bends. As the strip bends, the contacts move closer to each other. Once a predetermined temperature level has been exceeded, the contacts are closed and the circuit to ground is closed. When this happens, the warning lamp is turned on.

With normally open–type switches, the contacts are not closed when the ignition switch is turned to ON. In order to perform a bulb check on normally open switches, a **prove-out circuit** is included (**Figure 12-41**). A prove-out circuit completes the warning light circuit to ground through the ignition switch when it is in the START position. The warning light will be on during engine cranking to indicate to the driver that the bulb is working properly.

It is possible to have more than one sending unit connected to a single bulb. The illustration (**Figure 12-42**) shows a wiring circuit of a dual-purpose warning lamp. The lamp comes on whenever oil pressure is low or coolant temperature is too high.

Another system that is monitored with a warning lamp is the braking system. **Figure 12-43** shows a brake system combination valve. The center portion of the valve senses differences in the hydraulic pressures on both sides of the valve. With the

The prove-out function is also known as "Bulb Test" or "Bulb Check" position.

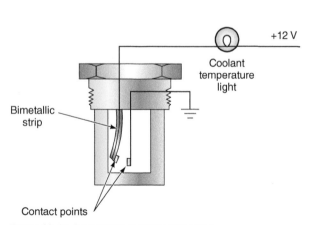

Figure 12-40 Temperature indicator light circuit.

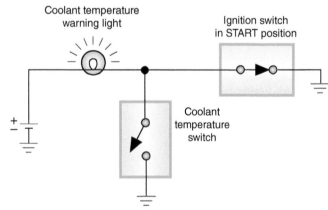

Figure 12-41 A prove-out circuit included in a normally open (NO) coolant temperature light system.

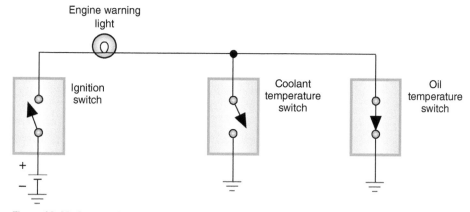

Figure 12-42 One warning lamp used with two sensors.

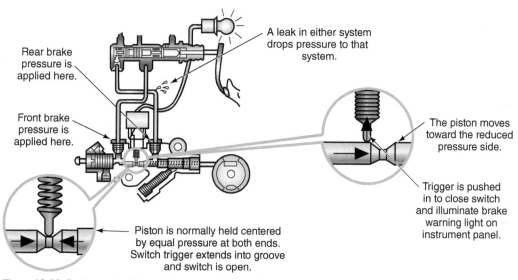

Figure 12-43 Brake warning light switch as part of the combination valve.

differential valve centered, the plunger on the warning lamp switch is in the recessed area of the valve. If the pressure drops in either side of the brake system, the differential valve will be forced to move by hydraulic pressure. When the differential valve moves, the switch plunger is pushed up and the switch contacts close.

Computer-Driven Warning Lamp Systems

The computer-driven warning lamp system uses either high-side or low-side drivers to illuminate the warning lamp. Usually the driver module (instrument cluster or BCM) receives a data bus message from the module, which monitors the affected system that the warning lamp needs to be turned on. The driver module then commands the lamp on. For example, the PCM monitors the engine temperature. If the engine temperature reaches the upper threshold, the PCM will send a data bus message to the instrument cluster to turn on the warning lamp. The instrument cluster will activate its driver to illuminate the lamp. Usually the instrument cluster must receive a bus message from the monitoring module at a set time interval. If the message is not received, the warning lamp is illuminated. Some systems illuminate a CHECK GAUGES lamp if the cluster uses a gauge and the gauge indicates a condition that the driver needs to be notified of.

Shop Manual
Chapter 12, page 611

SUMMARY

- Through the use of gauges and indicator lights, the driver is capable of monitoring several engine and vehicle operating systems.
- The gauges include speedometer, odometer, tachometer, oil pressure, charging indicator, fuel level, and coolant temperature.
- The most common types of electromechanical gauges are the d'Arsonval, three coil, two coil, and air core.
- Computer-driven quartz swing needle displays are similar in design to the air-core electromagnetic gauges used in conventional analog instrument panels.
- All gauges require the use of a variable resistance sending unit. Styles of sending units include thermistors, piezoresistive sensors, and mechanical variable resistors.

- Digital instrument clusters use digital and linear displays to notify the driver of monitored system conditions.
- The most common types of displays used on electronic instrument panels are: light-emitting diodes (LEDs), liquid crystal displays (LCDs), vacuum fluorescent displays (VFDs), and a phosphorescent screen that is the anode.
- A head-up display system displays visual images onto the inside of the windshield in the driver's field of vision.
- In the absence of gauges, important engine and vehicle functions are monitored by warning lamps. These circuits generally use an on/off switch–type sensor. The exception would be the use of voltage-controlled warning lights that use the principle of voltage drop.

REVIEW QUESTIONS

Short-Answer Essays

1. What are the most common types of electromagnetic gauges?

2. Describe the operation of the piezoresistive sensor.

3. What is a thermistor used for?

4. What is meant by *electromechanical?*

5. Describe the operation of the air-core gauge.

6. What is the basic difference between conventional analog and computer-driven analog instrument clusters?

7. Describe the operating principles of the digital speedometer.

8. Explain the operation of IC chip–type odometers.

9. Describe the operation of the electronic fuel gauge.

10. Describe the operation of quartz analog speedometers.

Fill in the Blanks

1. The purpose of the tachometer is to indicate _____.

2. A piezoresistive sensor is used to monitor _____ changes.

3. The most common style of fuel level sending unit is _____ variable resistor.

4. The brake warning light is activated by _____ pressure in the brake hydraulic system.

5. In a three-coil gauge, the _____ _____ produces a magnetic field that bucks or opposes the low-reading coil.
   The _____ _____ coil and the bucking coil are wound together, but in opposite directions.
   The _____ _____ coil is positioned at a 90° angle to the low-reading and bucking coils.

6. A _____ _____ circuit completes the warning light circuit to ground through the ignition switch when it is in the START position.

7. Digital instrument clusters use _____ and _____ displays to notify the driver of monitored system conditions.

8. _____ _____ is a calculation using the final drive ratio and the tire circumference to obtain accurate vehicle speed signals.

9. Most digital fuel gauges use a fuel level sender that _____ resistance value as the fuel level decreases.

10. Computer-driven quartz swing needle displays are similar in design to the _____ _____ electromagnetic gauges used in conventional analog instrument panels.

Multiple Choice

1. Odometer replacement is being discussed.

   Technician A says that it is permissible to turn back the reading on an odometer as long as the customer is notified.

   Technician B says that if an odometer is replaced, it must be set to the same reading as the original odometer.

   Who is correct?

   A. A only   C. Both A and B
   B. B only   D. Neither A nor B

2. Electromagnetic gauges are being discussed.

   Technician A says that the d'Arsonval gauge uses the interaction of a permanent magnet and an electromagnet, and the total field effect to cause needle movement.

   Technician B says that the three-coil gauge uses the interaction of three electromagnets and the total field effect upon a permanent magnet to cause needle movement.

   Who is correct?

   A. A only   C. Both A and B
   B. B only   D. Neither A nor B

3. *Technician A* says that the three-coil gauge uses the principle that the majority of current seeks the path of least resistance.

   Technician B says that the three coils used are the low-reading coil, a bucking coil, and a high-reading coil.

   Who is correct?

   A. A only   C. Both A and B
   B. B only   D. Neither A nor B

4. Warning light circuits are being discussed.

   Technician A says most oil pressure warning circuits use a normally closed switch.

   Technician B says most conventional coolant temperature warning light circuits use a normally open switch.

   Who is correct?

   A. A only   C. Both A and B
   B. B only   D. Neither A nor B

5. The brake failure warning system is being discussed.

   Technician A says if the pressure drops in either side of the brake system, the switch plunger is pushed up and the switch contacts close.

   Technician B says if the pressure is equal on both sides of the brake system, the warning light comes on.

   Who is correct?

   A. A only   C. Both A and B
   B. B only   D. Neither A nor B

6. The IC chip odometer is being discussed.

   Technician A says if the chip fails, some manufacturers provide for replacement of the chip.

   Technician B says depending on the manufacturer, the new chip may be programmed to display the last odometer reading.

   Who is correct?

   A. A only   C. Both A and B
   B. B only   D. Neither A nor B

7. Computer-driven quartz swing needle displays are being discussed.

   Technician A says the A coil is connected to system voltage and the B coil receives a voltage that is proportional to input frequency.

   Technician B says the quartz swing needle display is similar to air-core electromagnetic gauges.

   Who is correct?

   A. A only   C. Both A and B
   B. B only   D. Neither A nor B

8. *Technician A* says digital instrumentation displays an average of the readings received from the sensor.

   Technician B says conventional analog instrumentation gives more accurate readings but is not as decorative.

   Who is correct?

   A. A only   C. Both A and B

   B. B only   D. Neither A nor B

9. The microprocessor-initiated self-check of the electrical instrument cluster is being discussed.

   Technician A says during the first portion of the self-test all segments of the speedometer display are lit.

   Technician B says the display should not go blank during any part of the self-test.

   Who is correct?

   A. A only   C. Both A and B

   B. B only   D. Neither A nor B

10. *Technician A* says bar graph–style gauges do not provide for self-tests.

    Technician B says the digital instrument panel may display "CO" to indicate that the circuit is shorted.

    Who is correct?

    A. A only   C. Both A and B

    B. B only   D. Neither A nor B

CHAPTER 13

ACCESSORIES

Upon completion and review of this chapter, you should be able to understand and describe:

- The operation and function of the horn circuit.
- The operation of standard two- and three-speed wiper motors, both permanent magnet and electromagnetic field designs.
- The operation of intermittent wipers.
- How depressed-park wipers operate.
- The function of intelligent windshield wiper systems.
- The operation of windshield washer pump systems.
- The operation and methods used to control blower fan motor speeds.
- The operation of electric defoggers.
- Operational principles of power mirrors.
- The principles of operation for power windows, power seats, and power locks.

- The operating principles of the memory seat feature.
- The operation of climate-controlled seats.
- The purpose and operation of automatic door lock systems.
- The operation of the keyless entry system.
- The operation of common antitheft systems.
- The purpose of the cruise control system.
- The operating principles of the electronic cruise control system.
- The operating principles of adapter cruise control systems.
- The concepts of electronically controlled sunroofs.

Terms To Know

Adaptive cruise control (ACC)
Armed
Automatic door locks (ADL)
Child safety latch
Climate-controlled seats
Clockspring
Cruise control

Depressed-park
Diaphragm
Electrochromic mirror
Express down
Express up
Grid
Keyless entry

Negative logic
Park contacts
Peltier element
Resistor block
Sector gear
Trimotor
Window regulator

INTRODUCTION

Electrical accessories provide additional safety and comfort. Many electrical accessories can be installed into today's vehicles. This chapter explains the principles of operation for some of the most common electrical accessories. Systems not discussed here are similar in concept.

This chapter explores the operation of safety accessories, such as the horn, windshield wipers, and windshield washers. Comfort accessories explored in this chapter include the blower motor, electric defoggers, power mirrors, power windows, power seats, and power door locks.

In today's automobile there is no system that computers cannot control. The vehicle may be equipped with computer-controlled wipers, transmissions, locking differentials, brakes, suspensions, all-wheel drive systems, and so on. It would be impossible to cover the various operations of all these systems. It is important for today's technician to have an understanding of how electronics work and a basic knowledge of the control system. Whether you are working on a domestic or foreign-built automobile, electricity and electronics work the same. Always refer to proper service information to get an understanding of a system that may be new to you.

In addition to electrical accessories, this chapter discusses several of the electronic systems found in today's automobiles. Some of these systems are covered in greater detail in other *Today's Technician* series books. Refer to these textbooks for more information.

In this chapter, you will learn the operation of cruise control systems and the many electrical accessory systems that have electronic controls added to them for added features and enhancement. These accessories include memory seats, electronic sunroofs, antitheft systems, and automatic door locks.

The comfort and safety of the driver and/or passengers depend on the technician's proper diagnosis and repair of these systems. As with all electrical systems, the technician must have a basic understanding of the operation of these systems before attempting to perform any service.

HORNS

Shop Manual
Chapter 13, page 632

A horn is a device that produces an audible warning signal.

When two horns are used on a vehicle, one is called the low-note horn and the other is called the high-note horn.

The automotive electrical horn operates on an electromagnetic principle that vibrates a **diaphragm** to produce a warning signal. The diaphragm is a thin, flexible, circular plate that is held around its outer edge by the horn housing, allowing the middle to flex. Most electrical horns consist of an electromagnet, a movable armature, a diaphragm, and a set of normally closed contact points (**Figure 13-1**). The contact points are wired in series with the field coil. One of the points is attached to the armature. When current flows through the field coil, a magnetic field develops, which attracts the movable armature. The diaphragm is attached to and moves with the armature. Movement of the armature results in the contact points opening, which breaks the circuit. The diaphragm is released and returns to its normal position. The contact points close again and the cycle repeats. This vibration of the diaphragm is repeated several times per second.

As the diaphragm vibrates, it causes a column of air in the horn to vibrate. The vibration of the column of air produces the sound.

Most vehicles are equipped with two horns that are wired in parallel to each other and in series with the horn switch or relay. One of the horns will have a slightly lower pitch

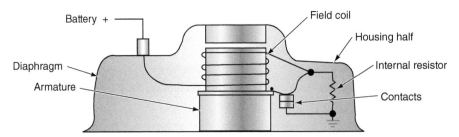

Figure 13-1 Horn construction. The internal resistor allows a weak magnetic field to remain after the contacts open, reducing the amount of time required to rebuild the field when the contacts close again.

Figure 13-2 Horn design determines sound quality.

Figure 13-3 Horn pitch adjustment screw.

than the other. The design and shape of the horn determines the frequency and tone of the sound (**Figure 13-2**). Pitch is controlled by the number of times the diaphragm vibrates per second. The faster the vibration, the higher the pitch. The horn pitch can be adjusted by changing the spring tension of the armature. This alters the magnetic pull on the armature and changes the rate of vibrations. The adjustment is made on the outside of the horn (**Figure 13-3**).

Horn Switches

Horn switches are installed either in the steering wheel or as a part of the multifunction switch. Most horn switches are normally open switches.

The steering wheel–mounted horn switch can be a single button in the middle of the steering wheel. Another design is to have multiple buttons in the horn pad. Switches mounted on the steering wheel require the use of a slip ring (**Figure 13-4**). The slip ring has contacts that provide continuity for the horn control in all steering wheel positions. The contacts consist of a circular contact in the steering wheel that slides against a

Shop Manual
Chapter 13, page 632

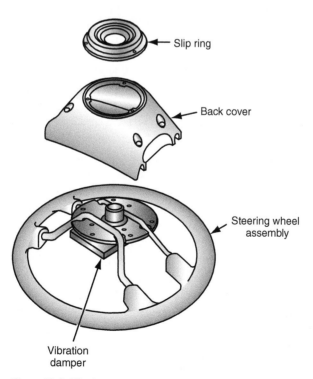

Slip ring

Back cover

Steering wheel assembly

Vibration damper

Figure 13-4 Slip ring contact provides horn continuity in all steering wheel positions.

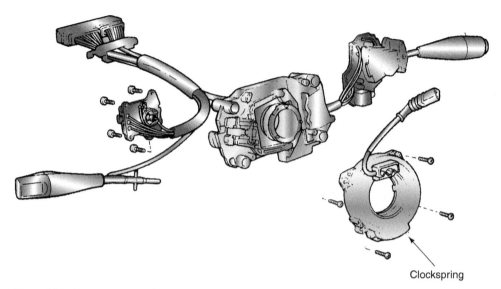

Clockspring

Figure 13-5 Most vehicle manufacturers now use a clockspring instead of sliding contacts.

spring-loaded contact in the steering column. Vehicles equipped with a steering wheel–mounted air bag use a **clockspring** (**Figure 13-5**) to provide continuity between the steering wheel components—horn switch, cruise control switches, air bag, and so on—and the steering column wiring harness. A clockspring is a winding of electrical conducting tape enclosed within a plastic housing. The clockspring maintains continuity between the steering wheel switches, the air bag, and the wiring harness in all steering wheel positions. The clockspring provides a more reliable connection than the sliding contacts.

Horn switches that are a part of the multifunction switch usually operate by a push button on the end of the lever.

HORN CIRCUITS

Shop Manual
Chapter 13,
pages 632, 635

There are two methods of circuit control: with or without a relay. If the horn circuit does not use a relay, the horns must be of low-current design because the horn switch carries the total current. Depressing the horn switch completes the circuit from the battery to the horns (**Figure 13-6**).

The most common type of circuit control is to use a relay (**Figure 13-7**). Most circuits have battery voltage present at the lower contact plate of the horn switch. When the switch is depressed, the contacts close and complete the circuit to ground. Only low current is required to operate the relay coil, so the horn switch does not have to carry the heavy current requirements of the horns. When the horn switch is closed, it energizes the relay core. The core attracts the relay armature, which closes the contacts and completes the horn circuit. Current flows from the battery to the grounded horns.

Shop Manual
Chapter 13, page 635

Many vehicle manufacturers now a module (or series of modules) to operate the horn. This reduces wiring and utilizes modules that can perform multiple functions. As a benefit to the use of modules, many are capable of performing diagnostic routines and setting diagnostic trouble codes to assist the technician in diagnostics. Consider the system illustrated in **Figure 13-8**. In this system, the steering wheel switches are used for radio and navigational system controls. The left steering switch is hard wired to the right switch. However, the right switch is more than just a switch. It is a slave microprocessor that puts inputs it receives onto the LIN bus to the master LIN module (steering column module). The steering column module then puts the message onto the CAN B bus network so other modules receive the input request made by the operator using the steering wheel switches.

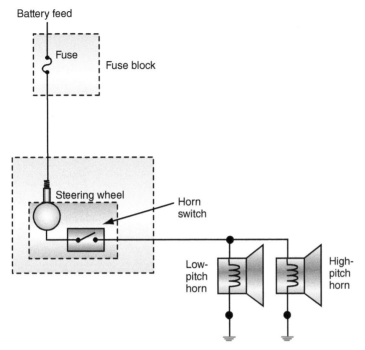

Figure 13-6 Insulated side switch without relay.

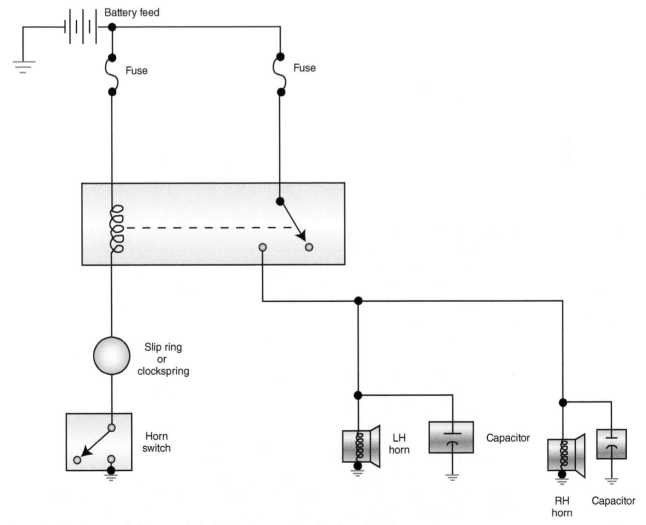

Figure 13-7 Relay-controlled horn circuit. The horn button completes the relay coil circuit.

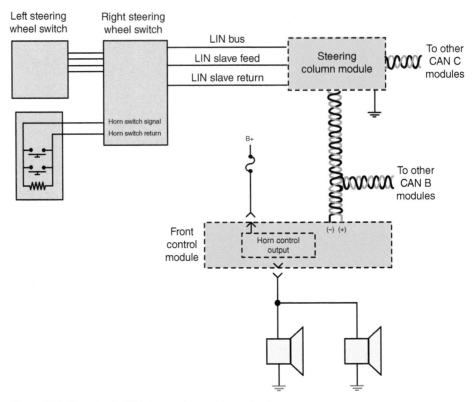

Figure 13-8 Horn circuit utilizing computers and bus networks.

The right steering wheel switch (LIN slave module) monitors the horn switch input. If the vehicle operator presses the horn switch, the right steering wheel switch sees a change in voltage on the signal circuit. It then broadcasts the horn request signal to the steering column module. The steering column module in turn broadcasts the request on the CAN B bus network to the front control module. The front control module then activates a high-side driver (HSD) to supply current to the horns. As you can see from this example, this system uses three microprocessors and two bus networks to operate the horns.

Shop Manual
Chapter 13, page 636

WINDSHIELD WIPERS

The windshield wiper system typically provides a two- or three-speed wiper system with an intermittent wipe feature. The wiper system may also provide a **depressed-park** feature. In systems equipped with depressed park, the wiper blades drop down below the lower windshield molding to hide them.

A single-speed rear window wiper and washer are used on many vehicles. In addition, headlight wipers are installed on many luxury vehicles that operate in union with the windshield wipers (**Figure 13-9**). The operation of these accessories is the same as the windshield wipers.

 A BIT OF HISTORY

Windshield wipers were introduced at the 1916 National Auto Show by several manufacturers. These wipers were hand operated. Vacuum-operated wipers were standard equipment by 1923. Electric wipers became common in the 1950s.

Figure 13-9 Headlight wipers.

Permanent Magnet Wiper Motors

Windshield wipers are mechanical arms that sweep back and forth across the windshield to remove water, snow, or dirt. Most windshield wiper motors use permanent magnet fields. Motor speed is controlled by the placement of the brushes on the commutator. Three brushes are used: common, high speed, and low speed. The common brush carries current whenever the motor operates. The most common placement of the low- and high-speed brushes is with the low-speed and common brushes opposite each other, and the high-speed brush offset or centered between them (**Figure 13-10**). Many wiper circuits use a circuit breaker to prevent temporary overloads from totally disabling the windshield wipers, such as would result if a fuse was to blow.

The placement of the brushes determines how many armature windings are connected in the circuit. When the wiper control switch is in the HIGH-SPEED position, battery voltage is supplied through wiper 1 to the high-speed brush (**Figure 13-11**). Wiper 2 moves with wiper 1 but does not complete any circuits. Current flows through the high-speed brush, through the armature, to the common brush, and to ground. There are fewer armature windings connected between the common and high-speed brushes. When battery voltage is applied to fewer windings, there is less magnetism in the armature and a lower counter electromotive force (CEMF). With less CEMF in the armature, there is greater armature current. The greater armature current results in higher speeds. Because the ground connection is after the park switch, the park switch position has no effect on motor operation.

When the switch is placed in the LOW-SPEED position, battery voltage is supplied through wiper 1 to the low-speed brush (**Figure 13-12**). Wiper 2 also moves to the LOW position, but does not complete any circuits. Current flows through the low-speed brush, the armature, and the common brush to ground. There are more armature

Shop Manual
Chapter 13,
pages 636, 638

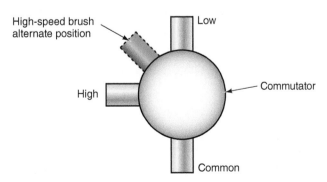

Figure 13-10 The most common brush arrangement is to place the low-speed brush opposite to the common brush.

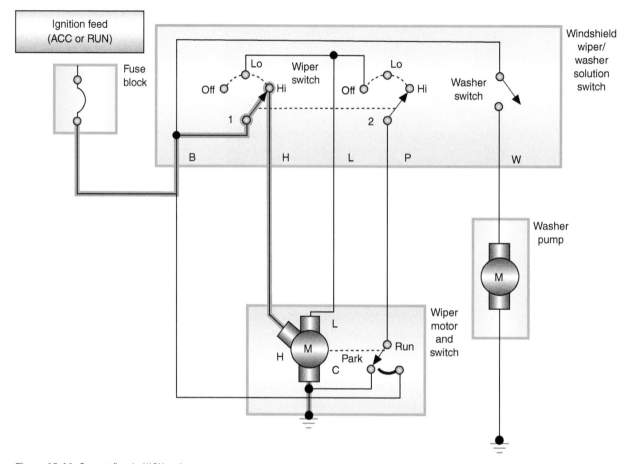

Figure 13-11 Current flow in HIGH, using a permanent magnet motor.

Shop Manual
Chapter 13, page 639

windings connected in the circuit between the common and low-speed brushes. With more windings, the magnetic field in the armature increases. This results in greater CEMF. The increased CEMF reduces the amount of current in the armature and slows the speed of the motor. Park switch position has no effect on motor operation.

A set of **park contacts** are incorporated into the motor assembly and operate off a cam or latch arm on the motor gear. Park contacts supply current to the motor after the wiper control switch has been turned to the OFF position. This allows the motor to continue operating until the wipers have reached their PARK position. The park switch changes position with each revolution of the motor. The switch remains in the RUN position for approximately 90% of the revolution. It is in the PARK position for the remaining 10% of the revolution. This does not affect the operation of the motor until the wiper control switch is placed in the OFF position. When the switch is returned to the OFF position, wiper 1 opens (**Figure 13-13**). Battery voltage is applied to the park switch. Wiper 2 is closed to allow current to flow to the low-speed brush, through the armature, and to ground. When the wiper blades are in their lowest position, the park switch is moved to the PARK position. This opens the circuit to the low-speed brush and the motor shuts off.

ELECTROMAGNETIC FIELD WIPER MOTOR CIRCUITS

Shop Manual
Chapter 13, page 636

Electromagnetic field motors use two fields that are wired in opposite directions, so their magnetic fields will oppose each other (**Figure 13-14**). The series field is wired in series with the motor brushes and commutator. The shunt field forms a separate circuit branch

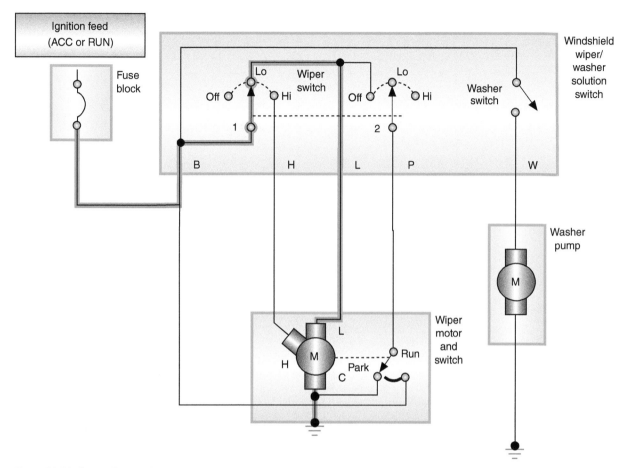

Figure 13-12 Current flow in LOW position.

off the series field to ground. The strength of the total magnetic field will determine the speed of the motor. Resistors are used in series with one of the fields to control the strength of the total magnetic field. The wiper control switch directs the current flow through the resistors to obtain the desired motor speed. Circuit operation varies between two- and three-speed systems.

Two-Speed Motors. The ground side switch will determine the current path. One path is directly to ground after the field coil; the other is through a 20-ohm (Ω) resistor (refer to Figure 13-14).

With the switch in the OFF position, switch wiper 1 breaks the circuit through the relay to ground. Because the relay is not energized, current is not supplied to the motor.

When the switch is placed in the LOW-SPEED position, wiper 1 completes the relay coil circuit to ground. The energized relay closes the contacts and applies current to the motor, through the series field and shunt field coils. Wiper 2 provides the ground path for the shunt field coil. Because there is less resistance to ground through wiper 2, the 20-ohm resistor is bypassed. With no resistance in the shunt field coil, the shunt field is very strong and bucks the magnetic field of the series field. The result is slow motor operation.

When the switch is located in the HIGH-SPEED position, wiper 1 completes the relay coil circuit to ground. This closes the relay contacts to the series field and shunt field coils. Wiper 2 opens the circuit to ground, and current must now pass through the 20-ohm resistor to ground. The resistor reduces the current flow and strength of the shunt field. With less resistance from the shunt field, the series field is able to turn the motor at an increased speed.

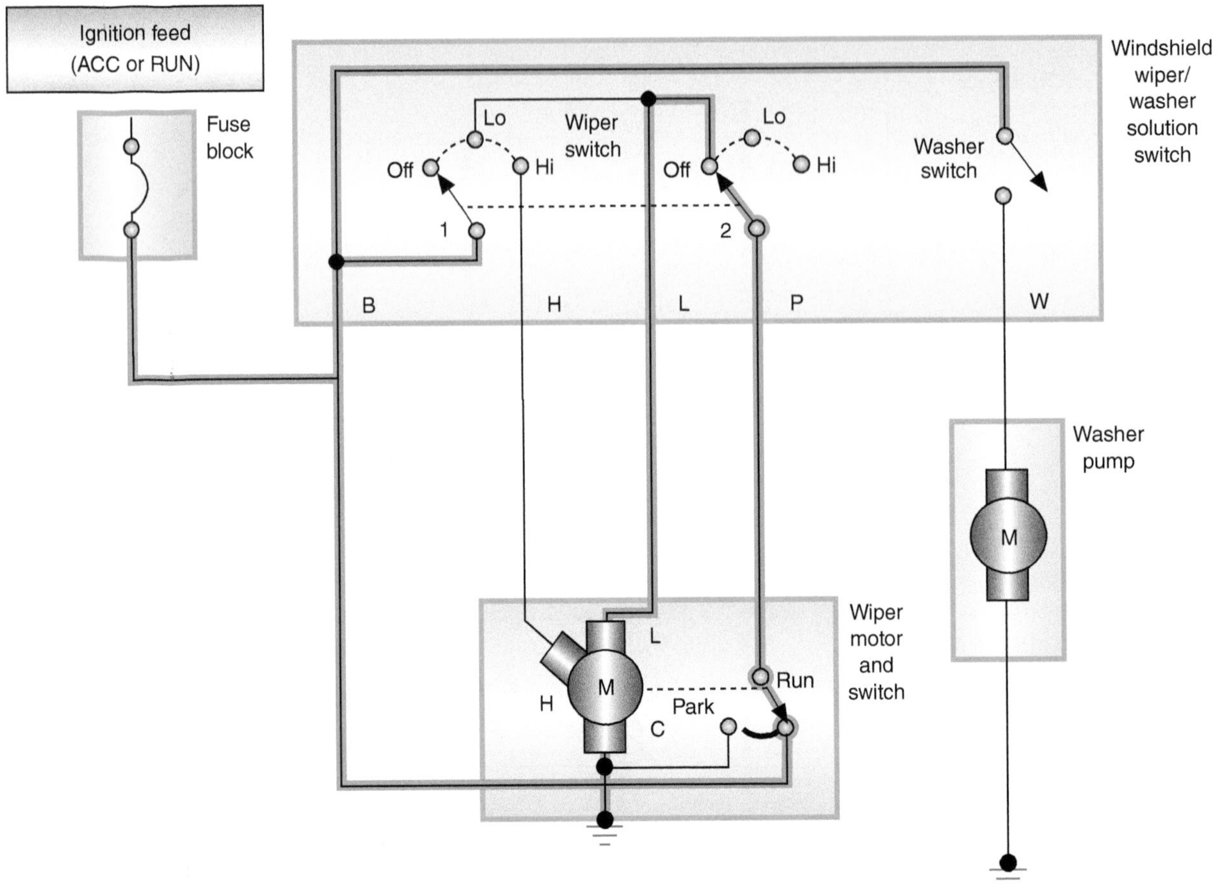

Ignition feed
(ACC or RUN)

Fuse
block

Windshield
wiper/
washer
solution
switch

Lo
Off   Hi   Wiper
switch

Lo
Off   Hi   Washer
switch

1   2

B   H   L   P   W

Washer
pump

M

L
Wiper
motor
and
switch

H   M   Run

C   Park

Figure 13-13 Current flow when the wipers are parking.

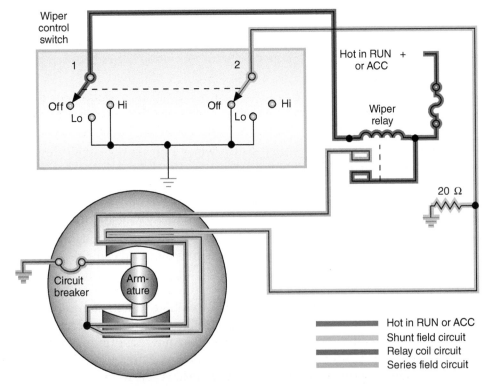

Wiper
control
switch

1   2

Hot in RUN   +
or ACC

Off   Hi   Off   Hi   Wiper
Lo   Lo   relay

20 Ω

Circuit
breaker   Arm-
ature

Hot in RUN or ACC
Shunt field circuit
Relay coil circuit
Series field circuit

Figure 13-14 Simplified diagram of an electromagnetic field, two-speed wiper system.

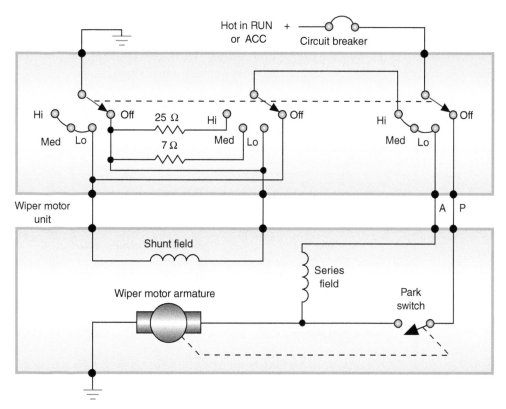

Figure 13-15 Three-speed wiper motor schematic.

When the switch is returned to the OFF position, wiper 1 opens the relay coil circuit. The relay is de-energized, but the contact points to the series and shunt field coils are manually held closed by the park switch. The park switch is closed until the wipers are in their lowest position. Wiper 2 closes the circuit path to ground. As long as the park switch is closed, current flows through the series field, shunt field, and wiper 2 to ground. Once the wipers are in their lowest point of travel, the park switch opens and the motor turns off.

Three-Speed Motors. The three-speed motor system offers a low-, medium-, and high-speed selection. The wiper control switch position determines what resistors, if any, will be connected to the circuit of one of the fields (**Figure 13-15**).

When the wiper control switch is placed in the LOW-SPEED position, both field coils receive equal current, so the total magnetic field is weak and the motor speed is slow.

When the switch is placed in the MEDIUM-SPEED position, the current flows through a resistor before flowing to the shunt field (**Figure 13-16**).The resistor weakens the strength of the shunt coil but strengthens the total magnetic field of the motor. The speed is increased over that of the LOW-SPEED position.

When the switch is placed in the HIGH-SPEED position, a resistor of greater value is connected into the shunt field circuit. The resistor weakens the magnetic field of the shunt coil, allowing a stronger total field to rotate the motor at a high speed.

Relay-Controlled, Two-Speed Wiper Systems

To reduce the size of the wires through the steering column harness, a wiper system may use relays to supply current to the wiper motor (**Figure 13-17**). With the ignition switch in the RUN or ACC position, battery voltage is supplied to the wiper motor and to the coils of the ON/OFF and HI/LO relays. If the wiper switch is placed in the LOW-SPEED position, the coil of the ON/OFF relay is energized. The contacts of the ON/OFF relay now provide ground for the wiper motor. Since the HI/LO relay coil is not energized, current flows through the motor's low-speed brush to ground.

Shop Manual
Chapter 13, page 636

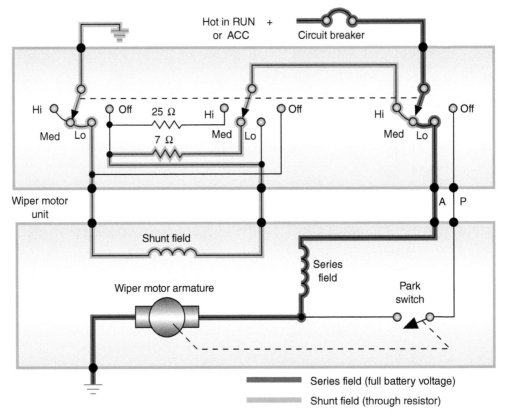

Figure 13-16 Current flow in MEDIUM position.

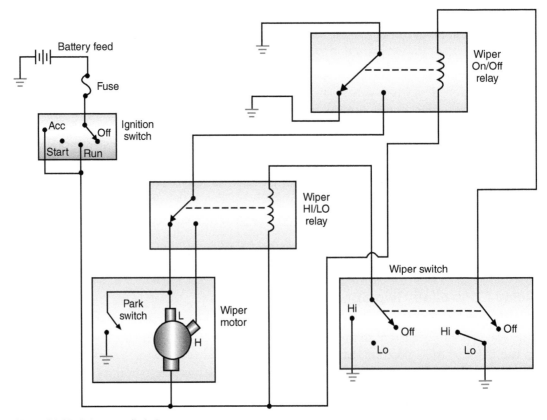

Figure 13-17 Relay-controlled wiper system.

When the wiper switch is placed in the HIGH-SPEED position, the switch provides ground for both the wiper ON/OFF and the wiper HI/LO relay coils. The wiper HI/LO relay is a circuit diverter and now provides ground for the wiper motor's high-speed brush through the contacts of the wiper ON/OFF relay (which remains energized).

Intermittent Wipers

Many wiper systems offer an intermittent mode that provides a variable interval between wiper sweeps. Many of these systems use a module located in the steering column. If the intermittent wiper mode is initiated when the wipers are in their parked position, the park switch is in the ground position. Current is sent to the solid-state module to the "timer activate" terminal. The internal timer unit triggers the electronic switch to close the circuit for the governor relay, which then closes the circuit to the low-speed brush (**Figure 13-18**). The wiper will operate until the park switch swings back to the PARK position.

The delay between wiper sweeps is determined by the amount of resistance the driver puts into the potentiometer control. By rotating the intermittent control knob, the resistance value is altered. The module contains a capacitor that is charged through the potentiometer. Once the capacitor is saturated, the electronic switch is triggered to

Shop Manual
Chapter 13, page 642

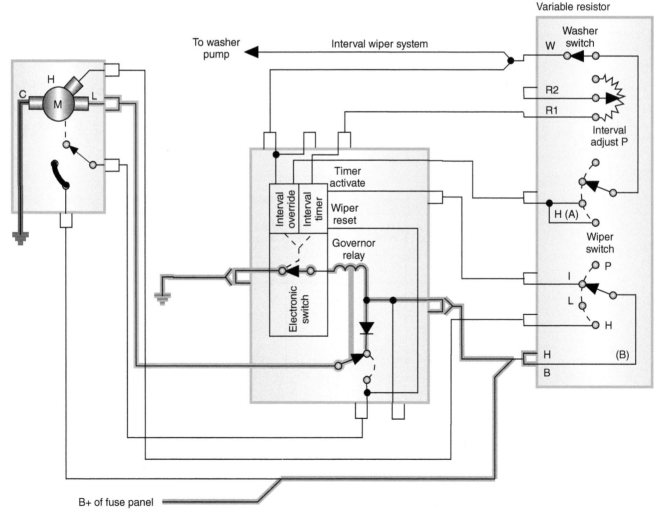

Figure 13-18 Current flow when intermittent wiper mode is initiated.

send current to the wiper motor. The capacitor discharge is long enough to start the wiper operation, and the park switch is returned to the RUN position. The wiper will continue to run until one sweep is completed and the park switch opens. The amount of time between sweeps is based on the length of time required to saturate the capacitor. As more resistance is added to the potentiometer, it takes longer to saturate the capacitor.

AUTHOR'S NOTE Many manufacturers are incorporating this function into the body computer. Also, some manufacturers are equipping their vehicles with speed-sensitive wiper systems. The delay between wiper sweeps is determined by the speed of the vehicle.

Depressed-Park Wiper Systems

Systems that have a depressed-park feature use a second set of contacts with the park switch. These contacts are used to reverse the rotation of the motor for about 15° after the wipers have reached the normal PARK position. The circuitry of the depressed circuit is different from that of standard wiper motors.

The operation of a depressed-park wiper system in the LOW-SPEED position is shown in **Figure 13-19**. Current flows through the number 3 wiper to the common brush. Ground is provided through the low-speed brush and switch wiper 2.

When the switch is placed in the OFF position, current is supplied through the park switch wiper B and switch wiper 3. Ground is supplied through the low-speed brush and switch wiper 1, and then to park switch wiper A.

When the wipers reach their PARK position, the park switch swings to the PARKING position (**Figure 13-20**). Current flow is through the park switch wiper A, to switch wiper 1. Wiper 1 directs the current to the low-speed brush. The ground path is through

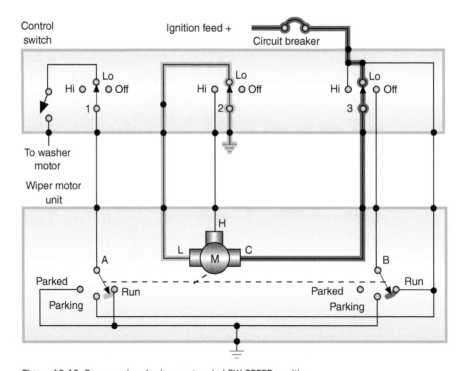

Figure 13-19 Depressed-park wiper system in LOW-SPEED position.

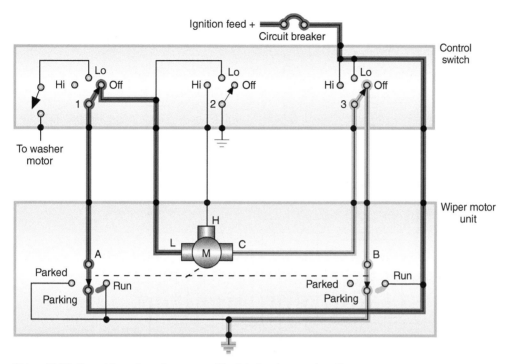

Figure 13-20 Current flow when wipers are parking into the depressed position.

the common brush, switch wiper 3, and park switch wiper B. This reversed current flow is continued until the wipers reach the depressed-park position, when park switch wiper A swings to the PARKED position.

AUTHOR'S NOTE Note the difference in operation between the depressed-park wiper system and the one shown in Figure 13-11.

COMPUTER-OPERATED WIPERS

For much the same reasons as going to computer control of the horn system, manufacturers are also using computers to operate the wiper system (**Figure 13-21**). In this system, the multifunction switch provides the driver's request for front wiper operation using a resistive multiplexed signal. The steering column module then broadcasts the request over the CAN B bus network to the front control module. This front control module operates the relays by low-side drivers.

The wiper motor is usually a permanent magnet motor. The relays are controlled to provide current to the appropriate brush of the motor. When the LOW-SPEED position of the multifunction switch is selected, the steering column module broadcasts a wiper switch LOW message to the front control module. The front control module energizes the wiper ON/OFF relay. This directs battery current through the closed contacts of the energized wiper ON/OFF relay and the normally closed contacts of the de-energized wiper HI/LO relay to the low-speed brush of the wiper motor, causing the wipers to cycle at low speed.

When the HIGH-SPEED position is selected, the steering column module broadcasts a wiper switch HIGH message to the front control module. The front control module energizes both the wiper ON/OFF relay and the wiper HI/LO relay. This directs battery current through the closed contacts of the energized wiper ON/OFF relay and the closed

Shop Manual
Chapter 13, page 644

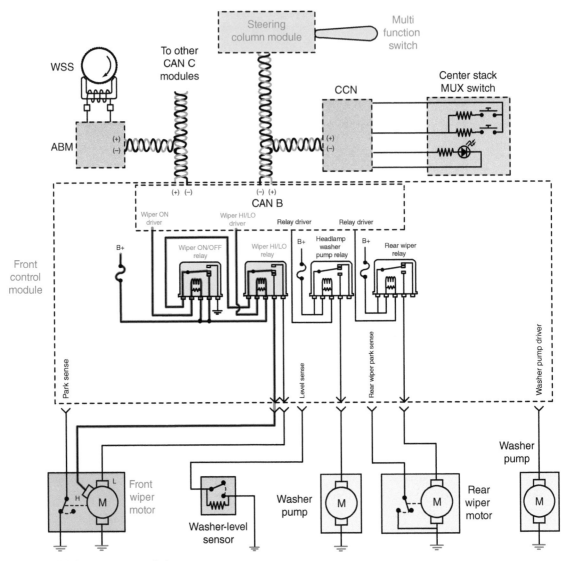

Figure 13-21 Computer-operated wiper system.

contacts of the energized wiper HI/LO relay to the high-speed brush of the wiper motor, causing the wipers to cycle at high speed.

When the OFF position of the multifunction switch is selected, the steering column module sends an electronic wiper switch OFF message to the front control module. Notice that the park switch does not operate the motor; instead it is used as an input. If the wipers are not in the PARK position on the windshield when the OFF request is broadcast, the front control module keeps the wiper ON/OFF relay energized, until the wiper blades are in the PARK position as indicated by the park switch input. If the wiper motor operates at high speed when the OFF request is broadcasted, the front control module de-energizes the wiper HI/LO relay, causing the wiper motor to return to low-speed operation before parking the wipers.

The computer-controlled wiper system also provides for intermittent wiper operation. When the multifunction switch is moved to one of the intermittent interval positions, the steering column module broadcasts the delay message to the front control module. The front control module uses an intermittent wipe logic circuit that calculates the correct length of time between wiper sweeps based upon the selected delay interval input. The front control module monitors the state of the wiper motor park switch to determine the proper intervals at which to energize and de-energize the wiper ON/OFF relay to operate the wiper motor intermittently for one low-speed cycle at a time.

The front control module can also provide vehicle speed sensitivity to the selected intermittent wipe delay intervals. The front control module monitors vehicle speed messages and doubles the selected delay interval whenever the vehicle speed is less than 10 mph (16 km/h).

With computer-controlled wiper operation, the driver can select to have the headlights turn on automatically whenever the wipers complete a minimum of five automatic wipe cycles within about 60 seconds. This meets the legal requirements in some states which stipulate that the headlights must be turned on whenever the wipers are in use. The headlights will also turn off automatically when the wipers are turned off and 4 minutes elapses without any wipe cycles.

INTELLIGENT WINDSHIELD WIPERS

To avoid making the driver select the correct speed of the windshield wipers according to the amount of rain, manufacturers have developed intelligent wiper systems. Two intelligent wiper systems will be discussed here: one senses the amount of rainfall and the other adjusts wiper speed according to vehicle speed.

Shop Manual
Chapter 13, page 647

The automatic wiper system selects the wiper speed needed to keep the windshield clear by sensing the presence and amount of rain on the windshield. The system relies on a series of LEDs that shine at an angle onto the inside of the windshield glass and an equal number of light collectors (**Figure 13-22**). The outer surface of a dry windshield will reflect the lights from the LEDs back into a series of collectors. The presence of water on the windshield will refract some of the light away from the collectors (**Figure 13-23**).

Figure 13-22 The rain sensor is mounted to the rearview mirror.

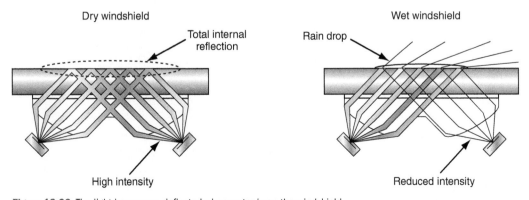

Figure 13-23 The light beams are deflected when water is on the windshield.

When this happens, the wipers are turned on. If the water is not cleared by one complete travel of the wipers, the wipers operate again. The frequency and speed of wiper operation are determined by the amount of water sensed on the windshield.

Speed-sensitive wipers do not require additional components to operate since most use body control module (BCM). Speed-sensitive wipers compensate for extra moisture that normally accumulates on the windshield at higher speeds in the rain. At higher speeds, the delay between wipers shortens when the wipers are operating in the interval mode. This delay is automatically adjusted at speeds between 10 and 65 mph. Basically, this system functions according to the input the computer receives about vehicle speed.

WASHER PUMPS

Shop Manual
Chapter 13, page 649

Windshield washers spray a washer fluid solution onto the windshield and work in conjunction with the wiper blades to clean the windshield of dirt. Some vehicles that have composite headlights incorporate a headlight washing system along with the windshield washer (**Figure 13-24**). Most systems have the washer pump motor installed into the reservoir (**Figure 13-25**). General Motors uses a pulse-type washer pump that operates off the wiper motor (**Figure 13-26**).

Figure 13-24 Headlight washer system may operate with the windshield washer or have a separate switch.

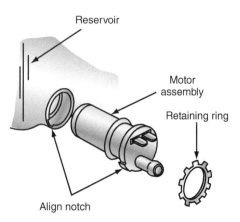

Figure 13-25 Washer motor installed into the reservoir.

Figure 13-26 General Motors' pulse-type washer system incorporates the washer motor into the wiper motor.

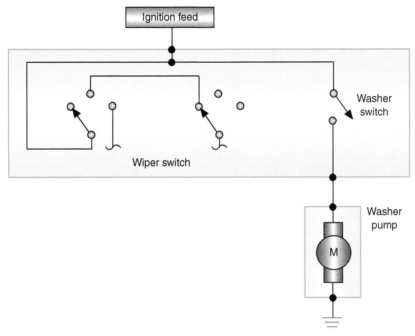

Figure 13-27 Windshield washer motor circuit.

The windshield washer system is activated by holding the washer switch (**Figure 13-27**). If the wiper/washer system also has an intermittent control module, a signal is sent to the module when the washer switch is activated (**Figure 13-28**). An override circuit in the module operates the wipers on low speed for a programmed length of time. The wipers will either return to the parked position or operate in intermittent mode, depending on system design.

The washer system includes plastic or rubber hoses to direct fluid flow to the nozzles and produce the spray pattern.

Computer-Operated Washer Systems

Referring to Figure 13-21, when the WASH request is made by depressing the control knob on the control stalk of the multifunction switch, the steering column module broadcasts a washer switch message to the front control module over the CAN B bus. The front control module then uses an HSD circuit that directs current to the washer pump. At the same time, the wipers are turned on and operate for about three wipes.

AUTHOR'S NOTE Some systems may use an H-gate driver circuit to operate the washer pump. The H-gate driver is used if the vehicle is also equipped with a rear window wiper system. This system uses only one washer pump. Based on the direction of current flow through the pump, the washer fluid will be directed to either the front windshield or the rear window.

If wash is requested while the wipers are already turned on and operating in one of the intermittent interval positions, the washer pump operation is the same. However, during this time, the front control module will abort the delay feature and will energize the wiper ON/OFF relay to operate the wiper motor in a continuous low-speed mode for as long as the WASH switch is closed. When the WASH request is no longer present, the front control module will resume the selected delay mode interval.

The headlamp washer system uses a separate high-pressure pump that is activated when the headlamps are turned on and the windshield washer switch is closed. The high-pressure pump will direct two-timed, high-pressure sprays onto the headlamp lens.

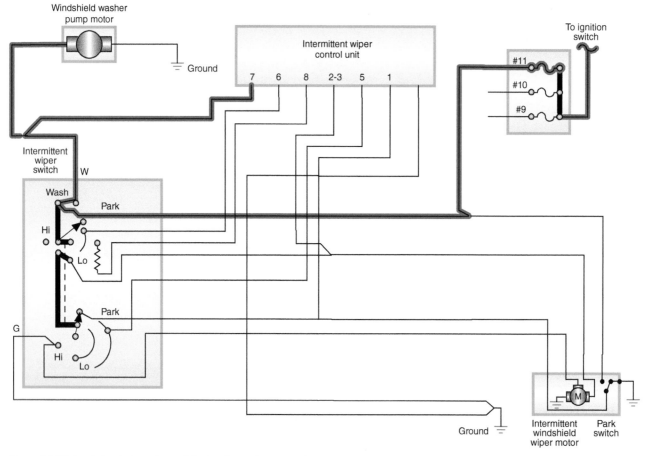

Figure 13-28 Input signal alerts the module that the washers are activated.

BLOWER MOTOR CIRCUITS

Shop Manual
Chapter 13, page 652

The blower motor is used to move air inside the vehicle for air conditioning, heating, defrosting, and ventilation. The motor is usually a permanent magnet, single-speed motor and is located in the heater housing assembly (**Figure 13-29**). A blower motor switch mounted on the dash controls the fan speed. The switch position directs current flow to

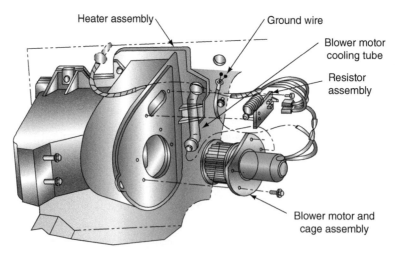

Figure 13-29 The blower motor is usually installed in the heater assembly. Mode doors control if vent, heater, or A/C-cooled air is blown by the motor cage.

a **resistor block** that is wired in series between the switch and the motor (**Figure 13-30**) and consists of two or three helically wound wire resistors wired in series.

The blower motor circuit includes the control assembly, blower switch, resistor block, and blower motor (**Figure 13-31**). This system uses an insulated side switch and a grounded motor. Battery voltage is applied to the control head when the ignition switch is in the RUN or ACC position. The current can flow from the control head to the blower switch and resistor block in any control head position except OFF.

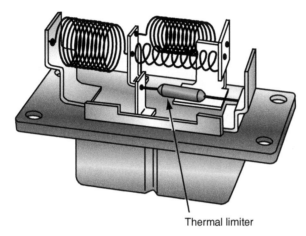

Thermal limiter

Figure 13-30 Fan motor resistor block.

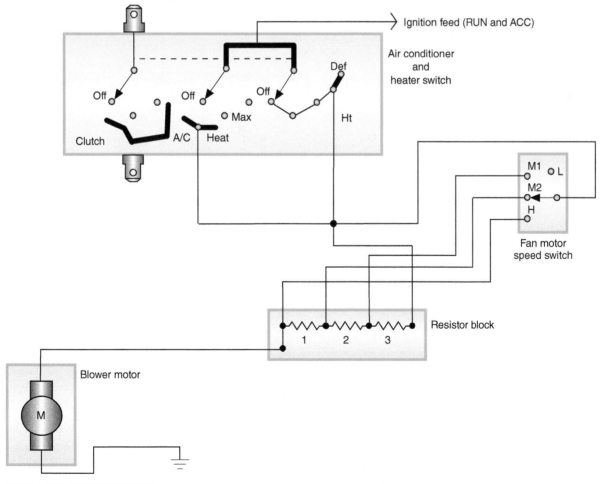

Figure 13-31 Blower motor circuit.

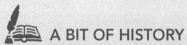

A BIT OF HISTORY

Automotive electric heaters were introduced at the 1917 National Auto Show. Hot water in-car heaters were introduced in 1926.

When the blower switch is in the LOW position, the blower switch wiper opens the circuit. Current can flow only to the resistor block directly through the control head. The current must pass through all the resistors before reaching the motor. When the voltage drops over the resistors, the motor speed slows (**Figure 13-32**).

When the blower switch is placed in the MED 1, MED 2, or HIGH position, the current flows through the blower switch to the resistor block. Depending on the speed selection, the current must pass through one, two, or none of the resistors. When more voltage is applied to the motor, the fan speed increases as the amount of resistance decreases.

Some manufacturers use ground side switching with an insulated motor. The switch completes the circuit to ground. The operating principles are identical to that of the insulated switch already discussed.

Shop Manual
Chapter 13, page 654

Many of today's vehicles use the BCM to control fan speed by pulse width modulation (PWM). Typically the BCM does not directly operate the motor. This is because of the high-amperage requirements of the motor. Instead a power module (**Figure 13-33**) controls the blower motor based on drive signals from the BCM. The BCM sends a PWM signal to the power module, and the power module amplifies the signals to provide variable fan speeds (**Figure 13-34**).

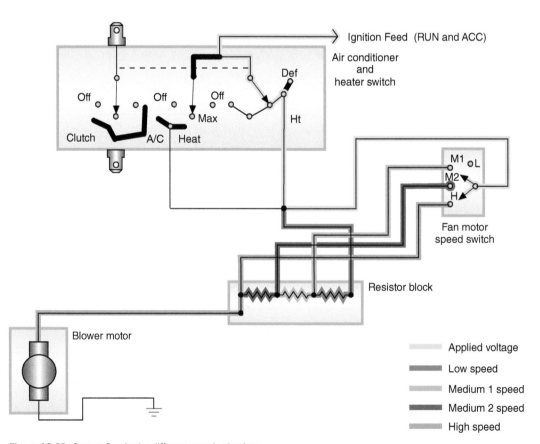

Figure 13-32 Current flow in the different speed selections.

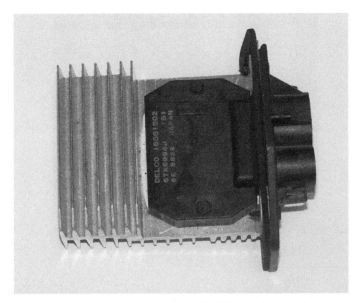

Figure 13-33 The power module amplifies the BCM signal to control motor speed.

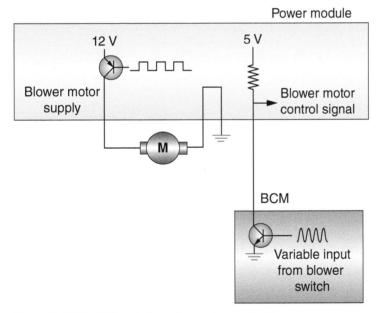

Figure 13-34 The BCM sends the motor speed request to the power module.

ELECTRIC DEFOGGERS

When electrons are forced to flow through a resistance, heat is generated. Rear window electric defoggers use this principle of controlled resistance to heat the glass. Electric defoggers heat the rear window to remove ice and/or condensation. Some vehicles use the same circuit to heat the outside side mirrors. The resistance is through a **grid** that is baked on the inside of the glass (**Figure 13-35**). The rear window defogger grid is a series of horizontal, ceramic silver–compounded lines. The terminals are soldered to the vertical bus bars. One terminal supplies the current from the switch; the other provides the ground (**Figure 13-36**).

Shop Manual
Chapter 13, page 655

Figure 13-35 Rear window defogger grid.

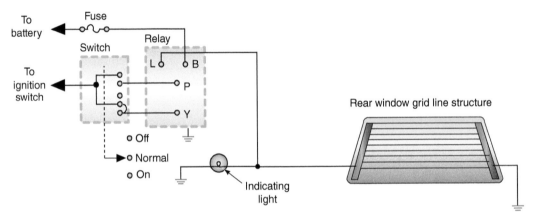

Figure 13-36 Rear window defogger circuit schematic.

Most systems incorporate a timer circuit to control the relay (**Figure 13-37**). The timer is used due to the high amount of current required to operate the system [approximately 30 amps (A)]. If this drain were allowed to continue for extended periods of time, battery and charging system failure could result. Because of the high-current draw, most vehicles equipped with a rear window defogger use a high-output AC generator.

The control switch may be a three-position, spring-loaded switch that returns to the center position after making momentary contact with the ON or OFF terminals. Activation of the switch energizes the electronic timing circuit, which energizes the relay coil. With the relay contacts closed, direct battery voltage is sent to the heater grid. At the same time, voltage is applied to the ON indicator. The timer is activated for 10 minutes. At the completion of the timed cycle, the relay is de-energized and the circuit to the grid and indicator light is broken. If the switch is activated again, the timer will energize the relay for 5 minutes.

The timer sequence can be aborted by moving the switch to the OFF position or by turning off the ignition switch. If the ignition switch is turned off while the timer circuit is activated, the rear window defogger switch will have to be returned to the ON position to activate the system again.

Ambient temperatures have an effect on electrical resistance; thus, the amount of current flow through the grid depends on the temperature of the grid. As the ambient temperature decreases, the resistance value of the grid also decreases. A decrease in resistance increases the current flow and results in quick warming of the window. The defogger system tends to be self-regulated to match the requirements for defogging.

The ON indicator can be either a bulb or a light-emitting diode (LED).

Many manufacturers refer to the electric rear window defogger as an electric backlight.

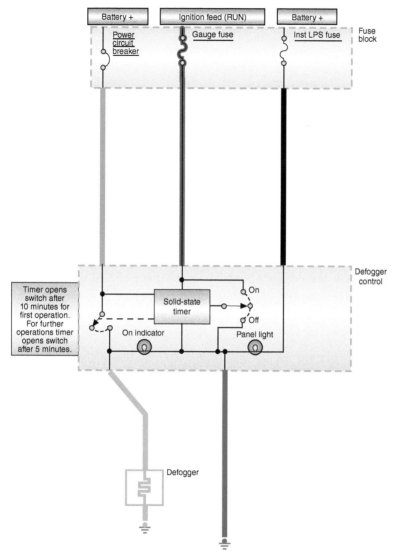

Figure 13-37 Defogger circuit using a solid-state timer.

Heated Mirrors

Outside heated mirrors are installed as an option on the vehicle to quickly remove snow, ice, or moisture from the mirror glass. Heated mirrors operate on the same principle as the rear electric defogger. Activation of the heated mirror is usually tied in with the rear electric defogger and when the electric defoggers are turned on the heated mirrors are also on. Some heated mirrors may come on automatically when the ignition is placed in the RUN position, while others are controlled by the BCM based on ambient temperature.

A heating grid is glued onto the back of the mirror and when activated will generate heat as current flows through it. The heat is transferred from the metallic backing of the mirror to the glass. Like the electric defogger, temperature affects the resistance of the grid. When the grid is cold, the resistance decreases and results in an increase in current flow. As the grid warms, the resistance increases and current flow is reduced. The heated mirror system tends to be self-regulated to match the requirements for heating.

POWER MIRRORS

Electrically controlled power mirrors allow the driver to position the outside mirrors by use of a switch. The mirror assembly will use built-in, dual-drive, reversible permanent magnet (PM) motors (**Figure 13-38**).

A single switch for controlling both the left and right side mirrors is used. On many systems, selection of the mirror to be adjusted requires positioning a switch. After the mirror is selected, movement of the power mirror switch (up, down, left, or right) moves the mirror in the corresponding direction. **Figure 13-39** shows a logic table for the mirror switch and motors.

Automatic Rearview Mirror

Some manufacturers have developed interior rearview mirrors that automatically tilt when the intensity of light that strikes the mirror is sufficient enough to cause discomfort to the driver.

The system has two photo cells mounted in the mirror housing. One of the photo cells is used to measure the intensity of light inside the vehicle. The second is used to measure the intensity of light the mirror receives. When the intensity of the light striking the mirror is greater than that of ambient light, by a predetermined amount, a solenoid is activated that tilts the mirror.

Electrochromic Mirrors

Electrochromic mirrors automatically adjust the amount of reflectance based on the intensity of glare (**Figure 13-40**). The electrochromic mirror uses forward- and rearward-facing photo sensors and a solid-state chip. Based on light intensity differences, the chip applies a small voltage to the silicon layer. As voltage is applied, the molecules of the layer rotate and redirect the light beams. Thus, the mirror reflection appears dimmer. If the glare is heavy, the mirror darkens to about 6% reflectivity. The electrochromic mirror has the advantage that it provides a comfort zone where the mirror will provide 20% to 30% reflectivity. When no glare is present, the mirror changes to the daytime reflectivity rating of up to 85%. The reduction of the glare by darkening of the mirror does not impair visibility.

Electrochromic mirrors can be installed as the outside mirror and/or inside mirrors. The mirror is constructed of a thin layer of electrochromic material that is placed between two plates of conductive glass. There are two photo cell sensors that measure

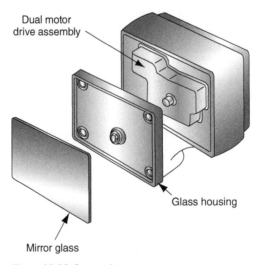

Figure 13-38 Power mirror motor.

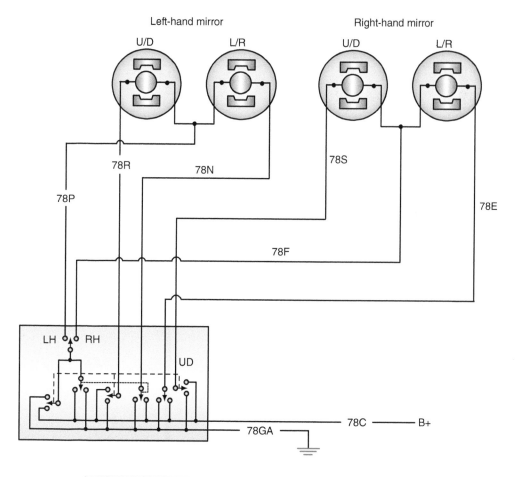

Left-hand mirror   Right-hand mirror

Switch function	Circuit function	
	Left mirror	Right mirror
Left	(78P+78C)(78N+78GA)	(78F+78C)(78E+78GA)
Right	(78N+78C)(78P+78GA)	(78E+78C) (78F+78GA)
Up	(78R+78C)(78P+78GA)	(78S+78C)(78F+78GA)
Down	(78P+78C)(78R+78GA)	(78F+78C)(78S+78GA)

Figure 13-39 Power mirror logic table.

Figure 13-40 Electrochromic mirror operation (A) day time; (B) mild glare; and (C) high glare.

light intensity in front and in back of the mirror. During night driving, the headlight beam striking the mirror causes the mirror to gradually become darker as the light intensity increases. The darker mirror absorbs the glare. Some systems allow for sensitivity of the mirror to be adjusted by the driver through a three-position switch (**Figure 13-41**).

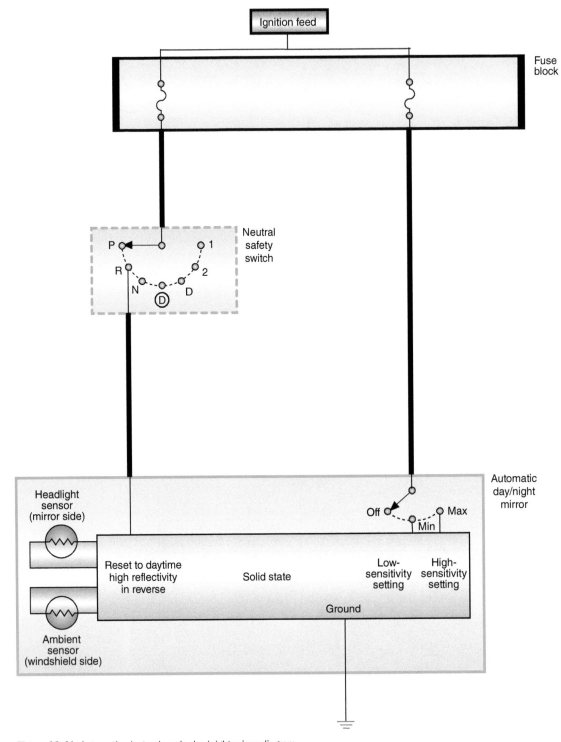

Figure 13-41 Automatic electrochromic day/night mirror diagram.

When the ignition switch is placed in the RUN position, battery voltage is applied to the three-position switch. If the switch is in the MIN position, battery voltage is applied to the solid-state unit and sets the sensitivity to a low level. The MAX setting causes the mirror to darken more at a lower glare level. When the transmission is placed in reverse, the reset circuit is activated. This returns the mirror to daytime setting for clearer viewing to back up.

The MIN position is used for city driving.

POWER WINDOWS

Most vehicle manufacturers have replaced the conventional window crank with electric motors that operate the side windows. In addition, most sport utility models are equipped with electric rear tailgate windows. The motor used in the power window system is a reversible PM or two-field winding motor.

Shop Manual
Chapter 13, page 658

The power window system usually consists of the following components:

1. Master control switch.
2. Individual control switches.
3. Individual window drive motors.
4. Lock-out or disable switch.

Another design is to use rack-and-pinion gears. The rack is a flexible strip of gear teeth with one end attached to the window (**Figure 13-42**).

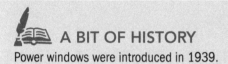

A BIT OF HISTORY
Power windows were introduced in 1939.

A **window regulator** converts the rotary motion of the motor into the vertical movement of the window. The motor operates the window regulator either through a cable or directly. On direct-drive motors, the motor pinion gear meshes with gear teeth on the regulator called the **sector gear** (**Figure 13-43**). As the window is lowered, the spiral spring is wound. The spring unwinds as the window is raised, to assist in raising the window. The spring reduces the amount of current that would be required to raise the window by the motor itself.

The master control switch provides the overall control of the system (**Figure 13-44**). Power to the individual switches is provided through the master switch. The master switch may also have a safety lock switch to prevent operation of the windows by the individual switches. When the safety switch is activated, it opens the circuit to the other switches

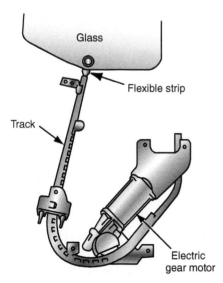

Figure 13-42 Rack-and-pinion–style power window motor and regulator.

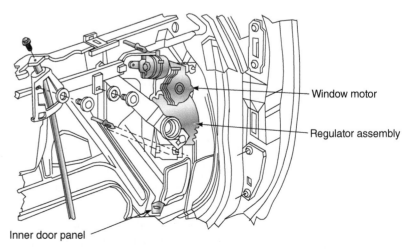

Figure 13-43 Window regulator.

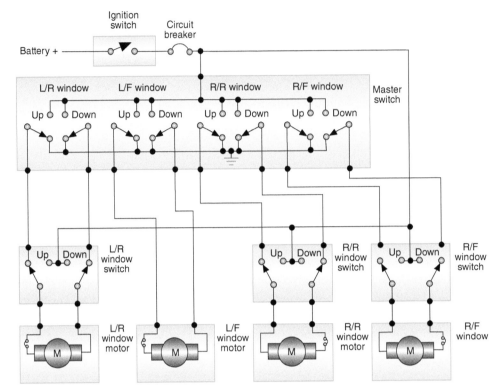

Figure 13-44 Typical power window circuit using PM motors.

The circuit breakers in the motor assemblies are used to turn the motors off once the window stalls. This prevents the window from operating if the switch is held after the window reaches its full travel.

Some motors have cam-operated limit switches that stop motor operation when the window reaches the end of its travel.

Shop Manual
Chapter 13, page 661

and control is only by the master switch. As an additional safety feature, some systems prevent operation of the individual switches unless the ignition switch is in the RUN or ACC position.

Wiring circuits depend on motor design. Most PM-type motors are insulated, with ground provided through the master switch (refer to Figure 13-44). When the master control switch is placed in the UP position, current flow is from the battery, through the master switch wiper, through the individual switch wiper, to the top brush of the motor. Ground is through the bottom brush and circuit breaker to the individual switch wiper, to the master switch wiper, and ground.

When the window is raised from the individual switch, battery voltage is supplied directly to the switch and wiper from the ignition switch. The ground path is through the master control switch.

When the window is lowered from the master control switch, the current path is reversed. In the illustration shown, current flows through the individual switch to lower the window.

Some manufacturers use a two-field coil motor that is grounded with insulated side switches. The two field coils are wired in opposite directions, and only one coil is energized at a time. Direction of the motor is determined by the coil that is activated (**Figure 13-45**).

Computer-Controlled Power Windows

Figure 13-46 illustrates a computer-controlled power window system. This system uses a driver door module (DDM) that receives the operator inputs and performs the output function. Notice that the DDM also controls many of the other electrical accessories, such as power mirrors, power door locks, and some lighting functions. The window switches are integral to the DDM and provide a direct input. When the window switch is moved to the UP or DOWN position, the module uses an H-gate circuit to provide power and

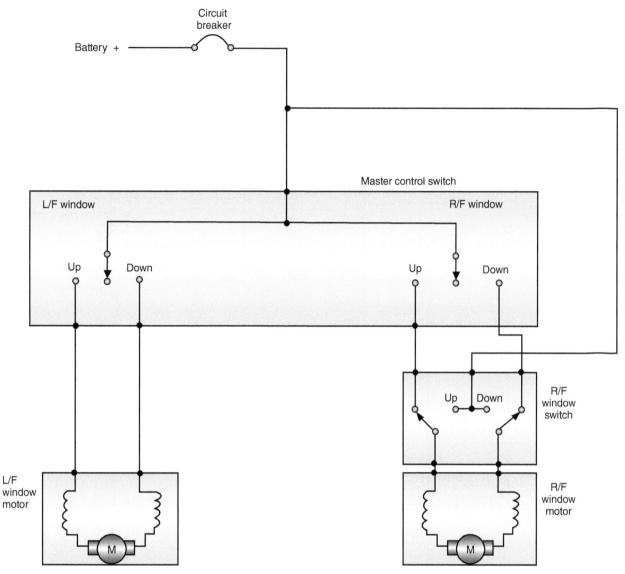

Figure 13-45 Wiring diagram of power window circuit using two-field coil motors.

ground to the window motor. When the driver operates one of the passenger side windows, the DDM sends a bussed message to the passenger door module (PDM) to operate that window motor.

In this example, the DDM operates the front and rear windows on the left side of the vehicle. The PDM controls the windows on the right side. The rear window switches are a MUX input to the door module. When a rear seat passenger operates a rear window, the door module controlling that side of the vehicle operates the window motor through an H-gate circuit.

The system can offer an **express down** feature. Express down allows the window to be lowered without having to hold the switch. When the second switch detent is actuated, the corresponding door module operates the power window to the open position until the circuit current increases. The increase in current indicates to the module that the window is at the end of its travel. An **express up** feature can also be included. The express up allows for the window glass to be raised without having to hold the switch. In this case the current is monitored to see if an obstruction is encountered. If an obstruction is detected, the window reverses direction.

Amperage or commutator pulses can be used to determine if the window has reached full travel. The control module will shut off motor operation regardless of the switch position.

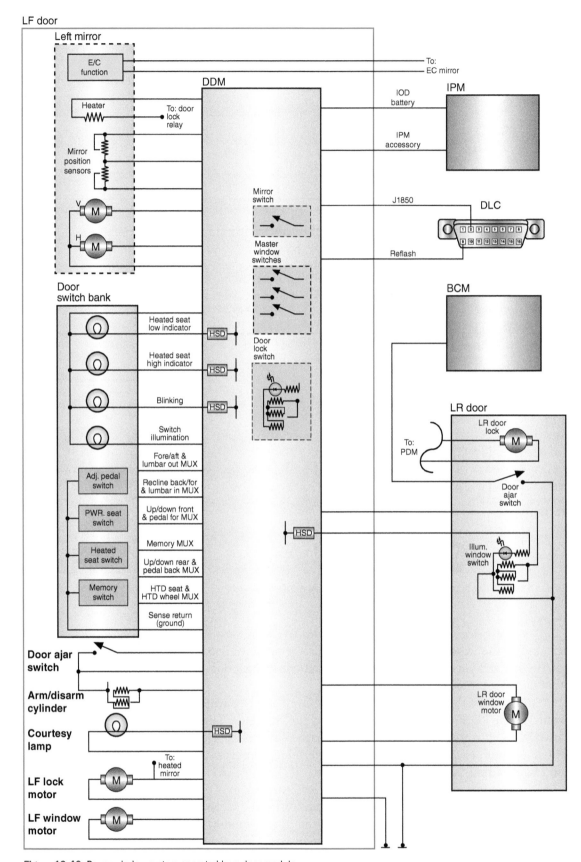

Figure 13-46 Power window system operated by a door module.

Vehicles with convertible tops may use an input from the top control switch to operate the windows. If the top is in the closed position and the switch is moved to lower it, the door module will lower the windows slightly. This releases the glass from the rubber weather strip molding to prevent the molding from being damaged when the top begins to move.

POWER SEATS

The power seat system is classified by the number of ways in which the seat moves. The most common classifications are as follows:

1. Two-way: Moves the seat forward and backward.
2. Four-way: Moves the seat forward, backward, up, and down.
3. Six-way: Moves the seat forward, backward, up, down, front tilt, and rear tilt.

Six-way power seats typically use a reversible, permanent magnet, three-armature motor called a **trimotor** (**Figure 13-47**). The motor may transfer rotation to a rack-and-pinion or to a worm gear drive transmission. A typical control switch consists of a four-position knob and a set of two-position switches (**Figure 13-48**). The four-position knob controls the forward, rearward, up, and down movements of the seat. The separate two-position switches are used to control the front tilt and rear tilt of the seat.

Shop Manual
Chapter 13, page 661

Some seat back latches use a solenoid to lock the seat unless the door is open. The solenoid is controlled by the door jamb switch.

AUTHOR'S NOTE Early General Motors' six-way power seats used a single motor. Solenoids were used to connect the motor to one of three transmissions that would move the seat.

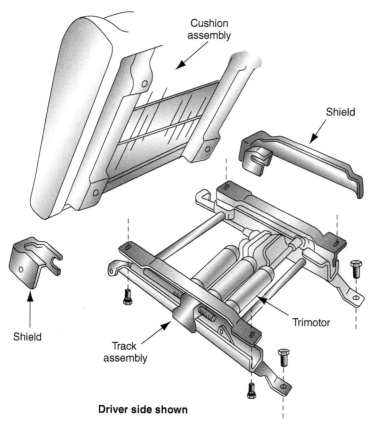

Figure 13-47 Trimotor power seat installation.

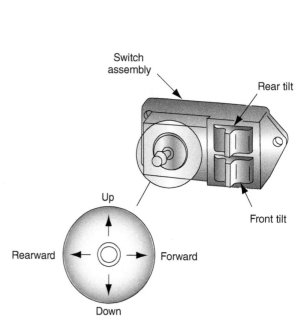

Figure 13-48 Power seat control switch.

This power seat system uses the circuit breakers in the motor assemblies as limit switches. When the end of travel is reached the higher current draw opens the circuit breaker and stops motor operation.

Current direction through the motor determines the rotation direction of the motor. The switch wipers control the direction of current flow. If the driver pushes the four-way switch into the down position, the entire seat lowers (**Figure 13-49**). Switch wipers 3 and 4 are swung to the left and battery voltage is sent through wiper 4 to wipers 6 and 8. These wipers direct the current to the front and rear height motors. The ground circuit is provided through wipers 5 and 7, to wiper 3 and ground.

Some manufacturers equip their seats with adjustable support mats that shape the seat to fit the driver (**Figure 13-50**). The lumbar support mat provides the driver with additional comfort by supporting the back curvature. Some systems use air that is pumped into the mats; others use a motor to roll the lumbar support.

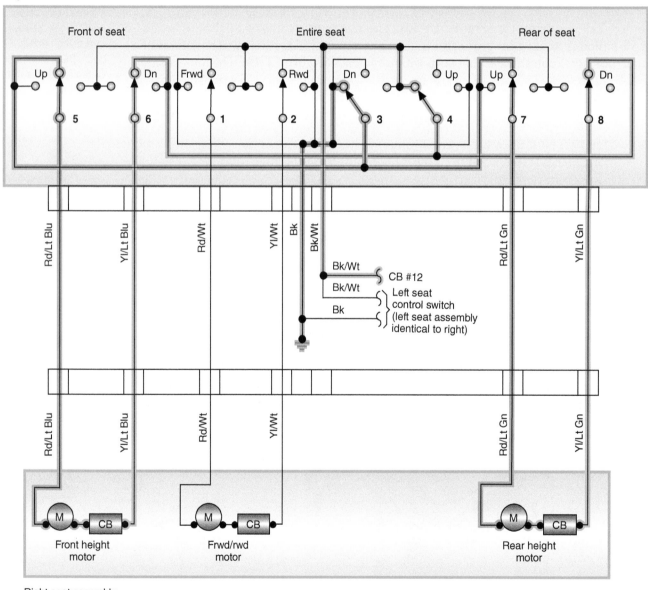

Figure 13-49 Current flow in the seat LOWER position.

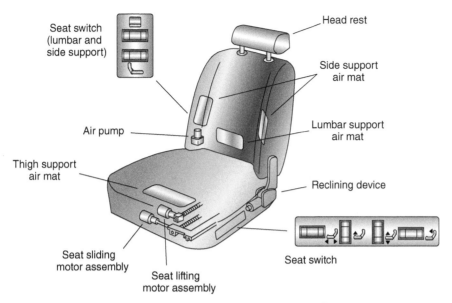

Seat switch
(lumbar and
side support)

Head rest

Side support
air mat

Air pump

Lumbar support
air mat

Thigh support
air mat

Reclining device

Seat sliding
motor assembly

Seat lifting
motor assembly

Seat switch

Figure 13-50 Adjustable seat cushions increase driver's comfort and safety.

MEMORY SEATS

The memory seat feature is an addition to the basic power seat system that allows the driver to program different seat positions that can be recalled at the push of a button. Most memory seat systems share the same basic operating principles. The difference is in programming methods and the number of positions that can be programmed.

Shop Manual
Chapter 13, page 663

The power seat system may operate in any gear position. However, the memory seat function will operate only when the transmission is in the PARK position. The purpose of the memory disable feature is to prevent accidental seat movement while the vehicle is being driven. In the PARK position, the seat memory module will receive a 12-volt (V) signal that will enable memory operation. In any other gear selection, the 12-volt signal is removed and the memory function is disabled. This signal can come from the gear selector switch or the neutral safety switch. Other systems can use a bussed data message concerning gear lever position from the transmission control module or instrument cluster.

Most systems provide for two-seat positions to be stored in memory. Some systems allow for three positions by pushing positions 1 and 2 buttons together. With the seat in the desired position, depressing the SET button and moving the memory select switch to either the memory 1 or 2 position will store the seat position into the module's memory (**Figure 13-51**).

When the seat is moved from its memory position, the seat memory module transmits the voltage applied from the switch to the motors. The module counts the pulses produced by motor operation and then stores the number of pulses and direction of movement in memory. When the memory switch is closed, the module operates the seat motors until it counts down to the preset number of pulses. Some systems use a potentiometer or a two-wire Hall sensor to monitor seat position and height instead of counting pulses.

Some systems offer an easy exit feature that is an additional function of the memory seat; the feature provides easier entrance and exit of the vehicle by moving the seat all the way back and down. Some systems also move the steering wheel up and to full retract. Some manufacturers use an easy exit switch as part of the power seat switch assembly. When the easy exit switch is closed, voltage is applied to both memory 1 and 2 inputs of the module. This signal is interrupted by the module to move the seat to its full down and

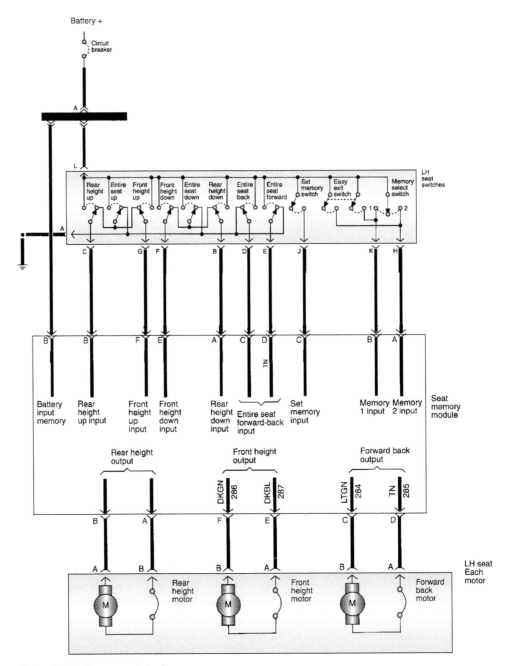

Figure 13-51 Memory seat circuit.

back positions. As the seat moves to the easy exit position, it counts the pulses and stores this information in memory. In some systems, the easy exit feature is activated when the key is removed from the ignition switch or when the driver's door is opened.

Memory is not lost when the ignition switch is turned off. However, it may be lost if the battery is disconnected. If memory is lost, the position of the seat at the time power is restored becomes set in memory for both positions.

The memory system may include many more features than just moving seats. In addition to the seat positions, some systems allow for two different groups of radio presets, two separate outside mirror positions, and two different steering wheel tilt positions to be recalled. All of these can be recalled by using the single switch and, on some systems, the remote keyless entry transmitter. Some systems even include electrically adjustable pedals as part of the memory system (**Figure 13-52**).

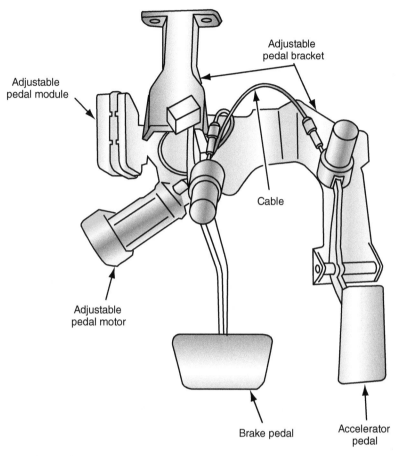

Figure 13-52 Electrically adjustable pedal assembly.

CLIMATE-CONTROLLED SEATS

Many manufacturers provide the option of **climate-controlled seats**. Climate-controlled seats provide addition comfort by heating and/or cooling the seat cushion and seat back.

    Figure 13-53 is a schematic of a heated seat system that uses heating element grids. This heated seat system uses four heated seat elements. Two elements are integral to each seat: one in the seat back and the other in the seat cushion. The heated seat module (HSM) contains the control logic and software for the heated seat system. Two heated seat switches are used, one for each heated seat. The cabin compartment node (CCN) is part of the instrument cluster on this vehicle and is the link between the heated seat switches and the HSM.

    The heated seat system operates on battery current received through a fused ignition switch output. Since ignition feed is used, the heated seat system will operate only when the ignition switch is in the RUN position. When either of the heated seat switches is depressed, a multiplexed resistance signal is received by the CCN. This requested heating level is then sent to the HSM over the data bus. The HSM controls the 12-volt output to the heating elements through the use of HSDs based on the heat level requested. The HSDs use PWM to control the current flow through the seat elements. The carbon fiber–heated seat elements consist of multiple heating circuits wired in parallel. The heated seat elements are located between the leather trim cover and the seat cushion. As electrical current passes through the heated seat element, the resistance of the wire used in the element converts electrical energy into heat energy. The heat is then radiated through

Shop Manual
Chapter 13, page 665

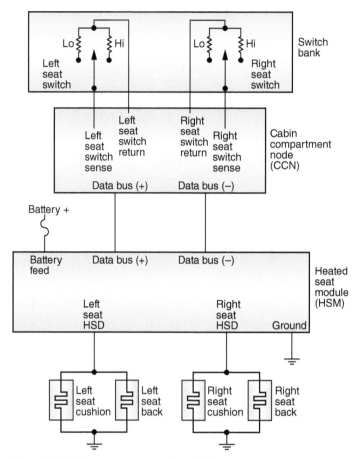

Figure 13-53 Heated seat system using grids in the cushion and seat back.

the seat cushion and seat back trim covers. Since the element grid is wired in parallel, if an open occurs in one or more of the individual carbon fiber circuits, the others will continue to operate.

When the switch selection indicates a high-temperature heating request, the HSM will provide a boosted heat level during the first 4 minutes of operation. During this time the HSM will use a 95% duty cycle. The heat output then drops to the normal high-temperature level of a 30% duty cycle. This will maintain a temperature of about 107.6°F (42°C). If high-level heating is activated, the HSM will automatically switch to the low level after 2 hours of continuous operation. The low heat level is maintained using a 15% duty cycle to hold the seat temperature to about 100.4°F (38°C). The HSM will turn off operation of the low heat level after 2 hours.

The HSDs in the HSM monitor the operation of the heater element circuits. The HSM will turn off the heating elements if it detects an open or a short in the heating element circuit.

Some systems use a thermistor to measure the seat temperature. This system will energize the seat elements at 100% duty cycle until the desired heat level is reached. Once it is reached, the HSM will stop current flow. If the temperature of the seat drops a programmed amount, the HSM will re-energize the seat elements. This process is repeated throughout the heated seat operation.

Some manufacturers use airflow to warm or cool the seat surface. A **Peltier element** operates similar to a bimetal switch. The element consists of two different types of metals, which are joined together. This joint area will generate or absorb heat when an electric current is applied to the element at a specified temperature. The Peltier element is integral to the climate controller (**Figure 13-54**). The climate controller cools or warms the air

Shop Manual
Chapter 13, page 667

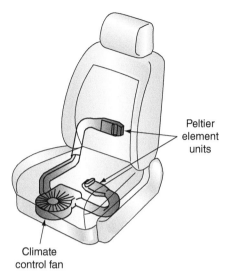

Figure 13-54 Climate-controlled seat using air heated or cooled by the Peltier element.

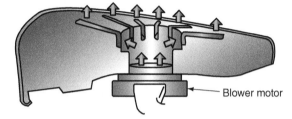

Figure 13-55 Airflow is directed through the seat cushion.

flow from the climate control fan motor based on the climate control electronic control unit (ECU) activation. The ECU output is based on the setting selection of the control switch. The control switches provide for cool air in three stages, airflow, and warm air in three stages. A temperature sensor is used to monitor the surface temperature of the seat back and cushion.

The climate control fan motor provides airflow to the seat cushion and seat back. The airflow passes through grooves in the seat pad, passes to the nonwoven cloth layer, and dissipates through the seat cover (**Figure 13-55**). The Peltier element will either heat or cool the airflow as it passes through the controller.

POWER DOOR LOCKS

Electric power locks use either a solenoid or a permanent magnet reversible motor. Due to the high-current demands of solenoids, most modern vehicles use PM motors (**Figure 13-56**). Depending on circuit design, the system may incorporate a relay (**Figure 13-57**). The relay has two coils and two sets of contacts to control current direction. In this system, the door lock switch energizes one of the door lock relay coils to send battery voltage to the motor. If the door lock switch is placed in the LOCK

Shop Manual
Chapter 13, page 669

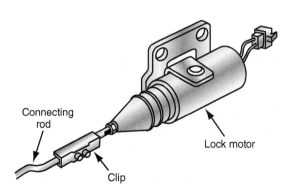

Figure 13-56 PM power door lock motor.

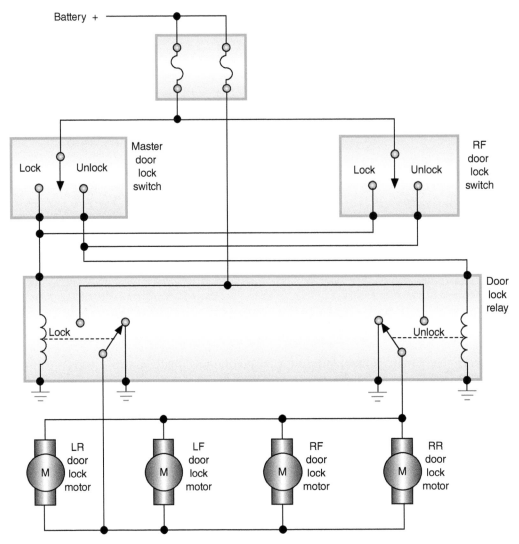

Figure 13-57 Power door lock circuit using a relay.

position, current flow is as shown in **Figure 13-58**. **Figure 13-59** shows current flow when the door lock switch is placed in the UNLOCK position.

Systems that do not use relays use the switch to provide control of current flow in the same manner as the power seat or power window systems.

A **child safety latch** in the door lock system prevents the door from being opened from the inside, regardless of the position of the door lock knob. The child safety latch is activated by a switch designed into the latch bellcrank (**Figure 13-60**). By placing the latch in the deactivated mode, the door operates as normal.

AUTHOR'S NOTE Like the power windows, the power locks can be operated by the BCM or door modules. In this case, the switch is nothing more than an input to the door module and the door module controls the relays or the solenoid directly.

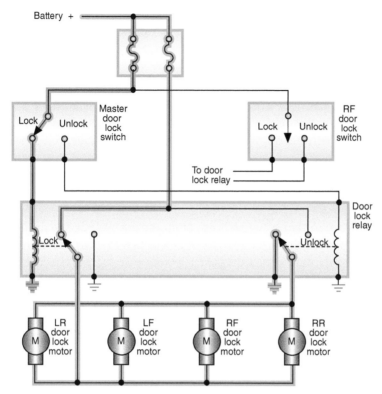

Figure 13-58 Current flow in the LOCK position.

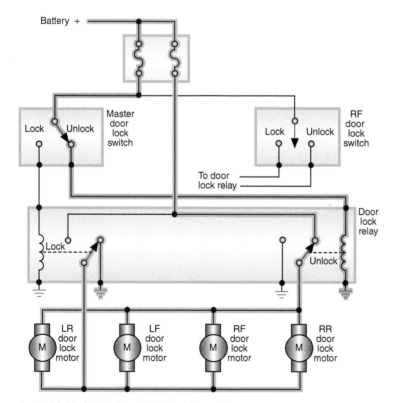

Figure 13-59 Current flow in the UNLOCK position.

Figure 13-60 Child safety latch.

AUTOMATIC DOOR LOCKS

Automatic door locks (ADL) is a passive system used to lock all doors when the required conditions are met. The ADL system is an additional safety and convenience system that uses the existing power door function. Most systems lock the doors when the ignition switch is in RUN, the gear selector is placed in drive, and all doors are shut. Some systems will lock the doors when the gear shift selector is passed through the reverse position; others do not lock the doors unless the vehicle is moving 8 mph (12.9 km/h) or faster.

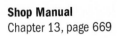

Shop Manual
Chapter 13, page 669

The system may use the body computer to control the door lock relays (**Figure 13-61**) or a separate controller. The controller (or body computer) takes the place of the door lock switches for automatic operation.

When all of the door jamb switches are open (doors closed), the ground is removed from the door jamb input circuit to the controller (**Figure 13-62**). This signals the controller to enable the lock circuit. When the gear selection moves from the PARK position, the neutral safety switch removes the power signal from the controller. The controller sends voltage through the LH seat switch to the lock relay coil. Current is sent through the motors to lock all doors.

When the gear selector is returned to the PARK position, voltage is applied through the neutral safety switch to the controller. The controller then sends power to the unlock relay coil to reverse current flow through the motors. Inputs concerning the gear position can be bussed instead of hardwired.

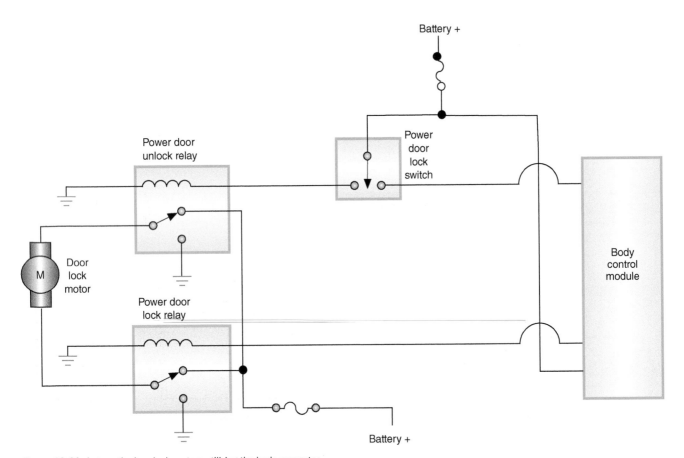

Figure 13-61 Automatic door lock system utilizing the body computer.

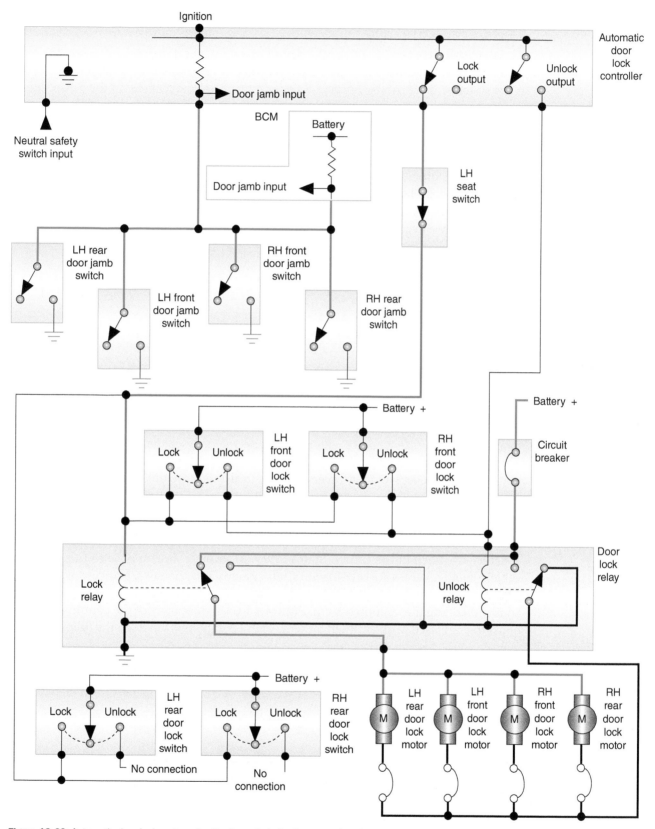

Figure 13-62 Automatic door lock system circuit schematic indicating operation during the lock procedure.

KEYLESS ENTRY

The **keyless entry** system allows the driver to unlock the doors or the deck lid (trunk) from outside the vehicle without the use of a key. The main components of the keyless entry system are the control module, a coded-button keypad located on the driver's door, and the door lock motors.

The keypad consists of five normally open, single-pole, single-throw switches. Each switch represents two numbers: 1-2, 3-4, 5-6, 7-8, and 9-0 (**Figure 13-63**).

The keypad is wired into the circuit to provide input to the control module. The control module is programmed to lock the doors when the 7-8 and 9-0 switches are closed at the same time. The driver's door can be unlocked by entering a five-digit code through the keypad. The unlock code is programmed into the controller at the factory. However, the driver may enter a second code. Either code will operate the system.

In addition to the aforementioned functions, the keyless entry system also:

1. Unlocks all doors when the 3-4 button is pressed within 5 seconds after the five-digit code has been entered.
2. Releases the deck lid lock if the 5-6 button is pressed within 5 seconds of code entry.
3. Activates the illuminated entry system if one of the buttons is pressed.
4. Operates in conjunction with the automatic door lock system and may share the same control module.

See the schematic (**Figure 13-64**) of the keyless entry system used by Ford. When the 7-8 and 9-0 buttons on the keypad are pressed, the controller applies battery voltage to all motors through the lock switch.

When the five-digit code is entered, the controller closes the driver's switch to apply voltage in the opposite direction to the driver's door motor. If the driver presses the 3-4 button, the controller will apply reverse voltage to all motors to unlock the rest of the doors.

ANTITHEFT SYSTEMS

Shop Manual
Chapter 13, page 670

A vehicle is stolen in the United States every 26 seconds. In response to this problem, vehicle manufacturers are offering antitheft systems as optional or standard equipment. These systems are deterrents designed to scare off would-be thieves by sounding alarms and/or disabling the ignition system. **Figure 13-65** shows many of the common components that are used in an antitheft system. These components include the following:

1. An electronic control module.
2. Door switches at all doors.
3. Trunk key cylinder switch.

Figure 13-63 Keyless entry system keypad.

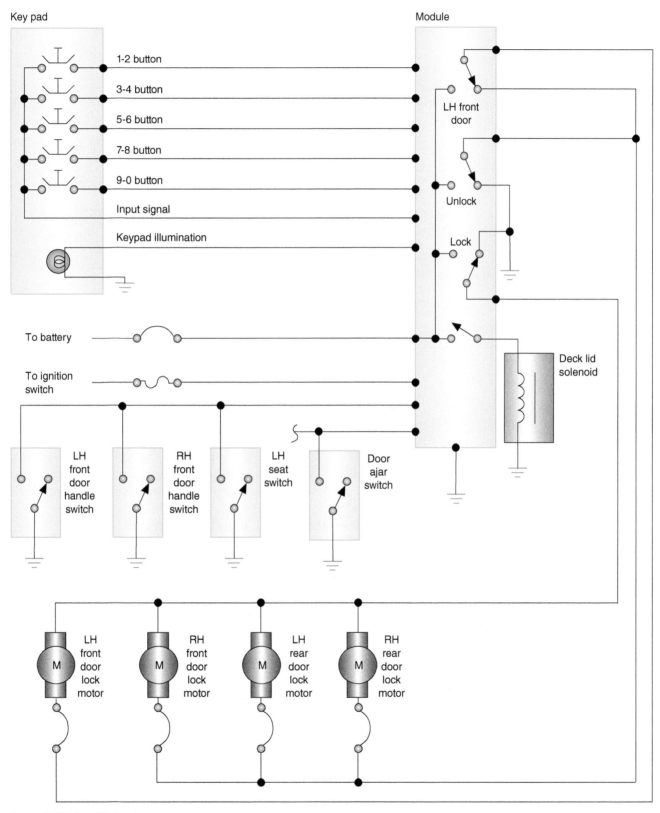

Figure 13-64 Simplified keyless entry system schematic.

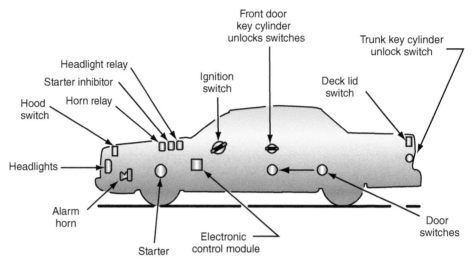

Figure 13-65 Typical components of an antitheft system.

4. Hood switch.
5. Starter inhibitor relay.
6. Horn relay.
7. Alarm.

In addition, many systems incorporate the exterior lights into the system. The lights are flashed if the system is activated.

For the system to operate, it must first be **armed**. This is done when the ignition switch is turned off and the doors are locked. When the driver's door is shut, a security light will illuminate for approximately 30 seconds to indicate that the system is armed and ready to function. If any other door is open, the system will not arm until it is closed. Once armed the system is ready to detect an illegal entry.

The control module monitors the switches. If the doors or trunk is opened or the key cylinders are rotated, the module will activate the system. The control module will sound the alarm and flash the lights until the timer circuit has counted down. At the end of the timer function, the system will automatically rearm itself.

Some systems use ultrasonic sensors that will signal the control module if someone attempts to enter the vehicle through the door or window. The sensors can be placed to sense the parameter of the vehicle and sound the alarm if someone enters within the protected parameter distance.

The system can also use current-sensitive sensors that will activate the alarm if there is a change in the vehicle's electrical system. The change can occur if the courtesy lights come on or if an attempt is made to start the engine.

The following systems are provided to give you a sample of the types of antitheft systems used. **Figure 13-66** illustrates an antitheft system that uses a separate control module and an inverter relay. If the system is triggered, it will sound the horn, flash the low-beam headlights, the taillights, and parking lamps, and disable the ignition system.

The arming process is started when the ignition switch is turned off. Voltage provided to the module at terminal K is removed. When the door is opened, a voltage is applied to the courtesy lamp circuit through the closed switch to terminal 2 of the inverter relay. This voltage energizes the inverter relay and provides a ground for module terminal J. This signal is used by the control module to provide an alternating ground at terminal D, causing the indicator lamp to blink. The flashing indicator light alerts the driver that the system is not armed. When the door lock switch is placed in the LOCK position, battery voltage is applied to terminal G of the module. The module uses this signal to apply a steady

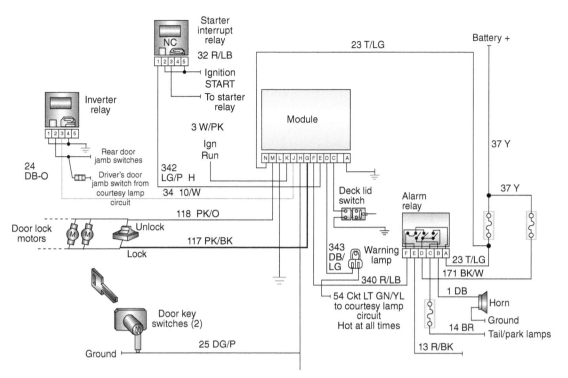

Figure 13-66 Circuit schematic of Ford's antitheft system.

ground at terminal D, causing the indicator light to stay on continuously. When the door is closed, the door switch is opened. The opened door switch de-energizes the inverter relay coil. Terminal J is no longer grounded, and the indicator light goes out after a couple of seconds.

To disarm the system, one of the front doors must be opened with a key or by pressing the correct code into the keyless entry keypad. Unlocking the door closes the lock cylinder switch and ground terminal H of the module. This signal disarms the system.

Once the system is armed, if terminals J and C receive a ground signal, the control module will trigger the alarm. Terminal C is grounded if the trunk tamper switch contacts close. Terminal J is grounded when the inverter relay contacts are closed. The inverter relay is controlled by the door jamb switches. If one of the doors is opened, the switch closes and energizes the relay coil. The contacts close and ground is provided to terminal J.

When the alarm is activated, a pulsating ground is provided at module terminal F. This pulsating ground energizes and de-energizes the alarm relay. As the relay contacts open and close, a pulsating voltage is sent to the horns and exterior lights.

At the same time, the start interrupt circuit is activated. The start interrupt relay receives battery voltage from the ignition switch when it is in the START position. When the alarm is activated, the module provides a ground through terminal E, causing the relay coil to be energized. The energized relay opens the circuit to the starter system, preventing starter operation.

Often the BCM controls the functions of the antitheft system. In this case the inputs and outputs are wired to the BCM (**Figure 13-67**). The BCM monitors the arming process and then triggers the alarm if an unauthorized entry is attempted. At this time the BCM controls the exterior lamps to cause them to flash; it also cycles the horn on and off. The BCM also sends a data bus message to the powertrain control module (PCM) not to start the engine. The PCM is programmed that it must receive a data bus message that it is alright to start. If the data bus circuit should fail, the engine may not start, since this message cannot be received.

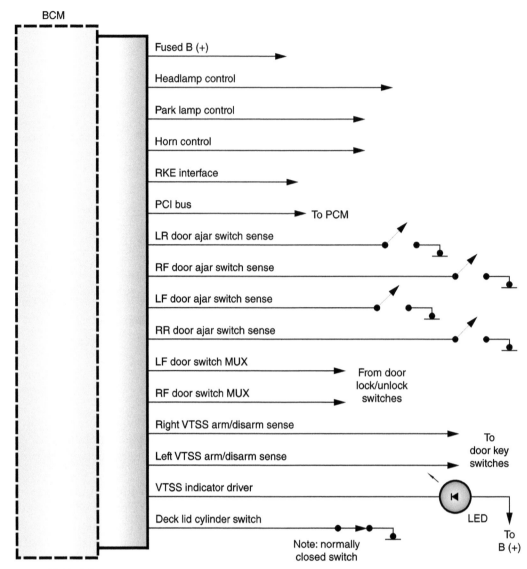

BCM

Fused B (+)

Headlamp control

Park lamp control

Horn control

RKE interface

PCI bus — To PCM

LR door ajar switch sense

RF door ajar switch sense

LF door ajar switch sense

RR door ajar switch sense

LF door switch MUX

RF door switch MUX

From door lock/unlock switches

Right VTSS arm/disarm sense

Left VTSS arm/disarm sense

To door key switches

VTSS indicator driver

Deck lid cylinder switch

LED

To B (+)

Note: normally closed switch

Figure 13-67 BCM-controlled antitheft system.

Systems that incorporate an intrusion monitor detect movement of a person or an object inside the passenger compartment. The intrusion monitor uses a single sensor that transmits 40-kHz ultrasonic sound waves. The sensor also receives the ultrasonic sound waves. If an object moves within the coverage area of the sensor, the received ultrasonic signals are distorted. This distortion is detected and processed by the module, which then signals the antitheft system to trigger.

Shop Manual
Chapter 13, page 674

Electromechanical cruise control receives its name from two subsystems: the electrical and the mechanical portion.

ELECTRONIC CRUISE CONTROL SYSTEMS

Cruise control is a system that allows a vehicle to maintain a preset speed with the driver's foot off the accelerator. Cruise control was first introduced in the 1960s for the purpose of reducing driver fatigue. When engaged, the cruise control system sets the throttle position to maintain the desired vehicle speed.

Most cruise control systems are a combination of electrical and mechanical components. The components used depend on the manufacturer and system design. However, the operating principles are similar.

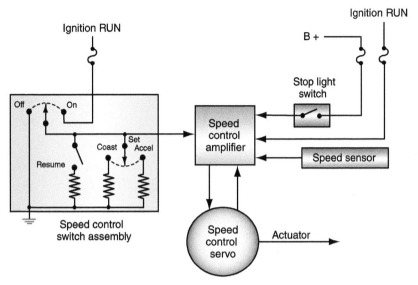

Figure 13-68 Block diagram of an electronic cruise control system.

![history icon] **A BIT OF HISTORY**

Ralph R. Teetor was born on August 17, 1890. At the age of 10, Ralph built a miniature dynamo. At the age of 12, he designed and built his own gasoline-powered automobile. Also at age 12, he built a generator to supply electricity not only to his home but for every house on the block. His most famous automotive invention was the Speedostat, which is now known as cruise control. He also designed and patented one of the first automatic gear shifts. However, the story of Ralph R. Teetor becomes more amazing once it is learned that he was totally blind from the age of 5. In 1902, a newspaper reporter wrote a story on Ralph but never noticed his blindness.

The electronic cruise control system uses an electronic module to operate the actuators that control throttle position (**Figure 13-68**). Other benefits include the following:

- More frequent throttle adjustments per second.
- More consistent speed increase/decrease when using the tap-up/tap-down feature.
- Greater correction of speed variation under loads.
- Rapid deceleration cutoff when deceleration rate exceeds programmed rates.
- Wheelspin cutoff when acceleration rate exceeds programmed parameters.
- System malfunction cutoff when the module determines there is a fault in the system.

Some manufacturers combine the transducer and servo into one unit. They usually refer to this unit as a servomotor.

Common Components

Common components of the electronic cruise control system include the following:

1. The control module: The module can be a separate cruise control module, the PCM, or the BCM. The operation of the systems is similar regardless of the module used.
2. The control switch (**Figure 13-69**): Depending on system design, the control switch contacts apply the ground circuit through resistors. Because each resistor has a different value, a different voltage is applied to the control module. In some systems, the control switch will send a 12-volt signal to different terminals of the control module.
3. The brake or clutch switch.

Shop Manual
Chapter 13, page 679

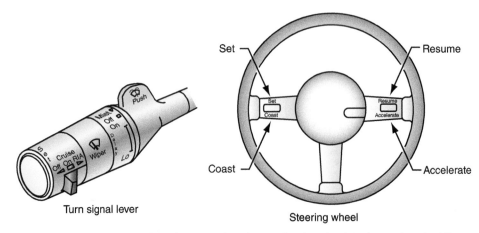

Figure 13-69 The control switch can be mounted on the turn signal stock or into the steering wheel. The switch is used to provide driver inputs for the system.

4. Vacuum release switch.
5. Servo: The servo operates on a vacuum that is controlled by supply and vent valves.

These operate from controller signals to solenoids.

Depending on system design, the sensors used as inputs to the control module include the vehicle speed sensor, servo position sensor, and throttle position sensor. Other inputs are provided by the brake switch, instrument panel switch, control switch, and the park-neutral switch.

The control module receives signals from the speed sensor and the control switch. When the speed of the vehicle is fast enough to allow cruise control operation and the driver pushes the SET button on the control switch, an electrical signal is sent to the controller. The voltage level received by the controller is set in memory. This signal is used to create two additional signals with values set at 1/4 mph (0.4 km/h) above and below the set speed. The module uses the comparator values to change vacuum levels at the servo to maintain set vehicle speed.

Three safety modes are operated by the control module:

1. Rapid deceleration cutoff: If the module determines that deceleration rate is greater than programmed values, it will disengage the cruise control system and return operation back over to the driver.
2. Wheelspin cutoff: If the control module determines that the acceleration rate is greater than programmed values, it will disengage the system.
3. System malfunction cutoff: The module checks the operation of the switches and circuits. If it determines there is a fault, it will disable the system.

The vacuum-modulated servo is the primary actuator. Vacuum to the servo is controlled by two solenoid valves: supply and vent. The vent valve is normally open and the supply valve is normally closed (**Figure 13-70**). The controller energizes the supply and vent valves to allow manifold vacuum or atmospheric pressure to enter the servo. The servo uses the vacuum and pressure to move the throttle to maintain the set speed. Vehicle speed is maintained by balancing the vacuum in the servo. The vacuum used to move the servo may be an engine vacuum or may be supplied by a vacuum pump.

Shop Manual
Chapter 13, page 674
If the voltage signal from the VSS drops below the low comparator value, the control module energizes the supply valve solenoid to allow more vacuum into the servo and increases the throttle opening. When the VSS signal returns to a value within the comparator levels, the supply valve solenoid is de-energized.

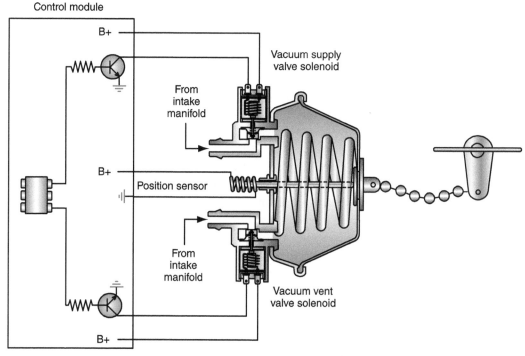

Figure 13-70 Servo valve operation in electronic control system. The servo position sensor informs the controller of servo operation and position.

If the VSS signal is greater than the high comparator value, the control module de-energizes the vent solenoid valve to release vacuum in the servo. The vehicle speed is reduced until the VSS signals are between the comparator values, at which time the control module will energize the vent valve solenoid again. This constant modulation of the supply and vent valves maintains vehicle speed.

During steady cruise conditions, both valves are closed and a constant vacuum is maintained in the servo.

Most manufacturers use electronic throttle control (ETC) systems instead of cables to operate the throttle body. The ETC throttle body (**Figure 13-71**) uses a motor to actuate the throttle plate. The driver's movement of the accelerator pedal is an input to the PCM. The PCM then directly controls the placement of the throttle body plates. When

Figure 13-71 An electronic throttle control throttle body.

ETC is used, cruise control servos are not necessary. Inputs and operation are the same as just described, except that the output is to the ETC motor instead of the servo solenoids.

Adaptive Cruise Control

Shop Manual
Chapter 13, page 680

The **adaptive cruise control (ACC)** system developed by TRW Automotive is similar to conventional cruise control systems, with the added function of adjusting and maintaining appropriate following distances between vehicles. This system uses laser or radar sensors to determine the vehicle-to-vehicle distances and relational speeds. Like other cruise control systems, the adaptive cruise control system maintains a fixed speed that has been set by the driver. However, the system also allows the driver to set a distance between their vehicle and any vehicle in front of them in the same lane, along with the set speed function. If a vehicle ahead is traveling slower and the distance is closing between the vehicles, then the ETC will close the throttle plate to decelerate the vehicle. If additional deceleration is necessary, then the transmission is shifted into a lower gear. If further deceleration is still necessary, the system controls the brake actuator to apply the brakes.

The system will continue to maintain the set distance between vehicles until the vehicle in front is no longer there (due to lane change). At this time, the vehicle is accelerated slowly until the set vehicle speed is reached, and then the system will maintain operation of driving at the fixed speed.

A steering wheel–mounted switch provides the driver with a selection of distances between his vehicle and the vehicle in front of him. Three distances are provided. The first button push sets the system to long, which is approximately 245 feet (75 meters) between vehicles. The second button push selects a middle range of approximately 165 feet (50 meters). A third button push sets the distance range to short, which is approximately 100 feet (30 meters).

The laser or radar sensor is usually mounted on the front grill (**Figure 13-72**). Radar is a system that uses electromagnetic waves to identify the range, direction, or speed of an object. Radar emits a radio wave reference signal and measures the time it takes to echo back. The waves are reflected by the object and detected by a receiver. Consider that radio waves travel at the speed of light (186,000 miles per second or 300,000,000 meters per second) and that the waves must travel to the object and echo back to the receiver; distance is calculated by measuring the length of the pulse multiplied by the speed of light, divided by two.

Laser is an acronym for light amplification by stimulated emission of radiation.

Figure 13-72 The laser or radar sensor is usually located behind the grill or in the front fascia.

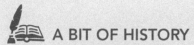

A BIT OF HISTORY

The term *RADAR* was first used in 1941 as an acronym for radio detection and ranging.

A laser is a device that controls the way that energized atoms release photons (light). Laser light has the following properties:

- Monochromatic—The light is only one color since it contains one specific wavelength of light.
- Coherent—The waves have the same wavelength and a fixed-phase relationship.
- Directional—The laser light produces a very tight beam.

The laser light uses stimulated emission to accomplish these properties. Stimulated emission results in the laser light atoms releasing photons in a very organized state. A typical laser emits light in a narrow, low-divergence beam and with a well-defined wavelength.

Laser distance sensors function on the same basic principle as the ultrasonic and radar distance sensors. They use a "time-of-flight" measurement that compares the outgoing and returning light waveform signals to determine the distance to an object. An internal clock measures the time it takes for the laser light waveform to be transmitted and then returned.

The laser sensor has three parts (**Figure 13-73**): the laser emitter that radiates laser rays forward, the laser-receiving portion that receives the laser beams as they are reflected back by the vehicle that is ahead, and the processing unit that determines the length of time it takes for the reflected beams to return to the sensor; the unit calculates the distance to the vehicle ahead and the relative speed. This data is then transmitted to the distance control ECU (**Figure 13-74**).

Regardless of the sensor type, the distance and relative speed information is sent to a control module (either a separate distance ECU or the PCM) that determines which vehicle to follow based on the information provided by the sensor. The control module will also calculate the target acceleration signals for following the vehicle. To maintain the set distance, the control module will send acceleration or deceleration requests. If necessary, it will also request a transmission downshift and brake application.

If brake control is necessary, the distance control module will determine a target deceleration rate. The maximum rate will be set at 0.30 G. The target deceleration rate is determined by current vehicle speed, the distance to the vehicle in front, and the relative speed. The brake apply request signal is sent to the antilock brake system control module,

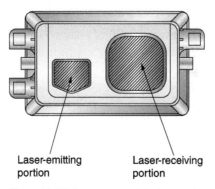

Laser-emitting portion   Laser-receiving portion

Figure 13-73 Components of the laser sensor.

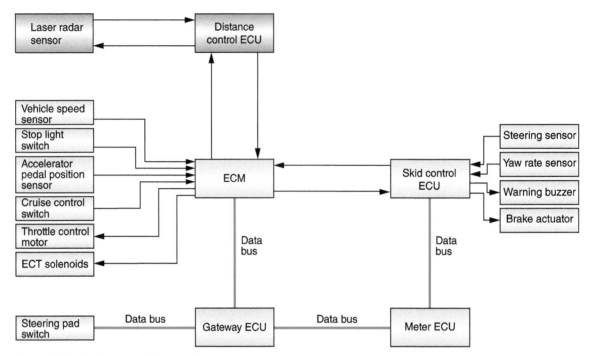

Figure 13-74 Block diagram of the adaptive cruise control system.

which controls the brake actuator to apply the brakes. If the vehicle does not decelerate at a fast enough rate to avoid a collision, a warning buzzer is sounded to prompt the driver to apply the brakes.

If the vehicle slows down due to a slower vehicle in front, then the system will enter follow mode and match the speed of the vehicle that is ahead. With the speeds of the two vehicles matched, the set distance is maintained. While in this mode, if either vehicle changes lanes, the system will request an acceleration mode. The throttle plates are opened to provide a gradual increase in speed until the set speed requested by the driver is reached. At this time, the system enters fixed speed mode and maintains the set speed.

ELECTRONIC SUNROOF CONCEPTS

Shop Manual
Chapter 13, page 686

Many manufacturers have introduced electronic control of their electric sunroofs. These systems incorporate a pair of relay circuits and a timer function into the control module. Although there are variations between manufacturers, the systems discussed here provide a study of the two basic types of systems.

Toyota Electronically Controlled Sunroof

Refer to the schematic in **Figure 13-75** of a sunroof control circuit used by Toyota. The movement of the sunroof is controlled by the motor that operates a drive gear. The drive gear either pushes or pulls the connecting cable to move the sunroof.

Motor rotation is controlled by relays that are activated according to signals received from the slide, tilt, and limit switches. The limit switches are operated by a cam on the motor (**Figure 13-76**).

The logic gates of this system operate on the principle of **negative logic**, which defines the most negative voltage as a logical 1 in the binary code. When the slide switch is moved to the OPEN position, either limit switch 1 or both limit switches are closed (**Figure 13-77**). Limit switches 1 and 2 provide a negative side signal to the OR gate labeled F. The output from gate F is sent to gate A. Gate A is an AND gate, requiring

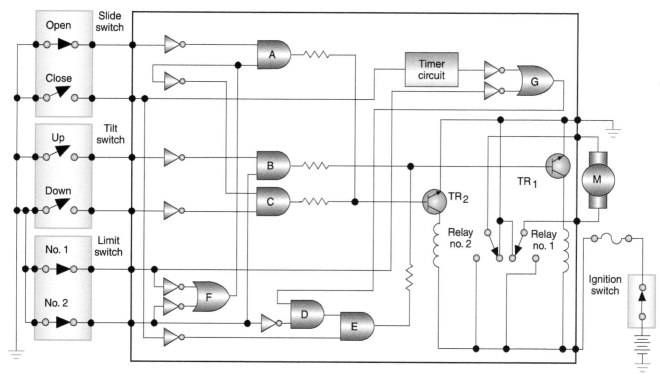

Figure 13-75 Toyota sunroof circuit using electronic controls.

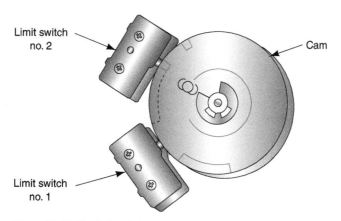

Figure 13-76 The limit switches operate off a cam on the motor.

input from gate F and the open slide switch. The output signal from gate A is used to turn on TR_2. This provides a ground path for the coil in relay 2. Battery voltage is applied to the motor through relay 2; the ground path is provided through the de-energized relay 1. Current is sent to the motor as long as the OPEN switch is depressed. If the OPEN switch is held in this position too long, a clutch in the motor disengages the motor from the drive gear.

AUTHOR'S NOTE The schematics used to explain the operation of the Toyota sunroof use logic gates. If needed, refer to Chapter 11 of this manual to review the operation of the gates.

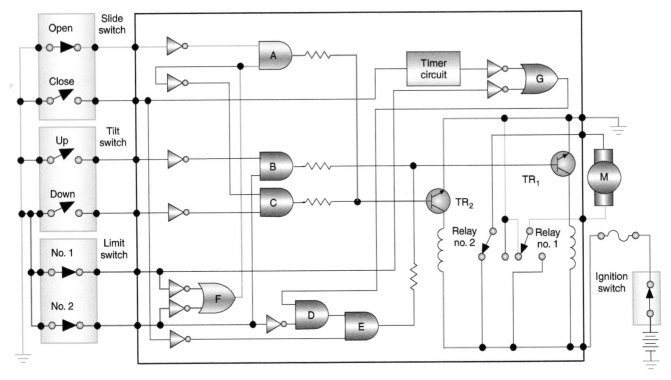

Figure 13-77 Circuit operation when the switch is in the OPEN position.

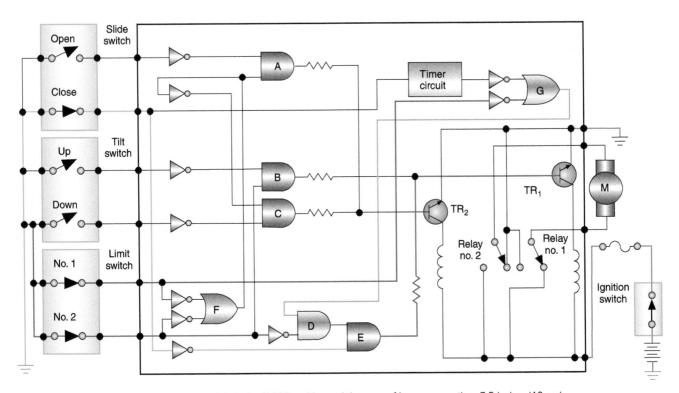

Figure 13-78 Circuit operation when the switch is in the CLOSE position and the sunroof is open more than 7.5 inches (19 cm).

Operation of the system during closing depends on how far the sunroof is open. If the sunroof is open more than 7.5 inches and the slide contact is moved to the CLOSE position, an input signal is sent to gate E (**Figure 13-78**). The other input signal required at gate E is received from the limit switches. The limit switch 1 signal passes through the OR gate G to the AND gate D. Limit switch 2 provides the second signal required by gate D. The output signal from D is the second input signal required by gate E. The output signal from E turns on TR_1. This energizes relay 1 and reverses the current flow through the motor. The motor will operate until the slide switch opens or limit switch 2 opens.

If the sunroof is open less than 7.5 inches and the slide switch is placed in the CLOSE position, the timer circuit is activated (**Figure 13-79**). The CLOSE switch signals the timer and provides an input signal to gate E. Limit switch 1 is open when the sunroof is open less than 7.5 inches. The second input signal required by gate D is provided by the timer. The timer is activated for 0.5 second. This turns on TR_1 and operates the motor for 0.5 second, or long enough for rotation of the motor to close limit switch 1. When limit switch 1 is closed, the operation is the same as described when the sunroof is closed after it is more than 7.5 inches open.

When the tilt switch is located in the UP position, a signal is imposed on gate B (**Figure 13-80**). This signal is inverted by the NOT gate and is equal to the value received from the opened number 2 limit switch. The output signal from gate B turns on TR_1, which energizes relay 1 to turn on the motor. The motor clutch will disengage if the switch is held in the closed position longer than needed.

When the tilt switch is placed in the DOWN position, a signal is imposed on gate C (**Figure 13-81**). The second signal to gate C is received from the limit switches (both are open) through gate F. The signal from gate F is inverted by the NOT gate and is equal to that from the DOWN switch. The output signal from gate C turns on TR_2 and energizes relay 2 to lower the sunroof. If the DOWN switch is held longer than necessary, limit switch 1 closes. When this switch is closed, the signals received by gate F are not opposite. This results in a mixed input to gate C and turns off the transistor.

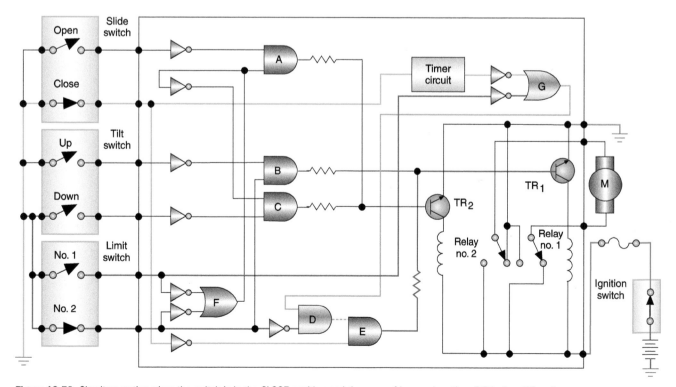

Figure 13-79 Circuit operation when the switch is in the CLOSE position and the sunroof is open less than 7.5 inches (19 cm).

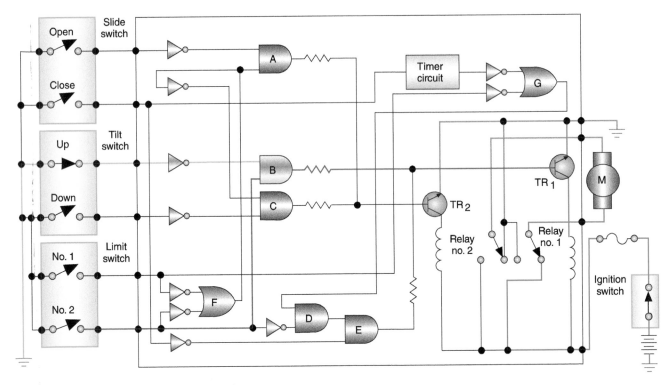

Figure 13-80 Circuit operation in the TILT UP position.

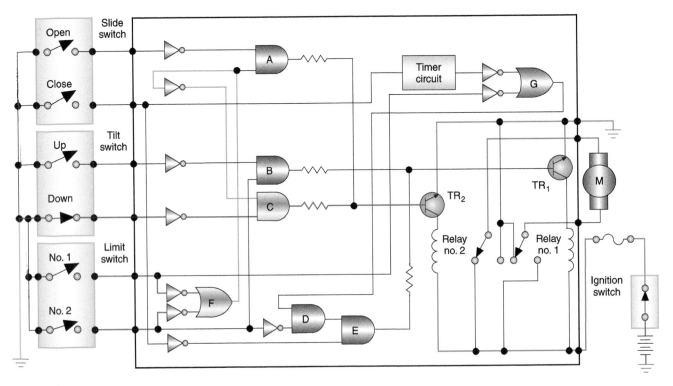

Figure 13-81 Circuit operation during TILT DOWN.

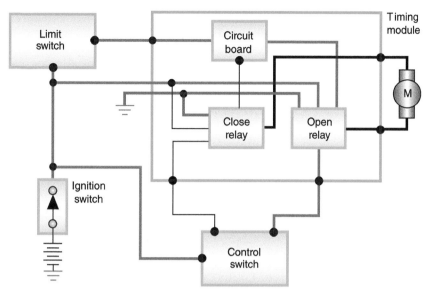

Figure 13-82 Block diagram of the GM sunroof.

General Motor Electronically Controlled Sunroof

See the schematic in **Figure 13-82** of the sunroof system used on some GM model vehicles. The timing module uses inputs from the control switch and the limit switches to direct current flow to the motor. Depending on the inputs, the relays will be energized to rotate the motor in the proper direction. When the switch is located in the OPEN position, the open relay is energized, sending current to the motor. The sunroof will continue to retract as long as the switch is held in the OPEN position. When the sunroof reaches its full open position, the limit switch will open and break the circuit to the open relay.

Placing the switch in the CLOSE position will energize the close relay. The current sent to the motor is in the opposite direction to close the sunroof. If the close switch is held until the sunroof reaches the full closed position, the limit switch will open.

SUMMARY

- Automotive electrical horns operate on an electromagnetic principle that vibrates a diaphragm to produce a warning signal.
- Horn switches are installed either in the steering wheel or as a part of the multifunction switch. Most horn switches are normally open switches.
- Horn switches that are mounted on the steering wheel require the use of sliding contacts. The contacts provide continuity for the horn control in all steering wheel positions.
- The most common type of horn circuit control is to use a relay.
- Most two-speed windshield wiper motors use permanent magnet fields whereby the motor speed is controlled by the placement of the brushes on the commutator.

- Some two-speed and all three-speed wiper motors use two electromagnetic field windings: series field and shunt field. The two field coils are wound in opposite directions so that their magnetic fields will oppose each other. The strength of the total magnetic field will determine at what speed the motor will operate.
- PARK contacts are located inside the wiper motor assembly and supply current to the motor after the switch has been turned to the PARK position. This allows the motor to continue operating until the wipers have reached the PARK position.
- Intermittent wiper mode provides a variable interval between wiper sweeps and is controlled by a solid-state module.
- Systems that have a depressed-park feature use a second set of contacts with the park switch, which

are used to reverse the rotation of the motor for about 15° after the wipers have reached the normal PARK position.

- Blower fan motors use a resistor block that consists of two or three helically wound wire resistors which are connected in series to control fan speed.
- The blower motor circuit includes the control assembly, blower switch, resistor block, and blower motor.
- Some blower motor systems are controlled by the BCM using a power module.
- Electric defoggers heat the rear window by means of a resistor grid.
- Electric defoggers may incorporate a timer circuit to prevent the high current required to operate the system from damaging the battery or charging system.
- The electrically controlled mirror allows the driver to position the outside mirrors by use of a switch that controls dual-drive, reversible PM motors.
- Power windows, seats, and door locks usually use reversible PM motors whereby motor rotational direction is determined by the direction of current flow through the switch wipers.
- Cruise control is a system that allows the vehicle to maintain a preset speed with the driver's foot off the accelerator.
- The cruise control module energizes the supply and vent valves to allow manifold vacuum to enter the servo. The servo moves the throttle to maintain

the set speed. The vehicle speed is maintained by balancing the vacuum in the servo.
- Adaptive cruise control uses laser or radar sensors to maintain a set distance between vehicles. They can also slow or stop the vehicle if the distance closes.
- The memory seat feature allows the driver to program different seat positions that can be recalled at the push of a button.
- The easy exit feature is an additional function of the memory seat that provides for easier entrance and exit of the vehicle by moving the seat all the way back and down.
- Seats can be climate controlled using heater grids and fans.
- Antitheft systems are deterrent systems designed to scare off would-be thieves by sounding alarms and/or disabling the ignition system.
- The antitheft control module monitors the switches. If the doors or trunk is opened or the key cylinders are rotated, the module will activate the system.
- Automatic door locks is a passive system used to lock all doors when the required conditions are met. Many automobile manufacturers are incorporating the system as an additional safety and convenience feature.
- The keyless entry system allows the driver to unlock the doors or the deck lid from outside the vehicle without using a key.

REVIEW QUESTIONS

Short-Answer Essays

1. Describe the function of the adaptive cruise control system.
2. Explain the basic operating principles of the electronic cruise control system.
3. Describe the operation of a relay-controlled horn circuit.
4. Explain how brush placement determines the speed of a two-speed, permanent magnet motor.
5. How do wiper motor systems that use a three-speed, electromagnetic motor control wiper speed?
6. List and describe the safety modes incorporated into electronic cruise control systems.
7. List the main components of common antitheft systems.
8. Explain two methods that the memory seat control module uses to determine seat position.
9. Describe the operation of electric defoggers.
10. Briefly explain the principles of operation for power windows.

Fill in the Blanks

1. Electrical accessories provide for additional_____and _____.
2. The _____ is a thin, flexible, circular plate that is held around its outer edge by the horn housing, allowing the middle to flex.

3. Horn switches that are mounted on the steering wheel require the use of _____ _____ to provide continuity in all steering wheel positions.

4. The_____ _____ feature is an additional function of the memory seat that provides for easier entrance and exit of the vehicle.

5. Adaptive cruise control uses _____ or _____ sensors to maintain a set distance between vehicles.

6. _____ is a system that uses electromagnetic waves to identify the range, direction, or speed of an object.

7. A _____ is a device that controls the way that energized atoms release photons (light).

8. Most blower motor fan speeds can be controlled through a_____ _____ that is wired in series to the fan motor.

9. Electric defoggers operate on the principle that when electrons are forced to flow through a _____, heat is generated.

10. Climate-controlled seats may use a _____ element to cool the seat cushion.

Multiple Choice

1. Sound from the horn is generated by:
   A. Heat causing the diaphragm to vibrate.
   B. Vibrating a column of air.
   C. Pulse width modulation of the horn relay.
   D. All of the above.

2. The horn circuit is being discussed.
   Technician A says if the circuit does not use a relay, the horn switch carries the total current requirements of the horns.
   Technician B says most systems that use a relay have battery voltage applied to the lower contact plate of the horn switch and the switch closes the path for the relay coil.
   Who is correct?
   A. A only   C. Both A and B
   B. B only   D. Neither A nor B

3. Electronic cruise control systems are being discussed.
   Technician A says if the voltage signal from the VSS drops below the low comparator value, the control module energizes the vent valve solenoid.
   Technician B says if the VSS signal is greater than the high comparator value, the control module energizes the supply solenoid valve.
   Who is correct?
   A. A only   C. Both A and B
   B. B only   D. Neither A nor B

4. Electronic cruise control is being discussed.
   Technician A says the electronic cruise control system offers more precise speed control than the electromechanical system.
   Technician B says the throttle position sensor is used to provide smooth throttle changes while the cruise control is engaged.
   Who is correct?
   A. A only   C. Both A and B
   B. B only   D. Neither A nor B

5. Memory seats are being discussed.
   Technician A says the power seat and memory seat functions can be operated only when the transmission is in the PARK position.
   Technician B says when the seat is moved from its memory position, the module stores the number of pulses and direction of movement in memory.
   Who is correct?
   A. A only   C. Both A and B
   B. B only   D. Neither A nor B

6. The operation of Toyota's electronically controlled sunroof is being discussed.
   Technician A says it is not necessary to understand logic gate operation to understand the control of the sunroof.
   Technician B says the movement of the sunroof is controlled by a motor that operates a drive gear.
   Who is correct?
   A. A only   C. Both A and B
   B. B only   D. Neither A nor B

7. The keyless entry system is being discussed.

    Technician A says an additional function of the system is that the deck lid lock can be released by pressing the 5-6 button.

    Technician B says a second code can be entered into the system.

    Who is correct?

    A. A only

    B. B only

    C. Both A and B

    D. Neither A nor B

8. The adaptive cruise control system is being discussed.

    Technician A says if the vehicle is approaching another vehicle in the same lane, ACC may activate the brake system to prevent a rear-end collision.

    Technician B says the radar or laser sensor is used to determine the distance between the vehicle and any other vehicle ahead of it.

    Who is correct?

    A. A only

    B. B only

    C. Both A and B

    D. Neither A nor B

9. Which statement about permanent magnet wiper motors is true?

    A. The more armature windings between the high-speed and common brushes result in less magnetism in the armature and lower CEMF.

    B. The lower the CEMF in the armature, the greater the armature current.

    C. The fewer windings between the low-speed and common brushes result in increased magnetism and higher CEMF.

    D. All of the above.

10. *Technician A* says the master control switch for power windows provides the overall control of the system.

    Technician B says current direction through the power seat motor determines the rotation direction of the motor.

    Who is correct?

    A. A only

    B. B only

    C. Both A and B

    D. Neither A nor B

CHAPTER 14

RADIO FREQUENCY AND INFOTAINMENT SYSTEMS

Upon completion and review of this chapter, you should be able to understand and describe:

- The generation of radio waves.
- The construction and operation of the antenna.
- The three common modulations used for radio waves.
- How radio waves are received.
- The operation of the remote keyless entry system.
- The operation of the keyless start system.
- The operation of the tire pressure monitoring system.
- The function of immobilizer security systems.

- The operating characteristics of speakers.
- The purpose and configuration of vehicle audio and entertainment systems.
- How a radio tuner operates.
- How radio frequency is used in vehicle audio entertainment systems.
- The operation of satellite radio systems.
- The operation of the telematics system including Wi-Fi connectivity.
- The operation of hands-free cell phone systems.
- The operation of the navigational system.

Terms To Know

Angular momentum
Antenna
Dead reckoning
Detector
Gyroscope
Immobilizer system
Infotainment
Keyless start system

Micro-Electro-Mechanical System (MEMS)
Multiplexer
Navigational systems
Radio choke
Reader
Receiver
Satellite radio

Speakers
Telematics
Tire pressure monitoring system
Transmitter
Tweeters
Wi-Fi
Woofers

INTRODUCTION

Radio wave transmitting and receiving devices are commonplace in the automobile. The first of these was the amplitude modulation (AM) radio system. Today, radio frequency is used in such systems as remote keyless entry, keyless start, immobilizer, and so forth. Examples of these systems are discussed to provide greater understanding of their operation.

Infotainment refers to a type of media system that provides a combination of information and entertainment. For example, the infotainment system can provide GPS directions and road conditions along with satellite radio, satellite TV, and Wi-Fi connectivity.

This chapter will explore how radio frequency waves are generated and how they are received. In addition, we will explore the operation of navigational systems and satellite receivers.

RADIO FREQUENCY GENERATION

Shop Manual
Chapter 14, page 734

Any system that is using radio frequency to broadcast data must have a **transmitter** and a **receiver**. The transmitter generates the radio frequency by taking the data and encoding it onto a sine wave. The data are then transmitted with radio waves. The receiver receives the radio waves and decodes the message from the sine wave.

Both components require an **antenna** to radiate and capture the radio wave signal. Antennas convert radio frequency electrical currents into electromagnetic waves and vice versa. The antenna is simply a wire or a metal stick that increases the amount of metal the transmitter's waves can interact with. The transmitting antenna will use the oscillating current that enters the antenna to create a magnetic field around the antenna. The magnetic field will induce an electric field (voltage and current) in space. This electric field in turn induces another magnetic field in space, which induces yet another electric field. This process is repeated as these electric and magnetic fields (electromagnetic fields) induce each other in space at the speed of light outward away from the antenna.

Radio wave transmission is actually a simple technology. It is based on the principle of induction. Radio waves are electromagnetic (EM) radiation. Remember that induced voltage occurs only when the magnetic field is in motion.

The transmitter induces a continuously varying AC electrical current (**Figure 14-1**). **Figure 14-2** illustrates a simple transmitter that can produce this type of sine wave using a capacitor and an inducer. Transistors are used to amplify the signal. Sending the signal to an antenna transmits the sine wave into space.

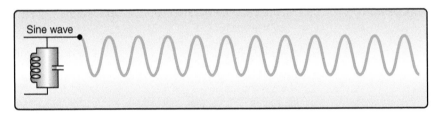

Figure 14-1 The transmitter produces an AC sine wave.

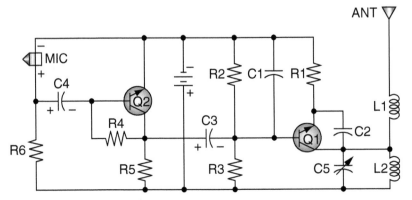

Figure 14-2 A simple transmitter using a capacitor and an inducer.

For actual data transmission, the sine-wave signal needs to be modulated. This encodes the data. There are three common ways to modulate a sine wave:

- *Pulse modulation.* This is accomplished by simply turning the sine wave on and off (**Figure 14-3**).
- *Amplitude modulation.* Used by AM radio stations and for the transmission of some video TV signals. In amplitude modulation, the amplitude of the sine wave (its peak-to-peak voltage) changes (**Figure 14-4**).

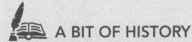 **A BIT OF HISTORY**

FM radio was invented by Edwin Armstrong, who desired to make high-fidelity and static-free music broadcasting possible. He built the first station in 1939, but FM did not become really popular until the 1960s.

- *Frequency modulation.* Used by frequency modulation (FM) radio stations and several wireless technologies, such as cell phones. This signal has the benefit of being almost immune to static. In FM, the transmitter's sine-wave frequency changes very slightly based on the data signal (**Figure 14-5**).

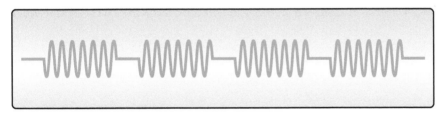

Figure 14-3 Pulse modulation (PM) sine wave.

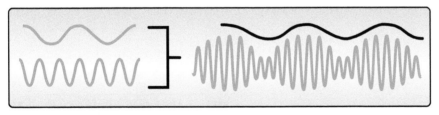

Figure 14-4 Amplitude modulation (AM) sine wave.

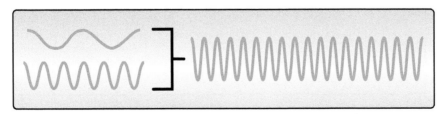

Figure 14-5 Frequency modulation (FM) sine wave.

RECEIVING RADIO FREQUENCIES

Shop Manual
Chapter 14, page 734

To illustrate the receiving of radio wave–transmitted signals, we will look at how the AM radio receives this data. In a simple crystal radio, a capacitor/inductor oscillator performs as the tuner for the radio. The oscillator is connected to an antenna and ground (**Figure 14-6**).

As the sine waves from different radio stations hit the antenna, the capacitor and inductor want to resonate at one particular frequency. The sine wave, which matches that particular frequency is amplified by the resonator, and all of the other frequencies are ignored.

In a radio, either the capacitor or the inductor in the resonator is adjustable, so as the tuner on the radio is changed, the resonant frequency of the resonator is changed. This changes the frequency of the sine wave that the resonator amplifies. **Figure 14-7** illustrates how the radio wave is received. For example, if the desired radio station is broadcasting at 910,000 Hz to receive this signal the radio dial is adjusted to 910. The data is modulated and amplified before it is sent out from the transmitting station's antenna. The radio receives the radio waves through its antenna. Since the antenna receives thousands of sine waves, the radio's tuner separates out the desired sine wave. In this case, the tuner is tuned to receive the 910,000 Hz signal.

The tuner resonates and amplifies the desired frequency and ignores all the others. The data (voices and music) is extracted from the sine wave using a **detector**. The detector is made with a diode so it creates a sine wave with only the upper-half wave cycles (**Figure 14-8**). Finally, the signal is amplified by a series of transistors and sent to the speakers.

> **AUTHOR'S NOTE** In an FM radio, the detector turns the changes in frequency into sound.

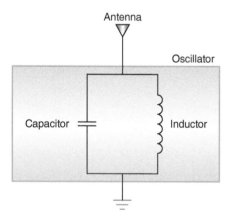

Figure 14-6 A capacitor/inductor oscillator performs as the tuner for the radio.

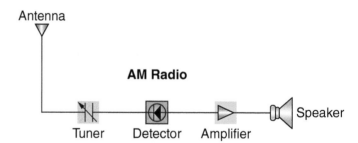

Figure 14-7 Reception of the radio wave.

Figure 14-8 The detector is made with a diode so only the upper half of the sine wave passes through.

REMOTE KEYLESS ENTRY

Many vehicles are equipped with a remote keyless entry system that is used to lock and unlock the doors, turn on the interior lights, and release the trunk latch. A small receiver is installed in the vehicle. The transmitter assembly is a handheld item attached to the key ring (**Figure 14-9**). Pressing a button on a handheld transmitter will allow operation of the system from a distance of 25 to 50 feet (7.6 to 15.2 m). When the UNLOCK button is pressed, the driver's door unlocks and the interior lights are illuminated. If a theft-deterrent system is installed on the vehicle, it is also disarmed when the unlock button is pressed. A driver exiting the vehicle can activate the door locks and arm the security system by pressing the lock button. Many transmitters also have a third button for opening the deck lid. Some systems will also include a PANIC button that, when pressed, signals the control module to honk the horns and flash the lights.

The system operates at a fixed radio frequency. If the unit does not work from a normal distance, check for two conditions: weak batteries in the remote transmitter or a stronger radio transmitter close by (radio station, airport transmitter, etc.).

Many keyless entry systems incorporate the ability to start the engine without entering the vehicle. A separate button on the FOB is used for this function. The engine start button usually requires multiple presses or for depressing it for a couple of seconds. When the remote keyless entry module receives the transmission of an engine start request, it will send bus messages to various modules to gather status information and to request activation.

When the transmission from the FOB is received, the parking lights will turn on (of flash) and the horn will chirp. The next command is to lock all doors. Some system will also raise the windows. Once the vehicle is secured, the request is sent to start the engine.

Shop Manual
Chapter 14, page 729

Remote keyless entry differs from the keyless entry system discussed in Chapter 13 since this system can perform the keyless entry functions from a distance.

Shop Manual
Chapter 14, page 730

Figure 14-9 Remote keyless entry system transmitter.

Messages over the bus system will confirm proper gear selection, proper parking brake position, and so forth. If all are within the correct parameters, the PCM will start the engine.

Most systems will only allow the engine to run for 5 to 10 minutes, and then the engine is shut off. Typically, the engine can only be started a couple of times without the ignition actually be placed in the RUN position. If the engine is started remotely too many times, the FOB request is denied and the horn will double chirp to indicate the engine will not be started. In order to reactivate the remote start function, the ignition must be placed in the RUN position.

Other inputs to the remote start system include door ajar and hood switches. If any door is not properly shut, or the hood is open, the remote start function is disabled.

The remote FOB can also be used to shut the engine off (if started remotely). This is done by a single press of the start button. The horn may chirp when the request is received, the engine shut down, and the parking lights are turned off. The doors will remain locked.

The remote start feature may be tied in with the automatic temperature control (ATC) system. When the start request is received, the ATC controller receives a bus message and will look at the in-car temperature sensor. Based on this input, the ATC system will activate A/C or heater/defrost modes. It is also possible that the climate-controlled seats and steering wheel function is included in the remote start feature.

To prevent the theft of the vehicle while it is running, the control module will monitor different inputs. For example, if the brake pedal is depressed prior to the ignition being placed in the RUN position the engine will shut off. On vehicles with manual transmissions, the clutch pedal switch is used for this input. In addition, if the hood is opened without the ignition in the RUN position, the engine is shut off. If the vehicle has an antitheft system, anything that triggers the system will cause the engine to shut down.

Shop Manual
Chapter 14, page 731

Aftermarket Kits. Vehicles that are factory equipped with power door locks can be upgraded to remote keyless entry and remote start by installing aftermarket kits. The kits will include a control module, antenna, transmitters, and wiring harnesses. In some cases, relays, diodes, and a program switch are also included. Depending on the system, add-on features may be installed or programmed with the remote keyless entry feature. Some of these add-on systems may require addition control units.

Typically, the aftermarket remote keyless system will operate very similar to the factory system. However, if remote start is also added the system can become very complex (**Figure 14-10**). The kit will usually require the installation of additional relays. The relays are used to bypass the operation of factory systems. This allows for original operation of the system without interruption from the add-on kit. For many systems, such as the horn, this is a better practice then attempting to wire the system using the existing relays. Keep in mind that these systems may now have two relays.

There are exceptions; for example, the factory door lock and unlock relays are used by splicing the wires from the control module to the vehicle's wiring harness for these circuits. If the vehicle does not have relays, these same circuits from the ECU are wired to provide a push/pull (H-gate) function of the actuators (if the manufacturer states that the circuits can handle the current). Some systems only have a 200-mA output to the door lock/unlock system, thus the installation of relays will be necessary.

There are different methods the aftermarket system can bypass the factory immobilizer system. One method is to connect one of the programmable output circuits of the ECU to control a bypass module. The bypass module allows for the add-on system to be installed, but maintains the integrity of the manufacturer's system. Vehicles equipped with the GM VATS system will have a resistor that matches the resistance of the key installed in a bypass circuit from the ignition switch to the VATS module (**Figure 14-11**). When

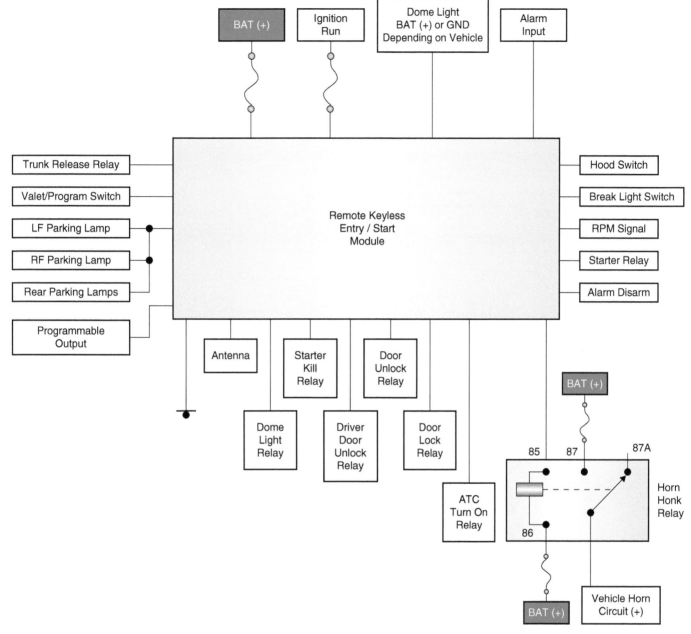

Figure 14-10 A typical aftermarket remote keyless entry system with remote start.

the relay is energized, the VATS module will see the correct resistance and allow the engine to start.

Some of the circuits from the control module will be wired to battery voltage or ground, depending on the vehicle. For example, the dome light input is connected to battery voltage if the door pin circuit is positive and to ground if the circuit is negative. The control module may be configurable to operate the parking lamps by either positive feed or ground. Changing the control of the lamps may be done by placement of a jumper harness or require a Bitwriter program.

Keyless Start

An enhancement to the remote keyless entry system is the **keyless start system**. This system allows the vehicle to be started without the use of an ignition key. Early versions of these systems use the key FOB to unlock the vehicle doors and disarm the antitheft

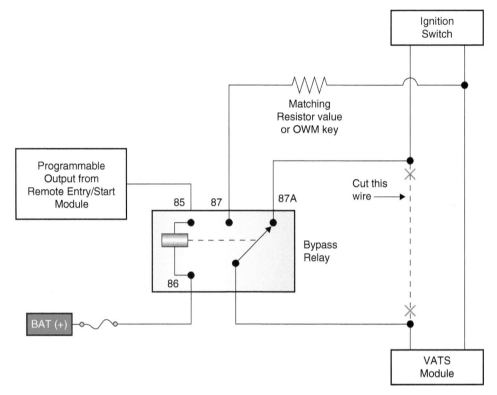

Figure 14-11 VATS bypass circuit.

system. At the same time, a message is sent that the engine can be started. Instead of the ignition key, the pointed end of the FOB is inserted into the ignition switch. This activates an infrared data exchange that unlocks the steering column and starts the engine.

A BIT OF HISTORY

The acronym "FOB" originally was the word *fob* referring to a small pocket in a waistcoat or pants that was used to hold a watch and chain. "FOB" is now used to mean a finger-operated button.

Newer versions have improved on this. They no longer require the use of a key FOB to unlock the doors or to start the engine. The following is the Keyless Go system used on some Mercedes-Benz and other vehicles.

This system uses a transmitter card or a FOB (**Figure 14-12**), signal acquisition and actuation modules (SAMs) that are installed throughout the vehicle, an electronic ignition switch (EIS), and a total of seven antennas. There are two antennas on the left side, two on the right side, and three in the trunk area.

The Keyless Go module communicates with the transmitter card at a frequency of 125 kHz. In order to receive messages, the transponder card must be within 5 feet (1.5 meters) of the vehicle. The electromagnetic field of the antennas causes the transmitter card to send its code by RF signal at a frequency of 433 MHz to the rear SAM. The antennas will determine if the transmitter card is located within the vehicle or outside of the vehicle.

When the driver approaches the vehicle with the transponder in his/her possession and attempts to open the door with the door handle, a capacitive sensor or microswitch signals the Keyless Go module to activate the left front door antenna. Messages are exchanged between the module and the card to determine authorized entry. If the entry is authorized, the doors are unlocked to allow the door to open.

Some manufacturers use the remote keyless entry transponder to do the function of the card being described here.

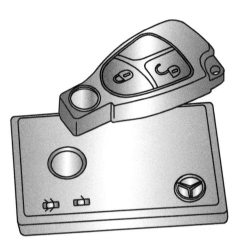

Figure 14-12 Transponder card used for entry and vehicle starting.

Figure 14-13 Start/Stop button.

When the vehicle doors are to be locked, the antennas determine if the card's signal is coming from inside or outside of the vehicle. This is determined by a decrease in antenna fields in the interior due to the sheet metal of the body. The radio range is limited to a defined range. Also, the ranges overlap between antennas further refines the location of the card. If it is determined that the card is in the vehicle or trunk, the doors cannot be locked. A warning message is displayed in the instrument cluster notifying the driver that the card is still in the vehicle.

The engine starting function can be performed only if the antenna determines that the transmitter is within the vehicle. Also, the transmission must be in PARK and the driver's foot on the brake pedal. Once these conditions are met, the driver simply presses the START/STOP button on the gear shift handle or mounted in the dash to start the engine (**Figure 14-13**). To shut off the engine, the driver can press the START/STOP button or simply exit the vehicle with the card in his/her possession. The module looks for the card every 10 seconds; if the card is no longer in the vehicle, the engine shuts off.

Not all manufacturers use the auto shut-off feature.

TIRE PRESSURE MONITORING SYSTEM

The **tire pressure monitoring system** is a safety system that notifies the driver if a tire is under- or overinflated. Most systems use a pressure sensor transmitter in each wheel. Wheel-mounted tire pressure sensors (TPSs) are attached to the rim by a sleeve nut on the valve stems (**Figure 14-14**). The TPS's internal transmitter uses a dedicated radio frequency signal to broadcast tire pressure and temperature information. In addition, it transmits its transmitter ID. Each TPS has an internal battery designed to last approximately 10 years. The battery is not serviceable.

Shop Manual
Chapter 14, page 736

Some systems only transmit tire inflation information when the vehicle is in motion, while others continue to monitor tire pressure even with the vehicle stationary. The TPSs transmit tire pressure data by RF once per minute at speeds over 20 mph (32 km/h) and up to once every hour when the vehicle is parked. A receiver module receives the tire inflation information and then sends the message over the data bus. If a tire pressure is out of the programmed parameters, a warning chime is sounded and a warning lamp is illuminated (**Figure 14-15**).

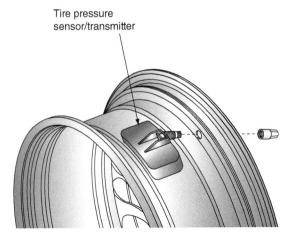

Tire pressure sensor/transmitter

Figure 14-14 The TPS is attached to the rim as part of the valve stem.

Figure 14-15 Low tire pressure warning symbol.

Although the receiver module stores each TPS's ID, some systems do not associate the ID with a particular location on the vehicle. If the pressure in a tire is outside the acceptable pressure limits, the warning would indicate a problem, but would not identify which tire. The actual tire causing the condition would need to be determined using a tire pressure gauge.

The TPS will send pressure and temperature data to the receiver module when any of the following occurs:

- If the vehicle is parked for more than 15 minutes, the TPS monitors the pressure every minute but only sends tire pressure data once every 13 hours, unless there is a change in pressure.
- While the vehicle is in motion, the TPS sends data once every 15 seconds for the first 30 data blocks and once per minute after that.
- If a drop in pressure greater than 1 psi (6.9 kPa) occurs when the vehicle is stationary, the TPS will send data once per minute.
- If a drop in pressure greater than 1 psi (6.9 kPa) occurs when the vehicle is in motion, the TPS will send data once every 5 seconds.

Logic in the receiving module prevents false warning displays to the driver. For example, since the signal transmission frequency is dependent on wheel speed, if a signal from one of the TPSs is not received as often as the other sensors, the receiving module will assume that the tire has been relocated and the spare is in use. Also, rapid tire pressure changes can occur due to changes in temperature. If a vehicle is moved from a heated garage and is driven on a cold day, a low-tire warning may be indicated. To prevent this, the system uses temperature information along with pressure data to filter the data and compensate for rapid pressure changes that occur as a result of temperature changes.

The receiver module can set a DTC when one of the transmitters fails to produce a signal. If a replacement transmitter is installed on the vehicle, the system learns the new transmitter when the vehicle is driven. When an unrecognized signal is received at the same transmission intervals as the recognized transmitters, the receiver module stores the ID and begins monitoring that transmitter. The receiver module will relearn the TPS IDs every time the vehicle is driven after being stopped for at least 15 minutes.

AUTHOR'S NOTE It may be necessary to drive the vehicle up to 10 miles before the replacement transmitter ID is learned.

Figure 14-16 Premium systems display the pressures of each tire.

Premium systems use the same basic TPSs as just described, but use additional components or programming to identify which tire is out of limits. The premium system records TPS ID locations, so this system is capable of displaying the actual tire pressures to the driver (**Figure 14-16**). Some systems require that the ID location be recorded into the receiving module. This is usually done by entering training mode and then, following the specified order of tire location, placing a magnet around the tire's valve stem. As each location is learned, the ID is locked to that position. This procedure needs to be performed each time the tires are rotated. When the magnet is placed around the valve stem, it pulls a reed switch closed and causes the TPS to send its data.

Some systems do not require ID location learning. These use a transponder (**Figure 14-17**) located behind the wheel splash shields at three locations. The fourth location is inferred when a fourth TPS ID is received while the vehicle is in motion. Tire position is relearned each time the vehicle is in motion.

The transponders use two ground terminals. One of the ground terminals is common for all transponders. The other three ground terminals are used to identify the transponder location. This arrangement allows the receiving module to determine the location of the transponder.

When the vehicle is in motion, all four TPSs send periodic signals to the receiving module. The receiving module then sends a signal to one of the transponders. This causes the transponder to emit a 125-kHz signal in the area surrounding the wheel. This signal

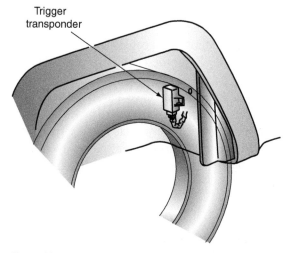

Figure 14-17 Transponder used to identify sensor location on the vehicle.

will excite the TPS at that location. Under this condition, the TPS will send a constant signal to the receiving module. The receiving module will then identify the TPS and its location. This procedure is repeated for the other two transponder locations. Once three wheel locations have been identified, the fourth location can be inferred and identified. This learning process occurs any time the vehicle is driven after being stationary for at least 15 minutes.

AUTHOR'S NOTE Because of this procedure to learn TPS locations, training magnets are not required.

Another method of tire location ID is done by locating the tire pressure monitor module in a specific location, such as in the left front fender well. The TPS broadcasts their pressures as described, but also broadcasts the direction of rotation. The module determines tire location by the strength of the signal from the TPS and its rotation direction.

A fourth method is to use the antilock brake systems (ABS) wheel speed sensors. Since the number of revolutions per mile is dependent upon the diameter of the tire, the wheel speed sensor data concerning tire speeds can be used to determine a tire with low pressures. As the tire pressure drops, the diameter of the tire effectively becomes smaller. This results in more revolutions per mile. When compared to the other wheel speed sensors, the system determines a low tire when it sees a wheel speed sensor reading that is higher than the others. Since the ABS system identifies the location of the wheel speed sensors, that data is used to identify the location of the low tire.

IMMOBILIZER SYSTEMS

The **immobilizer system** is designed to provide protection against unauthorized vehicle use by disabling the engine if an invalid key is used to start the vehicle or an attempt to hot-wire the ignition system is made. Manufacturers have several different names for this system. The following are examples of system operation.

Shop Manual
Chapter 14, page 731

The primary components of the system include the immobilizer module, ignition key with transponder chip, the powertrain control module (PCM), and an indicator lamp. The immobilizer module is composed of a magnetic coil as well as a charge/detect module (transceiver). The immobilizer module is either mounted to the steering column or in the instrument panel. It may include an integral antenna that surrounds the ignition switch lock cylinder (**Figure 14-18**). If the module is remote mounted in the instrument panel, then it is connected to an antenna by a cable. The signal communication that occurs between the transponder and the charge/detect circuit of the ECM is the heart of system operation (**Figure 14-19**)

The system includes special keys that have a transponder chip under the covering (**Figure 14-20**). Most systems are capable of recognizing up to eight different keys. Any additional keys that are to be used with the vehicle require programming to the immobilizer module. Most systems provide a procedure that allows the customer to program additional keys if they have two valid immobilizer keys that are programmed to the vehicle already.

System Operation

When the ignition key equipped with the radio frequency identification (RFID) transponder is inserted into the ignition switch cylinder and turned to the RUN position, the immobilizer module radiates an RF signal through its antenna to the key. The transponder's

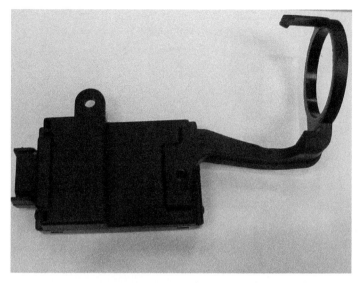

Figure 14-18 The immobilizer module with halo antenna.

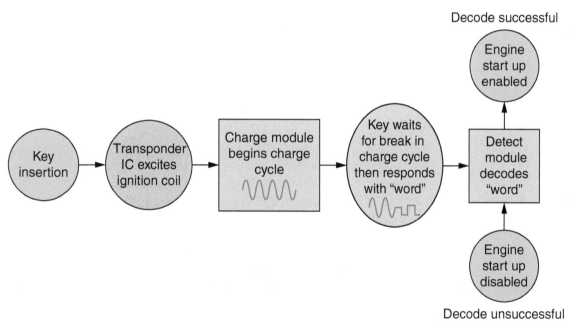

Figure 14-19 Flow chart of immobilizer system functionality.

integrated circuit induces a current in the magnetic coil and excites the coil, which feeds into a tuned circuit. The transponder signals back to the antenna its identification number. If the message properly identifies the key as being valid, the immobilizer module sends a message over the data bus to the PCM indicating that the engine may be started. If the response received from the key transponder is missing or identifies the key as invalid, the immobilizer module sends an "invalid key" message to the PCM over the data bus.

The Keyless GO system is incorporated within the immobilizer system.

The **reader** system serves as an interface between the transponder and the microprocessor.

AUTHOR'S NOTE The default condition in the PCM is "invalid key." If the PCM does not receive any messages from the immobilizer module, the engine is prevented from starting.

Figure 14-20 The immobilizer key has an internal transponder chip. The key cover was cut off for this illustration.

AUTHOR'S NOTE When the system is preventing the engine from running, it allows the engine to start and run for about 2 seconds and then shuts the engine off. Some systems prevent the starter from engaging after a set number of attempts with an invalid key are made. In this case, the vehicle will not recover until a valid key is used.

Within the immobilizer module is a **reader** that allows the module to read and process the identifier code from a transponder (**Figure 14-21**). The interface operates in two directions. In one direction, energy is transferred from the reader to the transponder. This is done by the reader creating a magnetic field through the antenna. The antenna is energized by using series resonance. The magnetic field induces a voltage in the transponder key's resonant circuit. This voltage is used to power the transponder IC. The current in the transponder coil generates a magnetic field that is superimposed to the reader's field. Provided the supply voltage to the transponder's IC is high enough, the transponder will transmit by damping the resonant circuit in accordance with the data signal.

In the other direction of interface operation, the data is transferred from the transponder to the microprocessor. The voltage signal from the transponder is smaller than that of the reader voltage. This results in a slight voltage modulation at the reader coil. Demodulation of the incoming signal is accomplished by a rectifier and decoupling capacitor (**Figure 14-22**). The signal is then amplified and conditioned into the appropriate digital output data.

The immobilizer module is programmed with a unique secret key code and also retains in memory the unique ID number of all keys that are programmed to the system. The secret key code is transmitted to the keys during the programming function and is stored in the transponder. In addition, the secret key code is programmed into the PCM.

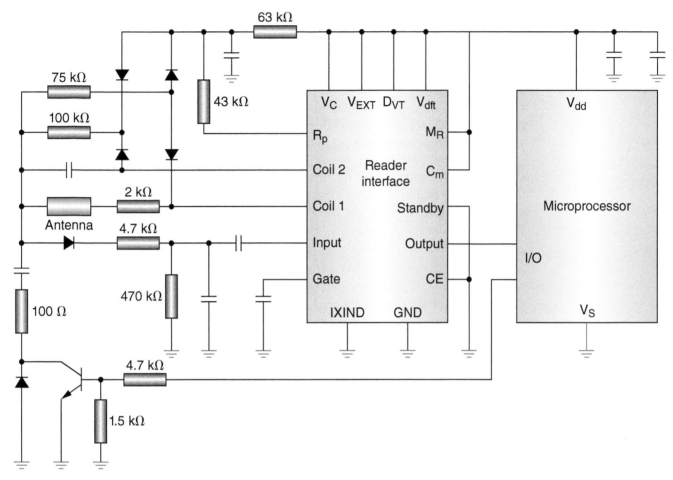

Figure 14-21 Reader interface circuit.

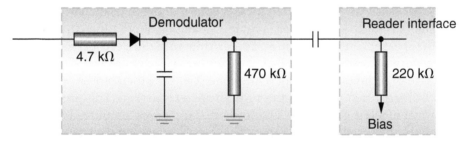

Figure 14-22 The signal from the transponder passes through a demodulation circuit to the reader.

Another identification code called a personal identification number (PIN) is used to gain secured access to the immobilizer module for service. The immobilizer module is also programmed with the vehicle identification number (VIN). All messages transmitted by the immobilizer module are scrambled to reduce the possibility of unauthorized immobilizer module access or disabling.

The immobilizer module also sends indicator lamp status messages. The indicator lamp will normally illuminate for 3 seconds for a bulb check when the ignition switch is first placed in the RUN position. After the bulb check is complete, the lamp should go out. If the lamp remains on after the bulb check, this indicates that the immobilizer module has detected a system malfunction or that the system has become inoperative. If the lamp flashes after the bulb check is completed, this indicates that an invalid key is detected or that a key-related fault is present.

Each key has a unique transponder identification code that is permanently programmed into it. When a key is programmed into the immobilizer module, the transponder identification code is then stored in the immobilizer's memory. In addition, the key learns the secret key code from the immobilizer module and programs it into its transponder memory. For the engine to start and run, all of the following must be in place:

- Each key's transponder ID must be programmed into the immobilizer module.
- The immobilizer's secret key code must be programmed into each key.
- The VIN number programmed in the immobilizer module must match the VIN in the PCM.
- The data bus network must be intact to allow messages to be sent and received between the immobilizer module and the PCM.
- The immobilizer module and the PCM must have properly functioning power and ground circuits.

AUTHOR'S NOTE   In 1996, GM introduced an immobilizer system that does not use a chip in the key. This system has a Hall-effect sensor in the key cylinder that measures the magnetic properties of the key as it is inserted into the cylinder. The cut pattern of every key has its own magnetic identity. If the wrong key is inserted into the lock cylinder, the car will not start, even if the lock cylinder turns.

SPEAKERS

Shop Manual
Chapter 14, page 744

Speakers turn the electrical energy from the radio receiver amplifier into acoustical energy. The acoustical energy moves air to produce sound. A speaker moves the air using a permanent magnet and electromagnet (**Figure 14-23**). The electromagnet is energized when the amplifier delivers current to the voice coil at the speaker. The coil forms magnetic poles that cause the voice coil and speaker cone to move in relation to the permanent magnet. The current to the speaker is rapidly changing alternating current (AC) that results in the speaker cone moving in and out in time with the electrical signal. When the cone moves forward, the air immediately in front of it is compressed. This will cause a slight increase in air pressure in front of the cone. The cone then moves back past its rest position, resulting in a reduction in the air pressure (rarefaction). This process continues so that a wave of alternating high- and low-pressure waves is radiated away from the speaker cone at the speed of sound. These changes in air pressure are actually sound.

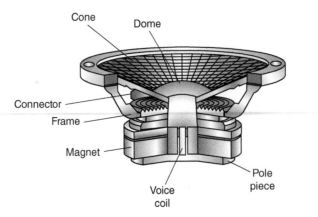

Figure 14-23 Speaker construction.

Since one speaker cannot reproduce the entire hearing frequency range of approximately 20 Hz to 20 kHz, speakers are designed to reproduce only parts of the desired frequency. Large speakers, called **woofers**, produce the low frequencies of midrange and bass better than small speakers. Smaller speakers, called **tweeters**, produce the high frequencies of treble better than large speakers. Coaxial speakers have two separate speakers combined in one speaker frame and cover a broader frequency range than a single-cone speaker. Sub-woofer speakers can be coupled with the coaxial speaker (or a separate tweeter and mid-range speaker) to cover the hearing range and maximize sound quality.

INFOTAINMENT SYSTEMS

When radios were first introduced to the automotive market, they produced little more than tinny noise and static. Today's audio sound systems produce music and sound that rivals the best that home sound systems can produce, and with nearly the same or even greater amounts of volume or sound power!

Shop Manual
Chapter 14, page 739

The most common sound system configuration is the all-in-one unit called a receiver. Housed in this unit are the radio tuner, amplifier, tone controls, and unit controls for all functions. These units may also include internal capabilities such as compact disc players, DVD players, digital audiotape players, and/or graphic equalizers. Most will be electronically tuned with a display that shows all functions being accessed or controlled and digital clock functions (**Figure 14-24**).

A BIT OF HISTORY

Radios were introduced in cars by Daimler in 1922. Cars were equipped with Marconi wireless receivers.

Recent developments by the manufacturers have been made to take individual functions (tape, disc, equalizer, control head, tuner, amplifier, etc.) and put them in individual boxes and call them components. This would allow owners greater flexibility

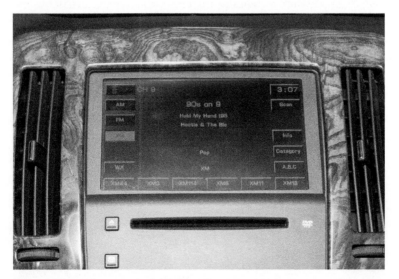

Figure 14-24 Radios are actually receivers. They contain the basic elements of a tuner, an amplifier, and a control assembly in one housing and can also contain a tape or CD player.

in selecting options to suit their needs and tastes (**Figure 14-25**). Componentizing has allowed greater dash design flexibility. Some components, such as multiple-CD changers, can be remotely mounted in a trunk area for greater security.

Wiring diagrams for component systems will be more complex (**Figure 14-26**). In addition to power and audio signal wires, note that some systems will have a serial data wire for microprocessor communication between components for the controlling of unit functions. Some functions are shared and integrated with factory-installed cell phones, such as radio mute. Some systems allow remote control of functions through a control assembly mounted in the steering wheel or alternate passenger compartment location (**Figure 14-27**).

The remote controls can either be resistive multiplexed switches or use a supplementary bus system, such as LIN. These inputs go to the controlling module. The switches send different voltage signals to the module, corresponding to which switch is pressed. The module responds by sending a request message via the data bus to the radio.

An antenna is needed to collect AM and FM radio signal waves. The radio station's broadcast tower transmits electromagnetic energy through air that induces an AC voltage that averages 50 microvolts in the antenna. The radio receiver processes this AC voltage signal and converts it to an audio output.

Some vehicles have the antenna incorporated into the rear window defogger grid. A rear window defogger/antenna module (**Figure 14-28**) separates the RF signal used by the radio from the electrical current used by the grid. When the radio is turned on, a 12-volt signal is sent to the defogger/antenna module (**Figure 14-29**). A coaxial cable from the module provides the AM and FM tuners the RF signal input. Usually the top grid lines are unheated. These are used to receive the AM signals. The heated grid lines are used to receive the FM signals.

Sound system amplifiers can be either remote mounted or integrated with the speakers. Most remote-mounted amplifiers are connected to the data bus. This allows for matching the configuration to the vehicle. Amplifier configuration includes amplification output power, the number of output channels, and the equalizer curve.

Shop Manual
Chapter 14, page 740

Figure 14-25 Components, like those shown here, allow for more options to the sound system.

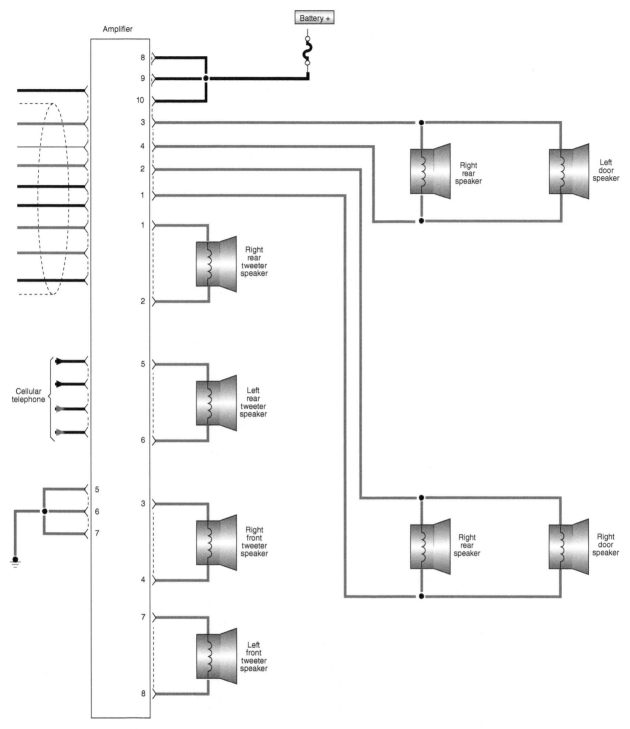

Figure 14-26 Schematic diagram showing the wiring hook-ups for remote-mounted components. Audio system with steering wheel–mounted remote control.

Functions that are supported by data bus inputs to the amplifier include speed-proportional volume increase, designating speakers for shared audio functions, and fixed audio output if an amplified speaker system is used.

Integrated amplifiers are mounted in the speakers. Power to the amplifiers is supplied through the amplifier relay. When the radio is turned on, 12 volts (V) are supplied to energize the relay.

Figure 14-27 Audio system with steering wheel–mounted remote control.

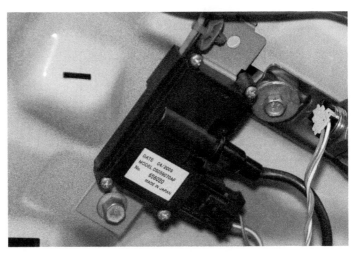

Figure 14-28 Defogger/antenna module.

Shop Manual
Chapter 14, page 743

Reception to the radio receiver from the antenna can be interfered with by noise resulting from radio frequency interference (RFI) and electromagnetic interference (EMI). This is especially true of the AM band. The noise is picked up by the radio receiver and amplified through the audio circuits. The FM band is susceptible to RFI and EMI, but usually is not as noticeable compared to AM. Control of RFI and EMI noise is done by proper radio antenna base, proper radio receiver, proper engine-to-body ground, and proper heater core grounds. In addition, resistor-type spark plugs and radio suppression secondary ignition wiring are used. The radio itself will also have internal suppression devices, such as capacitors that shunt AC noise to ground and slow sudden changes of voltage in a circuit. Some systems will use a **radio choke**. The choke is a winding of wire. In a direct current (DC) circuit the choke acts like a short, but in an AC circuit it represents high resistance. The choke blocks the noisy AC current but allows the DC current to pass normally.

Satellite Radio

Shop Manual
Chapter 14, page 740

Satellite radio (**Figure 14-30**) provides several music and talk show channels using orbiting satellites to provide a digital signal. An in-vehicle receiver receives the digital signal, which travels to the conventional vehicle radio as an auxiliary audio input and plays

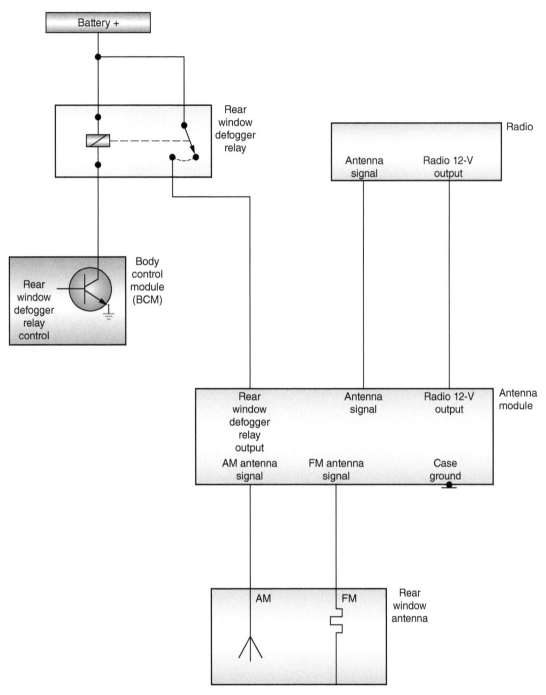

Figure 14-29 Rear window defogger/antenna circuit.

through the normal speaker system. In this case, the satellite radio function becomes an additional mode of the radio. Satellite radio operation is available only when the owner purchases the subscription service.

AUTHOR'S NOTE If all satellite signals are lost, the radio becomes totally silent, unlike the loss of an AM or FM signal where some static or hiss may be heard.

Figure 14-30 Satellite radio system.

The satellite digital audio receiver (SDAR) is an additional audio receiver that is installed separately from the vehicle's radio. The satellite signal is processed by the SDAR, which provides an input signal to the conventional radio over dedicated circuits for right and left channels. The usual radio controls for mode selection and tuning are bussed to the SDAR. This allows the radio controls to operate the SDAR. Any text information such as channel numbers, track, and artist that the stations transmit is sent to the radio over the data bus and is displayed on the radio screen.

The SDAR signals the satellite antenna located on the centerline of the roof (**Figure 14-31**), and the tuning of the antenna is calibrated to use the ground plane of the vehicle. Signal reception is possible only when the antenna and the satellite are in a direct line. Obstacles such as buildings, overpasses, and tunnels may temporarily disrupt the signal. A buffer is included that prevents brief interruptions when the vehicle passes under bridges.

Some radios have only one audio input. In order to add multiple audio sources such as SDAR, CD/DVD, and hands-free cell phone, a **multiplexer** may be used. The multiplexer acts as an electronic switch to change between the different audio sources. Depending on which source requires use of the system, the multiplexer will make the switch and send the output to the radio.

Figure 14-31 Satellite radio antenna

DVD Systems

There are several digital video disc/video entertainment systems (DVD/VES) available on today's vehicles. Most DVD systems display the video on a flip-down, roof-mounted monitor (**Figure 14-32**) or on a monitor attached to the back of the front seats (**Figure 14-33**). Most will play the audio through the vehicle's regular audio system. If a DVD is playing, the audio system automatically switches to a "surround sound" mode that is biased toward the rear of the vehicle. In addition, remote control and auxiliary input jacks that permit the display of video cameras and video games may also be included.

Shop Manual
Chapter 14, page 740

When headphones are supported, the speakers may play audio from one source while the headphones can play audio from another source. If the headphones and the radio are playing in the same mode, the radio is the master of the system. Headphones use either a 900-MHz signal or infrared. The headphones enable rear-seat occupants to listen to audio while front-seat occupants listen to CDs or the radio on the front speakers. The headphones receive their signals from the radio or the CD changer. Two separate channels are used to minimize interference. If another audio/video system is within range, the headphones automatically switch to the strongest channel.

Figure 14-32 Flip-down, roof-mounted DVD monitor.

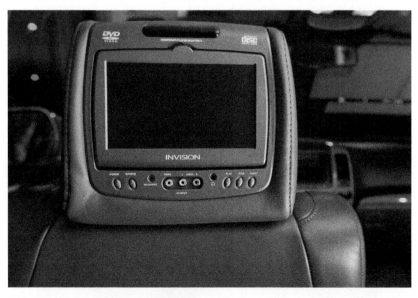

Figure 14-33 DVD monitor located in the seat back.

Shop Manual
Chapter 14, page 739

TELEMATICS

Telematics refers to the ability to connect the vehicle to the outside world. The vehicle telematics system can receive one-way transmissions from GPS satellites, XM Radio stations, HD Radio stations, and other sources. In addition, it can perform two-way communications to exchange data with cell phone networks by using the customer's smartphone. Several systems can be integrated into the operation of the telematics system. These may include the navigational system, smartphones, music players, emergency services, satellite and HD radio, and Wi-Fi connectivity. The vehicle's telematics system acts as the central hub where all the services connect.

WI-FI Connectivity

Shop Manual
Chapter 14, page 746

Wi-Fi is the wireless internet connection and network standard using the 2.4 GHz (12 cm) UHF and 5 GHz (6 cm) SHF ISM radio bands. Many vehicle manufacturers are offering Wi-Fi connectivity to the internet from the vehicle using a cellular connection. Most systems use Wi-Fi to create a Hotspot in the vehicle to allow for the connection of portable devices to the internet (**Figure 14-34**). The system can be expanded to include network connectivity to home or office computers for the sharing of information, folders, and files. Expansion of the Wi-Fi system can allow over-the-air vehicle computer software updates.

Like a home Wi-Fi system, the vehicle Wi-Fi requires a subscription and a cellular modem. In-vehicle systems utilize an external antenna that provides improved signal strength over that offered by portable Hotspot devices. The telematics system can use Wi-Fi to allow the use of smartphone apps to perform many functions not limited by distance. This includes the functions of the keyless entry FOB and remote start. From anywhere the owner can lock/unlock the doors, lower/raise the windows, start/shut off the engine, adjust the climate control systems, set or disarm the antitheft system. In actuality, any computer-controlled system can be accessed through the app from anywhere there is internet access.

Hands-Free Cell Phone System

Shop Manual
Chapter 14, page 744

Shop Manual
Chapter 14, page 748

The hands-free cell phone system can use different wireless technologies. Bluetooth™ wireless technology is the most common and is discussed here. As discussed in Chapter 10, this technology allows for communication between modules. In this case, the communication is between a compatible cell phone and the vehicle's on-board receiver. The system communicates with a cell phone that is anywhere within the vehicle.

Figure 14-34 The telematics system may offer Wi-Fi connectivity.

The system recognizes several cell phones. Each cell phone is given an identification number, name, and priority by the user during the setup process. The assigning of this information pairs the cell phone to the system. The pairing process stores the cell phone's IP address in the hands-free module (HFM) and the HFM's IP address into the cell phone.

The system uses voice recognition technology to control operation. When a programmed cell phone is within range of the HFM, communication between the two components is established. When a cell phone call is initiated, the voice is broadcast through the radio speakers. If any other audio device is using the speakers at that time, the hands-free system automatically overrides it.

When the HFM broadcasts a data bus message that hands-free operation is about to be initiated, the vehicle radio stores its current volume level and mode. The radio then switches to the stored hands-free volume level. The amplifier fades all of the speakers and then transmits the hands-free audio through the speakers. When the HFM broadcasts a data bus message that the cell phone call has been terminated, the radio stores the HFM volume level and returns to the previous radio mode and volume level. At this time, the amplifier returns all speakers to their previous audio level.

A dual-element microphone module (DEMM) is located in the dash, center console, or rearview mirror. The DEMM consists of microphone elements and electrical circuitry that includes a preamplifier network. The DEMM is capable of tuning the microphone frequency response to improve the voice recognition function.

Systems that incorporate the hands-free feature into the telematics system will typically use the cell phone to route audio and data to the telematics control head. However, some systems can access the local cell network directly to allow the vehicle to transmit and receive emergency voice communication and data even when no cell phone is present.

Navigational Systems

Many manufacturers offer **navigational systems** as an option on their vehicles, and are one of the fundamental components of the telematics system (**Figure 14-35**). These systems use satellites to direct drivers to their desired destinations. Many navigational systems are integrated in the vehicle radio. Regardless of the display method, most navigational systems use a global positioning system (GPS) antenna, to determine the vehicle's location by latitude and longitude coordinates, and a gyroscope to determine vehicle turns. Usually map data, provided on a DVD, and navigation information are displayed on a thin-film

Bluetooth is a wireless connection technology that uses the 2.4-GHz frequency band to connect a Bluetooth capable phone to the telematics system. This allows for hands-free function of the cellular phone.

Figure 14-35 Navigational system monitors and controls.

transistor (TFT), liquid crystal display (LCD) color screen. Voice prompts can be sent through the audio system speakers.

Most systems will provide at least some of the following features:

- Full-screen map display.
- Vehicle location and route guidance.
- Turn-by-turn distance in feet or meters.
- Points of interest.
- The storage of favorite routes and locations.
- The storage of recent routes.

During most conditions, the gyroscope, data bus information, and the map data locate the vehicle on the displayed map. If the vehicle is traveling in an unmapped area, the system uses the GPS data.

The GPS system can also be incorporated into a vehicle-tracking system. If the vehicle is stolen, its whereabouts can be tracked using the satellite. If the navigational system is tied to the hands-free cell phone system, the driver can get on-road assistance in the event of a problem. For example, if the driver locks the keys in the vehicle, he or she can call the assistance line and the representative can send a signal to unlock the doors. If the air bags deploy, the system can automatically send a signal of this event. A representative from the tracking subscription company will attempt to get in touch with the vehicle occupants to see if assistance is required. Emergency personnel can be dispatched because the satellite system informs the representative of the exact location of the vehicle.

Gyro Sensors

Differing types and names of X-Y-Z axis sensors are used in vehicle control, rollover mitigation, and navigational systems. They are used to detect yaw, lateral, and roll movements of the vehicle. Each of these measurements is done by individual sensors, but can be housed in a single unit.

Positions in space are given in X-Y-Z coordinates. These coordinates influence the forces that are acting upon a vehicle in motion. The yaw and lateral sensors can also be used in the navigational system to determine vehicle acceleration rate and turning.

A **gyroscope** is a device used to measure orientation in space (**Figure 14-36**). The operation of the gyroscope is founded on the principle of conservation of

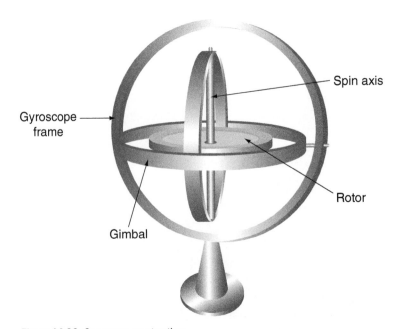

Figure 14-36 Gyroscope construction.

angular momentum. This is similar to inertia in that the angular momentum is the measure of the extent that an object rotating around a reference point will continue to rotate unless acted upon by an external torque. This "gyroscopic inertia" resists changes in orientation and makes the instrument a very good indicator of orientation. In fact, gyroscopes are often used as the element in high-quality compasses instead of the typical magnets. This is because a magnetic compass will point to magnetic north, which is actually some distance from true north. This variance needs to be compensated for when using a magnetic compass. The gyroscope possesses a much more accurate directional indicator.

Since gyroscopes of this type are not practical in an automotive application, it was not until **Micro-Electro-Mechanical System (MEMS)** gyros were developed that more advanced safety and navigational systems could be installed onto the vehicle. The vibrating structure gyroscope incorporates two proof masses that are induced to vibrate at a frequency along a plane within these small units. The Coriolis effect induces an acceleration on these masses. Due to the force of the Coriolis effect, lateral motion can be measured. This produces a signal that corresponds to the rate of rotation.

MEMS gyro sensors are typically used by navigational systems that rely on satellites and GPS to determine the position of the vehicle. The gyro informs the system of the speed and direction of the vehicle. If the vehicle is making a turn, the direction displayed on the monitor will indicate this change.

The gyro sensor measures the rotation of the vehicle around the Z-axis. This input, along with a lateral acceleration sensor and steering angle sensor, is used to indicate the vehicle's path of travel.

Micro-Electro-Mechanical System (MEMS) refers to the process of integrating mechanical elements, sensors, actuators, and electronics onto a common silicon substrate. This is done by a process called micro-fabrication technology. The electronic portion of the MEMS is fabricated using the integrated circuit (IC) process. The micromechanical components are fabricated using a process that etches away parts of the silicon wafer and then adds new layers to the structure that will form the desired mechanical and electromechanical device.

Dead Reckoning

In the event that GPS satellite signals are lost, the navigation system will invert to **dead reckoning**. This is sometimes evidenced on the monitor as the vehicle being shown in a location that it really is not—that is, showing the vehicle driving down railroad tracks instead of on the highway. Signal loss can occur when driving through a tunnel, in a canyon, or in parking garages. Using the last known position and direction of travel, the navigational system will use vehicle speed and gyro sensor inputs to calculate the approximate position of the vehicle. Many systems will display a message such as, "Satellite Signal Lost" to alert the driver that dead reckoning is being used and actual vehicle position may not be accurate.

Dead reckoning calculates position of the vehicle using the previously determined location and then uses vehicle speed, time, and gyro inputs to estimate current location.

SUMMARY

- Any system that is using radio frequency to broadcast data must have a transmitter and a receiver.
- The transmitter generates the radio frequency by taking the data and encoding it onto a sine wave. A receiver receives and decodes the sine wave.
- The transmitter and the receiver require an antenna to radiate and capture the radio wave signal. Antennas convert radio frequency electrical currents into electromagnetic waves and vice versa.

- There are three common ways to modulate a sine wave: pulse modulation, amplitude modulation, and frequency modulation.
- In a simple crystal radio, a capacitor/inductor oscillator performs as the tuner for the radio that resonates and amplifies one frequency and ignores all the others.
- A detector extracts the data (voices and music) from the sine wave.

- The electronics of the remote keyless entry (RKE) system are responsible for translating coded input from the key FOB into commands for driver outputs.
- Rolling code uses a counting system that requires a different signal every transmission to improve security. The RF generator modulates the carrier code, creating an output for the RKE receiver.
- The keyless start system is an enhancement to the remote keyless entry system that allows the vehicle to be started without the use of an ignition key.
- The keyless start system uses a transmitter card or a FOB, signal acquisition and actuation modules (SAMs) that are installed throughout the vehicle, an electronic ignition switch (EIS), and antennas. The electromagnetic field of the antennas causes the transmitter card to send its code by RF signal and determines if the transmitter card is located within the vehicle or outside of the vehicle.
- Keyless start systems will only perform the engine-starting function if the antenna determines that the transmitter is inside the vehicle.
- The tire pressure monitoring (TPM) system is a safety system that notifies the driver if a tire is under- or overinflated.
- Most TPM systems use a pressure sensor transmitter in each wheel. The TPS's internal transmitter uses a dedicated radio frequency signal to broadcast tire pressure and temperature information. In addition, it transmits its transmitter ID.
- Premium TPM systems use additional components or programming to identify which tire is out of limits. These systems can use a transponder located behind the wheel splash shields, signal strength and rotation, or the ABS wheel speed sensors to determine tire location.
- The immobilizer system is designed to provide protection against unauthorized vehicle use by disabling the engine if an invalid key is used to start the vehicle or an attempt to hot-wire the ignition system is made.
- The primary components of the immobilizer system include the immobilizer module, ignition key with transponder chip, the PCM, and an indicator lamp.
- When the ignition key equipped with the radio frequency identification (RFID) transponder is inserted into the ignition switch cylinder and turned to the RUN position, the immobilizer module radiates an RF signal through its antenna to the key. The transponder signals back to the antenna its identification number. If the message properly identifies the key as being valid, the immobilizer

- module sends a message over the data bus to the PCM indicating that the engine may be started.
- The immobilizer module is programmed with a unique secret key code and retains in memory the unique ID number of all keys that are programmed to the system. The secret key code is transmitted to the keys during the programming function and is stored in the transponder.
- Another identification code called a PIN is used to gain secured access to the immobilizer module for service. The immobilizer module also is programmed with the vehicle identification number (VIN).
- Infotainment refers to a type of media system that provides a combination of information and entertainment.
- Speakers turn the electrical energy from the radio receiver amplifier into acoustical energy. The acoustical energy moves air to produce sound.
- Vehicle audio entertainment systems are generally only a single component (receiver). Recently, manufacturers have been developing multiple components (tuner, amplifier, control head, etc.) to allow for greater flexibility.
- Some audio component systems utilize a serial data line to provide for communication and control of functions between components.
- The infotainment system remote controls can either be resistive multiplexed switches or use a supplementary bus system such as LIN.
- Sound system amplifiers can be either remote mounted or integrated with the speakers.
- Satellite radio provides several music and talk show channels using orbiting satellites to provide a digital signal. An in-vehicle receiver receives the digital signal, which travels to the conventional vehicle radio as an auxiliary audio input and plays through the normal speaker system.
- The satellite digital audio receiver (SDAR) is an additional audio receiver that is installed separate from the vehicle's radio. The satellite's signal is processed by an SDAR, which provides an input signal to the conventional radio over dedicated circuits for right and left channels.
- The multiplexer acts as an electronic switch to switch between the different audio sources. Depending on which source requires use of the system, the multiplexer makes the switch and sends the output to the radio.
- Most DVD systems display the video on a flip-down, roof-mounted monitor or on a monitor attached to the back of the front seats, and most play the audio through the vehicle's regular audio system.

- Hands-free cell phone system uses wireless technologies such as Bluetooth™, which allows for communication between a compatible cell phone and the vehicle's on-board receiver.
- Navigational systems use satellites to direct drivers to their desired destinations using a GPS antenna, to determine the vehicle's location by latitude and longitude coordinates, and a gyroscope to determine vehicle turns.
- The navigational system can also be incorporated into a vehicle-tracking system. If the vehicle is stolen, its whereabouts can be tracked using the satellite.

REVIEW QUESTIONS

Short-Answer Essays

1. Explain the operation of the immobilizer system.

2. Explain how radio waves are generated.

3. Explain the operation of the transmitting antenna.

4. How does the keyless start system determine if the transmitter is in the vehicle?

5. Explain the operating characteristics of speakers.

6. What methods are used to reduce RFI and EMI in the radio?

7. What are the three methods that can be used by the tire pressure monitoring system to determine the location of the tire on the vehicle?

8. Describe how the audio system changes outputs when a hands-free cell phone call is initiated.

9. Explain the principle of the MEMS gyro sensor.

10. How does the navigational system track vehicle movement using dead reckoning?

Fill in the Blanks

1. The _____ acts as an electronic switch to alternate between the different audio sources that travel to the conventional vehicle radio as an auxiliary audio input and play through the normal speaker system.

2. The tire pressure monitoring system uses a(n) _____ _____ signal to broadcast tire pressure and temperature information.

3. Satellite radio systems use an in-vehicle receiver that receives the _____ signal.

4. The _____ resonates and amplifies one frequency and ignores all the others.

5. The transmitter generates the radio frequency by taking the data and encoding it onto a(n) _____ wave.

6. _____ convert radio frequency electrical currents into electromagnetic waves.

7. Speakers turn the electrical energy from the radio receiver amplifier into _____ energy.

8. A tire pressure monitoring system can excite the tire pressure sensors to cause them to transmit a constant data stream by using a_____ Hz frequency.

9. The keyless start system will use_____ to determine the location of the transponder.

10. Any system that is using radio frequency to broadcast data must have a _____ and a _____.

Multiple Choice

1. The tire pressure monitoring system can identify the location of the tire on the vehicle by:
   A. Training magnets.
   B. Electronic transponders.
   C. Both A and B
   D. None of the above.

2. Transmitting antennas work by:
   A. Oscillating current to create a magnetic field.
   B. Magnetic field induction of an electric field.
   C. Both A and B
   D. Neither A nor B

3. What is the function of the radio's detector?
   A. Determines the frequency to amplify.
   B. Extracts the data from the sine wave.
   C. Receives the sine waves from space.
   D. All of the above

4. If a radio system has only one audio input but multiple audio sources, it may use a(n) _____ to switch between the sources.
   A. Decoder
   B. Translator
   C. Multiplexer
   D. Grid/antenna module

5. The speaker produces sound by:
   A. Creating a noise from a vibrating diaphragm.
   B. hanging air pressures in front of the speaker cone.
   C. Rubbing two dissimilar materials together at the speed of sound.
   D. Changing resistance of the permanent magnet.

6. Each key of the immobilizer system has a unique:
   A. Secret key code.
   B. Transponder ID code.
   C. Cover interface.
   D. TFT overlay.

7. The purpose of transmitting temperature information along with the tire pressure readings is to:
   A. Alert the driver of tire over temperature conditions.
   B. Prevent the tire pressure monitoring system from using the data sent by the tire pressure sensors when the tires are cold.
   C. To prevent false low tire pressure warning messages from being displayed.
   D. To alert the driver that ice may be on the road.

8. The satellite radio signal is processed by:
   A. The antenna.
   B. The radio mode selection transformer.
   C. Data bus interface chip.
   D. The satellite digital audio receiver.

9. A radio choke is a winding of wire that is used to:
   A. Allow the DC current to pass normally.
   B. Block AC current.
   C. Both A and B
   D. Neither A nor B

10. The vibrating structure gyroscope can measure lateral motion by:
   A. The force of the Coriolis effect on two elements.
   B. Stress on the Wheatstone bridge.
   C. Capacitance discharge between vibrating plates.
   D. All of the above.

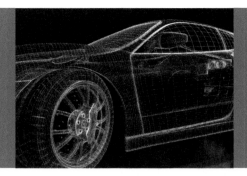

CHAPTER 15

PASSIVE RESTRAINT AND VEHICLE SAFETY SYSTEMS

Upon completion and review of this chapter, you should be able to:

- Explain the purpose of passive restraint systems.
- Describe the basic operation of passive seat belt systems.
- Describe the common components of an air bag system.
- List the components and explain the function of the air bag module.
- Describe the function of the clockspring.
- Explain the functions of the ORC used in air bag systems.
- List and describe the operation of the different types of air bag system sensors.
- List the sequence of events that occur during air bag deployment.
- Describe normal operation of the air bag system warning light.
- Describe the operation of a hybrid inflator module, and explain the advantages of this type of module.
- Explain the function of multistage air bags.
- Explain the function of the side-impact air bags and describe the locations of the modules and sensors.

- Describe the operation and purpose of factory-installed air bag on/off switches.
- Describe the purpose and operation of seat belt pretensioners.
- Describe the function of occupant classification systems (OCS).
- Describe the use of proximity sensors in today's automotive systems.
- Describe the operation of ultrasonic sensors.
- Describe the operation of visual imaging systems.
- Describe how the the infrared sensor is used to determine if an object is in the proximity and to detect obstructions.
- Describe how the the infrared sensor can be used in conjunction with a camera to assist in determining the shape and size of an object under low-light conditions.
- Explain the function of lane departure warning systems.
- Explain the functional characteristics of the park assist system.
- Describe the operation of rollover mitigation systems.

Terms To Know

Air bag
Belt tension sensor (BTS)
Carriers
Complementary metal oxide semiconductor (CMOS)
Crash sensor
Electronic roll mitigation (ERM)
Fifth percentile female
Fuel pump inertia switch
Haptic feedback

Hybrid air bag
Igniter
Inertia lock retractors
Driver-side inflatable knee blocker (IKB)
Lane departure warning (LDW)
Low voltage differential signaling (LVDS)
Multistage air bags
Park assist system

Piezoelectric accelerometers
Pretensioners
Proximity sensor
Safing sensor
Seat track position sensor (STPS)
Squib
Ultrasonic sensors
Ultrasound

INTRODUCTION

In a two-point system, the occupant must manually lock the lap belt.

Federal regulations have mandated the use of automatic passive restraint systems in all vehicles sold in the United States after 1990. Passive restraint systems operate automatically, with no action required on the part of the driver or occupant. Two- or three-point automatic seat belt and **air bag** systems are currently offered as a means of meeting this requirement.

In this chapter, you will learn the operation of the automatic passive restraint and air bag systems. The safety of the driver and/or passengers depends on the technician properly diagnosing and repairing these systems. As with all electrical systems, the technician must have a basic understanding of the operation of the restraint system before attempting to perform any service.

There are many safety cautions associated with working on air bag systems. Safe service procedures are accomplished through proper use of the service manual and by understanding the operating principles of these systems.

In addition, in 1998 the Federal Government mandated the Federal Motor Vehicle Safety Standard 201. These standards are known as the Interior Trim and Safety Refinements Acts. They regulate the requirements for occupant protection during accidents. In particular they address head injuries that may occur during an accident. As a result of this act, honeycombed plastic structures have been developed for use as trim and door materials in place of steel. The act also covers other structural components such as the instrument panel, the windshield, the rear window, and all side glass, along with all of the pillars, headers, roof, and side rails. It also includes the stitching of the seat belt loops. This means the technician must use all fasteners and proper service procedures whenever he/she is performing any service that requires the removal of any of these components.

This chapter will also introduce some of the newest vehicle safety features including vision systems, rollover protection, collision detection, lane-change departure, and rollover mitigation.

PASSIVE SEAT BELT SYSTEMS

Shop Manual
Chapter 15, page 780

The passive seat belt system automatically puts the shoulder and/or lap belt around the driver or occupant (**Figure 15-1**). The automatic seat belt system operates by means of direct current (DC) motors that move the belts by means of **carriers** on tracks (**Figure 15-2**). The carriers are attached to the shoulder anchor to move or carry the anchor from one end of the track to the other.

One end of the seat belt is attached to the carrier; the other end is connected to the **inertia lock retractors** (**Figure 15-3**). Inertia lock retractors use a pendulum mechanism to lock the belt tightly during sudden movement. When the door is opened, the outer end

Figure 15-1 Passive automatic seat belt system operation.

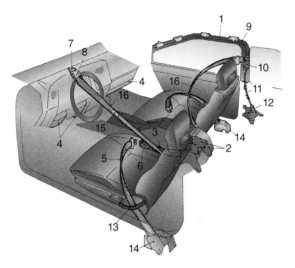

1 Rail and motor assembly
2 Emergency locking retractor assembly
3 Belt guide
4 Knee panel
5 Outer bell assembly (manual tap belt)
6 Inner belt assembly (manual lap belt)
7 Shoulder anchor
6 Emergency release buckle
9 Rail
10 Locking device
11 Tube
12 Motor
13 Belt holder
14 Emergency locking retractor assembly (manual lap belt)
15 Caution label
16 Shoulder belt

Figure 15-2 Passive seat belt restraint system uses a motor to put the shoulder harness around the occupant.

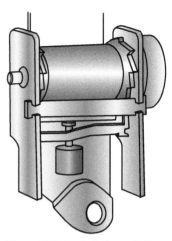

Figure 15-3 Inertia lock seat belt retractor.

of the shoulder harness moves forward (to the A-pillar) to allow for easy entry or exit (**Figure 15-4**). When the door is closed and the ignition switch is placed in the RUN position, the motor moves the outer end of the harness to the locked position in the B-pillar (**Figure 15-5**).

The automatic seat belt system uses a control module to monitor operation (**Figure 15-6**). The monitor receives inputs from door ajar switches, limit switches, and the emergency release switch.

A BIT OF HISTORY

In the 1920s, there was a movement by some American physicians to have the automobile manufacturers install padded dashboards and seat belts in cars. However, it took until the 1950s before vehicle manufacturers began to incorporate these recommendations. Volvo and Saab were the first manufacturers to install padded dashboards and seat belts.

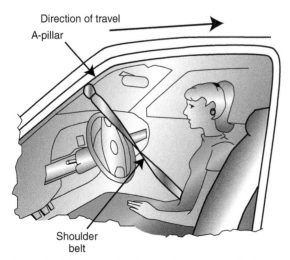

Figure 15-4 When the door is opened, the motor pulls the harness to the A-pillar.

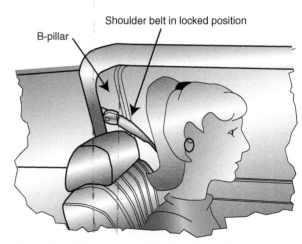

Figure 15-5 When the door is closed and the ignition switch is in the RUN position, the motor draws the harness to its lock position.

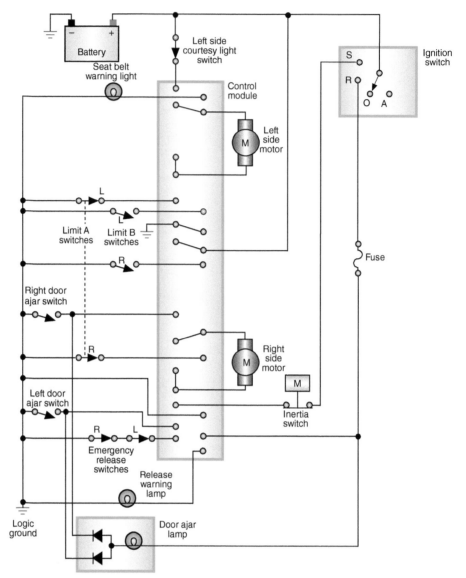

Figure 15-6 Typical circuit diagram of an automatic seat belt system using a control module.

The door ajar switches signal the position of the door to the module. The switch is open when the door is closed. This signal is used by the control module to activate the motor and move the harness to the lock point behind the occupant's shoulders. If the module receives a signal that the door is open, regardless of ignition switch position, it will activate the motor to move the harness to the FORWARD position.

The limit switches inform the module of the position of the harness. When the harness is moved from the FORWARD position, the front limit switch (limit A) closes. When the harness is located in the LOCK position, the rear limit switch (limit B) opens and the module turns off the power to the motor. When the door is opened, the module reverses the power feed to the motor until the A switch is opened.

An emergency release mechanism is provided in the event that the system fails to operate. The normally closed emergency release switch is opened whenever the release lever is pulled. The module turns on the warning lamp in the instrument panel and sounds a chime to alert the driver. The opened switch also prevents the harness retractors from locking.

Ford uses a **fuel pump inertia switch** that is a normally closed switch, which opens if the vehicle is involved in an impact at speeds over 5 mph or if it rolls over. When the

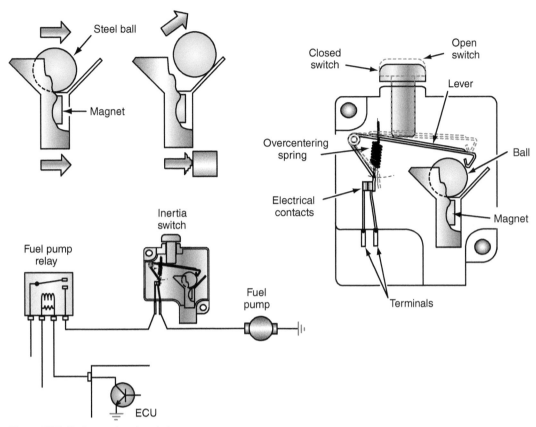

Figure 15-7 Fuel pump inertia switch.

switch opens, it turns off power to the fuel pump. This is a safety feature to prevent fuel from being pumped onto the ground or hot engine components if the engine dies. The inertia switch has been incorporated into the automatic seat belt system. If the seat belt module receives a signal that the inertia switch is open, it prevents the harness from moving to the forward position if the door opens. The switch has to be manually reset if it is triggered (**Figure 15-7**).

AIR BAG SYSTEMS

The need to supplement the existing restraint system during frontal collisions has led to the development of the supplemental inflatable restraint (SIR) or air bag systems (**Figure 15-8**).

Shop Manual
Chapter 15,
pages 780, 782

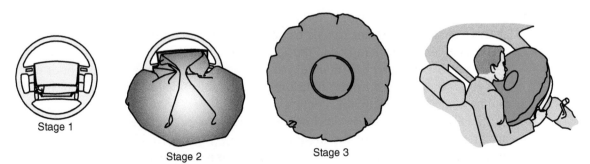

Figure 15-8 Air bag deployment sequence.

Today the most common name for the air bag system is supplemental restraint systems (SRS). The air bags are considered supplemental restraints because the seat belts must be worn at all times in order to provide maximum occupant protection. The air bag is a supplement and the seat belt is the primary restraint system. Seat belts must be worn in an air bag–equipped vehicle for the following reasons:

1. Seat belts hold the occupants in the proper position when the air bag inflates.
2. Seat belts reduce the risk of injury in less severe accidents in which the air bag does not deploy.
3. Seat belts reduce the risk of occupant ejection from the vehicle, thus reducing the possibility of injury.

 A BIT OF HISTORY

Although there were several early attempts at developing air bags, it was not until the mid-1980s that many manufacturers made air bags available as an option. In 1988, Chrysler was the first automotive manufacturer to offer the driver-side air bag as standard equipment.

The air bag system contains an inflatable air bag module that is designed into the steering wheel. Collapsible steering columns are used with the air bag system, and tilt steering wheels are still optional. If the vehicle is involved in a frontal collision, the air bag inflates rapidly to keep the driver's body from flying ahead and hitting the steering wheel or windshield. The frontal impact must be within 30° of the vehicle centerline to deploy the air bag. The air bag system helps to prevent head and chest injuries during a collision. The air bag system may be referred to as a passive restraint because it does not require active participation by the driver.

COMMON COMPONENTS

A typical air bag system consists of sensors, an occupant restraint controller, a clockspring, and an air bag module. See the typical location (**Figure 15-9**) of the common components of the SRS system.

> **AUTHOR'S NOTE** Many of the components used for driver-side air bags are similar to those used in passenger-side air bags. The basic operation of the two systems is the same.

Air Bag Module

The air bag module contains the air bag and inflator assembly packaged into a single module. This module is mounted in the center of the steering wheel (**Figure 15-10**).

The purpose of the air bag module is to inflate the air bag in a few milliseconds when the vehicle is involved in a frontal collision. A typical fully inflated driver-side air bag has a volume of 2.3 cubic feet (65 L).

The air bag module uses pyrotechnology (explosives) to inflate the air bag. The **igniter** is an integral component of the inflator assembly (**Figure 15-11**) because it starts a chemical reaction to inflate the air bag. The igniter is a combustible device that converts electrical energy into thermal energy to ignite the inflator propellant.

Shop Manual
Chapter 15, page 785

The air bag is made of neoprene-coated nylon.

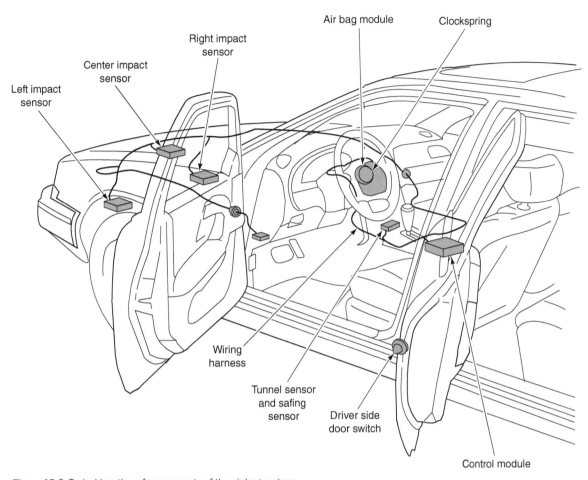

Figure 15-9 Typical location of components of the air bag system.

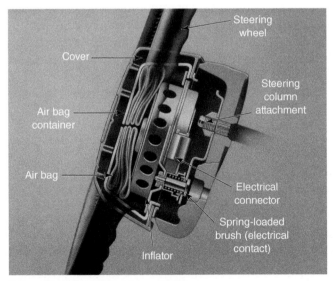

Figure 15-10 Air bag module components.

At the center of the igniter assembly is the **squib**, which contains zirconic potassium percolate (ZPP). The squib is similar to a blasting cap. Squib is a pyrotechnic term used for a fire cracker that burns but does not explode. The squib starts the process of air bag deployment. When as little as 400 mA is supplied through the squib, the air bag deploys.

The air bag module cannot be serviced. If it has been deployed, or is defective, it must be replaced.

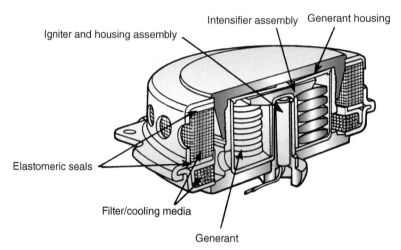

Figure 15-11 Igniter assembly.

Three components are required to create an explosion: fuel, oxygen, and heat. The squib and the igniter charge of barium potassium nitrate (a very fast-reacting explosive) provide the heat necessary for inflator module explosion. Fuel is supplied by the generant, which contains sodium azide and cupric oxide. The sodium azide provides hydrogen and the cupric oxide provides oxygen. When the chemicals in the inflator module explode, large quantities of hot, expanding nitrogen gas are produced very quickly. This expanding nitrogen gas flows through the igniter assembly diffuser, where it is filtered and cooled before inflating the air bag. Four layers of screen are positioned on each side of the ceramic in the filter. Sodium oxide dust is trapped by the filter. Sodium hydroxide is an irritating caustic. Therefore, automotive technicians are always warned to wear safety goggles and protective gloves when servicing deployed air bags. Within seconds after air bag deployment, the sodium hydroxide changes to sodium carbonate.

Tear seams in the steering wheel cover and in the instrument panel cover above the passenger-side air bag split easily and allow the air bag to exit from the module. Large openings under the air bag, where it attaches to the module, allow the air bag to deflate in 1 second so it does not block the driver's view or cause a smothering condition.

Combustion temperature in the inflator module reaches about 2,500°F (1,371°C), but the air bag will remain slightly above room temperature. Typical by-products from inflator module combustion are as follows:

1. Nitrogen—99.2%.
2. Water—0.6%.
3. Hydrogen—0.1%.
4. Sodium oxide—less than 1/10 of 1 part per million (ppm).
5. Sodium hydroxide—very minute quantity.

Many air bags pack corn starch into the inflator module. This, along with other combustion by-products, may appear as a white dust during and after air bag deployment.

Not all air bag systems use nitrogen gas to inflate the bag; some use compressed argon gas to inflate the air bag.

Clockspring. The clockspring conducts electrical signals to the module while allowing steering wheel rotation (**Figure 15-12**). The clockspring is a winding of special electric conductor tape housed in a plastic retainer. The clockspring maintains continuity between the air bar (and any steering wheel–mounted switches) and body wiring harness as the steering wheel is rotated. The clockspring is located between the column and the steering wheel. The clockspring electrical connector contains a long conductive ribbon. The wires from the air bag electrical system are connected through the underside of the clockspring

Shop Manual
Chapter 15, page 785

The clockspring is also known as the coil assembly, the cable reel assembly, the coil spring unit, and the contact reel.

Figure 15-12 The clockspring provides for electrical continuity in all steering wheel positions.

electrical connector to one end of the conductive ribbon. The other end of the conductive ribbon is connected through wires on the top side of the clockspring electrical connector to the air bag module. When the steering wheel is rotated, the conductive ribbon winds and unwinds, allowing steering wheel rotation while completing electrical contact between the system and the air bag module.

Occupant Restraint Controller (ORC). The air bag control module, called the occupant restraint controller (ORC), constantly monitors the readiness of the air SRS electrical system. If the ORC determines that there is a fault in the system, it illuminates the indicator light and stores a diagnostic trouble code. Depending on the fault, the SRS system may be disarmed until the fault is repaired.

The ORC also supplies back-up power to the air bag module in the event the battery or cables are damaged during an accident. The stored charge can last for up to 30 minutes after the battery is disconnected.

It is important for the technician to understand that not all SRS system ORCs are capable of turning off the system in the event of a fault. A typical ORC performs the following functions:

1. Controls the instrument panel warning lamp.
2. Continuously monitors all air bag system components.
3. Controls air bag system diagnostic functions.
4. Provides an energy reserve to deploy the air bag if battery voltage is lost during a collision.
5. Records sensor inputs and activation data whenever an air bag deployment is initiated.

On some systems, the ORC is responsible for deploying the air bag when appropriate signals are received from the sensors. These systems may be able to turn off the air bags if a fault is detected.

Most air bag sensors contain a resistor connected in parallel with the sensor contacts. When the ignition switch is in the RUN position, the ORC supplies a small amount of current through these resistors to monitor the system. If a short, ground, or open circuit occurs in the wiring or sensors, the current flow changes. When the ORC senses this condition, it illuminates the air bag warning light in the instrument panel.

Sensors. To prevent accidental deployment of the air bag, most systems require that at least two sensor switches be closed to deploy the air bag (**Figure 15-13**). The number of sensors used depends on the system design. Some systems use only a single sensor and others use up to five. The name used to identify the different sensors also varies among manufacturers.

Shop Manual
Chapter 15, page 782

The ORC is also called the air bag control module (ACM).

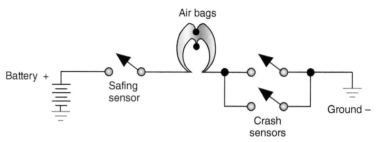

Figure 15-13 Typical sensor wiring circuit diagram.

On most three-sensor systems, the **crash sensors** are usually located in the engine compartment or below the headlights. Crash sensors are normally open electrical switches designed to close when subjected to a predetermined impact. A single **safing sensor** is usually located inside the ORC, which in turn is mounted on the centerline of the vehicle in the passenger compartment. The safing sensor determines if the collision is severe enough to inflate the air bag. When one of the crash sensors and the safing sensor closes, the electrical circuit to the igniter is complete. The igniter starts the chemical chain reaction that produces heat. The heat causes the generant to produce nitrogen gas, which fills the air bag.

There are several different types of sensors used. Common sensors include mass-type, roller-type, and accelerometers. The mass-type sensor contains a normally open set of gold-plated switch contacts and a gold-plated ball that acts as a sensing mass (**Figure 15-14**). The gold-plated ball is mounted in a cylinder coated with stainless steel. A magnet holds the ball about 1/8 inch (3.2 mm) away from the contacts. When the vehicle is involved in a frontal collision of sufficient force, the sensing mass (ball) moves forward in the sensor and closes the switch contacts.

For proper operation, sensors must be mounted with the forward marking (usually an arrow) on the sensor facing toward the front of the vehicle and in the original position designed by the manufacturer. Sensor brackets must not be bent or distorted.

Some mass-type air bag sensors contain a pivoted weight connected to a moving contact. When the vehicle is involved in a frontal collision with sufficient impact to deploy the air bag, the sensor weight moves in a circular path until the moving contact touches a fixed contact (**Figure 15-15**).

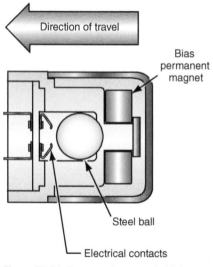

Figure 15-14 Some crash sensors hold the sensing mass by magnetic force. If the impact is severe enough to break the ball free, it will travel forward and close the electrical contacts.

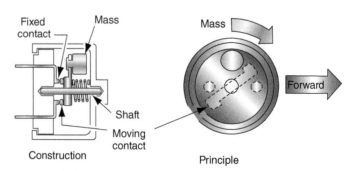

Figure 15-15 Mass-type air bag sensor with pivoted weight connected to a moving contact.

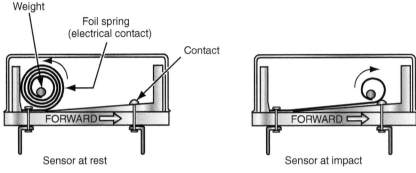

Figure 15-16 Roller-type air bag sensor.

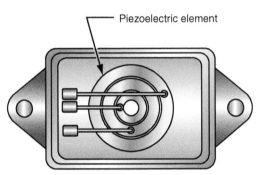

Figure 15-17 Accelerometer air bag sensor with piezoelectric element.

The roller-type sensor has a roller mass mounted on a ramp (**Figure 15-16**). One sensor terminal is connected to the ramp. The second sensor terminal is connected to a spring contact extending through an opening in the ramp without contacting the ramp. A 10,000-ohm (Ω) resistor is connected in parallel to the sensor contacts. The roller is held against a stop by small retractable springs on each side. These springs are similar to a retractable tape measure. If the vehicle is involved in a frontal collision at a high enough deceleration rate to deploy the air bag, the roller moves up the ramp and strikes the spring contact. In this position, the roller completes the circuit between the ramp and the spring contact.

In most current air bag systems, MEMS **piezoelectric accelerometers** are used to sense deceleration forces (**Figure 15-17**). The accelerometer contains piezoelectric element constructed of zinc oxide (ZnO) that is distorted during a high G force condition and generates an analog voltage in direct relationship to the amount of G force. The analog voltage is sent to a collision-judging circuit in the ACM. If the collision impact is great enough, the computer deploys the air bag.

Many MEMS-type piezoelectric sensors also contain a communication chip. This sensor will condition the analog voltage signal and send a pulsed digital signal to the ACM. The frequency of the pulses is proportional to the G force the sensor is subjected to. The MEMS sensor offers the advantages of being smaller than the piezoelectric sensor they replaced, and because these sensors have their own IC chip, they can perform self-diagnostic routines.

Air bag systems do not deploy the air bag based on vehicle speed information. Most systems do not even have vehicle speed input. The sensors close or provide an electrical signal based on the rate of deceleration. Most systems require about 30 G of force before the air bag is deployed.

Shorting Bars. The SRS electrical system is a dedicated system that is not interconnected with other electrical systems on the vehicle. All wiring harness connectors in the system are the same color for easy identification. Shorting bars are located in some of the component wiring harness connectors in the air bag system, such as the inflator module

The **piezoelectric accelerometer** generates an analog voltage proportional to a G force.

Shop Manual
Chapter 15,
page 788, 792

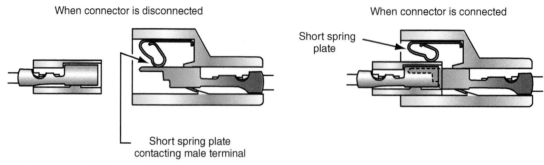

When connector is disconnected   When connector is connected

Short spring
plate

Short spring plate
contacting male terminal

Figure 15-18 Shorting bars on air bag system wiring connectors.

connector at the steering column base. The shorting bars connect terminals together when the wiring connectors are disconnected (**Figure 15-18**). Since the terminals are shorted, there is no way to have electric potential, which prevents accidental air bag deployment. If the ORC connector is disconnected, the shorting bars in the connector illuminate the air bag light.

AIR BAG DEPLOYMENT

Shop Manual
Chapter 15, page 784

The sequence of events occurring during an impact of a vehicle traveling at 30 mph (48 km/h) is as follows:

1. When an accident occurs, the arming sensor is the first to close. It closes due to sudden deceleration caused by braking or immediately upon impact. One of the crash sensors then closes. The amount of time required to close the switches is within 15 milliseconds.
2. Within 40 milliseconds, the igniter module burns the propellant and generates the gas to completely fill the air bag.
3. Within 100 milliseconds, the driver's body has stopped forward movement and the air bag starts to deflate. The air bag deflates by venting the nitrogen gas through holes in the back of the bag.
4. Within 2 seconds, the air bag is completely deflated.

AIR BAG WARNING LAMP

The air bag system warning lamp indicates the system condition to the driver. The warning lamp is operated by the ORC. Ignition on and crank signals are received by the ORC. When the ignition switch is placed in the RUN position, the air bag warning lamp should illuminate for a bulb check. In some systems, the lamp will flash seven to nine times and then remain steadily illuminated while the engine is cranking. Once the engine starts, the air bag warning lamp should be extinguished. An air bag system failure may be indicated by any of the following warning lamp conditions:

1. If the lamp remains on but does not flash when the ignition is turned on.
2. If the lamp flashes seven to nine times and then remains on when the ignition is turned on.
3. If the lamp comes on when the engine is running.
4. If the lamp does not come on at any time.
5. If the lamp does not come on steadily while the engine is cranking.

If any of these lamp conditions are present, the driver should have the air bag system checked.

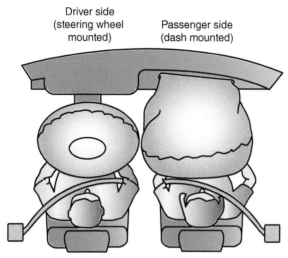

Driver side
(steering wheel
mounted)

Passenger side
(dash mounted)

Figure 15-19 Components of typical driver- and passenger-side air bag system.

PASSENGER-SIDE AIR BAGS

Federal law expanded to require that all passenger vehicles produced after 1995 be equipped with front passenger-side air bags (**Figure 15-19**). Since there is a greater distance between the passenger and the instrument panel compared to the distance between the driver and the steering wheel, the passenger-side air bag is much larger. A typical passenger-side air bag has a fully inflated volume of 7 cubic feet (198 L). In most systems, the passenger-side air bag deploys with the driver-side air bag.

AUTHOR'S NOTE Reference here to passenger-side air bags means the front-seat passenger air bag that is deployed from the instrument panel. This system is not to be confused with side-impact air bags, which will be discussed later.

HYBRID AIR BAG TYPES

Up to this point, the discussion of air bags has centered on the conventional sodium azide type. However, there are **hybrid air bag** systems that use compressed gas to fill the air bag. There are three common types of hybrid air bag modules.

Solid Fuel with Argon Gas

The first use of solid-fuel hybrid air bags was on the passenger-side air bags. They are now used for both driver- and passenger-side air bags. The hybrid inflator module contains an initiator similar to the squib in other inflator modules. However, the hybrid inflator module also has a container of pressurized argon gas (**Figure 15-20**). The same method is used to energize the initiator in the hybrid inflator as in conventional systems. When the initiator is energized, the propellant surrounding the initiator explodes and pushes out the burst disc. As the pressurized argon escapes through the exhaust holes and fills the air bag, the burning propellant heats the argon gas (**Figure 15-21**). Heating of the gas makes it expand quickly to fill the air bag.

The early version of the hybrid system used a pressure sensor mounted in the end of the argon gas chamber opposite from the initiator and propellant. This sensor sends a signal to the ORC in relation to the argon gas pressure. If the gas pressure decreases below a preset value, the module illuminates the air bag warning light.

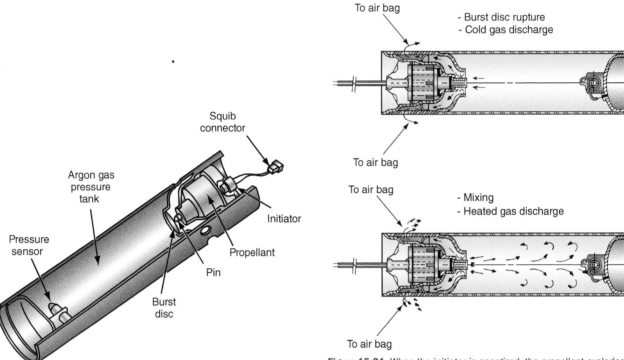

Figure 15-20 Hybrid inflator module with argon gas pressure chamber.

Figure 15-21 When the initiator is energized, the propellant explodes and punctures the container and allows pressurized argon gas to fill the air bag.

Liquid Fuel with Argon Gas

The liquid-fueled air bag module uses a small quantity of ethanol alcohol to blow out the burst disc (**Figure 15-22**). Although similar to the hybrid system previously discussed, there are some differences in methods. In this system, the fuel blows the burst disc, and as the fuel continues to burn, the heat expands the argon gas. The expanding gas pushes through an orifice cup, over the diffuser, and into the air bag.

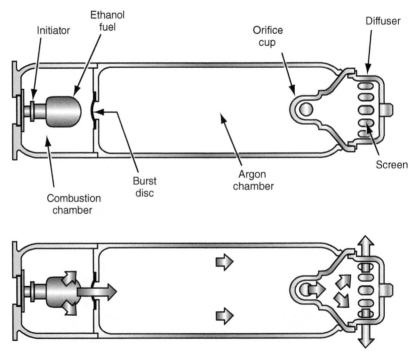

Figure 15-22 Liquid-fueled hybrid inflator operation.

Alcohol is used because it ignites at a lower temperature and does not leave a harmful residue.

Heated Gas Inflator

The heated gas inflator (HGI) is pressurized with a combustible mixture of 12% hydrogen gas and air. The mixture is ignited with a pyrotechnic squib.

MULTISTAGE AIR BAG DEPLOYMENT

One recent development to the air bag system is the development of **multistage air bags**. Multistage air bags are hybrid air bags that use two squibs to control the rate of inflation. Multistage air bags are used for both driver- and passenger-side air bags. These modules apply the principle of using heat to expand the argon gas to fill the air bag. The bag fills faster as the heat increases.

The air bag module of the multistage system uses two squibs (**Figure 15-23**). When one of the squibs is fired, the air bag begins to deploy. The second squib is then fired to generate more heat so the air bag fills faster (**Figure 15-24**). The length of time between the firing of the two squibs determines the rate of air bag deployment. In a minor accident that requires air bag deployment at a slow rate, only the first squib is ignited. The second squib may be ignited 160 milliseconds later to use up the other igniter charge, but this is too late to fill the air bag. As the severity of the deceleration forces indicate that faster air bag deployment is needed, the firing of the squibs will get closer together.

SIDE-IMPACT AIR BAGS

Many manufacturers are now offering side-impact air bag systems. Most of these are a single-stage hybrid design. The location of the air bag varies depending on the vehicle. Some are designed to come out of the door panel (**Figure 15-25**), from the seat back (**Figure 15-26**), between the A-pillar and the headliner (**Figure 15-27**), or from a roof-mounted curtain in the headliner that protects both the front- and rear-seat occupants (**Figure 15-28**).

The side-impact air bags deploy separately from the front air bags. The system may have a separate control module mounted in the B-pillars of the vehicle or sensors that relay impact information to the ORC. The ORC then deploys the side air bags.

Figure 15-23 The multistage air bag module uses two squibs.

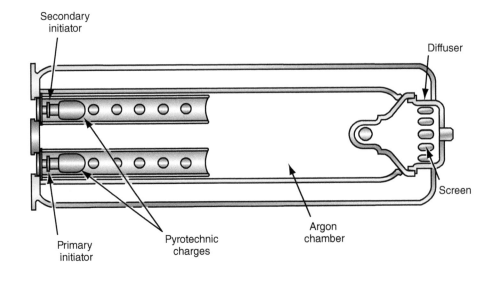

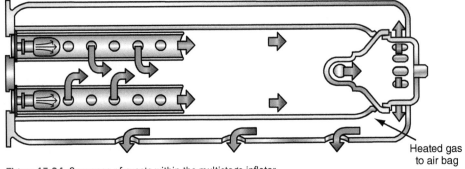

Figure 15-24 Sequence of events within the multistage inflator.

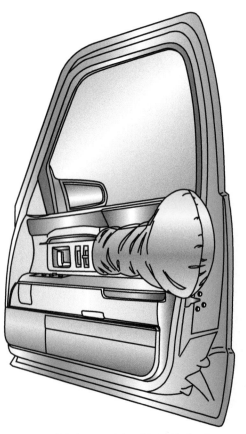

Figure 15-25 Side-impact air bag located in the door panel.

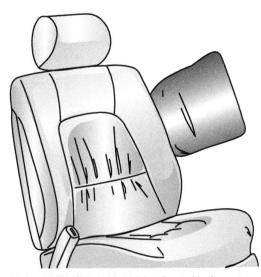

Figure 15-26 Side-impact air bag located in the seat back cushion.

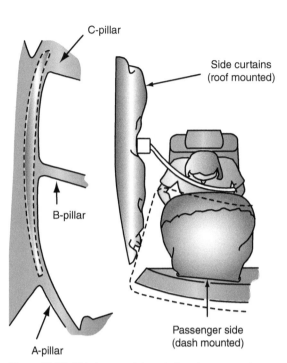

Figure 15-27 Side-impact air bag designed to protect the occupant's head from injury.

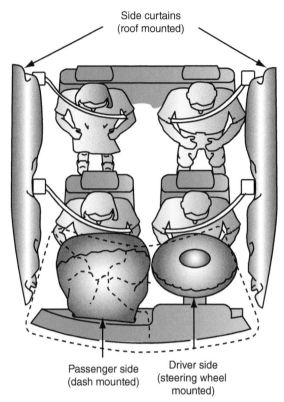

Figure 15-28 Side-impact air curtains protect both front- and rear-seat occupants.

To meet Federal Motor Vehicle Safety Standards (FMVSS) for vehicle-to-pole side impacts, some manufacturers utilize a pressure-sensitive sensor in the doors (**Figure 15-29**). The sensor is located between the door sheet metal or skin and the interior trim panel. The pressure sensor is an input to the ACM and does not deploy the air bag directly. The pressure sensor is used in conjunction with other accelerometer type impact sensors to indicate a side impact within the door area.

The ACM monitors all sensors for an indication of a deceleration rate or impact that exceeds a threshold. Once the threshold is exceeded, the air bag will be deployed. In the case of side-impact air bag systems, an impact in the door area may be absorbed

Shop Manual
Chapter 15, page 788

Figure 15-29 Impact pressure sensor.

by the door to a point that by the time the accelerometer-type impact sensor exceeds the threshold, air bag deployment may be a couple of milliseconds later than optimum. To make sure that the side air bag is deployed at the appropriate time, the pressure sensor was added. A sudden change in air pressure within the door indicates that an event may have happened. If the side-impact sensor also indicates an incident, the side air bag will deploy. If the pressure sensor input is above the threshold that indicates an impact in the door area, the threshold of the accelerometer-type impact sensor is modified so the side air bag is deployed a few milliseconds earlier. Basically, the pressure sensor provides a "heads-up" signal to the ACM. If the side-impact sensor agrees that the sudden pressure change is due to an impact, the side air bag is deployed.

AIR BAG ON/OFF SWITCHES

For the air bags to perform safely, there should be at least 10 inches (25 cm) between the air bag module and the occupant. Also, children should not sit in the front seat with an air bag, and a rear-facing infant seat (RFIS) should NEVER be used in the front seat with an air bag. Some vehicles with limited rear seating may be factory equipped to turn off the passenger air bag if it would not be safe to have it deploy.

Most systems rely on input from the driver or passenger to turn a switch (**Figure 15-30**). Early systems placed a resistor in the passenger-side air bag circuit when the switch was turned to the off position (**Figure 15-31**). Turning the switch opens the circuit between the ORC and the passenger-side air bag so it will not deploy. The resistor is used to trick the ORC into believing the circuit is still intact so it will not set false fault codes.

Newer systems use a MUX signal from an on/off module (**Figure 15-32**). The ORC provides a pulsed signal to the on/off module at a frequency of 10 Hz with a 3% duty cycle. The ORC monitors the voltage drop across the switch. If the switch is in the ON position, 4 volts (V) will be monitored. In the OFF position, about 10 volts will be monitored. A reading of 20 volts will be considered an open, while a reading of zero volts would set a short-circuit fault. If the switch is in the OFF position, the ORC will deactivate the passenger air bag internally. This system does not interrupt the circuit as earlier systems did.

Figure 15-30 Passenger-side air bag on/off switch.

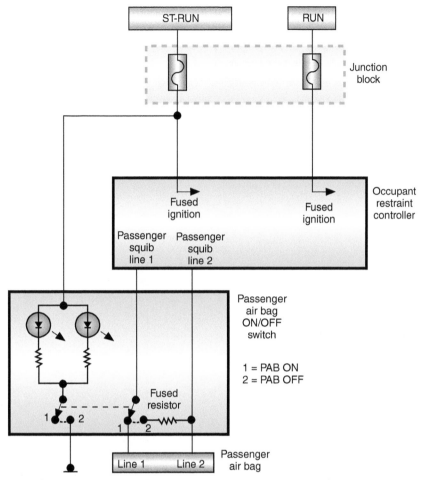

Figure 15-31 Wiring diagram of a hardwired passenger-side air bag on/off switch.

SEAT BELT PRETENSIONERS

To assure that the driver and/or passenger stay in position during an accident, some vehicles are equipped with seat belt **pretensioners**. There are two ways of mounting the pretensioner.

The first is to mount the pretensioner on the buckle side of the seat belt (**Figure 15-33**). At the same time the front air bags are deployed, the control module also deploys the pretensioner. A small piston is attached to a cable connected to the buckle. There is a pyrotechnic charge below the piston. When the pretensioner is fired, the expanding gas forces the piston to travel up the cylinder, pulling the buckle tight by the cable.

The pretensioner can also be mounted on the retractor side of the seat belt (**Figure 15-34**). The system type shown has a retractor assembly with a fan wheel–type unit attached to one end (**Figure 15-35**). When the pretensioner's pyrotechnic charge is fired, a series of balls shoot out of the channel and hit the fan wheel. As the balls hit the fan wheel, the retractor rotates and pulls the seat belt tight. The last ball is a little bigger than the others and lodges into the fan wheel, causing the seat belt to lock.

The seat belts used with the pretensioner systems usually use a special process of looping the end of the web called an energy management loop. The lower part of the seat belt webbing is looped and stitched with a special rip stitching (**Figure 15-36**) that will

Pretensioners are used to tighten the seat belt and shoulder harness around the occupant during an accident severe enough to deploy the air bag. Pretensioners can be used on all seat belt assemblies in the vehicle.

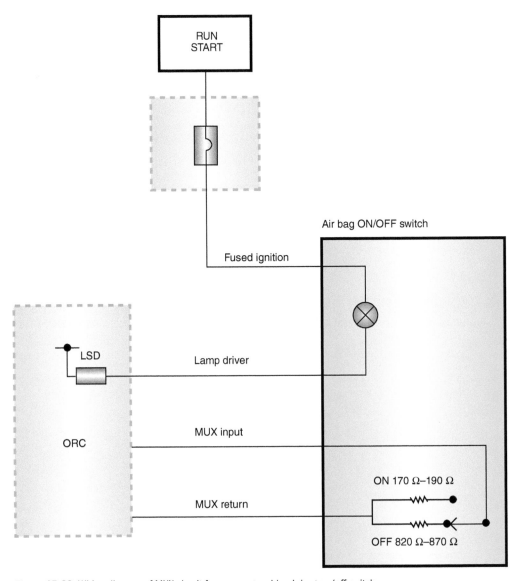

Figure 15-32 Wiring diagram of MUX circuit for passenger-side air bag on/off switch.

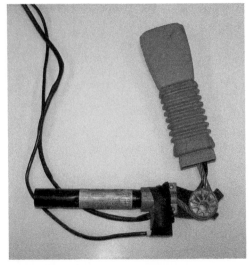

Figure 15-33 Buckle-mounted pretensioner.

Figure 15-34 Retractor-mounted pretensioner.

Figure 15-35 The balls are shot at the fan wheel, which causes the retractor to wind up the seat belt tight against the occupant.

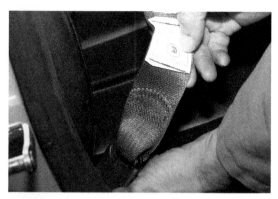

Figure 15-36 Special rip stitching is used for the seat belt loop.

release at a predetermined load. The seat belt does not come apart, but the ripping of the stitching allows the load being transmitted through the belt to the occupant to be reduced by giving the belt some additional "give." Anytime the vehicle is involved in an accident that the pretensioners have been deployed, the seat belt must be replaced.

INFLATABLE KNEE BLOCKERS

Manufacturers are now incorporating a **driver-side inflatable knee blocker (IKB)** into their air bag systems (**Figure 15-37**). The IKB is located on the driver side of the vehicle beneath the instrument panel cover and is attached to the instrument panel reinforcement. The IKB deploys simultaneously with the driver-side air bag to increase driver impact protection. The IKB provides upper-leg protection and positioning of the driver. When the IKB is deployed, it pushes a tethered plate against the driver's knees. This keeps the driver in the correct upright position during a collision, so the air bag is more efficient.

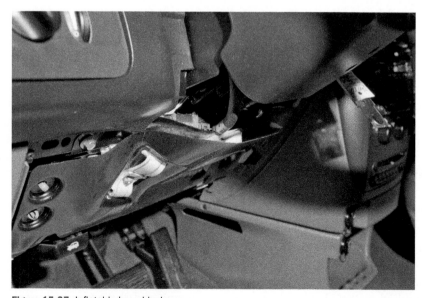

Figure 15-37 Inflatable knee blocker.

Figure 15-38 The PADL illuminates if the air bag is suppressed.

OCCUPANT CLASSIFICATION SYSTEMS

Shop Manual
Chapter 15, page 792

As the result of an amendment to the Federal Motor Vehicle Safety Standard 208, manufacturers have designed and installed air bag systems that reduce the risk of injuries resulting from air bag deployment. The goal of this amendment is to reduce injuries suffered by children and small adults that are in the **fifth percentile female** weight classification. The fifth percentile female is classified to be those who weigh less than 100 pounds (45 kg). The amendment mandates that the passenger-side air bag be suppressed when an infant in an RFIS occupies the front passenger seat. The amendment also mandated that a passenger air bag disable lamp (PADL) be illuminated whenever the passenger seat is occupied and the passenger-side air bag has been suppressed (**Figure 15-38**). If the front passenger seat is not occupied, the lamp is not illuminated.

AUTHOR'S NOTE The fifth percentile female is determined by averaging all potential occupants by size and then plotting the results on a graph. The middle of the bell curve on the graph would indicate the majority of occupants. At the far right of the bell curve would be those occupants who are very large, while on the left side of the curve would be occupants who are very small. At the fifth percentile range on the left of the curve, the majority of occupants would be female.

To meet these new requirements, manufacturers have taken different approaches. In this section, the Delphi and TRW systems are presented to provide two different methods of meeting this regulation. The Delphi system uses a bladder to determine weight, and the TRW system uses strain gauges. Both systems use an occupant classification module (OCM) that determines the weight classification of the front passenger and sends this information to the ORC. Also, both systems use multistage passenger-side air bags.

Delphi Bladder System

The bladder system uses a silicone-filled bladder that is positioned between the seat foam and the seat support (**Figure 15-39**). A pressure sensor is connected by a hose to the bladder (**Figure 15-40**). The three-wire pressure sensor operates similar to the way a manifold absolute pressure (MAP) sensor operates. When the seat is occupied, pressure that is applied to the bladder disperses the silicone and the pressure sensor reads the increase in pressure. The pressure reading is inputted to the ORC (**Figure 15-41**).

Shop Manual
Chapter 15,
pages 792, 799

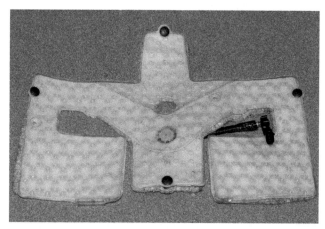

Figure 15-39 Bladder used to determine occupant classification.

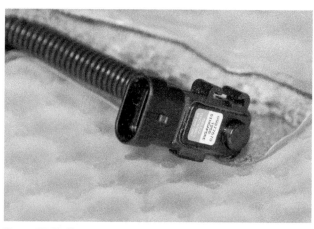

Figure 15-40 The pressure sensor changes voltage signals as weight is added to the bladder.

Since pressure is used to determine seat occupation and ultimately occupant classification, the system corrects for changes in atmospheric pressures. In addition, natural aging of the seat foam is also learned by monitoring gradual changes. The OCM stores seat aging and calibration information in the ORC. If a new OCM is installed, it retrieves this information from the ORC so the system continues to function properly.

When the seat is not occupied, the OCM compares the sensor voltage with the value stored in memory. The voltage values measured by the sensor change as weight is added to the seat. Based on the change of voltage from the sensor, the OCM can determine the weight of the occupant.

Based on the weight information that the OCM sends to the ORC, the following are determined:

- An empty seat. The PADL is off and the air bag is suppressed.
- Weight equivalent to or less than that of a six-year-old child. The PADL is illuminated and the air bag is suppressed.
- Weight equivalent to or greater than that of a fifth percentile female. The PADL light is off and the front passenger air bag is enabled. Deployment rate is based on the severity of the impact.

Since an infant seat that is securely strapped into the seat will cause an increase of downward pressures on the bladder, the system uses a **belt tension sensor (BTS)**. The BTS is a strain gauge–type sensor located on the seat belt anchor (**Figure 15-42**). The increase in pressures can be great enough to indicate a weight greater than that of a fifth percentile female in the seat. In this case, the air bag will not be suppressed. However, the BTS will indicate that the belt is tight around an object (about 24–26 lbs. [11–12 kg] of force). The reading will indicate that the belt is tighter than what it normally would be for a belt around a person.

Shop Manual
Chapter 15, page 800

As seat belt tension is increased, the sensor voltage changes. Based on the change of voltage from the BTS, the OCM can estimate how much of the sensed load results from the cinched seat belt. If the BTS indicates a cinched seat belt load over a certain threshold, the OCM determines a rear-facing infant classification.

TRW Strain Gauge System

The TRW system uses four strain gauges to determine weight classification (**Figure 15-43**). One strain gauge is located at each corner of the seat frame where the frame attaches to the seat riser. The strain gauges support the weight of the seat. Data from each sensor is sent to the OCM (**Figure 15-44**).

Shop Manual
Chapter 15, page 800

The OCM compares current voltage readings with the values stored in memory. The electrical resistance of the strain gauge changes based on the amount of strain against it. A circuit board is bonded to the frame of the gauge. The circuit board has a grid made of

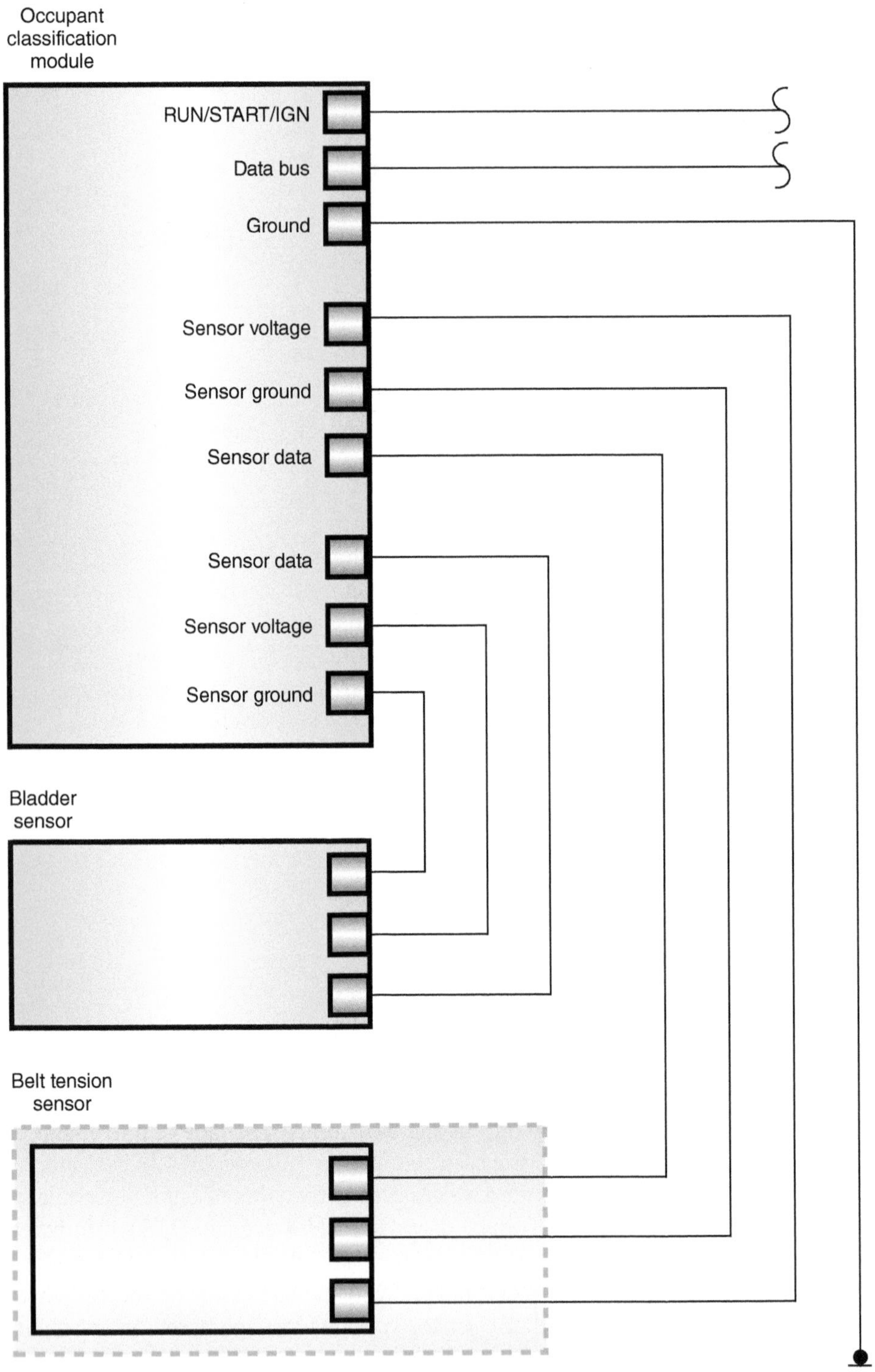

Figure 15-41 Schematic of bladder occupant classification system.

metallic foil that changes in resistance when strain is applied. As weight is added to the seat, the voltage values of the sensors change. Based on the change of voltage from each sensor, the OCM can determine the occupant's weight. The OCM will attempt to calibrate every key off if the seat is not occupied.

Based on the weight information that the OCM sends to the ORC, the following are determined:

■ An empty seat. The PADL is off and the air bag is suppressed.

Figure 15-42 Belt tension sensor.

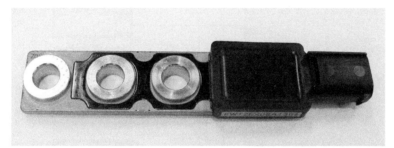

Figure 15-43 Strain gauge.

- Weight equivalent to or less than that of an RFIS. The PADL is illuminated and the air bag is suppressed.
- Weight equivalent to a child. The PADL is off and the air bag is enabled. Deployment is at the low-risk deployment level.
- Weight equivalent to or greater than that of a fifth percentile female. The PADL is off and the front passenger air bag is enabled. Deployment rate is based on the severity of the impact.

Either system can utilize a **seat track position sensor (STPS)** on both the driver and passenger seats. The STPS provides information to the OCM concerning the position of the seat in relation to the air bag. The OCM sends this information over the data bus to the ORC. The ORC modifies the deployment strategy based on this information. If the occupant's seat position is closer to the air bag, the deployment rate of the air bag will be slower than for an occupant's seat that is positioned farther away from the air bag.

The STPS uses a Hall-type sensor. The STPS is mounted onto a seat rail while a steel plate is mounted to the seat track. As the seat is moved, the steel plate covers or uncovers the sensor's magnetic field. This alters the current flow in the sensor.

AUTHOR'S NOTE Beginning in the 2006 model year, the use of classification systems was being reduced due to new technologies in air bag systems that protected an infant in the front seat as well as an adult driver. With the new technology air bags, the system is never suppressed and the air bag deploys anytime the system determines sufficient G-forces. However, the air bag will not deploy at a rate that would injure a properly restrained infant or adult.

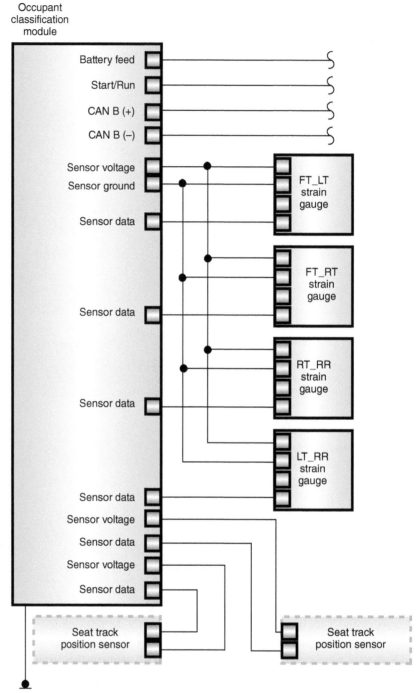

Figure 15-44 Schematic of strain gauge occupant classification system.

ACM OVERVIEW

As we have seen, in an accident, sensors measure the severity of the forces involved. The ACM is responsible for calculating the severity of the impact and for inflating up to eight air bags and the pretensioners. Some of these air bags can be deployed at different rates. Although it is more common to use hardwired outputs and actuators to the air bag modules, some systems use a high-speed optical bus system that controls the timing and extent of inflation of each air bag. Some systems will also deactivate the fuel pump and disengage the battery when deployment occurs.

Some ABMs will have two MEMS accelerometers that are surface mounted directly to the ACM board. The two MEMS accelerometers are mounted orthogonal to one another, one in the Y-axis for front collision detection and the other oriented in the X-axis to correspond to side-impact events. The board may also have an electromechanical sensor that shorts out contacts if sufficient deceleration is present. This sensor is used as confirmation and as a backup sensor to the MEMS.

A 32-bit microprocessor performs the algorithmic computations to determine if the air bag should be deployed and at the proper levels. The system will also incorporate a watchdog circuit to confirm the inputs from the MEMS and the conclusions. A communications chip handles data bus transmission and receipts with other modules. The capacitors are used to fire the air bags in the event the vehicle battery is disconnected or damaged during the accident.

ROLLOVER PROTECTION SYSTEMS

The MEMS accelerometer is also used for rollover detection. Rollover detection is an additional function of the air bag system that will deploy side air bag curtains and seat belt pretensioners if a rollover event is detected. This is a feature of the rollover protection systems (ROPS). Most of these systems use an additional low-G acceleration sensor and a gyroscopic sensor to detect if a turn is made too quickly or sense if the vehicle has swerved to avoid an obstacle. A MEMS inclinometer may also be used to make these determinations. The inclinometer is designed to measure lateral and vertical acceleration (vehicle speed and roll rate) to predict an impending rollover. The sensors may be located internal to the air bag control module. The inputs from the sensors will determine the rotation angle and rate around the X-axis (**Figure 15-45**). Based on this information, if a rollover event is determined to be imminent, the side-curtain air bags and the seat belt pretensioners are deployed to help protect the occupant from contact with the side of the vehicle interior and to help prevent occupant ejection as the vehicle rolls over. The ABM will determine the appropriate time to deploy the specific air bags or belt pretensioners. In addition, during a potential rollover event, enhanced traction or stability control can cut throttle and pulse the brakes to correct the vehicle's trajectory.

The ABM compares the tilt angle and rate-of-roll to determine if the vehicle is going to recover, or whether a roll is inevitable. If the vehicle is going to roll, the system deploys the side air bag curtain at a matched deployment speed to the event. A high-speed curb launch will trigger a faster deployment than a corkscrew roll on an embankment.

Shop Manual
Chapter 15, page 808

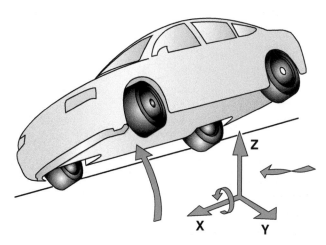

Figure 15-45 Sensors used for roll over protection measure vehicle movement on the X-Y-Z axis.

The side air bag curtains are a new style "safety canopy" curtain hidden above the head-liner. The canopy extends from the A-pillar to the C-pillar or D-pillar. Once the curtain has been deployed, tethers located on the bottom corners of the bag locks it in place to prevent the occupants from being ejected. The controlled deployment speed may happen much faster than with other air bag curtains. The deployment zone is from the roof rail to the bottom of the window, the full length of the vehicle, and the bag is about 5 inches (127 mm) thick.

The curtain is constructed of low-porous materials, so it retains its volume longer. The curtain is designed to remain inflated for up to 6 seconds. The curtain is folded inside the module using "roll fold technology." Rollover protection air bag curtains actually roll down between the window glass and the occupant. The rolling effect allows these curtains to deploy even if the occupant is out of position. In fact, the bag will actually set the occupant upright in the seat.

Convertibles that cannot use side air bag curtains use a ROPS system utilizing deployable roll bars that flip up or pop up (**Figure 15-46**). These are considered the flip-up type. The flip-up ROPS roll bar is concealed until activated; then it will flip up in less than three-tenths of a second. The flip-up roll bar is usually hydraulically deployed and then locked into position.

The pop-up system hides the roll bars within the rear seats (or directly behind them). When deployed, they pop straight up. These systems can be either hydraulically deployed or spring loaded and operated by an actuator latch device. The spring-loaded system uses an electromagnetic actuator system that releases the rollover protection system in fractions of a second in the event of an accident (**Figure 15-47**).

AUTHOR'S NOTE The Volvo C70 uses safety canopy air bags that deploy upward from the window sill and remain rigid to protect occupants in case the vehicle flips multiple times.

The roll-bar systems use an inclinometer to sense vehicle inclination and lateral acceleration, and a MEMS accelerometer to detect vehicle weightlessness in the event the vehicle becomes airborne. The control module determines when to deploy the roll

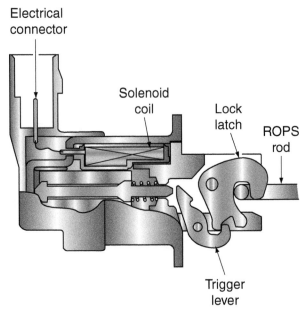

Figure 15-47 Spring-loaded ROPS may use a solenoid latch that releases the roll bar.

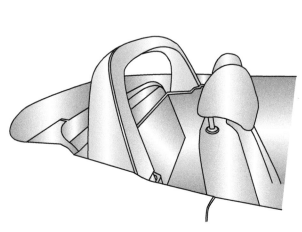

Figure 15-46 ROPS using pop-up roll bars.

bars based on this information. Certain criteria must be met before the control module will actuate deployment. For example, most systems will only deploy the roll bars in the following scenarios:

■ When the vehicle experiences a lateral acceleration of approximately 3 Gs.
■ When the vehicle approaches a lateral angle of 62°.
■ When the vehicle approaches its longitudinal angle of about 72°.
■ When a combination of longitudinal acceleration and longitudinal angle would cause the vehicle to roll over in the forward direction.
■ When the vehicle becomes airborne and for at least 80 milliseconds.

ACTIVE HEADRESTS

One of the largest factors contributing to injury during an accident results from the sudden, jerking movement of the head and neck. Active headrests incorporated into the seats support the head during an impact to prevent these types of injuries (**Figure 15-48**). Within 20 milliseconds of an impact being detected, the active headrests are pushed forward, reducing sudden head movement. The reduction of head movement helps to prevent neck injuries.

VISION SYSTEMS

Automotive electronics have made radical changes in the past few years. One of the newest technologies to emerge is that of the automotive vision system. This system covers a broad range of differing subsystems that include such automotive safety technologies as follows:

■ *Adaptive cruise control with collision mitigation system*—uses different sensors, along with laser or radar technologies, to control the throttle and brakes. The purpose is to maintain a safe following distance of the vehicle in front even under changes in traffic speeds or if another vehicle pulls into the lane. If the system determines that an accident is possible, it will automatically apply the brakes and some systems will tighten the seat belts.

Figure 15-48 Active headrests do not appear much different than standard headrests.

- *Blind-spot detection system*—is designed to alert the driver about vehicles or other objects that are in the "blind spots" of the driver. Usually, the system is activated when the turn signals are turned on. If an object is detected, a warning light located in the mirror will flash along with an audio warning tone. In addition, some manufacturers will cause the seat or steering wheel to vibrate to provide a warning indication.
- *Lane-departure warning*—similar to blind-spot detection, this system determines an approaching vehicle's speed and distance to warn the driver of potential danger if he or she should change lanes. This system will also provide a warning if it determines that the vehicle is wandering out of the lane.
- *Rearview camera*—designed to protect both people and property, this system provides the driver with a visual view of any objects directly behind the vehicle. This system provides a view of the area that cannot be seen by most mirrors.
- *Adaptive headlights system*—can be as simple as providing automatic headlight activation in low ambient light conditions. More advanced systems also provide automatic dimming based on speed and other conditions, and may also have headlights that follow the direction of the vehicle while it navigates around corners.
- *Night vision assist systems*—use infrared headlamps or thermal imaging cameras to allow the driver a better view of what is farther down the road during low ambient light conditions. An image of objects in the road is displayed for the driver to see.

The systems can use a variety of different sensors that will detect proximity, motion, and light intensity. The system can also use imaging cameras. This section discusses these types of components and details the operation of some of the systems.

Most vision systems rely upon a **proximity sensor** that can detect objects without physical contact. These sensors use several different technologies such as ultrasonic, radar, laser, camera, and infrared.

ULTRASONIC SENSORS

Ultrasound is a cycling sound pressure wave that is at a frequency greater than the upper limit of human hearing (about 20 kHz).

Ultrasonic sensors evaluate attributes of a target by interpreting the echoes from sound waves. These sensors are commonly used in the field of robotics for obstacle avoidance and as range finding for cameras. This technology is employed by the automotive industry for basically the same purpose. Today, vehicles may be equipped with sensors on the back of the vehicle that detect if an object is in the path while the driver is backing up (**Figure 15-49**). Some systems also include these sensors in the front of the vehicle to assist in parallel parking maneuvers (**Figure 15-50**).

Ultrasonic sensors emit a short burst of **ultrasound** waves that reflect or "echo" from a target. Most sensors operate in the range of about 40 kHz. If the waves hit an object,

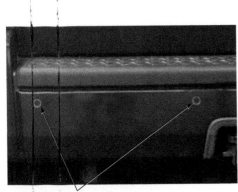

Ultrasonic sensor locations

Figure 15-49 Ultrasonic sensors used for back-up obstacle detection.

Figure 15-50 Front ultrasonic sensors used in park assist systems.

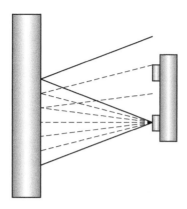

Figure 15-51 The echo effect of ultrasonic sound as it hits an object.

they are echoed back (**Figure 15-51**). The echo that is transmitted back from these high-frequency sound waves is then evaluated. A calculation of the distance the vehicle is from an object is made based on the time interval between sending the signal and receiving the echo and dividing that time value by the speed of sound.

Either the module or the sensor contains a transmitter that generates the ultrasonic sound waves by turning electrical energy into sound. The transducer circuit uses piezo-electric crystals that change size when a voltage is applied. This creates alternating voltage across the crystals that causes them to oscillate at very high frequencies. This in turn produces the high-frequency sound waves.

The echo is received by the detector circuit of the ultrasonic sensor, which converts the sound waves into electrical energy. A piezoelectric crystal that will generate a voltage when the sound wave pressure force is applied to it is used as the detector. Usually, the transmitter and detector are combined in a single piezoelectric transceiver.

Visual Imaging Systems

Many safety systems utilize a camera. The camera can be used to provide visual assistance for backing up the vehicle. However, it can also be used to determine light intensity and speed of oncoming vehicles for such systems as automatic headlight dimming, as discussed in Chapter 11.

Shop Manual
Chapter 15, page 804

Cameras can also be used to enhance the functionality of a system. For example, a camera can be used to provide a system the ability to recognize and learn patterns. Since radar sensors cannot determine if the object is a pedestrian, a vehicle, or road sign, camera-based pattern recognition provides the capability of recognizing these objects. Cameras are also used for night vision assistance systems. The use of the digital camera in public life has expanded greatly in the past decade; it is this technology that is also employed in the camera systems on today's vehicles. However, when used in the confines of the automobile, efficient processing of the captured image and displaying it is an obstacle. This is overcome by intelligent image capture systems.

Intelligent image capture systems perform local image processing. This processing could be to correct for lens "fisheye" or for full-object recognition. The information is transmitted over the bus network to the video controller. **Figure 15-52** illustrates a typical microprocessor-based system that performs the functions of controlling and processing the video data. The processing of the data conditions it for the type of media it will be sent to. These functions are performed before the video data is bussed to the display.

The captured data is sent by converting the captured parallel video data image into a serial stream and transmitting it over a **low voltage differential signaling (LVDS)** interface (**Figure 15-53**). At the display, the data is processed back to its original form. **Figure 15-54** provides an example of the system using an LCD display.

Low voltage differential signaling (LVDS) is a high-speed data bus that is used in point-to-point configurations for visual displays.

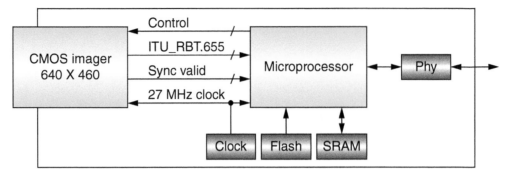

Figure 15-52 Intelligent image capture system.

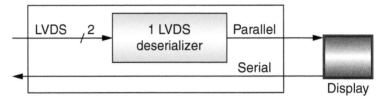

Figure 15-53 Transmission of captured video.

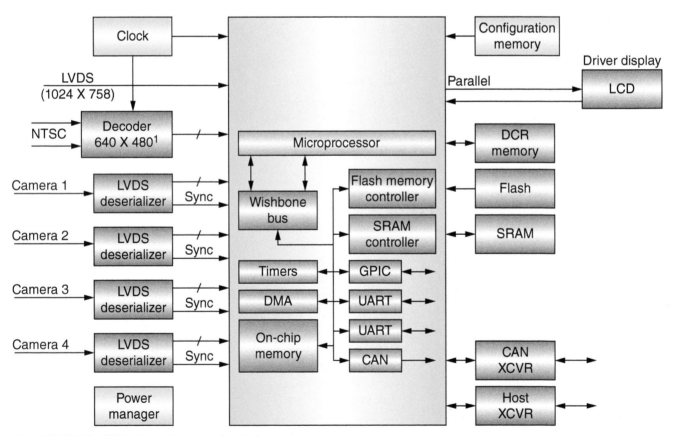

Figure 15-54 Using LVDS to transmit images to the display.

Infrared Sensors

Infrared can be used to measure the temperature of an object; however, it can also be used to determine if an object is in the proximity and detect obstructions. When used to measure temperature, the infrared sensor can measure the temperature of an object without contact. These sensors use the principle that any object emits an amount of energy that is a function of its temperature. This principle also establishes that the amount of energy emitted increases as the temperature of an object rises.

The infrared sensor collects energy from a target through a lens system. The lens focuses the energy onto a detector that generates a signal that is processed by the sensor's electronics. By programming, the sensor determines the relationship between the measured signal and corresponding temperature. When used as an obstruction sensor, the infrared light beams are monitored for distortions.

Infrared sensors can also be used in conjunction with a camera. This assists the camera in determining the shape and size of an object under low-light conditions. This can be used to see objects that the eye may not and is the technology used in some night vision systems. When used as a camera system to determine the identity of an object, the sensor uses the emitted energy to determine the object's shape.

PARK ASSIST OBSTACLE DETECTION

A **park assist system** is a parking aid that alerts the driver to obstacles located in the path immediately behind the vehicle. When an object is detected, the system may uses an LED display and warning chimes to provide the driver with visual and audible warnings of the object's presence.

Shop Manual
Chapter 15, page 802

The park assist system includes the following major components:

- *Instrument cluster*—used to display textual warnings and error messages related to the current operating status of the park assist system.
- *Park assist display*—mounted at the rear of the headlines just above the back glass, this display provides visual indication of the presence of an obstacle by use of LEDs. If the vehicle also has a front park assist system, a second display unit is mounted in the center of the instrument panel's top pad near the windshield.
- *Park assist module*—the central component of the system that supplies voltage to the park assist sensors and the park assist displays processes the data from the sensors, calculates the display information, and performs system diagnostics.
- *Park assist sensors*—ultrasonic sensors located behind the rear bumper fascia. In addition to the rear sensors, vehicles equipped with the front park assist system have ultrasonic sensors located behind the front bumper fascia.
- *Park assist switch*—allows the driver to manually disable or enable the park assist feature. This allows the driver to disable the system if towing a trailer.

The park assist system is active anytime the ignition switch is in the RUN position, the parking brake is released, and the vehicle speed is less than 10 mph (16 Km/h). Rear park assist is active only with the transmission gear selector in the REVERSE position. If the vehicle is also equipped with front park assist system, it is active with the transmission gear selector lever in the DRIVE or REVERSE position (automatic transmissions).

Ultrasonic transceiver park assist sensors in the bumpers locate and identify the proximity of obstacles in the path of the vehicle. The sensors are controlled by the module and generate ultrasonic sound pulses when triggered by the module. The sensor then signals the module when an echo of the reflected sound pulses is received.

Each sensor receives battery voltage and ground from the module. The voltage supply and ground circuit are wired in parallel from the module. However, each sensor has a dedicated serial bus communication circuit to the module. The module alternately oscillates and then quits the membrane of each sensor at the same time. When the sensor membrane is oscillated, it emits an ultrasonic signal. The ultrasonic signal echoes off of objects in the path of the vehicle.

When the membrane is quieted, each membrane will receive the echoes of the ultrasonic signals. These echoes may be from the sensor's own signals or from another sensor.

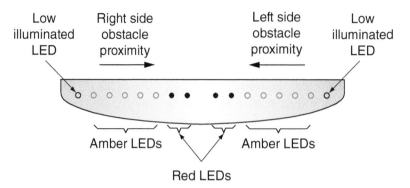

Figure 15-55 The display uses LEDs and audio tones to illustrate the proximity of an object.

The echo data is then sent over the serial bus to the module. The module uses the intervals between the ultrasonic transmission and reception data from the sensors to calculate the distance the vehicle is from any obstacles that may have been identified by the sensors.

The sensor area of detection from the vehicle is between 11.8 inches (0.3 meter) and 59.1 inches (1.5 meters), and the height from the ground is between 7.8 inch (0.2 meter), and 31 inches (0.8 meter). The detection area extends around the corners of the vehicle.

The module uses preprogrammed algorithms and calibrations to determine the appropriate outputs based on the data from the sensors. If a warning is needed, the module will send a message to the display unit(s) over a dedicated serial bus.

The displays contain two sets of LED indicators with eight on each side (**Figure 15-55**). Also, the display units house a chime tone transducer. While the park assist system is active, the number, position, and color of the illuminated LEDs and the frequency of the audible signal indicate the distance of obstacles from the vehicle. The visual indication of an object is by illuminating one or more LED indicators. The outer amber units will light first and will move toward the center of the display the closer the vehicle gets to the obstacle. Once the vehicle is within 31 inches (40 centimeters) of the obstacle, one of the red LED units is illuminated. Also, the display will emit a series of short, intermittent, audible beeps. If the vehicle comes within 12 inches (30.5 centimeters) of the obstacle, the second red LED is illuminated and the audible warning changes to a continuous tone.

When the park assist system is active and no obstacles are detected, the two outermost amber LED units are illuminated at a reduced intensity. This provides a visual confirmation that the system is operating.

An additional option with the park assist system is the rearview camera (RVC). A camera is mounted at the rear of the vehicle, usually in the license plate light bar (**Figure 15-56**). When the transmission is in REVERSE, backup light circuit will supply a voltage signal to the camera. The display can be in the radio (**Figure 15-57**), in the instrument cluster, or on a separate screen.

The RVC is a camera-on-chip device that utilizes **complementary metal oxide semiconductor (CMOS)** technology. CMOS technology is used for a wide variety of digital or analog circuits. In this application, it supports the analog image sensor of the RVC. The RVC function is to provide a wide-angle video image of the area behind the vehicle, including areas that might not be normally visible from the seated position of the vehicle operator, only while the transaxle gear selector is in the REVERSE position. This video image then displays to the vehicle operator within the display screen.

Advanced Parking Guidance System

With the advent of electric steering, ABS, proximity sensors, and CMOS cameras, it has become possible to develop systems that will parallel park or perform back-in parking of the vehicle without driver steering wheel input.

Shop Manual
Chapter 15, page 806

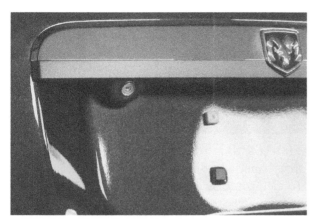

Figure 15-56 Rearview camera used for park assist and back-up monitoring.

Figure 15-57 Camera display shown on the radio screen.

The automatic parking system automatically operates the electric steering gear to maneuver the vehicle into a parking spot. To perform a parallel parking procedure the driver must activate the system with an input button. A message will be displayed that the system is active. As many as 12 ultrasonic sensors (six mounted in the rear fascia and six in the front fascia) will be activated to guide the vehicle into the space. The driver stops their vehicle just behind the parking space and activates the system. The front corner sensors will measure the clearance of the space from the front of the vehicle to front and far side of the parking space. The driver is then instructed to full just ahead of the parking space where the rear sensors will detect if the desired parking space has sufficient clearance. The guidance screen will display driver instructions, such as; "Shift Into Reverse," "Release Steering Wheel," or "Apply Brakes." Using the ultrasonic sensors, the system will control the steering gear to maintain sufficient clearance within the parking space. During the parking maneuver, if the sensors detect an obstacle that could result in a collision, the brakes are activated.

For back-in parking, the drive stops the vehicle perpendicular to the center of the target parking spot. The system is activated through button pushes, and the message center will display instructions to the driver. First, the system determines if there is sufficient space for the vehicle to enter. The system then controls the steering gear to back the vehicle into the space. If necessary, the system can use multi-turn maneuvering in order to back the vehicle into the parking spot.

LANE DEPARTURE WARNING

Lane departure warning (LDW) systems use a forward-facing camera that is often located behind the rearview mirror. The camera is capable of detecting white lines, yellow lines, or dots in the road and to track the road features. Latest systems use a camera that can operate even if a lane marker is detected on only one side of the vehicle. The system determines if the vehicle is about to leave its designated lane and then alerts the inattentive or drowsy driver. The LDW system can inform the driver of the potential that their vehicle is about to leave the lane by sounding warning buzzers, by **haptic feedback** in the steering wheel, or by tugging on the seat belt.

TRW's LDW system uses their electric power–assisted steering (EPAS) system to provide the feedback in the steering wheel. The system simulates the rumble strip sensation that is felt by the cuts placed in the shoulder of the road. It is also configurable to provide actual steering correction.

Haptic is from the Greek word meaning "pertaining to the sense of touch."

The system may also monitor the steering angle sensor to identify if the steering wheel has not been operated for an extended period of time. In addition, pressure sensors in the steering wheel can be used to determine if the wheel is being firmly gripped. If it is determined that a potential hazardous driving condition exists, a warning buzzer sounds.

ROLLOVER MITIGATION SYSTEM OVERVIEW

Shop Manual
Chapter 15, page 808

The purpose of **electronic roll mitigation (ERM)** is to attempt to prevent a vehicle rollover situation due to extreme lateral forces experienced during cornering or hard evasive maneuvers. This program is similar to electronic stability control (ESP), but is more aggressive in engine torque management. Rollover mitigation systems are typically used on SUV and other high-profile vehicles.

A majority of vehicle rollovers are the result of the vehicle being "tripped." Tripping occurs when a tire hits uneven ground and the vehicle's balance is disrupted. This can happen if the tire runs up on a soft shoulder or road obstruction. Usually, the ESP system is capable of correcting these types of events. However, the vehicle can also lose its balance as a result of excessive cornering forces. About 5° of rollovers occur as a result of high-speed lane-change maneuvers and overwhelming sideways momentum (**Figure 15-58**). The rollover mitigation feature uses additional sensors and actuators within the ESP system to prevent rollovers caused by these events.

Electronic roll mitigation (ERM) is also called proactive roll avoidance (PRA), roll stability control (RSC), active roll control (ARC), along with several other names.

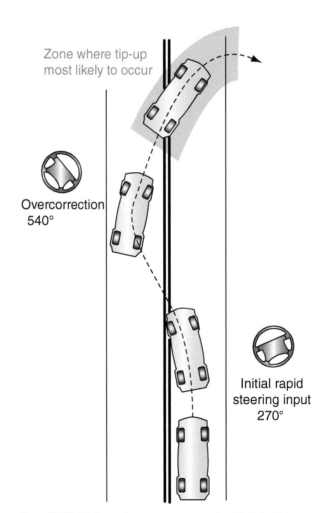

Figure 15-58 High-speed maneuvers cause about 5% of vehicle rollovers.

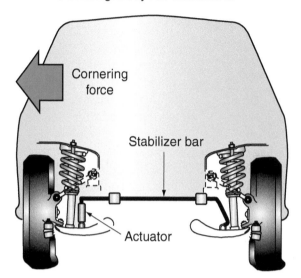

Cornering roll—No system.
Stabilizer bar deflects due to body
roll motion.

Cornering roll—with Active Roll Control
Actuator deflects stabilizer bar by
extending. Body roll eliminated.

Cornering
force

Cornering
force

Stabilizer bar

Stabilizer bar

Actuator

Figure 15-59 The actuator changes the stabilizer bar deflection to prevent body roll.

Some systems function by attempting to reduce the friction with the road. This is done by applying individual brake calipers to induce a skid. This allows the vehicle to follow in the current direction that its momentum is taking. By not attempting to change the vehicle's path and allowing the tires to skid on the road surface, it is less likely that the vehicle will overturn.

The rollover mitigation system can also be incorporated with the electronic suspension systems to control body roll and weight transfer. For example, the TRW system uses hydraulic actuators to actively alter the stiffness of the stabilizer bars by changing the length of the link to reduce body roll (**Figure 15-59**). Other systems change spring or shock absorber rates to prevent body roll.

SUMMARY

- Passive restraints operate automatically with no action required on the part of the driver or occupant.
- The automatic seat belt system uses a control module to monitor operation by receiving inputs from door ajar switches, limit switches, and the emergency release switch.
- The air bag is a supplemental restraint. The seat belt is the primary restraint system.
- The air bag module is composed of the air bag and inflator assembly. It is packaged in a single module and mounted in the center of the steering wheel.
- The ORC constantly monitors the readiness of the SRS electrical system. If the battery or cables are damaged during an accident, it supplies back-up power to the air bag module.

- The igniter is a combustible device that converts electrical energy into thermal energy to ignite the inflator propellant.
- Air bags will deploy if the vehicle is involved in a frontal collision of sufficient impact and the collision force is within 30° on either side of the vehicle centerline.
- The total air bag deployment time from the instant of impact until the air bag is inflated is less than 160 milliseconds.
- In most current air bag systems, MEMS piezoelectric accelerometers are used to sense deceleration forces and direction.
- The clockspring electrical connector maintains electrical contact between the inflator module and the air bag electrical system.

- The air bag warning light indicates an inoperative air bag system.
- A hybrid inflator module contains a pressurized argon gas cylinder, which is punctured by the exploding propellant to inflate the air bag.
- Shorting bars connect air bag system squib terminals together when the terminal is disconnected. This prevents accidental air bag deployment.
- Federal law expanded to require all passenger vehicles produced after 1995 be equipped with front passenger-side air bags.
- The air bag module of the multistage system uses two squibs. When one of the squibs is fired, the air bag begins to deploy. The second squib is then fired to generate more heat so the air bag is filled faster.
- Side-impact air bags can be designed to come out of the door panel, from the seat back, between the A-pillar and the headliner, or from a roof-mounted curtain in the headliner that protects both the front- and rear-seat occupants.
- To meet Federal Motor Vehicle Safety Standards (FMVSS) for vehicle-to-pole side impacts, some manufacturers utilize a pressure-sensitive sensor in the doors.
- Some vehicles with limited rear seating may be factory equipped to turn off the passenger air bag if it would not be safe to have it deploy.
- To assure the driver and/or passenger stay in position during an accident, some vehicles are equipped with seat belt pretensioners.
- There are two ways of mounting the pretensioner— on the buckle side or retractor side of the seat belt.
- The inflatable knee blocker (IKB) deploys simultaneously with the driver-side air bag to provide upper-leg protection and positioning of the driver.
- Occupant classification systems are a mandatory requirement designed to reduce the risk of injuries resulting from air bag deployment.
- The Delphi system uses a bladder to determine weight, and the TRW system uses strain gauges. Both systems use an occupant classification module (OCM) that determines the weight classification of the front passenger and sends this information to the ORC. Also, both systems use multistage passenger-side air bags.
- The belt tension sensor (BTS) is a strain gauge–type sensor located on the seat belt anchor that is used to indicate if an infant seat is cinched into the passenger-side front seat.
- The seat track position sensor (STPS) provides information to the OCM concerning the position of the seat in relation to the air bag.

- Some ABMs will have two MEMS accelerometers that are surface mounted directly to the ACM board, one in the Y-axis for front collision detection and the other oriented in the X-axis to correspond to side-impact events.
- A 32-bit microprocessor performs the algorithmic computations to determine if the air bag should be deployed and at the proper levels.
- The MEMS accelerometer is also used for rollover detection that provides an additional function of the air bag system that will deploy side air bag curtains and/or seat belt pretensioners if a rollover event is detected.
- The inclinometer is designed to measure lateral and vertical acceleration to predict an impending rollover.
- Convertibles that cannot use side air bag curtains will use a ROPS system utilizing deployable roll bars that flip up or pop up.
- Ultrasonic sensors evaluate attributes of a target by interpreting the echoes from sound waves.
- Many safety systems utilize a camera. The camera can be used to provide visual assistance for backing up the vehicle. However, it can also be used to determine light intensity and speed of oncoming vehicles for such systems as automatic headlight dimming.
- Since radar sensors cannot determine if the object is a pedestrian, a vehicle, or road sign, camera-based pattern recognition provides the capability of recognizing these objects.
- The infrared sensor can be used to determine if an object is in the proximity and detect obstructions by using the principle that any object emits an amount of energy that is a function of its temperature.
- When used as an obstruction sensor, the infrared light beams are monitored for distortions.
- Infrared sensors can also be used in conjunction with a camera to assist the camera in determining the shape and size of an object under low-light conditions.
- A park assist system is a parking aid using ultrasonic transceiver sensors to alert the driver of obstacles located in the path immediately behind the vehicle. When an object is detected, the system uses an LED display and warning chimes to provide the driver with visual and audible warnings of the object's presence.

  A. Lane departure warning (LDW) systems use a forward-facing camera to detect lane markers in the road and to track the road features. The system determines if the vehicle is about to leave its designated lane and then alerts the inattentive or drowsy driver.

REVIEW QUESTIONS

Short-Answer Essays

1. Define the term *passive restraint*.

2. Describe the basic operation of automatic seat belts.

3. List and describe the design and operation of three different types of air bag system sensors.

4. List and explain two of the functions of the OCM used in air bag systems.

5. List the sequence of events that occur during air bag deployment.

6. What is the purpose of the clockspring?

7. Describe how the infrared sensor can be used in conjunction with a camera to assist in determining the shape and size of an object under low-light conditions.

8. What is the purpose of multistage air bags?

9. Where is the side-impact air bag control module or sensor usually located?

10. List the common mounting locations of the seat belt pretensioner.

Fill in the Blanks

1. The _____ conducts electrical signals to the air bag module while permitting steering wheel rotation.

2. In the automatic seat belt system, the _____ switches inform the module of the position of the harness.

3. The _____ is a combustible device that converts electrical energy into thermal energy to ignite the inflator propellant.

4. The ORC supplies _____ to the air bag _____ in the event that the battery or cables are damaged during an accident.

5. To prevent accidental deployment of the air bag, most electromechanical systems require that at least _____ sensor switches be closed to deploy the air bag.

6. The frontal impact must be within _____ degrees of the vehicle centerline to deploy the air bag.

7. An accelerometer-type air bag sensor produces an analog voltage in relation to _____ _____.

8. The current flow through the squib required to deploy the air bag is approximately _____ amperes.

9. A hybrid inflator module contains a cylinder filled with compressed _____ gas.

10. The OCM system that uses a bladder may also use a _____ _____ _____ to determine if an infant seat is securely strapped into the seat.

Multiple Choice

1. In the ultrasonic sensor, the transmitter generates the ultrasonic sound waves by:

   A. Using a piezoelectric crystal with an alternating voltage across it that causes them to oscillate at very high frequencies.

   B. Using a piezoresistive bridge that varies the current flow through a sound generator.

   C. Converting the sound waves into electrical energy.

   D. Using a capacitive discharge to oscillate a thin-film transistor sheet at a high frequency.

2. All of the following are characteristics of the occupant classification system **EXCEPT:**

   A. The OCM is responsible for the deployment of the passenger-side air bag.

   B. The system is designed to determine if a rear-facing infant seat is being used in the front passenger seat.

   C. The PADL illuminates if the passenger-side air bag is suppressed.

   D. Weight classification of fifth percentile female and greater allows air bag deployment.

3. Air bag components are being discussed.

   Technician A says the igniter is a combustible device that converts electrical energy into thermal energy.

   Technician B says the inflation of the air bag is done through an explosive release of compressed air.

   Who is correct?

   A. A only
   B. B only
   C. Both A and B
   D. Neither A nor B

4. The air bag system is being discussed.

   Technician A says the clockspring is located at the bottom of the steering column.

   Technician B says the clockspring conducts electrical signals to the module while permitting steering wheel rotation.

   Who is correct?

   A. A only
   B. B only
   C. Both A and B
   D. Neither A nor B

5. The air bag system components are being discussed.

   Technician A says the ORC constantly monitors the readiness of the air SRS electrical system.

   Technician B says a crash sensor may be composed of a gold-plated ball held in place by a magnet.

   Who is correct?

   A. A only
   B. B only
   C. Both A and B
   D. Neither A nor B

6. Air bag sensors are being discussed.

   Technician A says the arrow on each sensor must face toward the rear of the vehicle.

   Technician B says air bag sensor brackets must not be bent or distorted.

   Who is correct?

   A. A only
   B. B only
   C. Both A and B
   D. Neither A nor B

7. Accelerometer-type air bag system sensors are being discussed.

   Technician A says an accelerometer senses collision force and direction.

   Technician B says an accelerometer produces a digital voltage.

   Who is correct?

   A. A only
   B. B only
   C. Both A and B
   D. Neither A nor B

8. Which of the following is a characteristic of the strain gauge–type occupant classification system?

   A. If the seat is empty, the PADL is illuminated.

   B. The system uses a belt tension sensor to determine the presence of an infant seat.

   C. The air bag will deploy if a child is determined to be sitting in the seat.

   D. None of the above.

9. The air bag deployment loop is being discussed.

   Technician A says if the arming sensor contacts close, this sensor completes the circuit from the inflator module to ground.

   Technician B says if the contacts close in two crash sensors, the air bag is deployed.

   Who is correct?

   A. A only
   B. B only
   C. Both A and B
   D. Neither A nor B

10. Hybrid inflator modules are being discussed.

    Technician A says a pressure sensor in the argon gas cylinder sends a signal to the ASDM in relation to gas pressure in the cylinder.

    Technician B says when the initiator is energized, the propellant explodes and pierces the propellant container, allowing the pressurized argon gas to escape into the air bag.

    Who is correct?

    A. A only
    B. B only
    C. Both A and B
    D. Neither A nor B

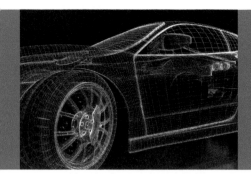

CHAPTER 16

VEHICLES WITH ALTERNATIVE POWER SOURCES

Upon completion and review of this chapter, you should be able to:

- Explain the basic operation of electric vehicles.
- Describe the typical operation of a hybrid vehicle.
- Explain the difference between parallel and series hybrids.
- Explain the purpose of regenerative braking.
- Describe the purpose of the 42-volt system.

- Explain the operating principles of integrated starter generator systems.
- Describe how a proton exchange membrane produces electricity in a fuel cell system.
- List and describe the different fuels that can be used in a fuel cell system.
- Describe the purpose of the reformer.
- Explain how different types of reformers operate.

Terms To Know

DC/DC converter
Dual-generator system
Dual-stator, dual-voltage system
Dual-voltage system
Electric vehicle (EV)
Electrolysis
Fuel cell
Full parallel hybrid

High-voltage ECU (HV ECU)
Hybrid electric vehicle (HEV)
Integrated starter generator (ISG)
Lean burn technology
Mild parallel hybrid
Parallel hybrid

Proton exchange membrane (PEM)
Reformer
Regenerative braking
Series hybrid
Single 42-volt system
Starter generator control module (SGCM)
Ultra capacitor

INTRODUCTION

Due to the increase in regulations concerning emissions, and the public's desire to become less dependent on foreign oil, most major automotive manufacturers have developed alternative fuel or alternative power vehicles. This chapter explores several alternative power sources. This includes a study of common hybrid systems. Also included is a discussion of the 42-volt (V) system and its influence on the new technology of integrated starter generator (ISG) idle stop systems. In addition, the final section of this chapter covers fuel cell theories and some of the methods that manufacturers are using to approach this alternative power source. These power sources are being sold in limited numbers, or they are still in the research and development stage.

ELECTRIC VEHICLES

Since the 1990s, most major automobile manufacturers have developed an **electric vehicle (EV)**. The EV powers its motor from a battery pack. The primary advantage of an EV is a drastic reduction in noise and emission levels. The California Air Resources Board (CARB) established a low-emission vehicles/clean fuel program to further reduce mobile source emissions in California during the late 1990s. This program established emission standards for five vehicle types (**Figure 16-1**): conventional vehicle (CV), transitional low-emission vehicle (TLEV), low-emission vehicle (LEV), ultra low-emission vehicle (ULEV), and zero emission vehicle (ZEV). The EV meets ZEV standards. **Figure 16-2** shows the basic components of an EV.

General Motors (GM) introduced the EV1 electric car to the market in 1996. The original battery pack in this car contained 26 12-volt batteries that delivered electrical energy to a three-phase 102-kilowatt (kW) AC electric motor. The electric motor is used to drive the front wheels. The driving range is about 70 miles (113 km) of city driving or 90 miles (145 km) of highway driving. Temperature, vehicle load, and speed affect this range. A 1.2-kW charger in the vehicle's trunk can be used to recharge the batteries. This charger takes about 15 hours to fully recharge the batteries. An external Delco Electronics's MAGNE CHARGE 6.6-kW inductive charger operating on 220 volts/30 amps can recharge the batteries in 3 to 4 hours (**Figure 16-3**). The weatherproof plastic paddle is inserted into the charging port located at the front of the vehicle. Power is transferred by magnetic fields.

If the batteries are fully charged, the EV1 accelerates from 0 to 60 mph (97 km/h) in 9 seconds and has a top speed of 80 mph (129 km/h). The EV1 is equipped with a Galileo electronic brake system that employs a computer and sensors at each wheel to direct power-assist braking, regenerative/friction brake blending, four-wheel antilock braking, traction assist, tire pressure monitoring, and system diagnostics. In 1998, GM installed nickel/metal/hydride batteries in the EV1 vehicles, which extended the driving range between battery charges to 160 miles (257 km).

The driving range of EVs is their biggest disadvantage. Much research continues to be done to extend the range and to decrease the required recharging times. Currently, the use of nickel/metal/hydride or lead-acid batteries and permanent magnet motors has

	CV	TLEV	LEV	ULEV	ZEV
NMOG	0.25*	0.125	0.075	0.040	0.0
CO	3.4	3.4	3.4	1.7	0.0
NOx	0.4	0.4	0.2	0.2	0.0

(*) Emission standards of NIMHC

Figure 16-1 California tailpipe emission standards in grams per mile at 50,000 miles.

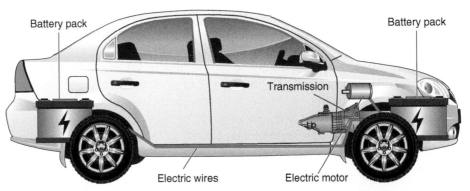

Figure 16-2 The electric vehicle is powered by an electric motor that receives its energy from battery packs.

extended the operating range. Another disadvantage is that the battery pack adds substantial weight to the vehicle (**Figure 16-4**). Other features, such as **regenerative braking** and highly efficient accessories (e.g., a heat pump for passenger heating and cooling), are also

Figure 16-3 Recharging the EV.

Figure 16-4 EV's battery pack.

being installed on EVs. Regenerative braking means that the braking energy is turned back into electricity instead of heat. Other disadvantages are the cost of replacement batteries and the danger associated with the high voltage and high frequency of the motors.

Currently, several manufacturers are offering EVs to the consumer market. These vehicles are providing some increases in range and performance. This new generation of EVs may prove to be the bridge to more advancements.

HYBRID VEHICLES

 A BIT OF HISTORY

The 1896 Armstrong is the world's first known gasoline/electric hybrid automobile. It was designed by Harry E. Dey and built in Bridgeport, Connecticut, by the American Horseless Carriage Company. There is only one known in existence and it was purchased in 2016 for $483,400 by Dutch collector Evert Louwman, who has it on display in his museum in The Hague, Netherlands. The vehicle has a 6,500 cc opposed, twin-cylinder gasoline engine with a dynamo wound flywheel. This design allowed the engine to charge the storage batteries for use by the ignition and lighting systems, but could also rotate the engine for starting. Solenoids installed into the intake valve housings would release compression while the engine was rotated electrically. The size of the flywheel dynamo allowed the vehicle to be propelled under electric power alone. Other innovations included automatic spark advance, an electric clutch, rear wheel brakes with regenerative electric motor assist, and a three-speed (with reverse) constant mesh semi-automatic transmission. A sliding key system engaged the transmission. Half of the gears were cut from rawhide to reduce noise.

Shop Manual
Chapter 16, page 849

The first alternative vehicle attempted was the EV, which had zero emissions and ran primarily on battery power. However, the battery has a limited energy supply and restricts the traveling distance. This limitation was a major stumbling block to many consumers. One method to improve the EV was the addition of an on-board power generator that is assisted by an internal combustion engine. The result was the **hybrid electric vehicle (HEV)**. An HEV has two different power sources. In most hybrid vehicles, the power source consists of a small displacement gasoline or diesel engine and an electric motor. The addition of the internal combustion engine meant that the vehicle could not be classified as a ZEV. However, it does reduce emission levels significantly and increases fuel economy.

Basically, the HEV relies on power from the electric motor, the engine, or both (**Figure 16-5**). When the vehicle moves from a stop and has a light load, the electric motor moves the vehicle. Power for the electric motor comes from stored electricity in the battery pack. During normal driving conditions, the engine is the main power source. Engine power is also used to rotate a generator that recharges the storage batteries (**Figure 16-6**). The output from the generator can also be used to power the electric motor, which is run to provide additional power to the powertrain (**Figure 16-7**). A computer controls the operation of the electric motor depending on the power needs of the vehicle. During full throttle or heavy load operation, additional electricity from the battery is sent to the motor to increase the output of the powertrain.

The components of a typical hybrid vehicle include the following:

Shop Manual
Chapter 16, page 854

- *Batteries.* Some types of batteries that are being used or experimented with now are the lead-acid battery, the nickel-metal hydride battery, and the lithium-ion battery. In the development of the battery, thermal management must be taken into consideration. The temperature can vary from module to module, so the performance of the battery is dependent on the temperature. An imbalance in temperature affects the power and capacity of the battery, the charge acceptance during regenerative braking,

Figure 16-5 Hybrid power system.

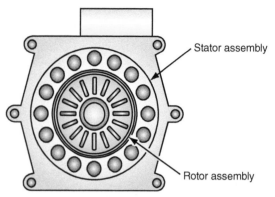

Figure 16-6 Engine power is also used to rotate a generator that recharges the storage batteries and drives the vehicle. The rotor assembly is a very powerful magnet that induces voltage into the stator windings as it is rotated.

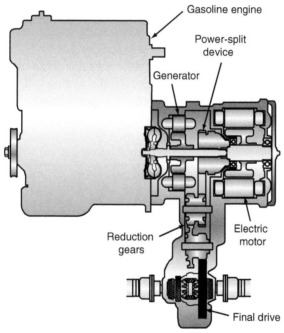

Figure 16-7 Hybrid power system with gasoline engine and electric propulsion motor.

and vehicle operation. Also, passenger safety is a major concern. The batteries must be kept in sealed containers in order to ensure complete protection. The Toyota Prius (**Figure 16-8**) seals its non-caustic, nonflammable nickel-metal hydride battery in a carbon composite case positioned in the rear of the vehicle (**Figure 16-9**).

■ *Electric motors.* One of the sources of power is the electric motor. The motor converts electrical energy to mechanical energy. This mechanical energy is what drives the wheels of the vehicle. This motor is designed to allow for maximum torque at low revolutions per minute (rpm). This gives the electric motor the advantage of having better acceleration than the conventional motor.

■ *Regenerative braking.* About 30% of the kinetic energy lost during braking is in heat. When decreasing acceleration, regenerative braking helps minimize energy loss by recovering the energy used to brake. It does this by converting rotational energy into electrical energy through a system of electric motors and generators. Regenerative braking assumes some of the stopping duties from the conventional friction brakes and uses the electric motor to help stop the car. To do this, the electric motor operates

Figure 16-8 Toyota Prius was the world's first mass-produced hybrid vehicle.

Figure 16-9 Battery pack.

as a generator when the brakes are applied, recovering some of the kinetic energy and converting it into electrical energy. The motor becomes a generator by using the kinetic energy of the vehicle to store power in the battery for later use.

■ *Ultra capacitors.* The **ultra capacitor** is a device that stores energy as electrostatic charge. It is the primary device in the power supply during hill climbing, acceleration, and the recovery of braking energy. To create a larger storage capacity for the ultra capacitors, the surface area must be increased, and in turn, the voltage is increased. However, because the voltage drops as energy is discharged, additional electronics are required to maintain a constant voltage.

Propulsion

There are two typical ways to arrange the flow of power in an HEV. If the combustion engine is capable of turning the drive wheels as well as the generator, then the vehicle is referred to as a **parallel hybrid** (**Figure 16-10**). In a parallel hybrid configuration, there is a direct mechanical connection between the engine and the wheels. Both the engine and the electric motor can turn the transmission at the same time.

There is a further distinction between a **mild parallel hybrid** and a **full parallel hybrid** vehicle. A mild parallel hybrid vehicle has an electric motor that is large enough to provide regenerative braking, instant engine start-up, and a boost to the combustion engine. A full parallel hybrid vehicle uses an electric motor that is powerful enough to propel the vehicle on its own.

Other configurations of the parallel hybrid vehicle include the use of an engine to power one axle and an electric motor to power the other (**Figure 16-11**). Another concept is to use a combination where the engine coupled with an electric motor powers one axle and another electric motor powers the other axle (**Figure 16-12**).

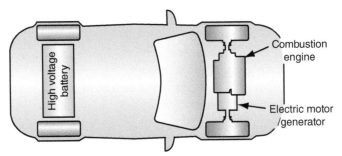

Figure 16-10 Parallel hybrid configuration.

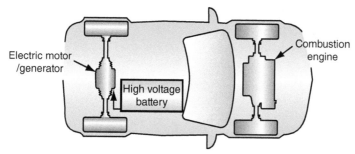

Figure 16-11 Parallel hybrid configuration using an engine to power one axle and an electric motor to power the other axle.

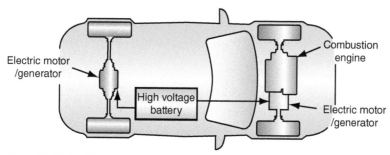

Figure 16-12 Parallel hybrid configuration using a combination where the engine coupled with an electric motor powers one axle and another electric motor powers the other axle.

Most parallel hybrid vehicles use the electric motor to accompany the engine to help drive the wheels. For example, the engine is used for long driving periods, while the electric motor is used for short, low-intensity drives. In other words, the engine is ideal for highway driving, and the electric motor is ideal for a trip around town. The electric motor also provides the vehicle with acceleration. This acceleration, however, is only sustained until the vehicle reaches a certain speed. After this speed is reached, the engine is started and replaces the electric motor. The parallel hybrid combines the alternator, starter, and wheels to create a system that starts the engine, electronically balance it, take power from the engine and turn it into electricity, and provide extra power to the driveline when a power-assist is needed for hill climbing or quick acceleration.

In the **series hybrid** vehicle, there is no mechanical connection between the engine and the wheels. The engine turns a generator, and the generator either charges the batteries or powers the electric motor, which in turn drives the transmission. Therefore, the engine never directly powers the automobile (**Figure 16-13**).

The power used to give the vehicle motion is transformed from chemical energy into mechanical energy, then into electrical energy, and finally back to mechanical energy to drive the wheels. This configuration is efficient in that it never idles. The automobile turns off completely at rest, such as at a stop sign or traffic light. This feature greatly reduces

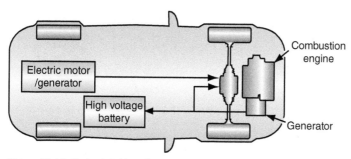

Figure 16-13 Series hybrid configuration.

emissions. There is a variety of options in the configuration and mounting of all the components. Some series hybrid vehicles do not use a transmission.

HEV Examples

Operating the engine independently of the vehicle speed means that even though the vehicle is traveling at highway speeds, the engine can be close to idle speed since it is only acting as a generator.

The Toyota Prius is considered a super ultra low-emissions vehicle (SULEV), meaning that it is 90% cleaner than an ULEV. The Prius uses a combined hybrid-electric structure to supply power. This system uses a combination of an internal combustion engine and an electric motor to turn the electrically controlled continuous variable transmission (ECCVT). However, the Prius can also accelerate using both the engine and the electric motor and can run solely on the electric motor. This is called a power split device (**Figure 16-14**).

Using a set of planetary gears (**Figure 16-15**), the vehicle can operate like a parallel vehicle in that either the electric motor or the gasoline engine powers the vehicle or they both do. However, the vehicle can also operate as a series hybrid where the engine can operate independently of the vehicle speed, either charging the batteries or providing power to the wheels when needed.

All the propulsion and auxiliary components of the Prius are packaged under the hood. The electric motor is rated at 33 kW and battery power is 21 kW. The nickel-metal hydride battery pack is located between the rear seatback and the trunk.

The Prius is equipped with a variable valve timing four-cylinder engine. The generator also works as the starter of this engine. Gear shifting is not required because the planetary system acts as a continuously variable transmission to keep engine rpm in the range of best efficiency. Instead of a normal transmission, the engine drives the planet carrier of a planetary gear set. The sun gear of that set connects to a motor/generator, and the ring gear drives both the front wheels and a second motor/generator.

The Prius is also equipped with a "drive-by-wire" accelerator. The driver inputs to the motor management how much speed is requested. The management system then decides where the necessary power should come from: the engine, the battery, or both. The same accounts for the brake-by-wire system: the driver calls for the appropriate amount of retardation, and the motor management coordinates this between the wheel brakes and the regenerative braking system. The computer also makes sure that the generator runs frequently enough to keep the battery charged.

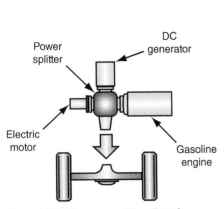

Figure 16-14 The power splitter allows for acceleration using both the engine and the electric motor and to be able to run solely on the electric motor.

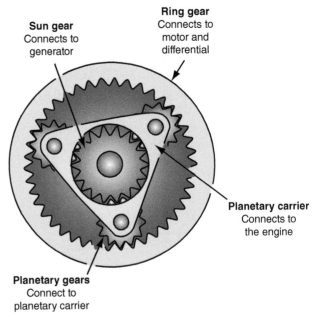

Figure 16-15 Planetary gears are used to transfer power to the drive wheels.

When the battery is fully charged and the engine temperature is acceptable, the engine can be shut off (if the vehicle speed is low enough). If the driver gently presses the accelerator, the vehicle is moved by battery power. If the driver requests a quicker acceleration, the engine is started and powers the vehicle. In normal driving, the Prius maintains the battery state-of-charge within a narrow window. However, driving the vehicle under heavy loads for an extended time may deplete the battery charge.

The Honda Insight is also a parallel hybrid vehicle. In this system, the gasoline engine provides the majority of the power. The electric motor is used to help the gasoline engine provide additional power during acceleration. The Insight uses regenerative braking technology to capture energy lost during braking. The Insight also has a lightweight engine that uses **lean burn technology** to maximize its efficiency. Lean burn technology, developed in the 1960s, uses high air-fuel ratios to increase fuel efficiency.

The Honda Civic Hybrid uses an integrated motor assist (IMA) system to power the vehicle. The system comprises a gasoline and electric motor combination. The electric motor is a source of additional acceleration and functions as a high-speed starter. The electric motor also acts as a generator for the charging system used during regenerative braking. This way the Civic Hybrid ensures efficiency by capturing lost energy by using regenerative braking, much like the Prius and the Insight.

42-VOLT SYSTEMS

The idea of 42-volt systems has been around for over 25 years. The first production vehicle that used the 42-volt system was the 2002 Toyota Crown Sedan, which was sold only in the Japanese market. Although the use of 42-volt systems is very limited at the current time, the technology learned has been applied to the hybrid systems. In this section, we will initially discuss the 42-volt system as it is used for vehicle electrical systems and then progress to the use of the ISG on some mild-hybrid vehicles.

As you have probably come to realize, the increased use of electrical and electronic accessories in today's vehicles has almost tapped the capabilities of the 12-volt/14-volt electrical system. Electronic content in vehicles has been rising at a rate of about 6% per year. It's been estimated that by the end of the decade the electronic content will be about 40% of the total cost of a high-line vehicle. The electrical demands on the vehicle have risen from about 500 watts in 1970 to about 7,000 watts in 2010. It is estimated that in 10 years the demand may reach 10,000 watts.

One way to meet the higher electrical demands would be to increase the amperage output of the generator. This has been the approach over the past 30 years. Years ago it was common for a generator to have a rating of 35 to 50 amps. Today, most manufacturers use a 150-amp generator. However, just increasing the amperage output of the generator will not suffice to meet the 10,000-watt demand. Using Watt's law, in order to obtain 10,000 watts with a 14-volt generator the output would have to be 714 amps. The more realistic alternative is to increase the voltage. As a result, vehicle manufacturers are developing a 36-volt/42-volt system that incorporates a 42-volt generator charging a 36-volt battery. To meet the 10,000-watt demand in a 42-volt system, 238 amps are required.

An additional benefit that may be derived from the use of a 42-volt system is that it allows manufacturers to electrify most of the inefficient mechanical and hydraulic systems that are currently used. The new technology will allow electro-mechanical intake and exhaust valve control, active suspension, electrical heating of the catalytic converters, electrically operated coolant and oil pumps, electric air-conditioning compressor, brake-by-wire, steer-by-wire, and so on to be utilized. Studies have indicated that as these mechanical systems are replaced, fuel economy will increase by about 10%. In addition, emissions will decrease.

The system is called 12 volt/14 volt because the battery is 12 volts but the charging system delivers 14 volts. The 42-volt system is actually a 36-volt/42-volt system.

Additional fuel savings can also be realized due to the more efficient charging system used for the 42-volt system. Current 14-volt generators have an average efficiency across the engine speed range band of less than 60%. This translates to about 0.5 gallons (1.9 L) of fuel for 65 miles (104.6 km) of driving to provide a continuous electrical load of 1,000 watts. With a 42-volt generator, the fuel consumption can be reduced the equivalent of up to 15% in fuel savings. The 42-volt generator will be discussed later in this chapter.

Changing to 42-volt system provides three times as much generator power as the current system and will deliver as much as 30,000 watts. Since the increase in voltage results in a two-thirds reduction in amperage, the size of components can be reduced. Also, a significant reduction in the vehicle wiring size and weight can be realized.

There are several methods that are being used and developed for the electrical architecture of the 42-volt system. One method is a simple **single 42-volt system** that is very similar to the current 12-volt/14-volt system (**Figure 16-16**). The challenge with this type of system is that all of the vehicle's electrical system will require redesigning to handle the 42 volts.

Due to the initial costs of implementing a single 42-volt system, a **dual-voltage system** is also being designed. With this architecture, the 42-volt system would power those electrical accessories that would require or benefit from the higher voltage. The remainder of the loads would remain on 14 volts. There are several ways to implement a dual-voltage system.

One method is to use a **dual-generator system** (**Figure 16-17**). One generator operates at 42 volts, while the other operates at 14 volts. Another system is the **dual-stator, dual-voltage system** (**Figure 16-18**). In this system, dual voltage is produced from a single alternator that has two output voltages.

Another design uses a **DC/DC converter** (**Figure 16-19**). The DC/DC converter is configured to provide a 14-volt output from the 42-volt input. The 14-volt output can be used to supply electrical energy to those components that do not require 42 volts.

As simple as it may seem to convert from a 12-volt/14-volt system to a 36-volt/42-volt system, many challenges need to be overcome. It is not as simple as adding a higher-voltage output generator and expecting the existing electrical components to work. One of the

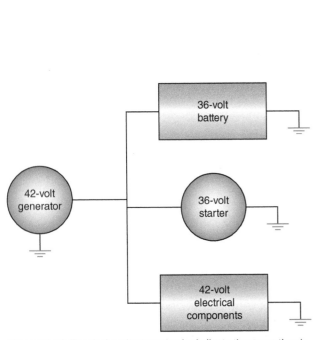

Figure 16-16 The single-voltage system is similar to the conventional 12-volt/14-volt system.

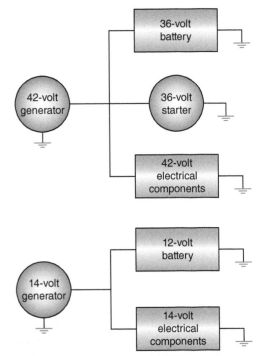

Figure 16-17 The dual-voltage system separates the electrical systems.

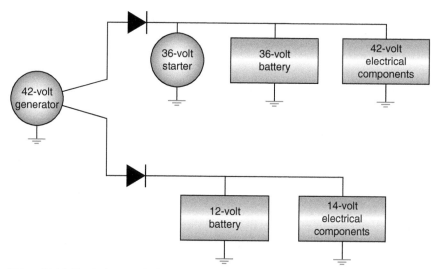

Figure 16-18 The dual-stator, dual-voltage system splits the electrical systems by using two outputs from a single generator.

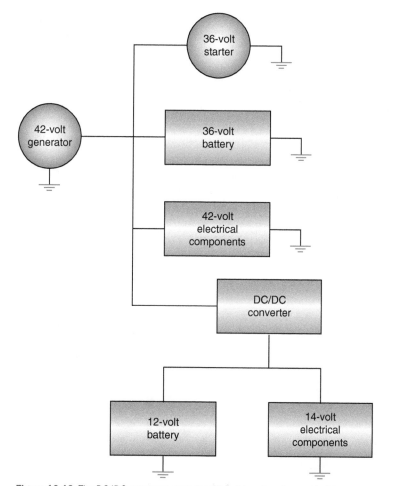

Figure 16-19 The DC/DC converter uses the 42-volt input and converts it to a 14-volt output.

biggest hurdles is the light bulb. Current 12-volt lighting filaments can't handle 42 volts. The dual-voltage systems can be used as a step toward full 42-volt implantation. However, dual-voltage systems are expensive to design. Another aspect of the 42-volt system is service technician training to address aspects of arcing, safety, and dual-voltage diagnostics.

A BIT OF HISTORY

In 1955, automobile manufacturers started to move from 6-volt electrical systems to the present 12-volt system. The change was due to the demand for increased power to accommodate a greater number of electrical accessories. In 1955, the typical car wiring harness weighed 8 to 10 pounds and required approximately 250 to 300 watts. In 1990, the typical car wiring harness weighed 15 to 20 pounds and required over 1,000 watts. In 2000, the typical car wiring increased to weigh between 22 and 28 pounds and required over 1,800 watts. The conventional 14-volt generator is capable of producing a maximum output of 2,000 watts.

Arcing is perhaps the greatest challenge facing the design and use of the 42-volt system. In fact, some manufacturers have abandoned further research and development of the system because of the problem with arcing. In the conventional 14-volt system, the power level is low enough that it is almost impossible to sustain an arc. Since there isn't enough electrical energy involved, the arcs collapse quickly and there is less heat buildup. Electrical energy in an arc at 42 volts is significantly greater and is sufficient to maintain a steady arc. The arc from a 42-volt system can reach a temperature of 6,000°F (3,316°C).

Voltage regulation also presents a challenge, expectably in dual-voltage systems. As mentioned earlier on method of dual-voltage control is the use of DC/DC converters. Another method is pulse width modulation (PWM). An advantage of this method is that it reduces most of the arcing problems associated with higher voltage. This is because there is not a true steady stream of current with PWM. This results in the arc collapsing quickly and in the reduction of heat.

One key contribution to the efficiency of the hybrid vehicle is its ability to automatically stop and restart the engine under different operating conditions. A typical hybrid vehicle uses a 14-kW electric induction motor or **integrated starter generator (ISG)** between the engine and the transmission (**Figure 16-20**). The ISG performs many functions such as fast, quiet-starting, automatic engine stops/starts to conserve fuel, recharging the vehicle batteries, smoothing driveline surges, and providing regenerative braking.

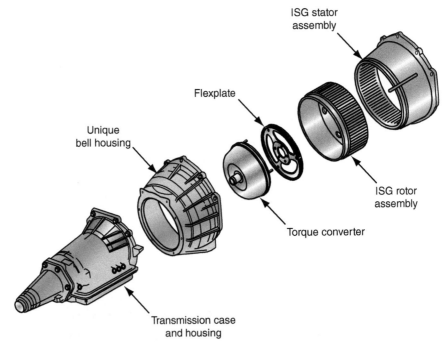

Figure 16-20 The ISG is usually located between the engine and the transmission in the bell housing.

Hybrid vehicles utilize the automatic stop/start feature to shut off the engine whenever the vehicle is not moving or when power from the engine is not required. Usually this feature is activated when the vehicle is stopped, no engine power is required, and the driver's foot is on the brake pedal. On manual transmission–equipped vehicles, this feature may be activated when the vehicle is stopped, no engine power is required, the transmission is in neutral, and the clutch pedal is released. Once the driver's foot is removed from the brake pedal (or the clutch is engaged), the starter automatically restarts the engine in less than one-tenth of a second. To further save fuel and reduce emissions, the engine is accelerated to idle speed by the starter-generator prior to the start of the combustion process and the injection of fuel.

The ISG can also convert kinetic energy to direct current (DC) voltage. When the vehicle is traveling downhill and there is zero load on the engine, the wheels can transfer energy through the transmission and engine to the ISG. The ISG then sends this energy to the battery for storage and use by the electrical components of the vehicle.

Some systems use a 42-volt ISG system since the power requirements of automatic stop/start are higher than that the 12-volt system can provide. Currently there are two main system designs.

The first design uses a belt alternator starter (BAS) that is about the same size as a conventional generator and is mounted in the same way (**Figure 16-21**). BASs have a maximum power output of around 5 kW. Two types of BASs are being designed: permanent magnet and induction BAS.

The second design is to mount the ISG at either end of the crankshaft. Most designs have the ISG mounted at the rear of the crankshaft between the engine and transmission. In some systems, the ISG may take the place of the engine flywheel. The ISG mounted in this method is larger than the BSG and is able to produce an output of 6 to 15 kW.

The ISG is a three-phase alternating current (AC) motor. At low vehicle speeds, the ISG provides power and torque to the vehicle. It also supports the engine, when the driver demands more power. During vehicle deceleration, ISG regenerates the power that is used to charge the traction batteries.

The ISG includes a rotor and a stator that are located inside the transmission bell housing (**Figure 16-22**). The stator is attached to the engine block and coils that are formed by laser-welding copper bars. Conventional generators use winding of copper wire for their stator. The rotor is bolted to the engine crankshaft.

Both the BAS and the ISG use the same principle to start the engine. Current flows through the stator windings and generates magnetic fields in the rotor. This causes the rotor to turn, thus turning the crankshaft and starting the engine. In addition, this same principle is used to assist the engine as needed when the engine is running.

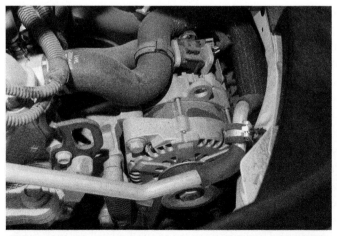

Figure 16-21 The belt starter generator looks very similar to a conventional generator.

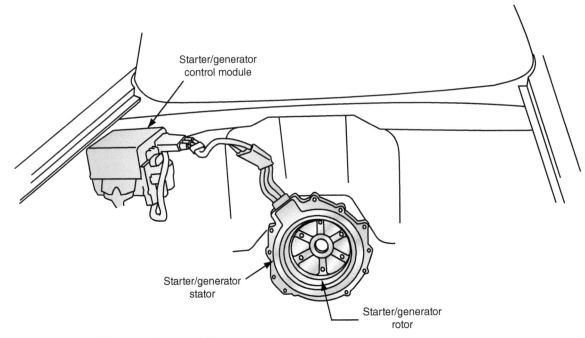

Starter/generator
control module

Starter/generator
stator

Starter/generator
rotor

Figure 16-22 The ISF stator and rotor assembly.

Shop Manual
Chapter 16, page 852

A **starter generator control module (SGCM)**, also called a **high-voltage ECU (HV ECU)**, is used to control the flow of torque and electrical energy. Remember that torque and energy can go into or out of the ISG. The function of the SGCM is to control the engine cranking, torque, speed, and active damping functions.

Our discussion up to this point has centered on the use of 42 volts for the ISG system. It is important to note that the hybrid system could work on 42 volts or 300 volts. Also, the voltage used to start the engine and to operate the electric motors is AC.

> **AUTHOR'S NOTE** See *Today's Technician: Advanced Automotive Electronic Systems* for more information on EVs and HEVs.

FUEL CELLS

A **fuel cell** produces current from hydrogen and aerial oxygen. Fuel cell–powered vehicles have a very good chance of becoming the vehicles of the future. They combine the reach of conventional internal combustion engines with high efficiency, low fuel consumption, and minimal or no pollutant emissions. At the same time, they are extremely quiet. Because they work with regenerative fuel such as hydrogen, they reduce the dependence on crude oil and other fossil fuels.

A fuel cell–powered vehicle (**Figure 16-23**) is basically an EV. Like the EV, it uses an electric motor to supply torque to the drive wheels. The difference is that the fuel cell produces and supplies electric power to the electric motor instead of the batteries. Most vehicle manufacturers and several independent laboratories are involved in fuel cell research and development programs. Manufacturers have produced a number of prototype fuel cell vehicles, with many being placed in fleets in North America and Europe.

Fuel cells electrochemically combine oxygen from the air with hydrogen to produce electricity. The oxygen and hydrogen are fed to the fuel cell as "fuel" for the electrochemical reaction. There are different types of fuel cells but the most common type is the **proton**

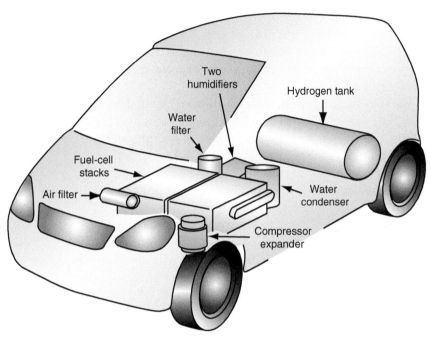

Figure 16-23 Fuel cell vehicle components. Technology has allowed engineers to design fuel cell vehicles without the loss of passenger and cargo space.

exchange membrane (PEM). Normally, hydrogen and oxygen bond with a loud bang but in fuel cells a special PEM impedes the oxyhydrogen gas reaction by ensuring that only protons (H+) and not elemental hydrogen molecules (H_2) react with the oxygen.

How the Fuel Cell Works

The PEM fuel cell is constructed like a sandwich (**Figure 16-24**). The electrolyte is situated between two electrodes of gas-permeable graphite paper. The electrolyte is a polymer membrane. Hydrogen is applied to the anode side of the PEM and ambient oxygen is applied to the cathode side (**Figure 16-25**). The membrane keeps the distance between the two gases and provides a controlled chemical reaction.

The anode is the negative post of the fuel cell. It conducts the electrons that are freed from the hydrogen molecules so that they can be used in an external circuit. It has channels etched into it that disperse the hydrogen gas evenly over the surface of the catalyst. The cathode is the positive post of the fuel cell. It also has channels etched into it that distribute the oxygen to the surface of the catalyst. It also conducts the electrons back from the external circuit to the catalyst where they can recombine with the hydrogen ions and oxygen to form water.

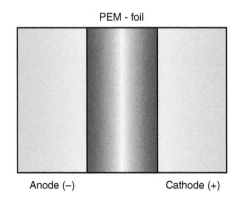

Figure 16-24 The PEM foil.

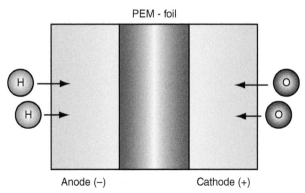

Figure 16-25 Hydrogen is applied to the anode side of the PEM, while oxygen is applied to the cathode side.

A fine coating of platinum is applied to the foil to act as a catalyst. This is used to accelerate the decomposition of the hydrogen atoms into electrons and protons (**Figure 16-26**). The catalyst is rough and porous so that the maximum surface area of the platinum can be exposed to the hydrogen or oxygen. The platinum-coated side of the catalyst faces the PEM.

The PEM is the electrolyte. This specially treated material (which looks similar to ordinary kitchen plastic wrap) allows only the protons to move across from the anode to the cathode (**Figure 16-27**). As a result, the anode has a surplus of electrons and the cathode has a surplus of protons. If the anode and cathode are connected outside of the cell, current flows through the conductor (**Figure 16-28**). The electrons move through the conductor to the cathode. The electrons then recombine with the protons and the oxygen to produce water.

The entire process is illustrated in **Figure 16-29**. Pressurized hydrogen gas (H_2) enters the fuel cell on the anode side. This gas is forced through the catalyst by the pressure. When an H_2 molecule comes in contact with the platinum on the catalyst, it splits into two H+ ions and two electrons (e−). The electrons are conducted through the

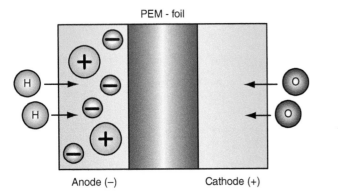

Figure 16-26 The catalyst breaks down H_2 into protons and electrons.

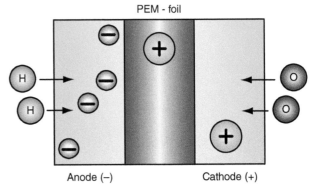

Figure 16-27 The PEM foil allows only the protons to migrate to the cathode, leaving the electrons on the anode.

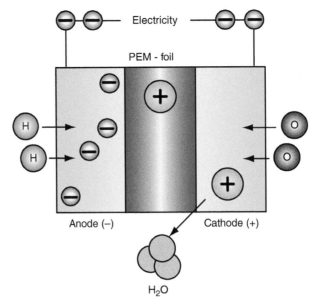

Figure 16-28 With an excess amount of electrons on the anode and an excess amount of protons on the cathode, current flows through an external conductor. The electrons then react with the protons and oxygen to produce water.

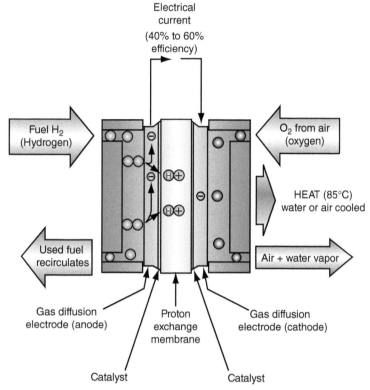

Figure 16-29 PEM fuel cell.

anode where they make their way through the external circuit (doing useful work such as turning a motor) and return to the cathode side of the fuel cell.

Meanwhile, on the cathode side of the fuel cell, oxygen gas (O_2) is being forced through the catalyst where it forms two oxygen atoms. Each of these atoms has a strong negative charge. This negative charge attracts the two H+ ions through the membrane, where they combine with an oxygen atom and two of the electrons from the external circuit to form a water molecule (H_2O).

This reaction in a single fuel cell produces only about 0.7 volt. For this voltage to become high enough to be used to move the vehicle, many separate fuel cells must be combined in series to form a fuel-cell stack (**Figure 16-30**).

Fuels for the Fuel Cell

A fundamental problem with fuel cell technology concerns whether to store hydrogen or convert it from other fuels on board the vehicle. All four principal fuels that automotive manufacturers are considering (hydrogen, methanol, ethanol, and gasoline) pose some challenges.

Hydrogen Fuel. One solution is to store hydrogen on board the vehicle. The ability to use hydrogen directly in a fuel cell provides the highest efficiency and zero tailpipe emissions. However, hydrogen has a low energy density and boiling point, thus on-board storage requires large, heavy tanks. There are three types of hydrogen storage methods under development: compressed hydrogen, liquefied hydrogen, and binding hydrogenate to solids in metal hydrides or carbon compounds.

Compressed hydrogen offers the least expensive method for on-board storage. However, at normal CNG-operating pressures of 3,500 psi (241 bar), reasonably sized, commercially available pressure tanks provide limited range for a fuel cell vehicle (about 120 miles or 193 km). Daimler and Hyundai are now using pressure tanks that are capable of 5,000 psi (345 bar). Quantum is conducting research of high-performance hydrogen

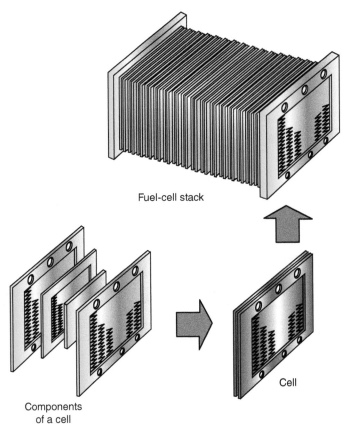

Fuel-cell stack

Cell

Components
of a cell

Figure 16-30 PEM fuel-cell stack.

storage systems, looking at pressure tanks that are capable of up to 10,000 psi (689 bar). This capability would permit a 400-mile (644-km) driving range.

Liquefied hydrogen can be stored in large cylinders containing a hydride material (something like steel wool). Liquefied hydrogen does not require the high storage capacity as that of compressed hydrogen for the same amount of driving range. However, the very low boiling point of hydrogen requires that the tanks have excellent insulation. Maintaining the extreme cold temperature of −423°F (−253°C) during refueling and storage is difficult. It is estimated that up to 25% of the liquid hydrogen may be boiled off during the refueling process. In addition, about 1% is lost per day in on-board storage. Storing liquid hydrogen on a vehicle also involves some safety concerns. As the fuel tank warms, the pressure increases and may activate the pressure relief valve. This action discharges flammable hydrogen into the atmosphere, creating a source of danger and pollution.

Methanol. Several automotive manufacturers are using methanol to power their fuel cells. It is believed that methanol fuel cells could bridge the gap over the next few decades while a hydrogen distribution infrastructure is being built. Using "methanolized" hydrogen as fuel has the advantage that it can be stored on the vehicle similarly to gasoline. For the reaction in the fuel cell, a **reformer** on board the vehicle produces hydrogen from the methanol fuel. A reformer is a high-temperature device that converts hydrocarbon fuels to carbon monoxide (CO) and hydrogen. To produce the hydrogen, the methanol fuel is mixed with water. When it evaporates, it is decomposed into hydrogen and carbon dioxide (CO_2). Prior to sending the hydrogen and CO_2 to the fuel cell, it is purified in additional steps.

Methanolized hydrogen contains more hydrogen atoms and has an energy density that is greater than that of liquid hydrogen. Like hydrogen, methanolized hydrogen is also independent of mineral oil. Compared with hydrogen, vehicles driven by methanol are not completely emissions-free, but they produce very little pollutants and much less CO_2 than internal combustion engines.

A special type of PEM fuel cell, called the direct methanol air fuel cell (DMAFC), utilizes methanol combined with water directly as a fuel and ambient air for oxygen. This technology enables the use of a liquid fuel without the need for an on-board reformer while still providing a zero-emissions system. However, current research has demonstrated that the power density is lower than for other PEM fuel cells.

Ethanol. Ethanol is considered less toxic than either gasoline or methanol. An ethanol system requires adding a reformer to the vehicle, similar to a methanol system. The fuel cell could use E100, E95, or E85.

Gasoline. Fuel cells can be driven with a special, more pure gasoline. Using reformers for on-board extraction of hydrogen from gasoline is one approach to commercialization of fuel cell vehicles since the gasoline infrastructure is already in place. However, producing hydrogen from gasoline in a vehicle system is much more difficult than producing hydrogen from methanol or ethanol. Gasoline reforming requires higher temperatures and more complex systems than the methanol or ethanol reforming. The reformation reactions occur at 1,562°F to 1,823°F (850°C to 995°C), making the devices slow to start and the chemistry temperamental. Thus, the drive would work less efficiently and produce more emissions. Moreover, the capabilities for cold starts would be restricted. The size of the reformer is also an issue, making it difficult to fit under the hood of a standard-sized vehicle. Furthermore, there is concern about the sulfur levels in current gasoline and CO in the reformer poisoning the fuel cell.

Reformer. As mentioned, some fuel cell systems may require the use of an on-board reformer to extract hydrogen from liquid fuels such as gasoline, methanol, or ethanol. On-board reformation of a hydrocarbon fuel into hydrogen allows the use of more established infrastructures, but adds additional weight and cost and reduces vehicle efficiency. In addition, the reformer does create some emissions.

PEM fuel cell reformers combine fuel and water to produce additional H_2 and convert the CO to CO_2. The CO_2 is then released to the atmosphere. Reformer technologies include steam reforming (SR), partial oxidation, and high-temperature electrolytes reforming.

SR uses a catalyst to convert fuel and steam to H_2, CO, and CO_2. The CO is further reformed with steam to form more H_2 and CO_2. A purification step then removes CO, CO_2, and any impurities to achieve a high hydrogen purity level (97% to 99.9%). SR of methanol is the most developed and least expensive method to produce hydrogen from a hydrocarbon fuel on a vehicle, resulting in 45% to 70% conversion efficiency.

Partial oxidation (POX) reforming is similar to stream reforming since both technologies combine fuel and steam, but this process adds oxygen in an additional step. The process is less efficient than SR, but the heat-releasing nature of the reaction makes it more responsive than SR to variable load. Heavier hydrocarbon (HC) can be used in POX, but it has lower carbon-to-hydrogen ratios, which limits hydrogen production.

Hydrogen can also be obtained by **electrolysis** from water. Electrolysis is the splitting of water into hydrogen and oxygen. The drawback to this process is that it requires a great deal of electrical energy. Recently, the development of high-temperature electrolytes that can operate at temperatures in excess of 212°F (100°C) has shown some positive results. The benefits of high-temperature electrolytes include the following:

- *Improved CO tolerance.* This allows the manufacturer to reduce or remove the need for an oxidation reactor and for air bleed. Since these requirements can be reduced, system efficiency is increased by 5% to 10%. There is also a considerable reduction in start-up time. The remaining CO is combusted in a catalytic tail-gas burner to prevent emissions of CO.
- *Facilitated stack cooling.* This reduces the size of the radiator and reduces the fuel-cell stack cooling plate requirements.
- *Humidity-independent operation.* Generally, high-temperature membranes require humidifiers and water recovery, whereas this system does not.

Solid Oxide Fuel Cells. Planar solid oxide fuel cells (SOFCs) operate at high temperatures of 932°F to 1,472°F (500°C to 800°C) and can use CO and H_2 fuel. SOFCs have a good tolerance to fuel impurities and use ceramic as an electrolyte. BMW is currently developing an auxiliary power unit (APU) with Delphi and Global Thermoelectric using SOFCs. Currently, SOFCs use gasoline fuel and require a reformer.

Sodium Borohydride. A joint venture between Daimler and Chrysler introduced a new idea in fuel technologies with a concept minivan called the Chrysler Town & Country Natrium (Latin for sodium). This research was driven by the lack of a safe, compact way to contain hydrogen. In addition, most fuel cell vehicles are not "on demand," meaning that there is a start-up time required before the reformer can begin producing hydrogen.

A fairly simple chemical process mixes sodium borohydride ($NaBH_4$) powder with water. The $NaBH_4$ powder produces free hydrogen for the fuel cell (**Figure 16-31**). Unlike the fuel cell systems that use methanol or gasoline, the Natrium produces no pollution and no CO_2.

The $NaBH_4$ powder holds more hydrogen than the most densely compressed hydrogen tank. The prototype minivan attains a top speed of 80 mph (129 km/h) and an operating range of almost 300 miles (487 km). The by-products of this process are water and sodium borate, which is basically a laundry detergent. After use, the spent powder goes into a storage tank where it can be pumped out and reclaimed.

As the name implies, $NaBH_4$ is a salt. The salt that powers this system is not ordinary table salt (sodium chloride—NaCl). Rather, it is a white salt whose molecules contain a relatively large amount of hydrogen. Through the use of a chemical catalyst, the $NaBH_4$ reaction with water results in elemental hydrogen. A slurry of sodium perborate ($NaBO_2$) forms during this reaction as well. This compound is chemically related to borax, which is used as a bleaching agent in conventional detergents. The $NaBH_2$ slurry is collected in a special tank and can be effectively recycled in a chemical process. In this reverse reaction, the $NaBH_2$ is reclaimed back to $NaBH_4$, which can then be reused as an energy source for the fuel cells.

The catalyst system is called Hydrogen on Demand™ and was developed by Millennium Cell (**Figure 16-32**). When stepping on the accelerator, the minivan driver is actually "throttling" a fuel pump. The $NaBH_4$ is pumped into the catalyst, which immediately generates hydrogen. This dynamic process slows or ceases entirely when the fuel pump is throttled or completely shut off. When the driver accelerates once more, hydrogen is immediately produced again and the fuel cells immediately generate electricity.

> The advantage of sodium borohydride is that it can be stored easily and transported in lightweight plastic tanks at normal temperatures and pressures when dissolved in water. Also, the borate solution is not poisonous, flammable, or explosive.

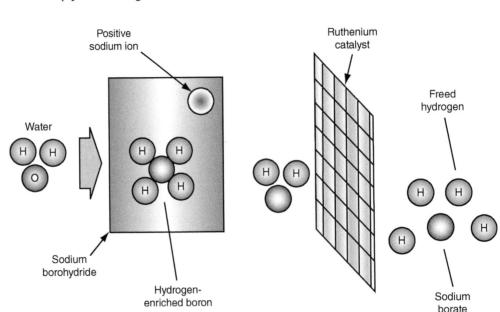

Figure 16-31 Process of extracting hydrogen from sodium borohydride.

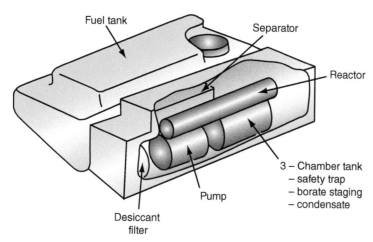

Figure 16-32 The Hydrogen on Demand™ catalyst system is the heart of the drivetrain. The hydrogen required for the fuel cell to generate power is extracted from sodium borohydride in a reactor.

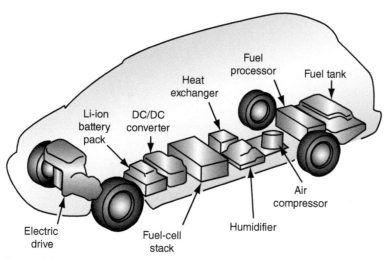

Figure 16-33 All components of the sodium borohydride fuel cell minivan are located between the frame rails.

An AC motor with an output of 35 kW powers the vehicle. A 55-kW lithium-ion battery serves as a storage unit for electrical energy. The battery is recharged by the fuel cell unit and by regenerative braking. If the fuel cell system should fail, the battery would be sufficient to drive the electric motor and move the vehicle. All fuel cell components are contained within the frame rail of the vehicle (**Figure 16-33**).

SUMMARY

- EVs powered by an electric motor run from a battery pack.
- The HEV relies on power from the electric motor, the engine, or both.
- The hybrid vehicle can be configured as a parallel, a series, or a combination hybrid system.
- Regenerative braking recovers the energy used to brake by converting rotational energy into electrical energy through a system of electric motors and generators. When the brakes are applied, the motor becomes a generator by using the kinetic energy of the vehicle to store power in the battery for later use.
- To meet the electrical demands of the automobile in the future, manufacturers are presently developing 42-volt systems.
- An additional benefit that may be derived from the use of a 42-volt system is it allows manufacturers to electrify most of the inefficient mechanical and hydraulic systems that are currently used.

- There are several methods that are being used and developed for the electrical architecture of the 42-volt system. One method is a simple single 42-volt system that is very similar to the current 12-volt/14-volt system. Another method is the dual-voltage system.
- The dual-generator system uses two generators: one that operates at 42 volts, while the other operates at 14 volts.
- In the dual-stator, dual-voltage system, voltage is produced from a single alternator that has two output voltages.
- Another design uses a DC/DC converter that is configured to provide a 14-volt output from the 42-volt input.
- Arcing is perhaps the greatest challenge facing the design and use of the 42-volt system.
- One of the newest technologies to emerge from the research and development of the 42-volt system is the ISG.
- Hybrid vehicles utilize the automatic stop (or idle stop) feature to shut off the engine whenever the vehicle is not moving or when power from the engine is not required.
- The ISG can also convert kinetic energy to storable electrical energy. When the vehicle is traveling downhill and there is zero load on the engine, the wheels can transfer energy through the transmission and engine to the ISG. The ISG then sends this energy to the battery for storage and use by the electrical components of the vehicle.
- The BAS is about the same size as a conventional generator and is mounted in the same way.
- The ISG is a three-phase AC motor. At low vehicle speeds, the ISG provides power and torque to the vehicle. It also supports the engine, when the driver demands more power.
- The ISG includes a rotor and a stator that are located inside the transmission bell housing and is formed by laser-welding copper bars. The rotor is bolted to the engine crankshaft.
- Both the BAS and the ISG use the same principle to start the engine. Current flows through the stator windings and generates magnetic fields in the rotor. This causes the rotor to turn, thus turning the crankshaft and starting the engine.
- A fuel cell–powered vehicle is basically an EV except the fuel cell produces and supplies electric power to the electric motor instead of batteries.
- Fuel cells electrochemically combine oxygen from the air with hydrogen to produce electricity.
- The most common type of fuel cell is the PEM.
- There are several methods being explored for fuel for the fuel cell. These include hydrogen, methanol, ethanol, and gasoline.
- Most fuel cell systems require the use of a reformer to extract hydrogen from liquid fuels such as gasoline, methanol, or ethanol.

REVIEW QUESTIONS

Short-Answer Essays

1. Explain the meaning of regenerative braking.
2. Describe the basic operation of a typical hybrid vehicle.
3. Explain the difference between parallel and series hybrids.
4. Briefly describe how the proton exchange membrane (PEM) fuel cell produces electrical energy.
5. What is the purpose of the reformer?
6. Explain why increasing amperage output of a generator is not able to meet the demands of the electrical system on future vehicles.
7. Explain the advantages that can be derived from the 42-volt electrical system.
8. Describe the basics of the three common dual-voltage systems.
9. What are some of the challenges of implementing the 42-volt system?
10. Describe the basic function of the integrated starter generator system.

Fill in the Blanks

1. _____ _____ recovers the heat energy used to brake by converting rotational energy into _____ energy through a system of electric motors and generators.
2. Fuel cells _____ combine oxygen from the air with hydrogen to produce electricity.
3. Most fuel cell systems require the use of a _____ to extract hydrogen from liquid fuels such as gasoline, methanol, or ethanol.
4. In a _____ hybrid configuration, there is a direct mechanical connection between the engine and the wheels.
5. In the _____ hybrid vehicle, there is no mechanical connection between the engine and the wheels.

6. The _____ is the negative post of the fuel cell.

7. In the ISG system, the _____ is attached to the engine block and coils that are formed by laser-welding copper bars.

8. The ISG is a _____ AC motor. At low vehicle speeds, the ISG provides power and torque to the vehicle. It also supports the engine, when the driver demands more power. During vehicle deceleration, ISG regenerates the power that is used to charge the traction batteries.

9. A _____ is used to control the flow of torque and electrical energy.

10. In a hybrid system, the _____ is a device that stores energy as electrostatic charge and is the primary device in the power supply during hill climbing, acceleration, and the recovery of braking energy.

Multiple Choice

1. *Technician A* says regenerative braking recovers the energy used to brake by converting rotational energy into electrical energy through a system of electric motors and generators.
Technician B says when the brakes are applied, the motor becomes a generator by using the kinetic energy of the vehicle to store power in the battery for later use.
Who is correct?
A. A only   C. Both A and B
B. B only   D. Neither A nor B

2. Electric vehicles power the motor by:
A. A generator.   C. An engine.
B. A battery pack.   D. None of the above.

3. *Technician A* says in a parallel hybrid vehicle, propulsion comes directly from the electric motor.
Technician B says in a series hybrid vehicle, both the engine and the electric motor can turn the transmission at the same time.
Who is correct?
A. A only   C. Both A and B
B. B only   D. Neither A nor B

4. *Technician A* says a fuel cell produces electrical energy by breaking down H_2 atoms into electrons and protons.
Technician B says the compressed hydrogen system requires a reformer to cool the fuel cell.
Who is correct?
A. A only   C. Both A and B
B. B only   D. Neither A nor B

5. The main advantage of the ISG system is that it:
A. Can be used to provide accessory power if the battery fails.
B. Allows for automatic stop/start functions.
C. Increases the life of the brake linings.
D. All of the above.

6. The splitting of water into hydrogen and oxygen is an example of:
A. Steam reforming.
B. Hydrocarbon reforming.
C. Partial oxidation reforming.
D. Electrolysis.

7. In the ISG, how does current flow to make the system perform as a starter?
A. Through the rotor to create an electromagnetic field that excites the stator, which causes the rotor to spin.
B. Through the rotor coils, which cause the magnetic field to collapse around the stator and rotate the crankshaft.
C. Through the stator windings, which generate magnetic fields in the rotor, causing the rotor to turn the crankshaft.
D. From the start generator control module to the rotor coils that are connected to the delta wound stator.

8. On which side of a PEM fuel cell does the pressurized hydrogen gas enter?
A. Anode.   C. Drain.
B. Cathode.   D. Gate.

9. All of the following are benefits of the 42-volt electrical system **EXCEPT:**
A. Allows for electrifying most of the mechanical and hydraulic systems.
B. Light bulbs last longer.
C. Reduction in size of electrical components and wires.
D. Increased efficiency of the charging system.

10. The dual-voltage system can use:
A. A dual-generator system.
B. A dual-stator, dual-voltage system.
C. A DC/DC converter.
D. All of the above.

Note: **Terms are highlighted in bold,** followed by Spanish translation in color.

Absorbed glass mat (AGM) battery A variation of the recombination batteries that hold their electrolyte in a moistened fiberglass matting instead of using a gel. The plates are made of high-purity lead and are tightly compressed into six cells.

Batería de malla de fibra de vidrio absorbente (AGM, por su sigla en inglés) Una variante de las baterías de recombinación que contienen sus electrolitos en una malla de fibra de vidrio humedecido en lugar de utilizar gel. Las placas están fabricadas con plomo de alta pureza y están bien comprimidas en seis celdas.

Accelerometer Generates an analog voltage in relation to the severity of deceleration forces. The accelerometer also senses the direction of impact force.

Acelerómetro Genera un voltaje análogo en relación a la severidad de las fuerzas de deceleración. El acelerómetro tambien detecta la dirección de la fuerza de un impacto.

A circuit A generator regulator circuit that uses an external grounded field circuit. In the A circuit, the regulator is on the ground side of the field coil.

Circuito A Circuito regulador del generador que utiliza un circuito inductor externo puesto a tierra. En el circuito A, el regulador se encuentra en el lado a tierra de la bobina inductora.

Actuators Devices that perform the actual work commanded by the computer. They can be in the form of a motor, relay, switch, or solenoid.

Accionadores Dispositivos que realizan el trabajo efectivo que ordena la computadora. Dichos dispositivos pueden ser un motor, un relé, un conmutador o un solenoide.

Adaptive brake lights A brake light system that selects different illumination levels or methods of display for the rear brake lights depending on conditions.

Luces de freno adaptables Un sistema de luces de freno que selecciona diferentes niveles de iluminación o métodos de visualización para las luces del freno traseras, de acuerdo con las condiciones.

Adaptive cruise control (ACC) A cruise control system using laser radar sensor to adjust and maintain appropriate following distances between vehicles. The sensor determines vehicle-to-vehicle distances and relational speeds.

Control de crucero adaptable (ACC) Sistema de control de crucero que utiliza un sensor de radar láser para ajustar y mantener las distancias de seguimiento apropiadas entre vehículos. El sensor determina las distancias de vehículo a vehículo y las velocidades de relación.

Adaptive headlight system (AHS) Headlight system that is designed to enhance night-time safety by turning the headlight beams to follow the direction of the road as the vehicle enters a turn.

Sistema de faros delanteros adaptables (AHS, por su sigla en inglés) Sistema de faros diseñado para mejorar la seguridad durante la noche, ya que las luces de los faros giran para seguir la dirección del camino al tomar una curva.

Adaptive memory The ability of the computer system to store changing values in order to correct operating characteristics.

Memoria adaptable La capacidad del sistema de computación de almacenar valores cambiantes para corregir las características operativas.

Adaptive strategy The capacity of the computer to make corrections to the normal strategy based on input signals.

Estrategia adaptable La capacidad de una computadora de realizar correcciones a la estrategia normal sobre la base de las señales recibidas.

Air bag Inflates in a few milliseconds when the vehicle is involved in a frontal collision. A typical fully inflated air bag has a volume of 2.3 cubic feet.

Bolsa de aire Infla en unos milisegundos cuando un vehículo se ha involucrado en una colisión delantera. Una bolsa de aire típica tiene un volúmen de 2.3 pies cúbicos al estar completamente inflada.

Air bag module Composed of the air bag and inflator assembly, which is packaged into a single module.

Unidad del Airbag Formada por el conjunto del Airbag y el inflador. Este conjunto se empaqueta en una sola unidad.

Air bag system Designed as a supplemental restraint that, in the case of an accident, will deploy a bag out of the steering wheel or passenger-side dash panel to provide additional protection against head and face injuries.

Sistema de bolsa de aire Diseñada como una restricción suplemental que, en el caso de un accidente, desplegará una bolsa del volante de dirección o del tablero lateral del pasajero para proveer la protección adicional contra los daños a la cabeza y a la cara.

Air-core gauge Gauge design that uses the interaction of two electromagnets and the total field effect upon a permanent magnet to cause needle movement.

Calibrador de núcleo de aire Calibrador diseñado para utilizar la interacción de dos electroimanes y el efecto inductor total sobre un imán permanente para generar el movimiento de la aguja.

Alternating current (AC) Electrical current that changes direction between positive and negative.

Corriente alterna Corriente eléctrica que recorre un circuito ya sea en dirección positiva o negativa.

Ambient temperature The temperature of the outside air.

Temperatura ambiente Temperatura del aire ambiente.

Ambient temperature sensor A thermistor used to determine the ambient temperature.

Sensor de la temperatura ambiente Termostato que se usa para determinar la temperatura ambiente.

American wire gauge (AWG) System used to determine wire sizes based on the cross-sectional area of the conductor.

Calibrador americano de alambres Sistema utilizado para determinar el tamaño de los alambres, basado en el área transversal del conductor.

Ammeter A test meter used to measure current draw.

Amperímetro Instrumento de prueba utilizado para medir la intensidad de una corriente.

Amortisseur winding The name given to the bars of a squirrel cage placed around the AC motor's rotor.

Amortisseur que enrolla El nombre dado a las barras de una jaula de ardilla colocó alrededor del rotor del motor de C.A.

Ampere-hour rating Indicates the amount of steady current a battery can supply for 20 hours.

Límite de amperio-hora Indica la cantidad de corriente fijo que un puede proveer una batería durante 20 horas.

Ampere (A) See current.

Amperios *Véase* corriente.

Analog A voltage signal that is infinitely variable or that can be changed within a given range.

Señal analógica Señal continua y variable que debe traducirse a valores numéricos discontinuos para poder ser trataba por una computadora.

Angular momentum The rotational analog of linear momentum. that remains constant unless acted on by an external torque.

Momento angular El análogo de rotación del momento lineal que permanece constante a menos que sea actuado por un esfuerzo de torsión externo.

Anode The positive charge electrode in a voltage cell.

Ánodo Electrodo de carga positiva de un generador de electricidad.

Antenna A wire or a metal stick that increases the amount of metal the transmitter's waves can interact with to convert radio frequency electrical currents into electromagnetic waves and vice versa.

Antena Un cable o varilla de metal que aumenta la cantidad de metal con la que pueden interactuar las ondas del transmisor para convertir las corrientes eléctricas de radio frecuencia en ondas electromagnéticas y viceversa.

Antilock brakes (ABS) A brake system that automatically pulsates the brakes to prevent wheel lockup under panic stop and poor traction conditions.

Frenos antibloqueo Sistema de frenos que pulsa los frenos automáticamente para impedir el bloqueo de las ruedas en casos de emergencia y de tracción pobre.

Antitheft system A system that prevents illegal entry or driving of a vehicle. Most are designed to deter entry.

Dispositivo a prueba de hurto Un dispositivo o sistema que previene la entrada o conducción ilícita de un vehículo. La mayoría se diseñan para detener la entrada.

Armature The movable component of an electric motor, which consists of a conductor wound around a laminated iron core and is used to create a magnetic field.

Armadura Pieza móvil de un motor eléctrico, compuesta de un conductor devanado sobre un núcleo de hierro laminado y que se utiliza para producir un campo magnético.

Armed The process of activating the antitheft system by shutting all doors and locking them.

Armado El proceso de activar el sistema antirrobo al cerrar todas las puertas y trabarlas con seguro.

Arming sensor A device that places an alarm system into "ready" to detect an illegal entry.

Sensor de armado Un dispositivo que pone "listo" un sistema de alarma para detectar una entrada ilícita.

Aspirator Tubular device that uses a venturi effect to draw air from the passenger compartment over the in-car sensor. Some manufacturers use a suction motor to draw the air over the sensor.

Aspirador Dispositivo tubular que utiliza un efecto Venturi para extraer aire del compartimiento del pasajero sobre el sensor dentro del vehículo. Algunos fabricantes utilizan un motor de succión para extraer el aire sobre el sensor.

Asynchronous Data that is sent on the bus network intermittently (as needed) rather than continuously.

Asíncronos Datos que se envían en el bus múltiple de modo intermitente (como sea necesario) en lugar de modo continuo.

Atom The smallest part of a chemical element that still has all the characteristics of that element.

Átomo Partícula más pequeña de un elemento químico que conserva las cualidades íntegras del mismo.

Audio system The sound system for a vehicle; can include radio, cassette player, CD player, amplifier, and speakers.

Sistema de audio El sistema de sonido de un vehículo; puede incluir el radio, el tocacaset, el toca discos compactos, el amplificador, y las bocinas.

Automatic door locks (ADL) A system that automatically locks all doors by activating one switch.

Cerrojos de compuertas automatizados Cerrojos de compuertas automatizados eléctricamente que utilizan o un solenoide o un motor reversible de imán permanente para cerrar y abrir las puertas.

Automatic headlight dimming An electronic feature that automatically switches the headlights from high beam to low beam under two different conditions: headlights from oncoming vehicles strike the photo cell-amplifier, or from the taillights of a vehicle being passed strikes the photo cell-amplifier.

Reducción automática de intensidad luminosa de los faros delanteros Característica electrónica que conmuta los faros delan teros automáticamente de luz larga a luz corta dadas las siguientes circunstancias: la luz de los vehículos que se aproximan alcanza el amplificador de fotocélula, o la luz de los faros traseros de un vehículo que se ha rebasado alcanza el amplificador de fotocélula.

Automatic on/off with time delay Turns on the headlights automatically when ambient light decreases to a predetermined level; also allows the headlights to remain on for a certain amount of time after the vehicle has been turned off. This system can be used in combination with automatic dimming systems.

Prendido/apagado automático con temporización Prende los faros automáticamente cuando la luz ambiental se oscurece a un nivel predeterminado. Tambien permite que los faros quedan prendidos por un tiempo determinado después de que se ha apagado el vehículo. Este sistema se puede utilizar en combinación con los sistemas de regulación de intensidad luminosas automáticos.

Automatic temperature control (ATC) A passenger comfort system that is capable of maintaining a preset temperature level as selected by the operator. Sensors are used to determine the present temperatures, and the system can adjust the level of heating or cooling as required using actuators to open and close air-blend doors to achieve the desired in-vehicle temperature.

Control automático de la temperatura (CAT) Un sistema de comodidad para el pasajero es capaz de mantener un nivel de temperatura previamente fijo tal como lo selecciona el operador. Los sensores se utilizan para determinar las temperaturas actuales, y el sistema puede ajustar el nivel de calentamiento o enfriamiento como se requiera al usar actuadores para abrir y cerrar las compuertas de recirculación para lograr la temperatura deseada dentro del vehículo.

Automatic traction control A system that prevents slippage of one of the drive wheels. This is done by applying the brake at that wheel and/ or decreasing the engine's power output.

Control Automático de Tracción Un sistema que previene el patinaje de una de las ruedas de mando. Esto se efectúa aplicando el freno en esa rueda y/o disminuyendo la salida de potencia del motor.

Avalanche diodes Diodes that conduct in the reverse direction when a reverse-bias voltage of about 6.2 volts or higher is applied. Causes the avalanche effect to occur when the reverse electric field moves across the PN junction, causing a wave of ionization that leads to a large current.

Diodo Zener Diodos que hacen conducción en dirección contraria cuando se aplica velocidad invertida de transmisión de baudios de alrededor de 6.2 voltios o más. Produce el efecto Zener o de avalancha cuando el campo eléctrico invertido se mueve a cruzar la unión PN, y causa así una onda de ionización que provoca una gran corriente.

B circuit A generator regulator circuit that is internally grounded. In the B circuit, the voltage regulator controls the power side of the field circuit.

Circuito B Circuito regulador del generador puesto internamente a tierra. En el circuito B, el regulador de tensión controla el lado de potencia del circuito inductor.

Balanced Refers to an atom that has the same number of electrons as the number of protons.

Balanceado Se refiere a un átomo que tiene el mismo número de electrones y de protones.

Balanced atom An atom that has an equal number of protons and electrons.

Átomo equilibrado Átomo que tiene el mismo número de protones y de electrones.

Ballast An electrical device that limits the current delivered through a circuit.

Balastra Un dispositivo eléctrico que limita la corriente que se distribuye a través de un circuito.

Base The center layer of a bipolar transistor.

Base Capa central de un transistor bipolar.

BAT The terminal identifier for the conductor from the generator to the battery positive terminal.

BAT El terminal que identifica el conductor del generador al terminal positivo de la batería.

Battery cables High-current conductors that connect the battery to the vehicle's electrical system.

Cables de batería Conductores de alta corriente que conectan la batería al sistema eléctrico del vehículo.

Battery cell The active unit of a battery.

Acumulador de batería Componente activo de una batería.

Battery holddowns Brackets that secure the battery to the chassis of the vehicle.

Portabatería Los sostenes que fijan la batería al chasis del vehículo.

Battery terminals Terminals of the battery to which the positive and the negative battery cables are connected. The terminals may be posts or threaded inserts.

Bornes de la batería Los bornes en la batería a los cuales se conectan los cables positivos y negativos. Los terminales pueden ser postes o piezas roscadas.

Baud rate   The measure of computer data transmission speed in bits per second.

Razón de baúd   Medida de la velocidad de la transmisión de datos de una computadora en bits por segundo.

Belt alternator starter (BAS)   A high-voltage starter/alternator combination that uses current flow through the stator windings to generate magnetic fields in the rotor, causing the rotor to turn, thus turning the crankshaft and starting the engine magnetic fields in the rotor, and starting the engine.

Arrancador del alternador por faja (AAF)   Una combinación de arrancador/alternador de alto voltaje que utiliza el flujo de corriente a través del devanado estatórico para generar campos magnéticos en el rotor, lo que provoca que el rotor dé vueltas, haciendo girar así el cigüeñal y arrancando el motor.

Belt tension sensor (BTS)   A strain gauge-type sensor located on the seat belt anchor of the passenger-side front seat, used to determine if an infant seat is being used.

Sensor de la tensión de la faja (STF)   Sensor de tipo medidor de tensiones localizado en el ancla del cinturón del asiento del lado del asiento frontal del pasajero y que se usa para determinar si se usa un asiento para bebés.

Bendix drive   A type of starter drive that uses the inertia of the spinning starter motor armature to engage the drive gear to the gears of the flywheel. This type of starter drive was used on early models of vehicles and is rarely seen today.

Acoplamiento Bendix   Un tipo del acoplamiento del motor de arranque que usa la inercia de la armadura del motor de arranque giratorio para endentar el engranaje de mando con los engranajes del volante. Este tipo de acoplamiento del motor de arranque se usaba en los modelos vehículos antiguos y se ven raramente.

Bias voltage   Voltage applied across a diode.

Tensión polarizadora   Tensión aplicada a través de un diodo.

Bimetallic strip   A metal contact wiper consisting of two different types of metals. One strip will react quicker to heat than the other, causing the strip to flex in proportion to the amount of current flow.

Banda bimetálica   Contacto deslizante de metal compuesto de dos tipos de metales distintos. Una banda reaccionará más rápido al calor que la otra, haciendo que la banda se doble en proporción con la cantidad de flujo de corriente.

Binary code   A series of numbers represented by 1s and 0s. Any number and word can be translated into a combination of binary 1s and 0s.

Código binario   Serie de números representados por unos y ceros. Cualquier número y palabra puede traducirse en una combinación de unos y ceros binarios.

Bipolar   The name used for transistors because current flows through the materials of both polarities.

Bipolar   Nombre aplicado a los transistores porque la corriente fluye por conducto de materiales de ambas polaridades.

Bit   A binary digit.

Bit   Dígito binario.

Bi-xenon headlamps   HID headlamps that use a single xenon lamp to provide both the high-beam and the low-beam operations. The full light output is used to produce the high beam, while the low beam is formed by moving a shutter between the bulb and the lens.

Faros delanteros de doble xenón   Faros de descarga de alta intensidad (HID, por su sigla en inglés) que utilizan una sola bombilla de xenón para cubrir las funciones de luces largas y luces cortas. La potencia de luz completa se utiliza para las luces largas, mientras que las luces bajas se forman moviendo un obturador entre la bombilla y el lente.

Blend-air door actuator   An electric motor that controls the position of the blend-air door, in order to supply the in-vehicle temperature the driver selected.

Actuador de puertas por aire mezclado   Un motor eléctrico que controla la posición de las puertas por aire mezclado para proporcionar la temperatura dentro del vehículo que seleccione el conductor.

Block diagrams   A wiring diagram that shows a single circuit.

Diagramas de bloque   Un diagrama de cableado que muestra un solo circuito.

Bluetooth   Technology that allows several modules from different manufacturers to be connected using a standardized radio transmission.

Bluetooth   Tecnología que permite que se conecten varios módulos de diferente manufactura por medio del uso de transmisión de radio estándar.

Brushes   Electrically conductive sliding contacts, usually made of copper and carbon.

Escobillas   Contactos deslizantes de conduccion eléctrica, por lo general hechos de cobre y de carbono.

Bucking coil   One of the coils in a three-coil gauge. It produces a magnetic field that bucks or opposes the low-reading coil.

Bobina compensadora   Una de las bobinas de un calibre de tres bobinas. Produce un campo magnético que es contrario o en oposición a la bobina de baja lectura.

Buffer   A buffer cleans up a voltage signal. Buffers are used with PM generator sensors to change AC voltage to a digitalized signal.

Separador   Un separador aguza una señal del tensión. Estos se usan con los sensores generadores PM para cambiar la tensión de corriente alterna a una señal digitalizado.

Buffer circuit   Changes AC voltage from the PM generator into a digitalized signal.

Circuito separador   Cambia el voltaje de corriente alterna del generador PM a una señal digital.

Bulkhead connector   A large connector that is used when many wires pass through the bulkhead or firewall.

Conectador del tabique   Un conectador que se usa al pasar muchos alambres por el tabique o mamparo de encendios.

Bus   Used in reference to data transmission since data is being transported from one place to another. The multiplex circuit is often referred to as the bus circuit.

Colectiva   Se usa en referir a la transmisión de datos que se estan transportando de un lugar a otro. El circuito multiplex suele referirse como el circuito colectivo.

Bus bar   A common electrical connection to which all of the fuses in the fuse box are attached. The bus bar is connected to battery voltage.

Barra colectora   Conexión eléctrica común a la que se conectan todos los fusibles de la caja de fusibles. La barra colectora se conecta a la tensión de la batería.

Bus (−)   The bus circuit that is most negative when the dominant bit is being transmitted.

Bus negativo (−)   Circuito del bus que es más negativo cuando se transmite el bit predominante.

Bus (+)   The bus circuit that is most positive when the dominant bit is being transmitted.

Bus positivo (+)   Circuito del bus que es más positivo cuando se transmite el bit predominante.

Buzzer   An audible warning device that is used to warn the driver of possible safety hazards.

Zumbador   Dispositivo audible de advertencia utilizado para prevenir al conductor de posibles riesgos a la seguridad.

Capacitance   The ability of two conducting surfaces to store voltage.

Capacitancia   Propiedad que permite el almacenamiento de electricidad entre dos conductores aislados entre sí.

Capacitance discharge sensor   A pressure sensor that uses a variable capacitor. As the distance between the electrodes changes, so does the capacity of the capacitor.

Sensor de descarga de capacitancia   Un sensor de presión que utiliza un capacitor variable. A medida que cambia la distancia entre los electrodos, también lo hace la capacidad del capacitor.

Capacitor   An electrical device made by wrapping two conductor strips around an insulating strip used to temporarily store electrical energy.

Capacitor   Un dispositivo eléctrico que se fabrica al envolver dos tiras conductoras alrededor de una tira de aislante y que se utiliza para almacenar energía eléctrica en forma temporal.

Carbon monoxide   An odorless, colorless, and toxic gas that is produced as a result of combustion.

Monóxido de carbono   Gas inodoro, incoloro y tóxico producido como resultado de la combustión.

Carriers   Attached to the shoulder anchor to move or carry the anchor from one end of the track to the other.

Portadoras   Conectados al reborde de anclaje para mudar o transportar el anclaje de una extremidad del carril a la otra.

Cartridge fuses   *See* maxi-fuse.

Fusibles cartucho   *Véase* maxifusible.

Cathode   Negatively charged electrode of a voltage cell.

Cátodo   Electrodo de carga negativa de un generador de electricidad.

Cathode ray tube Similar to a television picture tube. It contains a cathode that emits electrons and an anode that attracts them. The screen of the tube will glow at the points that are hit by the electrons.

Tubo de rayos catódicos Parecidos a un tubo de pantalla de televisor. Contiene un cátodo que emite los electrones y un ánodo que desde atrae. La pantalla del tubo iluminará en los puntos en donde pegan los electrones.

Caustic A material that has the ability to destroy or eat through something. Caustic materials are considered extremely corrosive.

Caustico Una materia que tiene la habilidad de destruir o carcomer algo. Las materias causticas se consideran extremadamente corrosivas.

Cell element The assembly of a positive and a negative plate in a battery.

Elemento de pila La asamblea de una placa positiva y negativa en una bateria.

Central gateway (CGW) A module on the CAN bus network that is the hub between the different networks.

Puerta central Un módulo de la red de bus CAN que está en el núcleo entre las diferentes redes.

Charging system Converts mechanical energy of the engine into electrical energy to recharge the battery and run the electrical accessories.

Sistema de carga Convierte la energía mecánica del motor en energía eléctrica para recargar la batería y hacer trabajar los accesorios eléctricos.

Child safety latch A switch that is designed into the door lock system that prevents the door from being opened from the inside, regardless of the position of the door lock knob.

Cerrojo de seguridad para niños Un interruptor diseñado en el sistema de bloqueo de las puertas para evitar que se abra la puerta desde el interior, sin importar la posición del botón de bloqueo de la puerta.

CHMSL The acronym for center high-mounted stop light, often referred to as the third brake light.

CHMSL El acrónimo de luz de freno central montada en la parte superior; se conoce comúnmente como la tercera luz de freno.

Choke An inductor in series with a circuit.

Reactancia Un inductor en serie con un circuit.

Choke coil Fine wire wound into a coil used to absorb oscillations in a switched circuit.

Bobina de inducción Alambre fino devanado en una bobina, utilizado para absorber oscilaciones en un circuito conmutado.

Chrysler Collision Detection (CCD) Chrysler's data bus network first used in 1988; uses a twisted pair of wires to transmit data.

Detección de colisión de Chrysler Red del bus de datos de Chrysler que se usó primero en 1988. Utilizó un par de hilos torcidos para transmitir datos.

Circuit The path of electron flow consisting of the voltage source, conductors, load component, and return path to the voltage source.

Circuito Trayectoria del flujo de electrones, compuesto de la fuente de tensión, los conductores, el componente de carga y la trayectoria de regreso a la fuente de tensión.

Circuit breaker A mechanical fuse that opens the circuit when amperage is excessive. In most cases, the circuit breaker resets when overload is removed.

Interruptor Un fusible mecánico que abre el circuito cuando la intensidad de amperaje es excesiva. En la mayoría de los casos, el interruptor se reengancha al eliminarse la sobrecarga.

Clamping diode A diode that is connected in parallel with a coil to prevent voltage spikes from the coil reaching other components in the circuit.

Diodo de bloqueo Un diodo que se conecta en paralelo con una bobina para prevenir que los impulsos de tensión lleguen a otros componentes en el circuito.

Climate-controlled seats Vehicle seats that provide additional comfort by heating and/or cooling the seat cushion and seat back.

Asientos climatizados Asientos del vehículo que brindan comodidad adicional al calentar y/o enfriar el cojín y el respaldo del asiento.

Clock circuit A crystal that electrically vibrates when subjected to current at certain voltage levels. As a result, the chip produces very regular series of voltage pulses.

Circuito de reloj Cristal que vibra electrónicamente cuando está sujeto a una corriente a ciertos niveles de tensión. Como resultado, el fragmento produce una serie sumamente regular de impulsos de tensión.

Clockspring A winding of electrical conducting tape enclosed within a plastic housing. The clockspring maintains continuity between the steering wheel, switches, the air bag, and the wiring harness in all steering wheel positions.

Muelle espiral Una bobina de cinta conductiva eléctrica encerrada en una caja de plástico. El muelle espiral mantiene la corriente continua entre el volante de dirección, los interruptores, la bolsa de aire, y el mazo de alambres en cualquier posición del volante de dirección.

Closed circuit A circuit that has no breaks in the path and allows current to flow.

Circuito cerrado Circuito de trayectoria ininterrumpida que permite un flujo continuo de corriente.

Coil Assembly An Assembly of coils that contains two coils or more.

Asamblea de bobina Una asamblea de bobinas que contiene dos bobinas o más.

Cold cranking rating Rating indicates the battery's ability to deliver a specified amount of current to start an engine at low-ambient temperatures.

Amperios de arranque en frío Tasa indicativa de la capacidad de la batería para producir una cantidad específica de corriente para arrancar un motor a bajas temperaturas ambiente.

Cold engine lock-out switch Prevents blower motor operation until the air entering the passenger compartment reaches a specified temperature.

Interruptor de cierre en motor inactivo Previene la operación del motor del ventilador hasta que el aire que entra en el compartimiento del pasajero alcanza su temperatura específica.

Collector The portion of a bipolar transistor that receives the majority of current carriers.

Dispositivo de toma de corriente Parte del transistor bipolar que recibe la mayoría de los portadores de corriente.

Common connection Splices in the wiring that are used to share a source of power or a common ground with different systems.

Conexión común Empalmes en el cableado que se utilizan para compartir una fuente de alimentación o una puesta a tierra común con distintos sistemas.

Common connector A connector that is shared by more than one circuit and/or component.

Conector común Un conector que se comparte entre más de un circuito y/o componente.

Commutator A series of conducting segments located around one end of the armature.

Conmutador Serie de segmentos conductores ubicados alrededor de un extremo de la armadura.

Complementary metal oxide semiconductor (CMOS) A technology for constructing integrated circuits using a metal gate electrode physical structure on top of an oxide insulator, that placed on top of a semiconductor material.

Semiconductor de óxido de metal complementario (CMOS) Una tecnología para construir circuitos integrados usando una estructura física de electrodo de compuerta de metal encima de un aislante de óxido, que se coloca encima de un material semiconductor.

Component locator A service manual that lists and describes the exact location of components on a vehicle.

Localizador de componentes Un manual de servicio que cataloga y describe la posición exacta de los componentes en un vehículo.

Composite bulb A headlight assembly that has a replaceable bulb in its housing.

Bombilla compuesta Una asamblea de faros cuyo cárter tiene una bombilla reemplazable.

Composite headlights A halogen headlight system that uses a replacement bulb.

Faros compuestos Un sistema de faros halógenos que usa un foco de recambio.

Compound motor A motor that has the characteristics of a series-wound and a shunt-wound motor.

Motor compuesta Un motor que tiene las características de un motor exitado en serie y uno en derivación.

Computer An electronic device that stores and processes data and is capable of operating other devices.

Computadora Dispositivo electrónico que almacena y procesa datos y que es capaz de ordenar a otros dispositivos.

Concealed headlight System used to help improve fuel economy and styling of the vehicle.

Faros ocultos Un sistema que se usa para mejorar el rendimiento del combustible y el estilo del vehículo.

Condenser A capacitor made from two sheets of metal foil separated by an insulator.

Condensador Capacitador hecho de dos láminas de metal separadas por un medio aislante.

Conduction Bias voltage difference between the base and the emitter has increased to the point that the transistor is switched on. In this condition, the transistor is conducting. Output current is proportional to that of the base current.

Conducción La diferencia de la tensión polarizadora entre la base y el emisor ha aumentado hasta el punto que el transistor es conectado. En estas circunstancias, el transistor está conduciendo. La corriente de salida está en proporción con la de la corriente conducida en la base.

Conductor A material in which electrons flow or move easily.

Conductor Una material en la cual los electrones circulen o se mueven fácilmente.

Contactors Heavy-duty relays that are connected to the positive and negative sides of the HV battery.

Contactos Relés para servicio pesado conectados al positivo y al negativo de la batería HV.

Continuity Refers to the circuit being continuous with no opens.

Continuidad Se refiere al circuito ininterrumpido, sin aberturas.

Control panel assembly Provides for driver input into the automatic temperature control microprocessor.

Asamblea de controles Permite la entrada del conductor al microprocesador del control de temperatura automático. La asamblea de control tambien se refiere como el tablero de instrumentos.

Controller area network (CAN) A two-wire bus network that allows the transfer of data between control modules.

CAN (Red del área del controlador) Red de bus de dos hilos que permite la transferencia de datos entre los módulos de control.

Conventional theory Electrical theory which states that current flows from a positive point to a more negative point.

Teoría convencional Teoría de electricidad la cual enuncia que el corriente fluye desde un punto positivo a un punto más negativo.

Cornering lights Lamps that illuminate when turn signals are activated. They burn steady when the turn signal switch is in a turn position, to provide additional illumination to the road in the direction of the turn.

Faros de viraje Los faros que iluminen cuando se prenden los indicadores de virajes. Quedan prendidos mientras que el indicador de viraje esta en una posición de viraje para proveer mayor iluminción del camino en la dirección del viraje.

Counter electromotive force (CEMF) An induced voltage that opposes the source voltage.

Fuerza cóntraelectromotriz Tensión inducida en oposición a la tensión fuente.

Courtesy lights Lamps that illuminate the vehicle's interior when the doors are open.

Luces interiores Lámparas que iluminan el interior del vehículo cuando las puertas están abiertas.

Covalent bonding The bonding where atoms share valence electrons with other atoms.

Enlace covalente Cuando los átomos comparten electrones de valencia con otros átomos.

Cranking amps A battery rating system that indicates the battery's ability to provide a cranking amperage at 32°F (0°C).

Amperios de arranque Un sistema de clasificación de baterías que indica la capacidad de una batería de suministrar un amperaje de arranque a 32°F (0°C).

Crash sensor Normally open electrical switch designed to close when subjected to a predetermined amount of jolting or impact.

Sensor de impacto Un conmutador normalmente abierto diseñado a cerrarse al someterse a un sacudo de una fuerza predeterminada o un impacto.

Cruise control A system that allows a vehicle to maintain a preset speed with the driver's foot off the accelerator.

Control crucero Un sistema que permite que el vehículo mantenga una velocidad predeterminada sin que el pie del conductor dceprime al accelerador.

Crystal A term used to describe a material that has a definite atom structure.

Cristal Término utilizado para describir un material que tiene una estructura atómica definida.

Current The aggregate flow of electrons through a wire. One ampere represents the movement of 6.25 billion billion electrons (or one coulomb) past one point in a conductor in one second.

Corriente Flujo combinado de electrones a través de un alambre. Un amperio representa el movimiento de 6,25 mil millones de mil millones de electrones (o un colombio) que sobrepasa un punto en un conductor en un segundo.

Cutoff The state where reverse-bias voltage is applied to the base leg of a transistor. In this condition, the transistor does not conduct and no current flows.

Corte Cuando se aplica tensión polarizadora inversa a la base del transistor. En estas circunstancias, el transistor no está conduciendo y no fluirá ninguna corriente.

Cycle Completed when the voltage has gone positive, returned to zero, gone negative, and returned to zero.

Ciclo Completado cuando el voltaje ha sido positivo, regresado al cero, ha sido negativo y regresado al cero.

Darlington pair An arrangement of transistors that amplifies current by one transistor acting as a preamplifier that creates a larger base current to the second transistor.

Par Darlington Conjunto de transistores que amplifica la corriente. Un transistor actúa como preamplificador y produce una corriente base más ámplia para el segundo transistor.

D'Arsonval gauge A gauge design that uses the interaction of a permanent magnet and an electromagnet and the total field effect to cause needle movement.

Calibrador d'Arsonval Calibrador diseñado para utilizar la interacción de un imán permanente y de un electroimán, y el efecto inductor total para generar el movimiento de la aguja.

Daytime running lamps (DRL) Generally use a high-beam or low-beam headlight system at a reduced intensity to provide additional visibility to the vehicle for other drivers and pedestrians.

Faros diurnos Generalmente usan el sistema de faros de alta y baja intensidad en una intensidad disminuada para proporcionar el vehículo más visibilidad para los otros conductores y los peatones.

DC/DC converter The DC/DC converter is configured to provide a 14-volt output from the high-voltage input. The 14-volt output can be used to supply electrical energy to those components that do not require the high voltage.

Convertidor continua-continua El convertidor continua-continua está configurado para proporcionar una salida de 14 V de una entrada de alto voltaje. La salida de 14 V puede usarse para suplir energía eléctrica a aquellos componentes que no requieran de alto voltaje.

Dead reckoning The process of calculating current position using previously determined positions, and advancing the current position based upon speeds over elapsed time and course.

Cuenta muerta El proceso de calcular la posición actual usando posiciones previamente determinadas, y avanzar de la posición actual basada en velocidades sobre el tiempo transcurrido y curso.

Deep cycling Discharging the battery completely before recharging it.

Operación cíclica completa La descarga completa de la batería previo al recargo.

Delta connection A connection that receives its name from its resemblance to the Greek letter delta (Δ).

Conexión delta Una conexión que recibe su nombre a causa de su aparencia parecida a la letra delta griega.

Delta stator A three-winding AC generator stator, with the ends of each winding connected to each other.

Estátor delta Estátor generador de corriente alterna de devanado triple, con los extremos de cada devanado conectados entre sí.

Depletion-type FET Cuts off current flow.

FET tipo agotamiento Corta el flujo del corriente.

Depressed-park A system in which the blades drop down below the lower windshield molding to hide them.

Limpiaparabrisas guardadas Un sistema en el cual los brazos se guardan abajo del borde inferior de la parabrisa para asi esconderlos.

Detector Extracts data (voices and music) from the sine wave.

Detector Extrae datos (voces y música) de la onda senoidal.

Diagnostic module Part of an electronic control system that provides self-diagnostics and/or a testing interface.

Módulo de diagnóstico Parte de un sistema controlado electronicamente que provee autodiagnóstico y/o una interfase de pruebas.

Diaphragm A thin, flexible, circular plate that is held around its outer edge by the horn housing, allowing the middle to flex.

Diagragma   Una placa redonda flexible y delgada sostenido en el cárter del claxon por medio de su borde exterior, así permitiendo flexionar la parte central.

Dielectric   An insulator material.

Dieléctrico   Material aislante.

Digital   A voltage signal is either on/off, yes/no, or high/low.

Digital   Una señal de tensión está Encendida-Apagada, es Sí-No o Alta-Baja.

Digital instrument clusters   Use digital and linear displays to notify the driver of monitored system conditions.

Grupo de instrumentos digitales   Usan los indicadores digitales y lineares para notificar el conductor de las condiciones de los sistemas regulados.

Dimmer switch   A switch in the headlight circuit that provides the means for the driver to select either high-beam or low-beam operation and to switch between the two. The dimmer switch is connected in series within the headlight circuit and controls the current path for high beam and low beam.

Conmutador reductor   Conmutador en el circuito para faros delanteros que le permite al conductor que elegir la luz larga o la luz corta, y conmutar entre las dos. El conmutador reductor se conecta en serie dentro del circuito para faros delanteros y controla la trayectoria de la corriente para la luz larga y la luz corta.

DIN   The abbreviation for Deutsche Institut füer Normung (German Institute for Standardization) and the recommended standard for European manufacturers to follow.

DIN   La abreviatura de Deutsche Institut füer Normung (Normas del Instituto Aleman) y se recomienda que los fabricantes europeos siguen estas normas.

Diode   An electrical one-way check valve that will allow current to flow only in one direction.

Diodo   Válvula eléctrica de retención, de una vía, que permite que la corriente fluya en una sola dirección.

Diode rectifier bridge   A series of diodes that are used to provide a reasonably constant DC voltage to the vehicle's electrical system and battery.

Puente rectificador de diodo   Serie de diodos utilizados para proveerles una tensión de corriente continua bastante constante al sistema eléctrico y a la batería del vehículo.

Diode trio   Used by some manufacturers to rectify the stator voltage of an AC generator current so the voltage can be used to create the magnetic field in the field coil of the rotor.

Trío de diodos   Utilizado por algunos fabricantes para rectificar el estátor de la corriente de un generador de corriente alterna y poder así utilizarlo para crear el campo magnético en la bobina inductora del rotor.

Direct current (DC)   Electric current that flows in one direction.

Corriente continua   Corriente eléctrica que fluye en una dirección.

Direct drive   A situation where the drive power is the same as the power exerted by the device that is driven.

Transmisión directa   Una situación en la cual el poder de mando es lo mismo que la potencia empleada por el dispositivo arrastrado.

Discrete devices   Electrical components that are made separately and have wire leads for connections to an integrated circuit.

Dispositivos discretos   Componentes eléctricos hechos uno a uno; tienen conductores de alambre para hacer conexiones a un circuito integrado.

Discriminating sensors   Part of the air bag circuitry; these sensors are calibrated to close with speed changes that are great enough to warrant air bag deployment. These sensors are also referred to as crash sensors.

Sensores discriminadores   Una parte del conjunto de circuitos de Airbag; estos sensores se calibran para cerrar con los cambios de la velocidad que son bastante severas para justificar el despliegue del Airbag. Estos sensores también se llaman los sensores de impacto.

Doping   The addition of another element with three or five valence electrons to a pure semiconductor.

Impurificación   La adición de otro elemento con tres o cinco electrones de valencia a un semiconductor puro.

Double-filament lamp   A lamp designed to execute more than one function. It can be used in the stop light circuit, taillight circuit, and the turn signal circuit combined.

Lámpara con filamento doble   Lámpara diseñada para llevar a cabo más de una función. Puede utilizarse en una combinación de los circuitos de faros de freno, de faros traseros y de luces indicadoras para virajes.

Double-start override   Prevents the starter motor from being energized if the engine is already running.

Sobremarcha de doble marcha   Previene que se excita el motor del encendido si ya esta en marcha el motor.

Drain   The portion of a field-effect transistor that receives the holes or electrons.

Drenador   Parte de un transistor de efecto de campo que recibe los agujeros o electrones.

Drive coil   A hollowed field coil used in a positive-engagement starter to attract the movable pole shoe of the starter.

Bobina de excitación   Una bobina inductora hueca empleada en un encendedor de acoplamiento directo para atraer la pieza polar móvil del encendedor.

Dual-climate control   Provides separate temperature settings for the driver and the front-seat passenger. This system is similar to previous systems except two blend doors are used to control separate temperature settings.

Control de clima doble   Provea la regulación de temperatura individual para el conductor ye el pasajero del asiento delantero. Este sistema es parecido a los sistemas anteriores menos que los dos compuertas de mezcla se usan para controlar la regulación individual de la temperatura.

Dual-generator system   Uses one generator that operates at 42 volts while another operates at 14 volts.

Sistema de doble voltaje   Arquitectura que utiliza dos voltajes separados. Uno es para el sistema de 42 voltios que enciende los accesorios eléctricos que podrían requerir o beneficiarse de mayor voltaje. El resto de las cargas continúan en 14 voltios.

Dual-range circuit   This circuit provides for a switch in the resistance values to allow the microprocessor to measure temperatures more accurately.

Circuitos de doble rango   Este circuito proporciona un interruptor en los valores de resistencia para permitir que el microprocesador mida las temperaturas de manera más precisa.

Dual-stator, dual-voltage system   Dual voltage (42 volts and 14 volts) is produced from a single alternator that has two output voltages.

Sistema de doble voltaje y doble estator   Un alternador sencillo que tiene voltajes de dos salidas y produce un doble voltaje (42 voltios y 14 voltios).

Dual-voltage system   Architecture that uses two separate voltages. One is for the 42-volt system that powers those electrical accessories that would require or benefit from the higher voltage. The remainder of the loads remain on 14 volts.

Sistema de doble voltaje   Arquitectura que utiliza dos voltajes separados. Uno es para el sistema de 42 voltios que enciende los accesorios eléctricos que podrían requerir o beneficiarse de mayor voltaje. El resto de las cargas continúan en 14 voltios.

Duty cycle   The percentage of on time to total cycle time.

Ciclo de trabajo   Porcentaje del trabajo efectivo a tiempo total del ciclo.

Easy exit   An additional function of the memory seat that provides for easier entrance and exit of the vehicle by moving the seat all the way back and down. Some systems also move the steering wheel up and to full retract.

Salida fácil   Una función adicional de la memoria del asiento que provee una entrada y salida más fácil del vehículo al mover el asiento hasta su posición más extrema hacia atrás y abajo. Algunos sistemas tambien muevan el volante de dirección hacia arriba y a su posición más alejada.

Eddy currents   Small induced currents.

Corriente de Foucault   Pequeñas corrientes inducidas.

Electric defoggers   Heat the rear window to remove ice and/or condensation. Some vehicles use the same circuit to heat the outside driver's-side mirror.

Desnebladoor eléctrica   Calientan la ventanilla trasera para remover el hielo y/o la condensación. Algunos vehículos usan el mismo circuito para calentar el espejo lateral del conductor.

Electric vehicle (EV)   A vehicle that powers its motor off of a battery pack.

Vehículo eléctrico (VE)   Vehículo que apaga su motor por medio de un paquete de baterías.

Electrical accessories   Electrical systems or components that provide for additional safety and comfort, including safety accessories such as the horn, windshield wipers, and windshield washers. Comfort accessories include the blower motor, electric defoggers, power mirrors, power windows, power seats, and power door locks.

Accesorios eléctricos   Sistemas o componentes eléctricos que proporcionan seguridad y comodidad adicionales, y que incluyen accesorios de seguridad tales como el claxon, limpia brisas y parabrisas. Los accesorios de comodidad incluyen un motor de aire, desnubilizador eléctrico, espejos mecánicos, asientos mecánicos y cierre mecánico de puertas.

Electrical load   The working device of the circuit.

Carga eléctrica   Dispositivo de trabajo del circuito.

Electrical symbols   Used to represent components in the wiring diagram.

Símbolos electrónicos   Se usan para representar los componentes en uyn esquema de conexiones.

Electrically erasable PROM (EEPROM)   Memory chip that allows for electrically changing the information one bit at a time.

Capacidad de borrado electrónico PROM   Fragmento de memoria que permite el cambio eléctrico de la información un bit a la vez.

Electrochemical   The chemical action of two dissimilar materials in a chemical solution.

Electroquímico   Acción química de dos materiales distintos en una solución química.

Electrochromic mirror   Automatically adjusts to light using forward- and rearward-facing photo sensors and a solid-state chip. Based on light intensity differences, the chip applies a small voltage to the silicon layer. As voltage is applied, the molecules of the layer rotate and redirect the light beams. Thus, the mirror reflection appears dimmer.

Espejo electrocrómico   Se ajusta automáticamente a la luz usando los sensores orientados hacia afrente y atrás juntos con un chip de estado sólido.

Electrolysis   The production of chemical changes by passing electrical current through an electrolyte; the splitting of water into hydrogen and oxygen.

Electrólisis   La producción de los cambios químicos al pasar un corriente eléctrico por un electrolito.

Electrolyte   A solution of 64% water and 36% sulfuric acid.

Electrolito   Solución de un 64% de agua y un 36% de ácido sulfúrico.

Electromagnetic gauge   Gauge that produces needle movement by magnetic forces.

Calibrador electromagnético   Calibrador que genera el movimiento de la aguja mediante fuerzas magnéticas.

Electromagnetic induction (EMI)   The production of voltage and current within a conductor as a result of relative motion within a magnetic field.

Inducción electromagnética   Producción de tensión y de corriente dentro de un conductor como resultado del movimiento relativo dentro de un campo magnético.

Electromagnetic interference (EMI)   An undesirable creation of electromagnetism whenever current is switched on and off.

Interferencia electromagnética   Fenómeno de electromagnetismo no deseable que resulta cuando se conecta y se desconecta la corriente.

Electromagnetism   A form of magnetism that occurs when current flows through a conductor.

Electromagnetismo   Forma de magnetismo que ocurre cuando la corriente fluye a través de un conductor.

Electromechanical   A device that uses electricity and magnetism to cause a mechanical action.

Electromecánico   Un dispositivo que causa una acción mecánica por medio de la electricidad y el magnetismo.

Electromotive force (EMF)   *See* voltage.

Fuerza electromotriz   *Véase* tensión.

Electron   Negatively charged particles of an atom.

Electrón   Partículas de carga negativa de un átomo.

Electronic roll mitigation (ERM)   A vehicle safety system that helps to prevent a vehicle rollover situation due to extreme lateral forces experienced during cornering or hard evasive maneuvers using brake and engine torque management.

Mitigación de rollos electrónicos (ERM)   Un sistema de seguridad para vehículos que ayuda a prevenir una situación de vuelco del vehículo debido a las fuerzas laterales extremas experimentadas durante las curvas o fuertes maniobras de evasión utilizando el freno y el esfuerzo de torción del motor.

Electron theory   Defines electrical movement from negative to positive.

Teoría del electrón   Define el movimiento eléctrico como el movimiento de lo negativo a lo positivo.

Electronic regulator   Uses solid-state circuitry to perform regulatory functions.

Regulador electrónico   Usa los circuitos de estado sólido para llevar a cabo los funciones de regulación.

Electronic stability control   An additional function of the antilock brake and traction control system that uses additional sensors and inputs to determine if the vehicle is actually moving in the direction intended by the driver, as indicated by steering wheel position sensors, yaw sensors, and lateral sensors. If the actual path is not the intended path, the module will apply the appropriate brake to bring the vehicle back onto the correct path.

Mando electrónico de estabilidad   Función adicional de un sistema de antibloqueo de frenos y de control de la tracción que utiliza sensores adicionales y entradas para determinar si efectivamente se mueve el vehículo en la dirección que desea el conductor, tales como indican los sensores de posición del volante, los sensores de guiñada y los sensores laterales. Si el recorrido real no es el corrido deseado, el módulo aplicará el freno apropiado para regresar el vehículo al recorrido correcto.

Electrostatic field   The field that is between two oppositely charged plates.

Campo electrostático   Campo que se encuentra entre las placas de carga opuesta.

Emitter   The outer layer of a transistor, which supplies the majority of current carriers.

Emisor   Capa exterior del transistor que suministra la mayor parte de los portadores de corriente.

Energy density   The amount of energy that is available for a given amount of space.

Densidad de la energía   La cantidad de energía disponible en un espacio dado.

Engine vacuum   Formed during the intake stroke of the cylinder. Engine vacuum is any pressure lower than atmospheric pressure.

Vacío del motor   Formado durante la carrera de entrada de un cilíndro. El vacío de motor es cualquier presión más baja de la presión atmosférica.

Enhancement-type FET   Improves current flow.

FET tipo de acrecentamiento   Mejor ael flujo del corriente.

Equivalent series load (equivalent resistance)   The total resistance of a parallel circuit, which is equivalent to the resistance of a single load in series with the voltage source.

Carga en serie equivalente (resistencia equivalente)   Resistencia total de un circuito en paralelo, equivalente a la resistencia de una sola carga en serie con la fuente de tensión.

Erasable PROM (EPROM)   Similar to PROM except that its contents can be erased to allow for new data to be installed. A piece of Mylar tape covers a window. If the tape is removed, the microcircuit is exposed to ultraviolet light and erases its memory.

Capacidad de borrado PROM   Parecido al PROM, pero su contenido puede borrarse para permitir la instalación de nuevos datos. Un trozo de cinta Mylar cubre una ventana; si se remueve la cinta, el microcircuito queda expuesto a la luz ultravioleta y borra la memoria.

Excitation current   Current that magnetically excites the field circuit of the AC generator.

Corriente de excitación   Corriente que excita magnéticamente al circuito inductor del generador de corriente alterna.

Express down   A power window feature that allows the operator to lower the window with a single press of the switch instead of having to hold the switch during window operation.

Bajada rápida   Una función de los levantavidrios automáticos que permite que el operador baje la ventanilla oprimiendo un interruptor una sola vez en lugar de mantener el interruptor oprimido durante el funcionamiento de la ventanilla.

Express up   A power window feature that allows the operator to raise the window with a single press of the switch instead of having to hold the switch during window operation.

Subida rápida   Una función de los levantavidrios automáticos que permite que el operador suba la ventanilla oprimiendo un interruptor una sola vez en lugar de mantener el interruptor oprimido durante el funcionamiento de la ventanilla.

Face shield   A clear plastic shield that protects the entire face.

Máscara protectora   Una máscara de plástico transparente que proteje la cara entera.

Feedback   1. Data concerning the effects of the computer's commands is fed back to the computer as an input signal; used to determine if the desired result has been achieved. 2. A condition that can occur when electricity seeks a path of lower resistance, but the alternate path operates a component other than that intended. Feedback can be classified as a short.

Realimentación   1. Datos referentes a los efectos de las órdenes de la computadora se suministran a la misma como señal de entrada. La realimentación se utiliza para determinar si se ha logrado el resultado deseado. 2. Condición que puede ocurrir cuando la electricidad busca una trayectoria de menos resistencia, pero la trayectoria alterna opera otro componente que aquel deseado. La realimentación puede clasificarse como un cortocircuito.

Fiber optics   A medium of transmitting light through polymethyl methacrylate plastic that keeps the light rays parallel even if there are extreme bends in the plastic.

Transmisión por fibra óptica   Técnica de transmisión de luz por medio de un plástico de polimetacrilato de metilo que mantiene los rayos de luz paralelos aunque el plástico esté sumamente torcido.

Field coils   Heavy copper wire wrapped around an iron core to form an electromagnet.

Bobina del campo   El alambre grueso de cobre envuelta alrededor de un núcleo de hierro para formar un electroimán.

Field current   The current going to the field windings of a motor or generator.

Corriente inductora   El corriente que va a los devanados inductores de un motor o generador.

Field-effect transistor (FET)   A unipolar transistor in which current flow is controlled by voltage in a capacitance field.

Transistor de efecto de campo   Transistor unipolar en el cual la tensión en un campo de capacitancia controla el flujo de corriente.

Field relay   The relay that controls the amount of current going to the field windings of a generator. This is the main output control unit for a charging system.

Relé inductor   El relé que controla la cantidad del corriente a los devanados inductores de un generador. Es la unedad principal de potencia de salida de un sistema de carga.

Fifth percentile female   The fifth percentile female is determined to be those who weigh less than 100 pounds (45 kg). The fifth percentile female is determined by averaging all potential occupants by size and then plotting the results on a graph.

Quinto percentil femenino   La determinan aquellos que pesan menos de 45 kg (100 lbs. Se determina al hacer el promedio de todos los posibles ocupantes por su tamaño, y luego al trazar los resultados en una gráfica.

Fixed resistors   Have a set resistance value and are used to limit the amount of current flow in a circuit.

Resistores fijos   Tienen un valor de resistencia fijo y se usan para limitar la cantidad de flujo del corriente en un circuito.

Flammable   A substance that supports combustion.

Inflamable   Una substancia que ampara la combustión.

Flasher   Used to open and close the turn signal circuit at a set rate.

Pulsador   Se usa para abrir y cerrar el circuito del indicador de vueltas en una velocidad predeterminada.

Floating   The movement of voltage levels when a switch is open.

Flotante   El movimiento de los niveles de voltaje cuando un interruptor se encuentra abierto.

Floor jack   A portable hydraulic tool used to raise and lower a vehicle.

Gato de pie   Herramienta hidráulica portátil utilizada para levantar y bajar un vehículo.

Flux density   The number of flux lines per square centimeter.

Densidad de flujo   Número de líneas de flujo por centímetro cuadrado.

Flux lines   Magnetic lines of force.

Líneas de flujo   Líneas de fuerza magnética.

Forward-biased   A positive voltage that is applied to the P-type material and negative voltage to the N-type material of a semiconductor.

Polarización directa   Tensión positiva aplicada al material P y tensión negativa aplicada al material N de un semiconductor.

Fuel cell   A battery-like component that produces current from hydrogen and aerial oxygen.

Celda de combustible   Componente tipo batería que produce corriente del hidrógeno y del oxígeno en el aire.

Fuel pump inertia switch   An NC switch that opens if the vehicle is involved in an impact at speeds over 5 mph or if it rolls over. When the switch opens, it turns off power to the fuel pump. This is a safety feature to prevent fuel from being pumped onto the ground or hot engine compartments if the engine dies. The switch has to be manually reset if it is triggered.

Interruptor inercia de la bomba de combustible   Un interruptor NC que se abre si el vehículo se involucra en un choque en una velocidad que exceda 5 millas por hora o si se invierte de arriba abajo. Cuando el interruptor se abre, corta la corriente a la bomba del combustible. Este es una precaución de seguridad para prevenir que la bomba vierte el combustibel en el suelo o sobre un compartimento caliente del motor si se muere el motor. El interruptor se tiene que reenganchar a mano si se acciona.

Full field   Maximum AC generator output.

Campo completo   Salida máxima de un generador de corriente alterna.

Full parallel hybrid   Uses an electric motor that is powerful enough to propel the vehicle on its own.

Híbrido en paralelo completo   Utiliza un motor eléctrico que es lo suficientemente potente para que el vehículo se impulse por sí mismo.

Full-wave rectification   The conversion of a complete AC voltage signal to a DC voltage signal.

Rectificación de onda plena   La conversión de una señal completa de tensión de corriente alterna a una señal de tensión de corriente continua.

Fuse   A replaceable circuit protection device that melts should the current passing through it exceed its rating.

Fusible   Dispositivo reemplazable de protección del circuito que se fundirá si la corriente que fluye por el mismo excede su valor determinado.

Fuse block   The term used to indicate the central location of the fuses contained in a single holding fixture.

Bloque de fusibles   El término que su usa para indicar la ubicación central de los fusibles contenidos en una fijación central.

Fuse box   A term used to indicate the central location of the fuses contained in a single holding fixture.

Caja de fusibles   Término utilizado para indicar la ubicación central de los fusibles contenidos en un solo elemento permanente.

Fusible link   A wire made of meltable material with a special heat-resistant insulation. When there is an overload in the circuit, the link melts and opens the circuit.

Cartucho de fusible   Alambre hecho de material fusible con aislamiento especial resistente al calor. Cuando ocurre una sobrecarga en el circuito, el cartucho se funde y abre el circuito.

G force   Term used to describe the measurement of the net effect of the acceleration that an object experiences and the acceleration that gravity is trying to impart to it. Basically G force is the apparent force that an object experiences due to acceleration.

Fuerza G   Término utilizado para describir la medida del efecto neto de la aceleración que experimenta un objeto y la aceleración que la gravedad está intentando impartirle. Básicamente, la fuerza G es la fuerza aparente que un objeto experimenta debido a la aceleración.

Gain   The ratio of amplification in an electronic device.

Ganancia   Razón de amplificación en un dispositivo electrónico.

Ganged switch   Refers to a type of switch in which all wipers of the switch move together.

Acoplado en tándem   Se refiere a un tipo de conmutador en el cual todos los contactos deslizantes del mismo se mueven juntos.

Gassing   The conversion of a battery's electrolyte into hydrogen and oxygen gas.

Burbujeo   La conversión del electrolito de una bateria al gas de hidrógeno y oxígeno.

Gate   The portion of a field-effect transistor that controls the capacitive field and current flow.

Compuerta   Parte de un transistor de efecto de campo que controla el campo capacitivo y el flujo de corriente.

Gauge   1. A device that displays the measurement of a monitored system by the use of a needle or pointer that moves along a calibrated scale. 2. The number that is assigned to a wire to indicate its size. The larger the number, the smaller the diameter of the conductor.

Calibrador   1. Dispositivo que muestra la medida de un sistema regulado por medio de una aguja o indicador que se mueve a través de una escala calibrada. 2. El número asignado a un alambre indica su tamaño. Mientras mayor sea el número, más pequeño será el diámetro del conductor.

Gear reduction   Occurs when two different-sized gears are in mesh and the driven gear rotates at a lower speed than the drive gear but with greater torque.

Desmultiplicación   Ocurre cuando dos engranajes de distinctos tamaños se endentan y el engrane arrastrado gira con una velocidad más baja que el engrane de mando pero con más par.

Grid   The frame structure of a battery that normally has connector tabs at the top. It is generally made of lead alloys.

Rejillas   La estructura encuadrador de una batería que normalmente tiene orejas de conexión en la parte superior. Generalmente se fabrica de aleaciones de plomo.

Grid growth   A condition where the grid grows little metallic fingers that extend through the separators and short out the plates.

Expansión de la rejilla   Una condición en la cual la rejilla produce protrusiones metálicas que se extienden por los separadores y causan cortocircuitos en las placas.

Ground   The common negative connection of the electrical system. It is the point of lowest voltage.

Tierra   Conexión negativa común del sistema eléctrico. Es el punto de tensión más baja.

Ground side   The portion of the circuit that is from the load component to the negative side of the source.

Lado a tierra   Parte del circuito que va del componente de carga al lado negativo de la fuente.

Grounded circuit   An electrical defect that allows current to return to ground before it has reached the intended load component.

Circuito puesto a tierra   Falla eléctrica que permite el regreso de corriente a tierra antes de alcanzar el componente de carga deseado.

Ground straps   Electrical conductors that complete the return path to the battery from components that are insulated from the vehicle's chassis. In addition, ground straps help suppress EMI conduction and radiation by providing a low-resistance circuit ground path.

Conectores de puesta a tierra   Conductores eléctricos que completan el circuito de regreso hacia la batería de los componentes que están aislados desde el chasis del vehículo. Además, los conectores de puesta a tierra ayudan a suprimir la conducción IEM y la radiación al proporcionar un recorrido de tierra del circuito de resistencia.

Gyroscope   A spinning wheel or disc that axis of rotation is capable assuming any orientation by itself.

Giroscopio   Una rueda giratoria o disco cuyo eje de rotación es capaz de asumir cualquier orientación por sí mismo.

Half-field current   The current going to the field windings of a motor or generator after it has passed through a resistor in series with the circuit.

Corriente de medio campo   El corriente que va a los devanados inductores de un motor o a un generador después de que haya pasado por un resistor conectado en serie con el circuito.

Half-wave rectification   Rectification of one half of an AC voltage.

Rectificación de media onda   Rectificación en la que la corriente fluye únicamente durante semiciclos alternados.

Hall-effect sensor   A sensor that operates on the principle that if a current is allowed to flow through a thin conducting material being exposed to a magnetic field, another voltage is produced.

Sensor de efecto Hall   Sensor que funciona basado en el principio de que si se permite el flujo de corriente a través de un material conductor delgado que ha sido expuesto a un campo magnético, se produce otra tensión.

Halogen   The term used to identify a group of chemically related nonmetallic elements. These elements include chlorine, fluorine, and iodine.

Halógeno   Término utilizado para identificar un grupo de elementos no metálicos relacionados químicamente. Dichos elementos incluyen el cloro, el flúor y el yodo.

Hand tools   Tools that use only the force generated from the body to operate. They multiply the force received through leverage to accomplish the work.

Herramientas manuales   Herramientas que para funcionar sólo necesitan la fuerza generada por el cuerpo. Para llevar a cabo el trabajo, las herramientas multiplican la fuerza que reciben por medio de la palancada.

Haptic feedback   The use of the sense of touch by generating vibrations.

Retroalimentación háptica   El uso del sentido del tacto mediante la generación de vibraciones.

Hazardous material   Materials that can cause illness, injury, or death; or pollute water, air, or land.

Material peligroso   Las materias que puedan causar la enfermedad, los daños, la muerte o que puedan contaminar el agua, el aire o la tierra.

Headlight leveling system (HLS)   Uses front lighting assemblies with a leveling actuator motor to allow the headlights to be adjusted into different vertical positions to compensate for headlight position that can occur when the vehicle is loaded.

Sistema de nivelación de los faros (SNF)   Utiliza el ensamblaje de los faros frontales con un motor accionador de nivelación para permitir el ajuste de los faros en diferentes posiciones verticales para compensar la posición que el faro pudiera tomar si se carga el vehículo.

Head-up display (HUD)   Displays images onto the inside of the windshield so the driver can see them without having to take his eyes off the road.

Presentación en pantalla (HUD)   Proyecta las imagenes en la parte interior de la parabrisas para que el conductor las pueda ver sin tener que tomar su atención de la pista.

Heat sink   An object that absorbs and dissipates heat to another object.

Dispersador térmico   Objeto que absorbe y disipa el calor de otro objeto.

Heated windshield system   A specially designed windshield that allows current flow through the glass without interfering with the driver's vision; it is capable of melting ice and frost from the windshield three to five times faster than conventional defroster systems.

Sistema de parabrisas térmico   Parabrisas especialmente diseñado para permitir el flujo de la corriente a través del vidrio sin interferir con la visión del conductor; está capacitado para derretir el hielo y la escarcha que haya en el parabrisas de 3 a 5 veces más rápido que los sistemas convencionales anticongelantes.

Heater core flow valve   Shuts off the coolant flow through the heater core when the A/C system is in the max air mode.

Válvula del flujo térmico del núcleo   Cierra el flujo del enfriador a través del núcleo del calentador cuando el sistema AC está en el mando de aire máximo.

H-gate   A set of four transistors that can reverse current.

Compuerta H   Juego de cuatro transistores que pueden invertir la corriente.

HID   High-intensity discharge; a lighting system that uses an arc across electrodes instead of a filament.

HID   Descarga de Alta Intensidad; un sistema de iluminación que utiliza un arco por dos electrodos en vez de un filamento.

High-intensity discharge (HID)   Uses an inert gas to amplify the light produced by arcing across two electrodes.

Descarga de alta intensidad (HID)   Usa un gas inerte para amplificar la luz producida al conectar dos electrodos con una arca.

High-reading coil   Position at 90° angle to the low-reading and bucking coils.

Bobina dfe lectura de alta tensión   Posiciona en un ángulo a las bobinas de lectura de tensión baja y las bobinas compensadoras.

High-side drivers   Control the output device by varying the positive (12-volt) side.

Impulsores del lado de alto potencial   Controlan el dispositivo de salida en variar el lado positivo (12 voltíos).

High-voltage ECU (HV ECU)   *See* starter generator control module (SGCM).

UCE de alto voltaje (UCE AV)   Vea el módulo de control del generador de arranque (MCGA)

Hoist   A lift that is used to raise the entire vehicle.

Elevador   Montacargas utilizado para elevar el vehículo en su totalidad.

Holddowns   Secure the battery to reduce vibration and to prevent tipping.

Portador   Aseguran la batería para disminuir la vibración y prevenir que se vierte.

Hold-in winding   A winding that holds the plunger of a solenoid in place after it moves to engage the starter drive.

Devanado de retención Un devanado que posiciona el núcleo móvil de un solenoide después de que mueva para accionar el acoplamiento del motor de arranque.

Hole The absence of an electron in an element's atom. These holes are said to be positively charged since they have a tendency to attract free electrons into the hole.

Agujero Ausencia de un electrón en el átomo de un elemento. Se dice que dichos agujeros tienen una carga positiva puesto que tienden a atraer electrones libres hacia el agujero.

Horn A device that produces an audible warning signal.

Claxon Un dispositivo que produce una señal de advertencia audible.

Hybrid air bag Modules use compressed gas to fill the air bag instead of burning a chemical to produce gas.

Bolsa de aire híbrido Los modulos usan el gas comprimido para llenar la bolsa de aire en vez de quemar una química para producir un gas.

Hybrid battery A battery that combines the advantages of low-maintenance and maintenance-free batteries.

Batería híbrida Una batería que combina las ventajas de las baterías de bajo mantenimiento y de no mantenimiento.

Hybrid electric vehicle (HEV) System that has two different power sources. In most HEV, the power sources are a small displacement gasoline or diesel engine and an electric motor.

Vehículo eléctrico híbrido (VEH) Sistema con dos diferentes fuentes de potencia. En la mayoría de los VEH, las fuentes de potencia son un motor de diesel o de gasolina de desplazamiento menor y un motor eléctrico.

Hydrometer A test instrument used to check the specific gravity of the electrolyte to determine the battery's state of charge.

Hidrómetro Instrumento de prueba utilizado para verificar la gravedad específica del electrolito y así determinar el estado de la carga de la batería.

Igniter A combustible device that converts electric energy into thermal energy to ignite the inflator propellant in an air bag system.

Ignitor Un dispositivo combustible que convierte la energía eléctrica a la energía termal para encender el propelente inflador en un sistema Airbag.

Ignition coil A step-up transformer that builds up the low-battery voltage of approximately 12.6 volts to a voltage that is high enough to jump across the spark plug gap and ignite the air–fuel mixture.

Bobina de encendido Transformador multiplicador que sube el bajo voltaje de la batería de aproximadamente 12.6 voltios a uno lo suficientemente alto para brincar sobre el huelgo de la bujía y encender la mezcla de aire y combustible.

Ignition switch The power distribution point for most of a vehicle's primary electrical systems.

Selector de encendido Punto de distribución de potencia para la mayoría de los sistemas eléctricos principales del vehículo.

Ignition system Responsible for delivering properly timed high-voltage surges to the spark plugs.

Sistema de encendido Es responsable de llevar subidas de alto voltaje reguladas apropiadamente a las bujías.

Ignitor Component of the HID headlight that is used to create a high-voltage arc across the lamp electrodes to ignite the gas.

Encendedor Componente del faro HID que se utiliza para crear un arco de alto voltaje a través de los electrodos de la lámpara para encender el gas.

Illuminated entry systems Turn on the courtesy lights before the doors are opened.

Sistemas de entrada iluminada Enciendan las luces de cortesía antes de que se abren las puertas.

Immobilizer system Designed to provide protection against unauthorized vehicle use by disabling the engine if an invalid key is used to start the vehicle or if an attempt to hot-wire the ignition system is made.

Sistema inmovilizante Diseñado para proporcionar protección contra el uso no autorizado del vehículo al desactivar el motor si una llave que no es válida se utiliza para encender el vehículo o para intentar "hacerle el puente" al vehículo.

Incandescence The process of changing energy forms to produce light.

Incandescencia Proceso a través del cual se cambian las formas de energía para producir luz.

Induced voltage Voltage that is produced in a conductor as a result of relative motion within magnetic flux lines.

Tensión inducida Tensión producida en un conductor como resultado del movimiento relativo dentro de líneas de flujo magnético.

Induction The magnetic process of producing a current flow in a wire without any actual contact to the wire. To induce 1 volt, 100 million magnetic lines of force must be cut per second.

Inducción Proceso magnético a través del cual se produce un flujo de corriente en un alambre sin contacto real alguno con el alambre. Para inducir 1 voltio, deben producirse 100 millones de líneas de fuerza magnética por segundo.

Induction motor An AC motor that generates its own rotor current as the rotor cuts the magnetic flux lines of the stator field.

Motor de inducción Un motor de CA que genera su propia corriente de rotor a medida que el rotor corta las líneas de flujo magnético del campo del estator.

Inductive reactance The result of current flowing through a conductor and the resultant magnetic field around the conductor that opposes the normal flow of current.

Reactancia inductiva El resultado de un corriente que circule por un conductor y que resulta en un campo magnético alrededor del conductor que opone el flujo normal del corriente.

Inductive reluctance A statement of a material's ability to strengthen the magnetic field around it.

Reluctancia a la inducción Una indicación de la habilidad de una materia en reenforzar el campo que la rodea.

Inertia The tendency of an object that is at rest and an object that is in motion to stay in motion.

Inercia La tendencia de un objeto que esta en descanso quedarse en descanso y un objeto en movimiento de quedarseen movimiento.

Inertia engagement A type of starter motor that uses rotating inertia to engage the drive pinion with the engine flywheel.

Conexión por inercia Tipo de motor de arranque que utiliza inercia giratoria para engranar el piñón de mando con el volante de la máquina.

Inertia lock retractors Use a pendulum mechanism to lock the belt tightly during sudden movement.

Retractores de cierre tipo inercia Usan un mecanismo de péndulo para enclavar fuertemente la cinta durante un movimiento repentino.

Inflatable knee blocker (IKB) A small air bag that deploys simultaneously with the driver's-side airbag to provide upper-leg protection and positioning of the driver.

Bloqueante inflable de la rodilla (BIR) Pequeña bolsa de aire que se despliega simultáneamente con la bolsa de aire lateral del conductor para proporcionar protección a la parte superior de las piernas y el posicionamiento del conductor.

Infotainment A type of media system that provides a combination of information and entertainment.

Infotenimiento Un tipo de sistema de medios que brinda una combinación de información y entretenimiento.

Infrared temperature sensor A senor that measures the surface temperature of an object or person by measuring the intensity of the energy given off by an object.

Sensor infrarrojo para la temperatura Sensor que mide la temperatura de la superficie de un objeto o persona al medir la intensidad de la energía que desprende un objeto.

Esquemas de instalación Proveen una duplicación más precisa de donde se encuentran el cableado preformado, los conectores, y los componentes en el vehículo.

Instrument panel dimming System in which the headlight switch dimming control is used as an input to the computer instead of having direct control of the illumination lights.

Reducción luminosa del tablero de instrumentos Un sistema en el cual el control del interruptor de luminosidad se usa como una entrada a la computadora en vez de tener control directo al luminosidad.

Instrument voltage regulator (IVR) Provides a constant voltage to the gauge, regardless of the voltage output of the charging system.

Instrumento regulador de tensión Le provee tensión constante al calibrador, sin importar cual sea la salida de tensión del sistema de carga.

Insulated side The portion of the circuit from the positive side of the source to the load component.

Lado aislado   Parte del circuito que va del lado positivo de la fuente al componente de carga.

Insulator   A material that does not allow electrons to flow easily through it.

Aislador   Una material que no permite circular fácilmente los electrones.

Integrated circuit (IC)   A complex circuit of thousands of transistors, diodes, resistors, capacitors, and other electronic devices that are formed into a small silicon chip. As many as 30,000 transistors can be placed on a chip that is 1/4 inch (6.35 mm) square.

Circuito integrado (Fragmento CI)   Circuito complejo de miles de transistores, diodos, resistores, condensadores, y otros dispositivos electrónicos formados en un fragmento pequeño de silicio. En un fragmento de 1/4 de pulgada (6,35 mm) cuadrada, pueden colocarse hasta 30.000 transistores.

Integrated starter generator (ISG)   A combination starter generator in one unit that attaches directly to the crankshaft to allow for the automatic stop/start function of an HEV. It can also convert kinetic energy to DC voltage when the vehicle is traveling downhill and there is zero load.

Generador de arranque integrado   Combinación de generador de arranque en una unidad que se adhiere directamente al cigüeñal para permitir la función automática de encendido y apagado de un VEH. También puede convertir la energía cinemática a voltaje de corriente continua cuando el vehículo va de bajada y no lleva carga.

Intelligent windshield wipers   A wiper system that uses a monitoring system to detect if water is present on the windshield and that automatically turns on the wiper system.

Limpiaparabrisas inteligente   Sistema de limpiadores que utiliza un sistema de monitoreo para detectar si hay agua en el parabrisas, y esto hace que automáticamente se encienda el sistema de los limpiadores.

Interface   Used to protect the computer from excessive voltage levels and to translate input and output signals.

Interfase   Utilizada para proteger la computadora de niveles excesivos de tensión y traducir señales de entrada y salida.

International Standards Organization (ISO)   Symbols used to represent the gauge function.

Organización Internacional de Normas (ISO)   Los símbolos que se usan para representar la función del indicador.

In-vehicle sensor   The in-vehicle sensor contains a temperature-sensing NTC thermistor to measure the average temperature inside the vehicle.

Sensor en el vehículo   El sensor dentro del vehículo contiene un termostato de coeficiente de temperatura negativo (CTN) para percibir la temperatura que mide la temperatura promedio dentro del vehículo.

Ion   An atom or group of atoms that has an electrical charge.

Ion   Átomo o grupo de átomos que poseen una carga eléctrica.

Ionize   To electrically charge.

Ionizar   Cargar eléctricamente.

ISO   An abbreviation for International Standards Organizations.

ISO   Una abreviación de las Organizaciones de Normas Internacionales.

ISO 14230-4   A bus data protocol that uses a single-wire bidirectional data line to communicate between the scan tool and the nodes. This data bus is used only for diagnostics and maintains the ISO 9141 protocol with a baud rate of 10.4 kb/s.

ISO 14230-4   Protocolo de un bus de datos que utiliza una línea de datos en dos direcciones en un hilo sencillo para comunicarse entre el instrumento de exploración y los nodos. Este bus de datos se utiliza solamente para diagnósticos y mantiene el protocolo de ISO 9141 con una velocidad de transmisión de baudios de 10.4 kb/s.

ISO 9141-2   A class B system with a baud rate of 10.4 kb/s used only for diagnostic purposes between the nodes on the data bus and an OBD II standardized scan tool.

ISO 9141-2   Sistema B de clase A con una velocidad de transmisión de baudios de 10.4 kb/s que se utiliza sólo con un propósito de diagnóstico entre los nodos en el bus de datos y un instrumento de exploración estandarizado del sistema de diagnóstico a bordo II o DAB II.

ISO-K   An adoption of the ISO 9141-2 protocol that allows for bidirectional communication on a single wire. Vehicles that use the ISO-K bus require that the scan tool provide the bias voltage to power up the system.

ISO-K   La adopción del protocolo del ISO 9141-2 es que permite la comunicación en dos direcciones en un hilo sencillo. Los vehículos que utilizan el bus de ISO-K requieren que el instrumento de exploración proporcione la tensión de polarización para hacer funcionar el sistema.

ISO relays   Conform to the specifications of the International Standards Organization (ISO) for common size and terminal patterns.

Relés ISO   Conforman a las especificaciones de la Organización Internacional de Normas (ISO) en tamaño normal y conformidades de terminales.

J1850   The bus system that is the class B standard for OBD II. The J1850 standard allows for two different versions based on baud rate. The first supports a baud rate of 41.6 kb/s, which is transmitted by a pulse-width–modulated (PWM) signal over a twisted pair of wires. The second protocol supports a baud rate of 10.4 kb/s average, which is transmitted by a variable pulse width (VPW) data bus over a single wire.

J1850   El sistema de bus que es el estándar de clase B para el sistema de diagnóstico a bordo II o DAB II. El estándar J1850 permite dos diferentes versiones que se basan en la velocidad de transmisión de baudios. El primero respalda una velocidad de transmisión de baudios de 41.6 kb/s que transmite una señal de modulación de duración de impulsos (MDI o PWM) mediante un par torcido de hilos. El segundo protocolo respalda una velocidad de transmisión de baudios de 10.4 kb/s promedio que transmite un bus de datos de anchura variada entre impulsos (AVI o VPW) mediante un hilo sencillo.

Keyless entry system   A lock system that allows for locking and unlocking of a vehicle with a touch keypad instead of a key.

Entrada sin llave   Un sistema de cerradura que permite cerrar y abrir un vehículo por medio de un teclado en vez de utilizar una llave.

Keyless start system   System that allows an engine to start without using an ignition key.

Sistema de arranque sin llave   Sistema que permite arrancar un motor sin utilizar una llave de encendido.

Kirchhoff's current law   Electrical law which states that the total flow of current in a parallel circuit will be the sum of the currents through all the circuit branches.

Ley de corriente de Kirchhoff   Ley eléctrica que afirma que el flujo total de corriente en un circuito paralelo será la suma de las corrientes a través de todas las ramas del circuito.

Kirchhoff's voltage law   Electrical law which states that the sum of the voltage drops in a series circuit must equal the source voltage.

Ley de voltaje de Kirchhoff   Ley eléctrica que afirma que la suma de las caídas de voltaje en un circuito en serie debe ser igual al voltaje fuente.

K-line   One circuit of the ISO 9141-2 data bus that is used for transmitting data from the module to the scan tool. The scan tool provides the bias voltage onto this circuit and the module pulls the voltage low to transmit its data.

Línea K   Un circuito del bus de datos ISO 9141-2 que se utiliza para transmitir datos de un módulo a un instrumento de exploración. El instrumento de exploración proporciona la tensión de polarización sobre este circuito, y el módulo baja el voltaje para transmitir sus datos.

Laminated construction   Construction of the armature from individual stampings.

Construcción laminada   La armadura esta construida de un matrizado individual.

Lamination   The process of constructing something with layers of materials that are firmly connected.

Laminación   El proceso de construir algo de capas de materiales unidas con mucha fuerza.

Lamp   A device that produces light as a result of current flow through a filament. The filament is enclosed within a glass envelope and is a type of resistance wire that is generally made from tungsten.

Lámpara   Dispositivo que produce luz como resultado del flujo de corriente a través de un filamento. El filamento es un tipo de alambre de resistencia hecho por lo general de tungsteno, que es encerrado dentro de una bombilla.

Lamp outage module   A current-measuring sensor that contains a set of resistors, wired in series with the power supply to the headlights, taillights, and stop lights. If the sensor indicates that a lamp is burned out, the module will alert the driver.

Unidad de avería de la lámpara   Sensor para medir corriente que incluye un juego de resistores, alambrado en serie con la fuente de alimentación a los faros delanteros, traseros y a las luces de freno. Si el sensor indica que se ha apagado una lámpara, la unidad le avisará al conductor.

Lane departure warning (LDW)   A vehicle safety system that uses a forward-facing camera, warning buzzers, haptic feedback in the steering wheel, and/or tugging on the seatbelt if the vehicle is about to leave its designated lane.

Advertencia de salida de carril (LDW) Un sistema de seguridad del vehículo que utiliza una cámara orientada hacia delante, zumbadores de advertencia, retroalimentación háptica en el volante y/o tirón en el cinturón de seguridad si el vehículo está a punto de abandonar su carril designado.

Leading edge The edges of the rotating blade that enter the switch in a Hall-effect switch.

Borde anterior Los bordes de la ala giratorio que entran al interruptor en un terruptor efecto Hall.

Lean burn technology Uses lean air–fuel ratios to increase fuel efficiency.

Tecnología de quema limpia Determina las relaciones aire-combustible limpios para aumentar la eficacia del combustible.

Light-emitting diode (LED) A gallium-arsenide diode that converts the energy developed when holes and electrons collide during normal diode operation into light.

Diodo emisor de luz Diodo semiconductor de galio y arseniuro que convierte en luz la energía producida por la colisión de agujeros y electrones durante el funcionamiento normal del diodo.

Lighting system Electrical system that consists of all of the lights used on the vehicle, including headlights, front and rear park lights, front and rear turn signals, side marker lights, daytime running lights, cornering lights, brake lights, back-up lights, instrument cluster backlighting, and interior lighting.

Sistema de iluminación Sistema eléctrico que consta de todas las luces que usa el vehículo, incluyendo los faros, las luces frontales y traseras de estacionamiento, las luces intermitentes frontales y traseras, luces de posición, luces de marcha diurna, luces de esquina?, luces de freno, luces de marcha atrás, iluminación trasera de tablero de controles e iluminación interior.

Limit switch A switch used to open a circuit when a predetermined value is reached. Limit switches are normally responsive to a mechanical movement or temperature changes.

Disyuntor de seguridad Un conmutador que se emplea para abrir un circuito al alcanzar un valor predeterminado. Los disyuntores de seguridad suelen ser responsivos a un movimiento mecánico o a los cambios de temperatura.

Linearity Refers to the sensor signal being as constantly proportional to the measured value as possible. It is an expression of the sensor's accuracy.

Linealidad Significa que la variación del valor de una magnitud es lo más proporcional posible a la variación del valor de otra magnitud. Expresa la precisión del sensor.

Liquid crystal display (LCD) A display that sandwiches electrodes and polarized fluid between layers of glass. When voltage is applied to the electrodes, the light slots of the fluid are rearranged to allow light to pass through.

Visualizador de cristal líquido Visualizador digital que consta de dos láminas de vidrio selladas, entre las cuales se encuentran los electrodos y el fluido polarizado. Cuando se aplica tensión a los electrodos, se rompe la disposición de las moléculas para permitir la formación de caracteres visibles.

L-line One circuit of the ISO 9141-2 data bus that is used by the module to receive data from the scan tool. The module provides the bias onto this circuit and the scan tool pulls the voltage low to communicate.

Línea L Un circuito del bus de datos del ISO 9141-2 que utiliza un módulo que recibe datos de un instrumento de exploración. El módulo proporciona la tensión sobre este circuito y el instrumento de exploración baja el voltaje para comunicarse.

Load device The component that performs some form of work.

Dispositivo de carga El componente que lleva a cabo algun forma de trabajo.

Local interconnect network (LIN) A bus network that was developed to supplement the CAN bus system. The term *local interconnect* refers to all of the modules in the LIN network being located within a limited area.

LIN (Red local de interconexiones) Una red de bus que se desarrolló para complementar el sistema de bus CAN. El término "interconexión local" se refiere a todos los módulos en la red de LIN que se encuentran dentro de un área.

Logic gates Electronic circuits that act as gates to output voltage signals depending on different combinations of input signals.

Compuertas lógicas Circuitos electrónicos que gobiernan señales de tensión de salida, dependiendo de las diferentes combinaciones de señales de entrada.

Look-up tables The part of a microprocessor's memory that indicates how a system should perform in the form of calibrations and specifications.

Tablas de referencia La parte de la memoria de una microprocesora que indica como debe ejecutar las calibraciones y las especificaciones la sistema.

Low-reading coil Wound together with the bucking coil but in the opposite direction.

Bobina de lectura de baja tensión Envueltas juntas con la bobina compensadora pero en una dirección opuesta.

Low-side drivers Used to complete the path to ground to turn on an actuator.

Impulsorer del lado a tierra Usados para completar el circuito a tierra para activar un actuador.

Low voltage differential signaling (LVDS) A signaling method used for high-speed bus transmission of binary data that is used in point-to-point configurations for visual displays.

Señalización diferencial de baja tensión (LVDS) Método de señalización utilizado para la transmisión de datos binarios de buses a alta velocidad que se utiliza en configuraciones punto a punto para pantallas visuales.

LUX The International System unit of measurement of the intensity of light. It is equal to the illumination of a surface 1 meter away from a single candle (1 lumen per square meter).

LUX (lumen por metro cuadrado) Unidad del sistema internacional de medida de la intensidad de la luz. Es igual a la iluminación de una superficie a un metro de distancia de una vela sencilla (un lumen por metro cuadrado).

Magnetic field The area surrounding a magnet where energy is exerted due to the atoms aligning in the material.

Campo magnético Espacio que rodea un imán donde se emplea la energía debido a la alineación de los átomos en el material.

Magnetic flux density The concentration of the magnetic lines of force.

Densidad de flujo magnético Número de líneas de fuerza magnética.

Magnetic pulse generator Sensor that uses the principle of magnetic induction to produce a voltage signal. Magnetic pulse generators are commonly used to send data to the computer concerning the speed of the monitored component.

Generador de impulsos magnéticos Sensor que funciona según el principio de inducción magnética para producir una señal de tensión. Los generadores de impulsos magnéticos se utilizan comúnmente para transmitir datos a la computadora relacionados a la velocidad del componente regulado.

Magnetically coupled linear sensors Sensors that can be used to measure movement by use of a moveable magnet, a resistor card, and a magnetically sensitive comb. Changes in the location of the magnet on the card provide a variable output.

Sensores lineales acoplados magnéticamente Sensor que se puede utilizar para medir el movimiento mediante el uso de un imán móvil, una tarjeta de resistor y un peine sensible al magnetismo. Los cambios en la ubicación del imán en la tarjeta proporcionan información variable.

Magnetism An energy form resulting from atoms aligning within certain materials, giving the materials the ability to attract other metals.

Magnetismo Forma de energía que resulta de la alineación de átomos dentro de ciertos materiales y que le da a éstos la capacidad de atraer otros metales.

Magnetoresistive effect The increase in resistance of a current-carrying magnetic substance when an external magnetic field is applied perpendicular to its surface.

Efecto magneto resistivo El aumento en la resistencia de una substancia magnética transportadora de corriente cuando se aplica un campo magnético externo, perpendicular a su superficie.

Magnetoresistive (MR) sensors Speed detection sensor consisting of a permanent magnet and an integrated signal conditioning circuit to change resistance due to the relationship of the tone wheel and magnetic field surrounding the sensor. The change in resistance results in a digital reading of current levels by the control module.

Sensores magneto resistivos (MR) Sensor de detección de velocidad que está compuesto por un imán permanente y un circuito de acondicionamiento de señales integradas para cambiar la resistencia debido a la relación entre la rueda de virado y el campo magnético que rodea al sensor. El cambio en la resistencia tiene como resultado una lectura digital de los niveles actuales por parte del módulo de control.

Maintenance-free battery A battery that has no provision for the addition of water to the cells. The battery is sealed.

Sin mantención Que no tiene provisión para añadir el agua a la células. Es una batería sellada.

Master module Controller on the network that translates messages between different network systems.

Instancia maestra Controlador o entidad única, dentro de una red distribuida, que traduce los mensajes entre los diferentes sistemas de la red.

Material expanders Fillers that can be used in place of the active materials in a battery. They are used to keep the cost of manufacturing low.

Expansores de materias Los rellenos que se pueden usar en vez de las materiales activas de una batería. Se emplean para mantener bajos los costos de la fabricación.

Matrix A rectangular array of grids.

Matriz Red lógica en una rejilla de forma rectangular.

Maxi-fuse A circuit protection device that looks similar to a blade-type fuse except that it is larger and has a higher amperage capacity. Maxi-fuses are used because they are less likely to cause an under-hood fire when there is an overload in the circuit. If the fusible link burned in two, it is possible that the hot side of the fuse could come into contact with the vehicle frame and the wire could catch fire.

Maxifusible Dispositivo de protección del circuito parecido a un fusible de tipo de cuchilla, pero más grande y con mayor capacidad de amperaje. Se utilizan maxifusibles porque existen menos probabilidades de que ocasionen un incendio debajo de la capota cuando ocurra una sobrecarga en el circuito. Si el cartucho de fusible se quemase en dos partes, es posible que el lado "cargado" del fusible entre en contacto con el armazón del vehículo y que el alambre se encienda.

Media-oriented system transport (MOST) A data bus system based on standards established by a cooperative effort between automobile manufacturers, suppliers, and software programmers that resulted in a data system specifically designed for the data transmission of media-oriented data. MOST uses fiber optics to transmit data at a rate up to 25 Mb/s.

MOST Sistema de bus de datos basado en estándares establecidos por un esfuerzo cooperativo entre los fabricantes de vehículos, los proveedores y los programadores de software que resultó en un sistema de datos específicamente diseñado para la transmisión de datos informativos. MOST utiliza fibras ópticas para transmitir datos a una velocidad de 25 megabitios por segundo (25Mb/s).

Memory effect The battery failing to fully charge because it remembers its previous charge level. This results in a low-battery charge due to a battery that is not completely discharged before it is recharged.

Efecto memoria Fallas en la carga completa de la batería debido a que "recuerda" su nivel de carga previo. En consecuencia, se produce una baja carga de la batería ya que ésta no está completamente descargada antes de la recarga.

Memory seat Power seats that can be programmed to return or adjust to a point designated by the driver.

Asientos con memoria Los asientos automáticos que se pueden programar a regresar o ajustarse a un punto indicado por el conductor.

Metri-pack connector Special wire connectors used in some computer circuits. They seal the wire terminals from the atmosphere, thereby preventing corrosion and other damage.

Conector metri-pack Los conectores de alambres especiales que se emplean en algunos circuitos de computadoras. Impermealizan los bornes de los alambres, así previniendo la corrosión y otros daños.

Microprocessor The brains of the computer where most calculations take place.

Microprocesador El cerebro de la computadora en donde se realizan la mayoría de los cálculos.

Micro-Electro-Mechanical System (MEMS) A technology that combines miniaturized mechanical and electro-mechanical elements using the techniques of microfabrication.

Sistema Micro-Electro-Mecánico (MEMS) Una tecnología que combina elementos mecánicos y electromecánicos miniaturizados utilizando las técnicas de microfabricación.

Mild parallel hybrid Uses an electric motor that is large enough to provide regenerative braking, instant engine start-up, and a boost to the combustion engine.

Híbrido de medio paralelo Utiliza un motor eléctrico que es lo suficientemente grande para proveer freno regenerativo, encendido instantáneo del motor y un aumento a la combustión del motor.

Mode door actuator An electric motor that is linked to the mode door to supply air flow to the floor ducts, A/C panel ducts, or defrost ducts.

Actuador de mando puerta Motor eléctrico que está unido al mando puerta para suministrar flujo de aire a los conductos del piso, los conductos del panel de corriente alterna o a los conductos de descongelamiento.

Momentary contact A switch type that operates only when held in position.

Contacto momentáneo Tipo de conmutador que funciona solamente cuando se mantiene en su posición.

Multiplexer An electronic switch that switches between the different audio sources.

Multiplexor Interruptor electrónico que se usa para hacer cambios entre las diferentes fuentes de audio.

Multiplexing (MUX) A means of transmitting information between computers. It is a system in which electrical signals are transmitted by a peripheral serial bus instead of conventional wires, allowing several devices to share signals on a common conductor.

Multiplexaje Medio de transmitir información entre computadoras. Es un sistema en el cual las señales eléctricas son transmitidas por una colectora periférica en serie en vez de por líneas convencionales. Esto permite que varios dispositivos compartan señales en un conductor común.

Multistage air bags Hybrid air bags that use two squibs to control the rate of inflation.

Bolsas de aire de etapas múltiples Las bolsas de aire híbridas que usan dos petardos para controlar la velocidad de la inflación.

Mutual induction An induction of voltage in an adjacent coil by changing current in a primary coil.

Inducción mutua Una inducción de la tensión en una bobina adyacente que se efectúa al cambiar la tensión en una bobina primaria.

MUX Common acronym for multiplexing.

MUX Una sigla común del proceso de multiplex.

Navigational systems Use satellites to direct drivers to desired destinations.

Sistema de navegación Usa los satélites para dirigir el conductor a la destinación deseada.

Negative logic Defines the most negative voltage as a logical 1 in the binary code.

Lógica negativa Define la tensión más negativa como un 1 lógico en el código binario.

Negative temperature coefficient (NTC) thermistors Thermistors that reduce their resistance as the temperature increases.

Termistores con coeficiente negativo de temperatura Termistores que disminuyen su resistencia según aumenta la temperatura.

Nematic Describes a fluid that is a liquid crystal with a threadlike form. It has light slots that can be rearranged by applying small amounts of voltage.

Nemático Describe un flúido que es un cristal líquido con una forma de filamento. Tiene aberturas de luz que se pueden reorganizar por medio de la aplicación de pequeñas cantidades de voltaje.

Neon lights A light that contains a colorless, odorless inert gas called neon. These lamps are discharge lamps.

Luces de neón Una luz que contiene un gas inerto sin color, inodoro llamado neón. Estas lámparas son lámparas de descarga.

Network Incorporating the vehicle's electrical systems together through computers so information gathered by one system can be used by another.

En red Incorporar los sistemas eléctricos del vehículo mediante el uso de computadoras para que la información que obtenga un sistema pueda usarla otro sistema.

Network architecture Describes how modules are connected on a network. Common types include linear, stub, ring, and star.

Arquitectura de red Describe la forma en que se conectan los módulos en una red. Los tipos comunes son: lineal, aislada, anillo y estrella.

Neutral atom See balanced atom.

Átomo neutro Véase átomo equilibrado.

Neutral junction The center connection to which the common ends of a Y-type stator winding are connected.

Empalme neutro Conexión central a la cual se conectan los extremos comunes de un devanado del estátor de tipo Y.

Neutral safety switch A switch used to prevent the starting of an engine unless the transmission is in PARK or NEUTRAL.

Disyuntor de seguridad en neutral Un conmutador que se emplea para prevenir que arranque un motor al menos de que la transmisión esté en posición PARK o Neutral.

Neutrons Particles of an atom that have no charge.

Neutrones Partículas de un átomo desprovistas de carga.

Night vision Describes the ability to see objects at night by use of infrared cameras.

Visión nocturna   Describe la capacidad de ver objetos en la noche mediante el uso de cámaras infrarrojas.

Node   A computer that is connected to a data bus network and capable of sending or receiving messages.

Nodo   Computadora conectada a una red de bus de datos y con capacidad de mandar o recibir mensajes.

Nonvolatile   RAM memory that will retain its memory if battery voltage is disconnected. NVRAM is a combination of RAM and EEPROM into the same chip. During normal operation, data is written to and read from the RAM portion of the chip. If the power is removed from the chip, or at programmed timed intervals, the data is transferred from RAM to the EEPROM portion of the chip. When the power is restored to the chip, the EEPROM will write the data back to the RAM.

Memoria de acceso aleatorio no volátil [NV RAM]   Memoria de acceso aleatorio (RAM) que retiene su memoria si se desconecta la carga de la batería. La NV RAM es una combinación de RAM y EEPROM en el mismo fragmento. Durante el funcionamiento normal, los datos se escriben en y se leen de la parte RAM del fragmento. Si se remueve la alimentación del fragmento, o si se remueve ésta a intervalos programados, se transfieren los datos de la RAM a la parte del EEPROM del fragmento. Cuando se restaura la alimentación en el fragmento, el EEPROM volverá a escribir los datos en la RAM.

Normally closed (NC) switch   A switch designation denoting that the contacts are closed until acted upon by an outside force.

Conmutador normalmente cerrado   Nombre aplicado a un conmutador cuyos contactos permanecerán cerrados hasta que sean accionados por una fuerza exterior.

Normally open (NO) switch   A switch designation denoting that the contacts are open until acted upon by an outside force.

Conmutador normalmente abierto   Nombre aplicado a un conmutador cuyos contactos permanecerán abiertos hasta que sean accionados por una fuerza exterior.

N-type material   When there are free electrons, the material is called an N-type material. The N means negative and indicates that it is the negative side of the circuit that pushes electrons through the semiconductor and the positive side that attracts the free electrons.

Material tipo N   Al material se le llama material tipo N cuando hay electrones libres. La N significa negativo e indica que el lado negativo del circuito empuja los electrones a través del semiconductor y el lado positivo atrae los electrones libres.

Nucleus   The core of an atom that contains protons and neutrons.

Núcleo   Parte central de un átomo que contiene los protones y los neutrones.

Occupant classification systems   A mandated requirement to reduce the risk of injuries resulting from air bag deployment by determining the weight classification of the front-seat passenger.

Sistemas de clasificación de los ocupantes   Mandato para reducir el riesgo de daños que resulten del desarrollo de la bolsa de aire al determinar la clasificación del peso del pasajero del asiento de enfrente.

Occupational safety glasses   Eye protection that is designed with special high-impact lens and frames and provides for side protection.

Gafas de protección para el trabajo   Gafas diseñadas con cristales y monturas especiales resistentes y provistas de protección lateral.

OCS service kit   Special kit that consists of the seat foam, the bladder, the pressure sensor, the occupant classification module (OCM), and the wiring.

Kit de servicio SCO   Kit especial que consiste en hule-espuma del asiento, el depósito, el sensor de presión, el módulo de clasificación del ocupante (MCO) y el alambrado.

OCS validation test   A test that confirms that the system can properly classify the occupant.

Prueba de revalidación del SCO   Prueba que confirma que el sistema puede clasificar apropiadamente al ocupante.

Odometer   A mechanical counter in the speedometer unit indicating total miles accumulated on the vehicle.

Odómetro   Aparato mecánico en la unidad del velocímetro con el que se cuentan las millas totales recorridas por el vehículo.

Ohms   Unit of measure for resistance. One ohm is the resistance of a conductor such that a constant current of 1 amp in it produces 1 volt between its ends.

Ohmio   Unidad de resistencia eléctrica. Un ohmio es la resistencia de un conductor si una corriente constante de 1 amperio en el conductor produce una tensión de 1 voltio entre los dos extremos.

Ohmmeter   A test meter used to measure resistance and continuity in a circuit.

Ohmiómetro   Instrumento de prueba utilizado para medir la resistencia y la continuidad en un circuito.

Ohm's law   Defines the relationship between current, voltage, and resistance.

Ley de Ohm   Define la relación entre la corriente, la tensión y la resistencia.

Open   Term used to describe a break in the circuit. The break can be from a switch in the off position or a physical break in the wire.

Abierto   Término utilizado para describir una interrupción en el circuito. La interrupción puede ser debido a un interruptor en la posición de apagado o a una rotura física en el cable.

Open circuit   A term used to indicate that current flow is stopped. By opening the circuit, the path for electron flow is broken.

Circuito abierto   Interrupción en el circuito eléctrico que causa que pare el flujo de corriente.

Optical horn   A name Chrysler uses to describe their flash-to-pass headlamp system.

Claxón óptico   Un nombre que usa Chrysler para describir su sistema de faros "relampaguea para rebasar."

Oscillator   Creates a rapid back-and-forth movement of voltage.

Oscilador   Crea un movimiento de oscilación rápido de voltaje.

Output driver   A transistor circuit that controls the operation of an actuator. It can send a voltage to the actuator or provide the ground for the actuator.

Controlador de salida   Un circuito de transistores que controla la operación de un actuador. Puede enviar un voltaje al actuador o proporcionar la puesta a tierra para el actuador.

Output signal   A command signal sent from the computer to an actuator.

Señal de salida   Una señal de comando que se envía desde la computadora hacia un actuador.

Overload   Excess current flow in a circuit.

Sobrecarga   Flujo de corriente superior a la que tiene asignada un circuito.

Overrunning clutch   A starter drive that uses a roller clutch to transmit torque in one direction and freewheels in the other direction.

Embrague de sobremarcha   Una asamblea de embrague en un acoplamiento del motor de arranque que se emplea para prevenir que el volante del motor dé vueltas al armazón del motor de arranque.

Oversteer   The tendency of the back of the vehicle to turn on the vehicle's center of gravity and come around the front of the vehicle.

Tener la dirección muy sensible   Tendencia de la parte trasera de un vehículo de dar vuelta en el centro de gravedad del vehículo y de doblar al frente del vehículo.

Oxygen sensor   A voltage-generating sensor that measures the amount of oxygen present in an engine's exhaust.

Sensor de oxígeno   Un sensor generador de tensión que mide la cantidad del oxígeno presente en el gas de escape de un motor.

Parallel circuit   A circuit that provides two or more paths for electricity to flow.

Circuito en paralelo   Circuito que provee dos o más trayectorias para que circule la electricidad.

Parallel hybrid   A hybrid vehicle configuration that has a direct mechanical connection between the engine and the wheels. Both the engine and the electric motor can turn the transmission at the same time.

Híbrido en paralelo completo   Utiliza un motor eléctrico que es lo suficientemente potente para que el vehículo se impulse por sí mismo.

Parasitic loads   Electrical loads that are still present when the ignition switch is in the OFF position.

Cargas parásitas   Cargas eléctricas que todavía se encuentran presente cuando el botón conmutador de encendido está en la posición OFF.

Park contacts   Located inside the motor assembly and supply current to the motor after the wiper control switch has been turned to the PARK position. This allows the motor to continue operating until the wipers have reached their PARK position.

Contactos de Park   Ubicado dentro de la asamblea del motor y proveen el corriente al motor después de que el interruptor de control de la limpiaparabrisa se ha puesto en la posición de estacionamiento. Esto permite que el motor continua operando hasta que los brazos de la limpiaparabrisas hayan llegado a su posición de estacionamiento.

Park switch   Contact points located inside the wiper motor assembly that supply current to the motor after the wiper control switch has been turned to the PARK position. This allows the motor to continue operating until the wipers have reached their PARK position.

Conmutador PARK   Puntos de contacto ubicados dentro del conjunto del motor del frotador que le suministran corriente al motor después de que el conmutador para el control de los frotadores haya sido colocado en la posición PARK. Esto permite que el motor continue su funcionamiento hasta que los frotadores hayan alcanzado la posición original.

Park assist system   A parking aid that alerts the driver to obstacles located in the path immediately behind or in front of the vehicle.

Sistema de ayuda al aparcamiento   Ayuda al aparcamiento que avisa al conductor de los obstáculos que se encuentran en el camino inmediatamente detrás o delante del vehículo.

Pass key   A specially designed vehicle key with a coded resistance value. The term *pass* is derived from personal automotive security system.

Llave maestra   Una llave vehícular de diseño especial que tiene un valor de resistencia codificado. El termino pass se derive de las palabras Personal Automotive Security System (sistema personal de seguridad automotriz).

Passive restraints systems   A passenger restraint system that automatically operates to confine the movement of a vehicle's passengers.

Correas passivas   Un sistema de resguardo del pasajero que opera automaticamente para limitar el movimiento de los pasajeros en el vehículo.

Passive suspension systems   Use fixed spring rates and shock valving.

Sistemas pasivos de suspensión   Utilizan elasticidad de muelle constante y dotación con válvulas amortigadoras.

Peak reverse voltage (PRV)   Indicates the maximum reverse-bias voltage that may be applied to a diode without causing junction breakdown.

Voltaje inverso pico (PRV, por su sigla en inglés)   Indica el voltaje de polarización inversa máximo que se puede aplicar a un diodo sin causar una avería a un empalme.

Peltier element   An element consisting of two different types of metals that are joined together. The area of the joint will generate or absorb heat when an electric current is applied to the element at a specified temperature.

Elemento Peltier   Un elemento compuesto por dos tipos diferentes de metales unidos. El área de la unión generará o absorberá calor cuando se aplique una corriente eléctrica al elemento a una temperatura específica.

Permanent magnet gear reduction (PMGR)   A starter that uses four or six permanent magnet field assemblies in place of field coils.

Reducción de engranaje de imán permanente (PMGR)   Un arrancador que usa cuatro o seis asambleas permanentes de campo magnético en vez de las bobinas de campo.

Permeability   Term used to indicate the magnetic conductivity of a substance compared with the conductivity of air. The greater the permeability, the greater the magnetic conductivity and the easier a substance can be magnetized.

Permeabilidad   Término utilizado para indicar la aptitud de una sustancia en relación con la del aire, de dar paso a las líneas de fuerza magnética. Mientras mayor sea la permeabilidad, mayor será la conductividad magnética y más fácilmente se comunicará a un cuerpo propiedades magnéticas.

Piezoelectric accelerometers   A piezoelectric element constructed of zinc oxide that is distorted during a high G force condition and generates an analog voltage in direct relationship to the amount of G force. Used to sense deceleration forces.

Acelerómetros piezoeléctricos   Un elemento piezoeléctrico construido de óxido de zinc que se distorsiona durante una condición de alta fuerza G y genera un voltaje análogo en relación directa con la cantidad de fuerza G. Se utiliza para detectar fuerzas de desaceleración.

Photo cell   A variable resistor that uses light to change resistance.

Fotocélula   Resistor variable que utiliza luz para cambiar la resistencia.

Photoconductive mode   Operating mode of the photodiode when an external reverse-bias is applied that results in a small reverse current through the diode.

Modo fotoconductor   Modo de operación del fotodiodo cuando se aplica una polarización inversa externa, la cual produce una pequeña corriente inversa a través del diodo.

Photodiode   Allows current to flow in the opposite direction of a standard diode when it receives a specific amount of light.

Fotodiodo   Permite que fluye el corriente en la dirección opuesta de él de un diodo normal al recibir una cantidad específica de luz.

Photoresistor   A passive light-detecting device composed of a semiconductor material that changes resistance when its surfaced is exposed to light.

Fotorresistor   Un dispositivo detector de luz pasivo, compuesto por un material semiconductor que cambia su resistencia cuando su superficie se expone a la luz.

Phototransistor   A light-detecting transistor that uses the application of light to generate carriers to supply the base leg current.

Fototransistor   Un transistor detector de luz que utiliza la aplicación de luz para generar portadores que suministren corriente a la pata base.

Photovoltaic diodes   Diodes capable of producing a voltage when exposed to radiant energy.

Diodos fotovoltaicos   Diodos capaces de generar una tensión cuando se encuentran expuestos a la energía de radiación.

Photovoltaic mode   An operating mode of the photodiode with no bias applied where the generated current or voltage is in the forward direction.

Modo fotovoltaico   Un modo de operación del fotodiodo sin aplicar polarización en donde la corriente o voltaje que se genera es en dirección positiva.

Photovoltaics (PV)   One of the forces that can be used to generate electrical current by converting solar radiation through the use of solar cells.

Fotovoltaica (PV)   Una de las fuerzas que puede utilizarse para generar corriente eléctrica al convertir la radiación solar a través del uso de celdas solares.

Pickup coil   The stationary component of the magnetic pulse generator consisting of a weak permanent magnet that has fine wire wound around it. As the timing disc rotates in front of it, the changes in magnetic lines of force generate a small voltage signal in the coil.

Bobina captadora   Componente fijo del generador de impulsos magnéticos compuesta de un imán permanente débil devanado con alambre fino. Mientras gira el disco sincronizador enfrente de él, los cambios de las líneas de fuerza magnética generan una pequeña señal de tensión en la bobina.

Piconets   Small transmission cells that assist in the organization of data.

Picoredes   Pequeñas células de transmisión que ayudan a organizar los datos.

Piezoelectric device   A voltage generator with a resistor connected in series that is used to measure fluid and air pressures.

Dispositivo piezoeléctrico   un generador de voltaje con un resistor conectado en series que se utiliza para medir las presiones de fluido y aire.

Piezoelectricity   Voltage produced by the application of pressure to certain crystals.

Piezoelectricidad   Generación de polarización eléctrica en ciertos cristales a consecuencia de la aplicación de tensiones mecánicas.

Piezoresistive device   Similar to a piezoelectric device except that it operates like a variable resistor. Its resistance value changes as the pressure applied to the crystal changes.

Dispositivo piezoresistivo   Similar a uno piezoeléctrico excepto porque operan como un resistor variable. Su valor de resistencia cambia a medida que lo hace la presión que se aplica al cristal.

Piezoresistive sensor   A sensor that is sensitive to pressure changes.

Sensor piezoresistivo   Sensor susceptible a los cambios de presión.

Ping (or denotation)   A knocking sound that occurs as two flame fronts collide.

Golpeteo (detonación)   Un ruido de impacto que ocurre cuando hay colisión entre dos bordes térmicos.

Pinion factor   A calculation using the final drive ratio and the tire circumference to obtain accurate vehicle speed signals.

Factor de piñón   Una calculación que usa la relación de impulso final y la circunferencia de la llanta para obtener unas señales precisad de la velocidad del vehículo.

Pinion gear   A small gear; typically refers to the drive gear of a starter drive assembly or the small drive gear in a differential assembly.

Engranaje de piñón   Un engranaje pequeño; tipicamente se refiere al engranaje de arranque de una asamblea de motor de arranque o al engranaje de mando pequeño de la asamblea del diferencial.

Plate straps   Metal connectors used to connect the positive or negative plates in a battery.

Abrazaderas de la placa   Los conectores metálicos que sirven para conectar las placas positivas o negativas de una batería.

Plates The basic structure of a battery cell; each cell has at least one positive plate and one negative plate.

Placas La estructura básica de una celula de batería; cada celula tiene al menos una placa positiva y una placa negativa.

P-material Silicon or germanium that is doped with boron or gallium to create a shortage of electrons.

Material-P Boro o galio añadidos al silicio o al germanio para crear una insuficiencia de electrones.

PMGR An abbreviation for permanent magnet gear reduction.

PMGR Una abreviación de desmultiplicación del engrenaje del imán permanente.

Pneumatic tools Power tools that are powered by compressed air.

Herrimientas neumáticas Herramientas mecánicas accionadas por aire comprimido.

PN junction The point at which two opposite kinds of semiconductor materials are joined together.

Unión pn Zona de unión en la que se conectan dos tipos opuestos de materiales semiconductores.

Polarizers Glass sheets that make light waves vibrate in only one direction. This converts light into polarized light.

Polarizadores Las láminas de vidrio que hacen vibrar las ondas de luz en un sólo sentido. Esto convierte la luz en luz polarizada.

Polarizing The process of light polarization or of setting one end of a field as a positive or negative point.

Polarizadora El proceso de polarización de la luz o de establecer un lado de un campo como un punto positivo o negativo.

Pole The number of input circuits.

Poste El número de los circuitos de entrada.

Pole shoes The components of an electric motor that are made of high-magnetic permeability material to help concentrate and direct the lines of force in the field assembly.

Expansión polar Componentes de un motor eléctrico hechos de material magnético de gran permeabilidad para ayudar a concentrar y dirigir las líneas de fuerza en el conjunto inductor.

Positive-engagement starter A type of starter that uses the magnetic field strength of a field winding to engage the starter drive into the flywheel.

Acoplamiento de arranque positivo Un tipo de arrancador que utilisa la fuerza del campo magnético del devanado inductor para accionar el acoplamiento del arrancador en el volante.

Positive plate The plate connected to the positive battery terminal.

Placa positiva La placa conectada al terminal positivo de la batería.

Positive temperature coefficient (PTC) thermistors Thermistors that increase their resistance as temperature increases.

Termistores con coeficiente positivo de temperatura Termistores que aumentan su resistencia según aumenta la temperatura.

Potential The ability to do something; typically voltage is referred to as the potential. If you have voltage, you have the potential for electricity.

Potencial La capacidad de efectuar el trabajo; típicamente se refiere a la tensión como el potencial. Si tiene tensión, tiene la potencial para la electricidad.

Potentiometer A variable resistor that acts as a circuit divider to provide accurate voltage drop readings proportional to movement.

Potenciómetro Resistor variable que actúa como un divisor de circuito para obtener lecturas de perdidas de tensión precisas en proporcion con el movimiento.

Potentiometric pressure sensor Sensor used to measure pressure by use of a Bourdon tube, a capsule, or bellows to move a wiper arm on a resistive element. The movement of the wiper across the resistive element records a different voltage reading.

Sensor de presión potenciométrica Sensor que se utiliza para medir la presión mediante un tubo Bourdon, una cápsula o un tubo flexible ondulado para mover el brazo de un contacto deslizante en un elemento resistivo. El movimiento del contacto deslizante contra el elemento resistivo registrará una lectura de voltaje diferente.

Power (P) The rate of doing electrical work.

Potencia La tasa de habilidad de hacer el trabajo eléctrico.

Power door locks Electric power door locks use either a solenoid or a permanent magnet reversible motor to lock and unlock doors.

Cerradura mecánica de puertas Las cerraduras electro-mecánicos de puertas utilizan o un solenoide o un motor reversible de imanes permanentes para abrir y cerrar la puerta.

Power formula A formula used to calculate the amount of electrical power a component uses. The formula is $P = I \times E$, where P stands for power (measured in watts), I stands for current, and E stands for voltage.

Formula de potencia Una formula que se emplea para calcular la cantidad de potencia eléctrica utilizada por un componente. La formula es $P = I \times E$, en el que el P quiere decir potencia (medida en wats), I representa el corriente y el E representa la tensión.

Power mirrors Outside mirrors that are electrically positioned from the inside of the driver's compartment.

Espejos eléctricos Los espejos exteriores que se ajusten eléctricamente desde el interior del compartimiento del conductor.

Power tools Tools that use forces other than those generated from the body. They can use compressed air, electricity, or hydraulic pressure to generate and multiply force.

Herramientas mecánicas Herramientas que utilizan fuerzas distintas a las generadas por el cuerpo. Dichas fuerzas pueden ser el aire comprimido, la electricidad, o la presión hidráulica para generar y multiplicar la fuerza.

Power windows Windows that are raised and lowered by use of electrical motors.

Ventanillas eléctricas Las ventanillas que se suban y se bajan por medio de los motores eléctricos.

Pressure control solenoid A solenoid used to control the pressure of a fluid, commonly found in electronically controlled transmissions.

Solenoide de control de la presión Un solenoide que controla la presión de un fluido, suele encontrarse en las transmisiones controladas electronicamente.

Pretensioners Used to tighten the seat belt and shoulder harness around the occupant during an accident severe enough to deploy the air bag. Pretensioners can be used on all seat belt assemblies in the vehicle.

Pretensadores Se usan para apretar la cinta de seguridad y el arnés del cuerpo alrededor del ocupante durante un accidente bastante severo como para activar la bolsa de aire. Los pretensadores se pueden usarse en cualquier asamblea de cinta de seguridad del vehículo.

Primary circuit All the components that carry low voltage through the system.

Circuito pirmario Todos los componentes que llevan un voltaje bajo dentro del sistema.

Primary coil winding The second set of winding in the ignition coil. The primary winds will have about 200 turns to create a magnetic field to induce voltage into the secondary winding.

Devanado primario El grupo segundo de devanados en la bobina del encendido. Los devandos primarios tendrán unos 200 vueltas para crear un camp magnético para inducir el voltaje al devanado secundario.

Primary wiring Conductors that carry low voltage and current. The insulation of primary wires is usually thin.

Hilos primarios Hilos conductores de tensión y corriente bajas. El aislamiento de hilos primarios es normalmente delgado.

Printed circuit boards Made of thin phenolic or fiberglass board with copper deposited on it to create current paths. These are used to simplify the wiring of circuits.

Circuito impreso Un circuito hecho de un tablero de fenólico delgado o de fibra de vidrio el cual tiene depósitos del cobre para crear los trayectorios para el corriente. Estos se emplean para simplificar el cableado de los circuitos.

Prism lens A light lens designed with crystal-like patterns, which distort, slant, direct, or color the light that passes through it.

Lente prismático Un lente de luz con diseños cristalinos que distorcionan, inclinan, dirigen o coloran la luz que lo atraviesa.

Prisms Redirect the light beam and create a broad, flat beam.

Prismas Dirigen un rayo de luz y crean un rayo ancho y plano.

Program A set of instructions that the computer must follow to achieve desired results.

Programa Conjunto de instrucciones que la computadora debe seguir para lograr los resultados deseados.

Program number Represents the amount of heating or cooling required to obtain the temperature set by the driver.

Número de programa Representa la cantidad de calefacción o enfriamiento requerido para obtener la temperatura indicado por el conductor.

Programmable Communication Interface (PCI)   A single-wire, bidirectional communication bus where each module supplies its own bias voltage and has its own termination resistors. As a message is sent, a variable pulse width modulation (VPWM) voltage between 0 and 7.75 volts is used to represent the 1 and 0 bits.

Interfaz de comunicación programable (PCI)   Bus de comunicación en dos direcciones en un hilo sencillo en donde cada módulo proporciona su propia velocidad de transmisión de baudios y tiene sus propias resistencias de unión. Mientras se envía un mensaje, un voltaje de anchura variada entre impulsos entre 0 y 7.75 voltios se utiliza para representar los bits 1 y 0.

Programmer   Controls the blower speed, air mix doors, and vacuum motors of the SATC system. Depending on manufacturer, they are also called servo assemblies.

Programador   Controla la velocidad del ventilador, las puertas de mezcla de aire y los motores de vacío de un sistema SATC. Según el fabricante, tambien se llaman asambleas servo.

Projector headlight   A headlight system using a HID or halgen bulb to produce a very focused light beam.

Faro proyector   Un sistema de faro que utiliza un HID o bulbo de halógeno para producir un haz de luz muy concentrado.

PROM (programmable read only memory)   Memory chip that contains specific data which pertains to the exact vehicle in which the computer is installed. This information may be used to inform the CPU of the accessories that are equipped on the vehicle.

PROM (memoria de sólo lectura programable)   Fragmento de memoria que contiene datos específicos referentes al vehículo particular en el que se instala la computadora. Esta información puede utilizarse para informar a la UCP sobre los accesorios de los cuales el vehículo está dotado.

Protection device   Circuit protector that is designed to turn off the system that it protects. This is done by creating an open to prevent a complete circuit.

Dispositivo de protección   Protector de circuito diseñado para "desconectar" el sistema al que provee protección. Esto se hace abriendo el circuito para impedir un circuito completo.

Protocol   A language used by computers to communicate with each other over a data bus.

Protocolo   Lenguaje que se utiliza en computadoras para comunicarse entre sí sobre un mando de bus.

Proton   Positively charged particles contained in the nucleus of an atom.

Protón   Partículas con carga positiva que se encuentran en el núcleo de todo átomo.

Proton exchange membrane (PEM)   Impedes the oxyhydrogen gas reaction in a fuel cell by ensuring that only protons (H+), and not elemental hydrogen molecules (H_2), react with the oxygen.

Membrana de intercambio de protones (MIP)   Impide la reacción de gas de oxigeno-hidrógeno en una célula de combustible al asegurar que sólo los protones reaccionen con el oxígeno y no las moléculas de hidrógeno elemental.

Prove-out circuit   A function of the ignition switch that completes the warning light circuit to ground through the ignition switch when it is in the START position. The warning light will be on during engine cranking to indicate to the driver that the bulb is working properly.

Circuito de prueba   Función del botón conmutador de encendido que completa el circuito de la luz de aviso para que se ponga a tierra a través del botón conmutador de encendido cuando éste se encuentra en la posición START. La luz de aviso se encenderá durante el arranque del motor para avisarle al conductor que la bombilla funciona correctamente.

Proximity sensor   A sensor able to detect the presence of nearby objects without any physical contact.

Sensor de proximidad   Un sensor capaz de detectar la presencia de objetos cercanos sin ningún contacto físico.

Pull-down circuit   Closes the switch to ground.

Circuito de bajada   Cierra el interruptor a tierra.

Pull-down resistor   A current-limiting resistor used to assure a proper low-voltage reading by preventing float when the switch is open.

Resistor de bajada   Un resistor que limita la corriente y que se utiliza para garantizar una lectura de voltaje bajo adecuada al evitar la flotación cuando el interruptor se encuentra abierto.

Pull-in windings   An electrical coil internal to a solenoid that is energized to create a magnetic field used to move the solenoid plunger to the engaged position.

Devanados de puesta en trabajo   Una bobina eléctrica que es íntegra a un solenoide que se excita para crear un campo magnético que sirve para mover el relé de solenoide a la posición de engranaje.

Pull-up circuit   Closes the switch to voltage.

Circuito de subida   Cierra el interruptor al voltaje.

Pull-up resistor   A current-limiting resistor that is used to assure proper high-voltage reading of a voltage sense circuit by connecting the voltage sense circuit to an electrical potential that can be removed when the switch is closed.

Resistor de subida   Un resistor que limita la corriente y que se utiliza para garantizar una lectura de voltaje alto adecuada de un circuito de sensores de voltaje al conectar el circuito de sensores de voltaje a un potencial eléctrico que se puede quitar cuando el interruptor se encuentra cerrado.

Pulse width   The length of time in milliseconds that an actuator is energized.

Duración de impulsos   Espacio de tiempo en milisegundos en el que se excita un accionador.

Pulse-width modulation (PWM)   On/off cycling of a component. The period of time for each cycle does not change; only the amount of on time in each cycle changes.

Modulación de duración de impulsos   Modulación de impulsos de un componente. El espacio de tiempo de cada ciclo no varía; lo que varía es la cantidad de trabajo efectivo de cada ciclo.

Radial   A design pattern that branches out from a common center, such as spokes of a wheel.

Radial   Un patrón de diseño que se ramifica a partir de un centro común, como los rayos de una rueda.

Radial grid   A type of battery grid that has its patterns branching out from a common center.

Rejilla radial   Un tipo de rejilla de bateria cuyos diseños extienden de un centro común.

Radio choke   Absorbs voltage spikes and prevents static in the vehicle's radio.

Impedancia del radio   Absorba los impulsos de la tensión y previene la presencia del estático en el radio del vehículo.

Radio frequency interference (RFI)   Radio and television interference caused by electromagnetic energy.

Interferencia de frecuencia radioeléctrica   Interferencia en la radio y en la televisión producida por energía electromagnética.

RAM (random access memory)   Stores temporary information that can be read from or written to by the CPU. RAM can be designed as volatile or nonvolatile.

RAM (memoria de acceso aleatorio)   Almacena datos temporales que la UCP puede leer o escribir. La RAM puede ser volátil o no volátil.

Ratio   A mathematical relationship between two or more things.

Razón   Una relación matemática entre dos cosas o más.

Reactivity   A statement of how easily a substance can cause or be a part of a chemical.

Reactividad   Una indicación de cuan fácil una sustancia puede causar o ser parte de una química.

Reader   an interface between the transponder and the microprocessor that allows the module to read and process the identifier code from a transponder.

Lector   una interfaz entre el transpondedor y el microprocesador que permite al módulo leer y procesar el código identificador de un transpondedor.

Receiver   The receiver receives radio waves from the transmitter and decodes the message from the sine wave.

Receptor   El receptor recibe las ondas de radio del transmisor y decodifica el mensaje de la onda senoidal.

Recirc/air inlet door actuator   An electric motor that is linked to the recirculation door to provide either outside air or in-vehicle air into the A/C heater case.

Actuador de la compuerta de entrada de recirculación del aire   Motor eléctrico que está unido a una compuerta de recirculación para proporcionar ya sea aire de fuera o que provenga del vehículo, dentro de la caja del calentón de corriente alterna.

Recombination battery   A type of battery that is sometimes called a dry-cell battery because it does not use a liquid electrolyte solution.

Batería de recombinación   Un tipo de batería que a veces se llama una pila seca porque no requiere una solución líquida de electrolita.

Rectification   The conversion of AC current to DC current.

Rectificación   Proceso a través del cual la corriente alterna es transformada en una corriente continua.

Reflectors   A device whose surface reflects or radiates light.

Reflectores   Un dispositivo cuyo superficie refleja o irradia la luz.

Reformer   A high-temperature device that converts hydrocarbon fuels to CO and H_2.

Reformador   Dispositivo de alta temperatura que convierte los combustibles de hidrocarburo a monóxido de carbono CO y a H_2.

Regenerative braking   Braking energy is converted back into electricity instead of heat.

Frenado regenerativo   La energía de frenado se convierte nuevamente en electricidad en lugar de calor.

Relay   A device that uses low current to control a high-current circuit. Low current is used to energize the electromagnetic coil, while high current is able to pass over the relay contacts.

Relé   Dispositivo que utiliza corriente baja para controlar un circuito de corriente alta. La corriente baja se utiliza para excitar la bobina electromagnética, mientras que la corriente alta puede transmitirse a través de los contactos del relé.

Reluctance   A term used to indicate a material's resistance to the passage of flux lines.

Reluctancia   Término utilizado para señalar la resistencia ofrecida por un circuito al paso del flujo magnético.

Reserve capacity   The amount of time a battery can be discharged at a certain current rate until the voltage drops below a specified value.

Capacidad de reserva   La cantidad de tiempo que puede descargarse una batería a una cierta proporción de corriente, hasta que el voltaje caiga debajo de un valor especificado.

Reserve-capacity rating   An indicator, in minutes, of how long a vehicle can be driven, with the headlights on, if the charging system should fail. The reserve-capacity rating is determined by the length of time, in minutes, that a fully charged battery can be discharged at 25 amps before battery cell voltage drops below 1.75 volts per cell.

Clasificación de capacidad en reserva   Indicación, en minutos, de cuánto tiempo un vehículo puede continuar siendo conducido, con los faros delanteros encendidos, en caso de que ocurriese una falla en el sistema de carga. La clasificación de capacidad en reserva se determina por el espacio de tiempo, en minutos, en el que una batería completamente cargada puede descargarse a 25 amperios antes de que la tensión del acumulador de la batería disminuya a un nivel inferior de 1,75 amperios por acumulador.

Resistance   Opposition to current flow.

Resistencia   Oposición que presenta un conductor al paso de la corriente eléctrica.

Resistor block   Consists of two or three helically wound wire resistors wired in series.

Bloque de resistencia   Consiste de dos o tres cables helicoilades rostáticos intercalados en serie.

Resistive multiplex switch   Provides multiple inputs over a single circuit. Since each switch position has a different resistance value, the voltage drop is different. This means a switch can have one power supply wire and one ground wire instead of a separate wire for each switch position.

Interruptor multiplex resistente   Provee múltiples entradas de energía sobre un circuito simple. Debido a que la posición de cada interruptor tiene un valor diferente de resistencia, el voltaje de la caída de tensión es diferente. Esto significa que un interruptor puede tener un cable de abastecimiento de energía y un cable conductor a tierra, sin que sea necesario un cable individual para cada posición del interruptor.

Reverse-biased   A positive voltage is applied to the N-type material and a negative voltage is applied to the P-type material of a semiconductor.

Polarización inversa   Tensión positiva aplicada al material N y tensión negativa aplicada al material P de un semiconductor.

Rheostat   A two-terminal variable resistor used to regulate the strength of an electrical current.

Reóstato   Resistor variable de dos bornes utilizado para regular la resistencia de una corriente eléctrica.

Right-hand rule   Identifies the direction of the lines of force of an electromagnet.

Regla de la mano derecha   Identifica la dirección de las líneas de fuerza de un electroimán.

ROM (read only memory)   Memory chip that stores permanent information. This information is used to instruct the computer on what to do in response to input data. The CPU reads the information contained in ROM, but it cannot write to it or change it.

ROM (memoria de sólo lectura)   Fragmento de memoria que almacena datos en forma permanente. Dichos datos se utilizan para darle instrucciones a la computadora sobre cómo dirigir la ejecución de una operación de entrada. La UCP lee los datos que contiene la ROM, pero no puede escribir en ella o puede cambiarla.

Rotating magnetic field   The magnetic field of the stator windings in an AC motor. The stator is stationary, but the field rotates from pole to pole.

Campo magnético rotativo   El campo magnético de los giros de un estator en un motor de CA. El estator permanece fijo, pero el campo rota de polo en polo.

Rotor   The component of the AC generator that is rotated by the drive belt and creates the rotating magnetic field of the AC generator.

Rotor   Parte rotativa del generador de corriente alterna accionada por la correa de transmisión y que produce el campo magnético rotativo del generador de corriente alterna.

Safety goggles   Eye protection device that fits against the face and forehead to seal off the eyes from outside elements.

Gafas de seguridad   Dispositivo protector que se coloca delante de los ojos para preservarlos de elementos extraños.

Safety stands   Support devices used to hold the vehicle off the floor after it has been raised by the floor jack.

Soportes de seguridad   Dispositivos de soporte utilizados para sostener el vehículo sobre el suelo después de haber sido levantado con el gato de pie.

Safing sensor   Determines if the collision is severe enough to inflate the air bag.

Monitor de seguridad (safing sensor)   Determina si el impacto del choque es lo suficientemente grave para inflar las bolsas de aire.

Satellite radio   Provide several commercial-free music and talk show channels using orbiting satellites to provide a digital signal.

Radios por satélite   Proporcionan música variada sin comerciales y canales de programas de entrevistas al usar satélites en órbita que dan una señal digital.

Saturation   1. The point at which the magnetic strength eventually levels off, and where an additional increase of the magnetizing force current no longer increases the magnetic field strength. 2. The point where forward-bias voltage to the base leg is at a maximum. With bias voltage at the high limits, output current is also at its maximum.

Saturación   1. Máxima potencia posible de un campo magnético, donde un aumento adicional de la corriente de fuerza magnética no logra aumentar la potencia del campo magnético. 2. La tensión de polarización directa a la base está en su máximo. Ya que polarización directa ha alcanzado su límite máximo, la corriente de salida también alcanza éste.

Schematic   An electrical diagram that shows how circuits are connected, but not details such as color codes.

Esquemático   Diagrama eléctrico que muestra cómo se conectan los circuitos, pero no los detalles tales como las claves por colores.

Schmitt trigger   An electronic circuit used to convert analog signals to digital signals or vice versa.

Disparador de Schmitt   Un circuito electrónico que se emplea para convertir las señales análogas en señales digitales o vice versa.

Sealed-beam headlight   A self-contained glass unit that consists of a filament, an inner reflector, and an outer glass lens.

Faro delantero sellado   Unidad de vidrio que contiene un filamento, un reflector interior y una lente exterior de vidrio.

Seat track position sensor (STPS)   A Hall-effect sensor that provides information to the occupant classification module (OCM) concerning the position of the seat in relation to the air bag.

Sensor de la posición del carril de asiento (SPCA)   Sensor de efecto de Hall que proporciona información al MCS (Módulo de control de salida) que concierne a la posición del asiento en relación con la bolsa de aire.

Secondary wiring   Conductors, such as battery cables and ignition spark plug wires, that are used to carry high voltage or high current. Secondary wires have extra thick insulation.

Hilos secundarios   Conductores, tales como cables de batería e hilos de bujías del encendido, utilizados para transmitir tensión o corriente alta. Los hilos secundarios poseen un aislamiento sumamente grueso.

Sector gear The section of gear teeth on the regulator.

Engranaje de cables La sección de los dientes de engranaje en el regulador.

Self-induction The generation of an electromotive force by a changing current in the same circuit.

Autoinducción Este es el generamiento de la fuerza electromotriz cuando la corriente cambia en el mismo circuito.

Semiconductor An element that is neither a conductor nor an insulator. Semiconductors are materials that conduct electric current under certain conditions, yet will not conduct under other conditions.

Semiconductor Elemento que no es ni conductor ni aislante. Los semiconductores son materiales que transmiten corriente eléctrica bajo ciertas circunstancias, pero no la transmiten bajo otras.

Sending unit The sensor for the gauge. It is a variable resistor that changes resistance values with changing monitored conditions.

Unidad emisora Sensor para el calibrador. Es un resistor variable que cambia los valores de resistencia según cambian las condiciones reguladas.

Sensing voltage Input voltage to the AC generator.

Detección del voltje Determina la tensión de entrada de energía al generador de Corriente Alterna AC.

Sensitivity control A potentiometer that allows the driver to adjust the sensitivity of the automatic dimmer system to surrounding ambient light conditions.

Controles de sensibilidad Un potenciómetro que permite que el conductor ajusta la sensibilidad del sistema de intensidad de iluminación automático a las condiciones de luz ambientales.

Sensor Any device that provides an input to the computer.

Sensor Cualquier dispositivo que le transmite información a la computadora.

Sentry key Describes a sophisticated antitheft system that prevents the engine from starting unless a special key is used.

Llave guardiante centinela Describe un sofisticado sistema de guardia anti-robo el que evita prender el motor, a menos que se haga con una llave específicamente creada.

Separators Normally constructed of glass with a resin coating. These battery plates offer low resistance to electrical flow but high resistance to chemical contamination.

Separadores Normalmente se construyen del vidrio con una capa de resina. Estas placas de la batería ofrecen baja resistencia al flujo de la electricidad pero alta resistencia a la contaminación química.

Sequential logic circuits Flip-flop circuits in which the output is determined by the sequence of inputs. A given input affects the output produced by the next input.

Circuitos de lógica secuenciales Cambia los circuitos en los cuales la salida de energía es determinada según la secuencia de las entradas de corriente. Una entrada de energía afecta la salida de corriente que va ser producida en una próxima entrada de energía.

Sequential sampling The process that the MUX and DEMUX operate on. This means the computer will deal with all of the sensors and actuators one at a time.

Muestreo secuencial La forma en que funcionan los sistemas de las abreviaturas (MUX y DEMUX). Esto significa que la computadora se encargará de que todos los monitores y actuadores funcionen uno por uno.

Series circuit A circuit that provides a single path for current flow from the electrical source through all the circuit's components and back to the source.

Circuito en serie Circuito que provee una trayectoria única para el flujo de corriente de la fuente eléctrica a través de todos los componentes del circuito, y de nuevo hacia la fuente.

Series hybrid Hybrid configuration where propulsion comes directly from the electric motor.

Híbrido en serie Configuración del híbrido en donde la propulsión llega directamente del motor eléctrico.

Series-parallel circuit A circuit that has some loads in series and some in parallel.

Circuito en series paralelas Circuito que tiene unas cargas en serie y otras en paralelo.

Series-wound motor A type of motor that has its field windings connected in series with the armature. This type of motor develops its maximum torque output at the time of initial start. Torque decreases as motor speed increases.

Motor con devanados en serie Un tipo de motor cuyos devanados inductores se conectan en serie con la armadura. Este tipo de motor desarrolla la salida máxima de par de torsión en el momento inicial de ponerse en marcha. El par de torsión disminuye al aumentar la velocidad del motor.

Servomotor An electrical motor that produces rotation of less than a full turn. A feedback mechanism is used to position itself to the exact degree of rotation required.

Servomotor Motor eléctrico que genera rotación de menos de una revolución completa. Utiliza un mecanismo de realimentación para ubicarse al grado exacto de la rotación requerida.

Shell The electron orbit around the nucleus of an atom.

Corteza Órbita de electrones alrededor del núcleo del átomo.

Short An unwanted electrical path; sometimes this path goes directly to ground.

Corto Una trayectoria eléctrica no deseable; a veces este trayectoria viaja directamente a tierra.

Shorted circuit A circuit that allows current to bypass part of the normal path.

Circuito corto Este circuito permite que la corriente pase por una parte del recorrido normal.

Shunt More than one path for current to flow.

Desviación Más de una derivación para que la corriente pueda fluir.

Shunt circuits The branches of the parallel circuit.

Circuitos en derivación Las ramas del circuito en paralelo.

Shunt-wound motor A type of motor whose field windings are wired in parallel to the armature. This type of motor does not decrease its torque as speed increases.

Motor con devanados en derivación Un tipo de motor cuyos devanados inductores se cablean paralelos a la armadura. Este tipo de motor no disminuya su par de torsión al aumentar la velocidad.

Shutter wheel A metal wheel consisting of a series of alternating windows and vanes. It creates a magnetic shunt that changes the strength of the magnetic field from the permanent magnet of the Hall-effect switch or magnetic pulse generator.

Rueda obturadora Rueda metálica compuesta de una serie de ventanas y aspas alternas. Genera una derivación magnética que cambia la potencia del campo magnético, del imán permanente del conmutador de efecto Hall o del generador de impulsos magnéticos.

Sine wave A waveform that shows voltage changing polarity.

Onda senoidal Una forma de onda que muestra un cambio de polaridad en la tensión.

Single 42-volt system An electrical system that uses 42 volts for all circuit operations including starter, lighting, and accessories.

Sistema único de 42 voltios Un sistema eléctrico que utiliza 42 voltios para todas las operaciones del circuito, incluyendo el arrancador, la iluminación y los accesorios.

Single-phase voltage The sine wave voltage induced in one conductor of the stator during one revolution of the rotor.

Tensión monofásica La tensión en forma de onda senoidal inducida en un conductor del estator durante una revolución del rotor.

Slave module Controller on the network that must communicate through a master controller.

Módulo esclavo Controlador en la red que debe comunicarse por medio del control maestro.

Slip The difference between the synchronous speed and actual rotor that is directly proportional to the load on the motor.

Slip La diferencia entre la velocidad síncrona y rotor real es directamente proporcional a la carga en el motor.

Slip rings Rings that function much like the armature commutator in the starter motor; however, they are smooth.

Anillos colectores Estos funcionan como casi el conmutador inducido en el arranque del motor, excepto que estos anillos son lisos.

Smart sensors Sensors that are capable of sending digital messages on the data bus.

Sensor inteligente Sensores capaces de enviar mensajes digitales en el bus de datos.

Solenoid An electromagnetic device that uses movement of a plunger to exert a pulling or holding force.

Solenoide Dispositivo electromagnético que utiliza el movimiento de un pulsador para ejercer una fuerza de arrastre o de retención.

Sound generator *See* buzzer.

Generador de sonido Vea "timbre."

Source The portion of a field-effect transistor that supplies the current-carrying holes or electrons.

Fuente Terminal de un transistor de efecto de campo que provee los agujeros o electrones portadores de corriente.

Spark plug Electrodes provide gaps inside each combustion chamber across which the secondary current flows to ignite the air–fuel mixture in the combustion chambers.

Enchufe de chispa Electrodos para proveer intervalos dentro de cada cámara de combustión a través de la corriente secundaria que fluye para encender la mezcla de aire/combustible en las cámaras de combustión.

Speakers Convert electrical energy provided by the radio receiver amplifier into acoustical energy using a permanent magnet and an electromagnet to move air to produce sound.

Altavoces Convierten la energía eléctrica que suministra el amplificador receptor de radio en energía acústica mediante el uso de un imán permanente y un electroimán, los cuales mueven el aire para producir sonido.

Specific gravity The weight of a given volume of a liquid divided by the weight of an equal volume of water.

Gravedad específica El peso de un volumen dado de líquido dividido por el peso de un volumen igual de agua.

Speedometer An instrument panel gauge that indicates the speed of the vehicle.

Velocímetro Calibrador en el panel de instrumentos que marca la velocidad del vehículo.

Splice A common connection point for the electrical circuit; used to eliminate the need for multiple connection points.

Empalme Un punto de conexión común para el circuito eléctrico; se utiliza para eliminar la necesidad de varios puntos de conexión.

Squib A pyrotechnic term used for a fire cracker that burns but does not explode. The squib starts the process of air bag deployment.

Mecha Un término pirotécnico usado para prender una pólvora que se quema pero no explota. La mecha inicia el proceso de la salida de la bolsa de aire.

Start/clutch interlock switch Used on vehicles equipped with manual transmissions.

Interruptor de seguridad de embrague Usado en los vehículos equipados con transmisiones manuales.

Starter drive The part of the starter motor that engages the armature to the engine flywheel ring gear.

Transmisión de arranque Parte del motor de arranque que engrana la armadura a la corona del volante de la máquina.

Starter generator control module (SGCM) Also called a high-voltage ECU (HV ECU). Used to control the flow of torque and electrical energy into and from the motor generator of the HEV.

Módulo de control del generador de encendido (MCGE) También se llama UCE de alto voltaje (UCE AV). Se utiliza para controlar el flujo del par motor y la energía eléctrica dentro y fuera del generador del motor del VHE.

Starting system A combination of mechanical and electrical parts that work together to start the engine by changing the electrical energy that is being supplied by the battery into mechanical energy by use of a starter or cranking motor.

Sistema de encendido Combinación de partes mecánicas y eléctricas que trabajan unidas para encender el motor al cargar la energía eléctrica que proporciona la batería, en energía mecánica mediante el uso de un encendedor o motor de arranque.

State of charge The condition of a battery's electrolyte and plate materials at any given time.

Estado de carga Condición del electrolito y de los materiales de la placa de una batería en cualquier momento dado.

Static electricity Electricity that is not in motion.

Electricidad estática Electricidad que no está en movimiento.

Static neutral point The point at which the fields of a motor are in balance.

Punto neutral estático El punto en que los campos de un motor estan equilibrados.

Stator The stationary coil of the AC generator where current is produced.

Estátor Bobina fija del generador de corriente alterna donde se genera corriente.

Stator neutral junction The common junction of Wye stator windings.

Unión de estátor neutral La unión común de los devanados de un estátor Y.

Stepped resistor A resistor that has two or more fixed resistor values.

Resistor de secciones escalonadas Resistor que tiene dos o más valores de resistencia fija.

Stepper motor An electrical motor that contains a permanent magnet armature with two or four field coils; can be used to move the controlled device to the desired location. By applying voltage pulses to selected coils of the motor, the armature will turn a specific number of degrees. When the same voltage pulses are applied to the opposite coils, the armature will rotate the same number of degrees in the opposite direction.

Motor de pasos Contiene una armadura magnética permanente con dos, cuatro o más bobinas del campo.

Strain gauge A sensor that determines the amount of applied pressure by measuring the strain a material experiences when subjected to pressure.

Extensómetro de resistencia eléctrica Un sensor que determina la cantidad de presión aplicada al medir la resistencia eléctrica que experimenta un material cuando se lo somete a presión.

Stranded wire A conductor comprised of many small solid wires twisted together. This type conductor is used to allow the wire to flex without breaking.

Cable trenzado Un conductor que comprende muchos cables sólidos pequeños trenzados. Este tipo de conductor se emplea para permitir que el cable se tuerza sin quebrar.

Sulfation A condition in a battery that reduces its output. The sulfate in the battery that is not converted tends to harden on the plates, resulting in permanent damage to the battery.

Sulfatación Una condición en una batería que disminuya su potencia de salida. El sulfato en la batería que no se convierte suele endurecerse en las placas y resulta en daños permanentes en la batería.

Supplemental bus networks Bus networks that are on the vehicle in addition to the main bus network.

Redes de bus complementarios Redes de bus que hay en el vehículo aparte de la red del bus principal.

Synchronous motor Type of AC motor that operates at a constant speed regardless of load. The rotor always rotates at the speed of the rotating magnetic field.

Motor sincrónico Tipo de motor de CA que opera a velocidad constante sin importar la carga. El rotor siempre gira a la velocidad del campo magnético rotativo.

Synchronous speed The speed at which the magnetic field of the stator rotates in an AC motor.

Velocidad sincrónica La velocidad a la cual el campo magnético del estator gira en un motor de CA.

Tachometer An instrument that measures the speed of the engine in revolutions per minute (rpm).

Tacómetro Instrumento que mide la velocidad del motor en revoluciones por minuto (rpm).

Telematics A system that encompasses telecommunications, instrumentation, wireless communications, road transportation, multimedia, and internet connectivity.

Telemática Sistema que abarca telecomunicaciones, instrumentación, comunicaciones inalámbricas, transporte por carretera, multimedia y conectividad a internet.

Terminals Provide a means of connecting the battery plates to the vehicle's electrical system.

Terminales Provee los medios de conección de las placas de batería al sistema eléctrico del vehículo.

Termination resistors Used to control induced voltages. Since voltage is dropped over resistors, the induced voltage is terminated.

Resistores de terminación Usados para controlar la conducción de los voltajes. Como el voltaje cae sobre los resistores, es así como se determina el voltaje.

Thermistor A solid-state variable resistor made from a semiconductor material that changes resistance in relation to temperature changes.

Termistor Resistor variable de estado sólido hecho de un material semiconductor que cambia su resistencia en relación con los cambios de temperatura.

Three-coil gauge   A gauge design that uses the interaction of three electromagnets and the total field effect upon a permanent magnet to cause needle movement.

Calibrador de tres bobinas   Calibrador diseñado para utilizar la interacción de tres electroimanes y el efecto inductor total sobre un imán permanente para producir el movimiento de la aguja.

Throw   Term used in reference to electrical switches or relays referring to the number of output circuits from the switch.

Posición activa   Término utilizado para conmutadores o relés eléctricos en relación con el número de circuitos de salida del conmutador.

Thyristor   A semiconductor switching device composed of alternating N and P layers. It can also be used to rectify current from AC to DC.

Tiristor   Dispositivo de conmutación del semiconductor compuesto de capas alternas de N y P. Puede utilizarse también para rectificar la corriente de corriente alterna a corriente continua.

Timer circuit   Uses a bimetallic strip that opens as a result of the heat being generated by the current flow.

Circuito Sincronizador   Consiste de una cinta bimetálica la que se abre, debido al calor generado por el flujo de corriente.

Timer control   A potentiometer that is part of the headlight switch in some systems. It controls the amount of time the headlights stay on after the ignition switch is turned off.

Control temporizador   Un potenciómetro que es parte del conmutador de los faros en algunos sistemas. Controla la cantidad del tiempo que quedan prendidos los faros después de apagarse la llave del encendido.

Timing disc   Known as armature, reluctor, trigger wheel, pulse wheel, or timing core. It is used to conduct lines of magnetic force.

Disco medidor de tiempo   Se conoce como una armadura de inducción, rueda disparadora, rueda de pulsación o un núcleo de tiempo. Este disco es usado para conducir líneas de fuerza magnética.

Tire pressure monitoring system   A safety system that notifies the driver if one or more tires are underinflated or overinflated.

Sistema de monitoreo de la presión de las llantas   Sistema de seguridad que le hace saber al conductor si una llanta o más llantas están desinfladas o sobre infladas.

Torque converter   A hydraulic device found on automatic transmissions. It is responsible for controlling the power flow from the engine to the transmission; works like a clutch to engage and disengage the engine's power to the drive line.

Convertidor de par   Un dispositivo hidráulico en las transmisiones automáticas. Se encarga de controlar el flujo de la potencia del motor a la transmisión; funciona como un embrague para embragar y desembragar la potencia del motor con la flecha motríz.

Total reflection   A phenomenon wherein a light wave reflects off the surface 100% when the light wave advances from a medium of high index of refraction to a medium of low index of refraction.

Reflexión total   Fenómeno en el que una onda de luz se refleja 100% de una superficie cuando la onda de luz avanza de un índice medio de refracción de uno alto a uno índice medio de refracción de uno bajo.

Tracer   A thin or dashed line of a different color than the base color of the insulation.

Traza líneas   Una línea delgada o instrumento de color diferente al color básico de la insulación.

Trailing edge   In a Hall-effect switch, the edges of the rotating blade that exit the switch.

Borde de salida   Un interruptor de efecto Hall indica los bordes de la paleta giratoria que sale del interruptor.

Transducer   A device that changes energy from one form into another.

Transductor   Dispositivo que cambia la energía de una forma a otra.

Transistor   A three-layer semiconductor used as a very fast switching device.

Transistor   Semiconductor de tres capas utilizado como dispositivo de conmutación sumamente rápido.

Transmitter   Generates a radio frequency by taking data and encoding it onto a sine wave that is tranmitted to a receiver.

Transmisor   Genera una radiofrecuencia al tomar los datos y codificarlos en una onda senoidal que se transmite a un receptor.

Trimotor   A three-armature motor.

Trimotor   Motor de tres armaduras.

Turn-on voltage   The voltage required to jump the PN junction and allow current to flow.

Voltaje de conección   El voltaje requerido para hacer funcionar el cable de empalme PN y permitir que la corriente fluya.

TVRS   An abbreviation for television-radio-suppression cable.

TVRS   Una abreviación del cable de supresíon del televisión y radio.

Tweeters   Smaller speakers that produce the high frequencies of treble.

Tweeters   Pequeñas bocinas que producen altas frecuencias agudas.

Two-coil gauge   A gauge design that uses the interaction of two electromagnets and the total field effect upon an armature to cause needle movement.

Calibrador de dos bobinas   Calibrador diseñado para utilizar la interacción de dos electroimanes y el efecto inductor total sobre una armadura para generar el movimiento de la aguja.

Ultra-capacitor   A device that stores energy as electrostatic charge. It is the primary device in the power supply during hill climbing, acceleration, and the recovery of braking energy.

Ultra capacitor   Dispositivo que guarda energía como carga electrostática. Es el dispositivo primario en la fuente de energía durante una subida, una aceleración y el recobro de la energía de frenado.

Ultrasonic sensors   A proximity detector that measures the distances to nearby objects using acoustic pulses and a control unit that measures the return interval of each reflected signal and calculates the object's distance.

Sensores ultrasónicos   Detector de proximidad que mide las distancias a objetos cercanos usando pulsos acústicos y una unidad de control que mide el intervalo de retorno de cada señal reflejada y calcula la distancia del objeto.

Ultrasound   A cycling sound pressure wave that is at a frequency greater than the upper limit of human hearing (about 20 kHz).

Ultrasonido   Una onda de presión de sonido cíclico que está en una frecuencia mayor que el límite superior de la audición humana (aproximadamente 20kHz).

Vacuum distribution valve   A valve used in vacuum-controlled concealed headlight systems. It controls the direction of vacuum to various vacuum motors or to vent.

Válvula de distribución al vacío   Válvula utilizada en el sistema de faros delanteros ocultos controlado al vacío. Regula la dirección del vacío a varios motores al vacío o sirve para dar salida del sistema.

Vacuum fluorescent display (VFD)   A display type that uses anode segments coated with phosphor and bombarded with tungsten electrons to cause the segments to glow.

Visualización de fluorescencia al vacío   Tipo de visualización que utiliza segmentos ánodos cubiertos de fósforo y bombardeados de electrones de tungsteno para producir la luminiscencia de los segmentos.

Valence ring   The outermost orbit of the atom.

Anillo de valencia   Órbita más exterior del átomo.

Valve body   A unit that consists of many valves and hydraulic circuits. This unit is the central control point for gear shifting in an automatic transmission.

Cuerpo de la válvula   Una unedad que consiste de muchas válvulas y circuitos hidráulicos. Esta unedad es el punto central de mando para los cambios de velocidad en una transmisión automática.

Valve-Regulated Lead–Acid (VRLA) battery   Another variation of the recombination battery that uses lead–acid. The oxygen produced on the positive plates is absorbed by the negative plate causing a decrease in the amount of hydrogen produced at the negative plate which is then combined with the oxygen to produce water.

Batería de plomo-ácido de válvula regulada (VRLA, por su sigla en inglés)   Otra variante de la batería de recombinación que utiliza plomo-ácido. El oxígeno que se produce en las placas positivas es absorbido por la placa negativa, lo que causa una disminución en la cantidad de hidrógeno que se produce en la placa negativa y que luego se combina con el oxígeno para formar agua.

Vaporized aluminum   Gives a reflecting surface that is comparable to silver.

Aluminio vaporizado   Da a la superficie un reflejo brillante como si fuera plata.

Variable resistor   A resistor that provides for an infinite number of resistance values within a range.

Resistor variable   Resistor que provee un número infinito de valores de resistencia dentro de un margen.

Vehicle instrumentation systems   A system that monitors the various vehicle operating systems and provides information to the driver of their correct operation.

Sistemas de instrumentación del vehículo Sistema que monitorea varios sistemas de operación del vehículo y proporciona información de su operación correcta al conductor.

Volatile Easily vaporizes or explodes.

Volátil Vaporiza o explota fácilmente.

Volatile RAM memory that is erased when it is disconnected from its power source. Also known as Keep Alive Memory.

Volátil Memoria RAM cuyos datos se perderán cuando se la desconecta de la fuente de alimentación. Conocida también como memoria de entretenimiento.

Volt The unit used to measure the amount of electrical force.

Voltio Unidad práctica de tensión para medir la cantidad de fuerza eléctrica.

Voltage The difference or potential that indicates an excess of electrons at the end of the circuit the farthest from the electromotive force. It is the electrical pressure that causes electrons to move through a circuit. One volt is the amount of pressure required to move 1 amp of current through 1 ohm of resistance.

Tensión Diferencia o potencial que indica un exceso de electrones al punto del circuito que se encuentra más alejado de la fuerza electromotriz. La presión eléctrica genera el movimiento de electrones a través de un circuito. Un voltio equivale a la cantidad de presión requerida para mover un amperio de corriente a través de un ohmio de resistencia.

Voltage drop A resistance in the circuit that reduces the electrical pressure available after the resistance. The resistance can be the load component, conductors, any connections, or unwanted resistance.

Caída de tensión Resistencia en el circuito que disminuye la presión eléctrica disponible después de la resistencia. La resistencia puede ser el componente de carga, los conductores, cualquier conexión o resistencia no deseada.

Voltage limiter Connected through the resistor network of a voltage regulator. It determines whether the field will receive high, low, or no voltage. It controls the field voltage for the required amount of charging.

Limitador de tensión Conectado por el red de resistores de un regulador de tensión. Determina si el campo recibirá alta, baja o ninguna tensión. Controla la tensión de campo durante el tiempo indicado de carga.

Voltage regulator Used to control the output voltage of the AC generator, based on charging system demands, by controlling field current.

Regulador de tensión Dispositivo cuya función es mantener la tensión de salida del generador de corriente alterna, de acuerdo a las variaciones en la corriente de carga, controlando la corriente inductora.

Voltmeter A test meter used to read the pressure behind the flow of electrons.

Voltímetro Instrumento de prueba utilizado para medir la presión del flujo de electrones.

Wake-up signal An input signal used to notify the body computer that an engine start and operation of accessories is going to be initiated soon. This signal is used to warm up the circuits that will be processing information.

Señal despertadora Señal de entrada para avisarle a la computadora del vehículo que el arranque del motor y el funcionamiento de los accesorios se iniciarán dentro de poco. Dicha señal se utiliza para calentar los circuitos que procesarán los datos.

Warning lamp A lamp that is illuminated to warn the driver of a possible problem or hazardous condition.

Luz de aviso Lámpara que se enciende para avisarle al conductor sobre posibles problemas o condiciones peligrosas.

Watchdog circuit Supplies a reset voltage to the microprocessor in the event that pulsating output voltages from the microprocessor are interrupted.

Circuito de vigilancia Proporciona un voltaje de reposición para el microprocesador en caso de que se interrumpan los voltajes de potencia útil de pulsaciones del microprocesador.

Watt The unit of measure of electrical power, which is the equivalent of horsepower. One horsepower is equal to 746 watts.

Watio Unidad de potencia eléctrica, equivalente a un caballo de vapor. 746 watios equivalen a un caballo de vapor (CV).

Wattage A measure of the total electrical work being performed per unit of time.

Vataje Medida del trabajo eléctrico total realizado por unidad de tiempo.

Watt-hour rating Equals the battery voltage times ampere-hour rating.

Contador de voltaje-hora nominal Su función es igualar el tiempo del voltaje de la batería con el voltaje-hora nominal.

Weather-pack connector A type of connector that seals the terminal's ends. This type connector is used in electronic circuits.

Conectador impermeable Un tipo de conectador que sella las extremidades de los terminales. Este tipo de conectador se emplea en los circuitos electrónicos.

Wheatstone bridge A series–parallel arrangement of resistors between an input terminal and ground.

Puente de Wheatstone Conjunto de resistores en series paralelas entre un borne de entrada y tierra.

Wi-Fi A wireless local area networking technology based on the IEEE 802.11 standards.

Wi-Fi Tecnología de red local inalámbrica basada en los estándares IEEE 802.11.

Window regulator Converts rotary motion of the motor into the vertical movement of the window.

Regulador del vidrio parabrisas Convierte la acción rotatoria del motor en un movimiento vertical sobre el vidrio parabrisas.

Windshield wipers Mechanical arms that sweep back and forth across the windshield to remove water, snow, or dirt.

Limpiador del parabrisas Funciona con brazos mecánicos, de un extremo al otro del vidrio, para remover agua, nieve y suciedad.

Wiper Moveable contact of a variable resistor.

Limpiador Consiste de un movimiento deslizable de resistores variables.

Wireless networks Connection of modules together to transmit information without the use of physical connection by wires.

Redes inalámbricas Conexión entre los módulos para transmitir información sin el uso de conexiones físicas por medio de alambres.

Wiring diagram An electrical schematic that shows a representation of actual electrical or electronic components and the wiring of the vehicle's electrical systems.

Esquema de conexiones Esquema en el que se muestran las conexiones internas de los componentes eléctricos o electrónicos reales y las de los sistemas eléctricos del vehículo.

Wiring harness A group of wires enclosed in a conduit and routed to specific areas of the vehicle.

Cableado preformado Conjunto de alambres envueltos en un conducto y dirigidos hacia áreas específicas del vehículo.

Woofers Large speakers that produce low frequencies of midrange and bass.

Woofers Bocinas grandes que producen bajas frecuencias de media distancia y bajo.

Worm gear A type of gear whose teeth wrap around the shaft. The action of the gear is much like that of a threaded bolt or screw.

Engranaje de tornillo sin fin Un tipo de engranaje cuyos dientes se envuelven alrededor del vástago. El movimiento del engranaje es muy parecido a un perno enroscado o una tuerca.

Wye-wound connection A type of stator winding in which one end of the individual windings is connected at a common point. The structure resembles the letter Y.

Conexión Y Un tipo de devanado estátor en el cual una extremidad de los devanados individuales se conectan en un punto común. La estructura parece la letra "Y."

Yaw The tendency of a vehicle to rotate around its center of gravity.

Giro longitudinal Tendencia de un vehículo a girar sobre su propio eje de gravedad.

Y-type stator A three-winding AC generator that has one end of each winding connected at the neutral junction.

Estátor de tipo Y Generador de corriente alterna de devanado triple; un extremo de cada devanado se conecta al empalme neutro.

Zener diode A diode that allows reverse current to flow above a set voltage limit.

Diodo Zener Diodo que permite que el flujo de corriente en dirección inversa sobrepase el límite de tensión determinado.

Zener voltage The voltage that is reached when a diode conducts in reverse direction.

Tensión de Zener Tensión alcanzada cuando un diodo conduce en una dirección inversa.

INDEX